# Legionella Contamination in Water Environment

# Legionella Contamination in Water Environment

Editors

**Silvia Bonetta**
**Sara Bonetta**

MDPI • Basel • Beijing • Wuhan • Barcelona • Belgrade • Manchester • Tokyo • Cluj • Tianjin

*Editors*

Silvia Bonetta
Department of Life Sciences and Systems Biology, University of Torino
Italy

Sara Bonetta
Department of Public Health and Pediatrics, University of Torino
Italy

*Editorial Office*
MDPI
St. Alban-Anlage 66
4052 Basel, Switzerland

This is a reprint of articles from the Special Issue published online in the open access journal *Pathogens* (ISSN 2076-0817) (available at: https://www.mdpi.com/journal/pathogens/special_issues/Legionella_contamination).

For citation purposes, cite each article independently as indicated on the article page online and as indicated below:

LastName, A.A.; LastName, B.B.; LastName, C.C. Article Title. *Journal Name* **Year**, *Volume Number*, Page Range.

**ISBN 978-3-0365-0694-4 (Hbk)**
**ISBN 978-3-0365-0695-1 (PDF)**

# Contents

# About the Editors

**Silvia Bonetta** graduated in Biological Science in 1999 at the University of Piemonte Orientale "A. Avogadro". In 2006, she obtained her Ph.D. in "Environmental Science" at the University of Piemonte Orientale "A. Avogadro". From 2000 to 2017, she undertook Post-Doc activity at the Environmental Hygiene Laboratory of the Department of Environmental and Life Sciences at the University of Piemonte Orientale "A. Avogadro" and the Department of Public Health and Pediatrics, University of Torino. In 2018, she was a Researcher (Junior Fixed-Term Researcher—RTDA) at the Department of Public Health and Pediatrics, University of Torino. Since 2020, she has been a Researcher (Senior Fixed-Term Researcher- RTDB) at the Department of Life Sciences and Systems Biology, University of Torino and lecturer in the courses of "Hygiene and preventive medicine", "Epidemiology", "Laboratory of Environmental Health and Toxicology" and "General Microbiology and Food&Health Safety". She is also a member of the Istituto Superiore di Sanità Commission for "Biological and microbiological methods". Her research is focused on the development and application of methods for pathogen detection in the environment and food to highlight emerging public health issues in the environment (ex. emerging pathogen detection, evaluation of antibiotic resistance, evaluation of antibacterial activity using emerging compounds or techniques). She is the author/co-author of 41 scientific publications in international journals, a member of the editorial board of *Pathogens*, and the Guest Editor of some Special Issues from MDPI concerning the topic "Environmental Hygiene".

**Sara Bonetta** graduated in Biological Sciences in 1999 at the University of Piemonte Orientale "A. Avogadro". In 2005, she obtained her Ph.D. in Environmental Sciences from the University of Piemonte Orientale "A. Avogadro". From 2000 to 2012, she undertook a Post-Doc at the Environmental Hygiene Laboratory of the Department of Environmental and Life Sciences, University of Piemonte Orientale. From 2012 to 2019, she was a Researcher (Junior Fixed-Term Researcher- RTDA and Senior Fixed-Term Researcher- RTDB) at the Department of Public Health and Pediatrics, University of Torino and lecturer in the courses of "Microbiology and Hygiene", "Laboratory of environmental health and general toxicology", "Occupational health and risk communication" and "Epidemiology". Since 2019, she has been an Associate Professor at the University of Torino. Her area of research focuses on the evaluation of methods for the detection of Legionella in water, study of disinfection methods against waterborne pathogens, and evaluation of microbiological air quality. Her other topics of interest include evaluation of antibiotic resistance in the environment and in vitro and in vivo genotoxicity of particulate matter and environmental pollutants in occupational and life environment. She is the author/co-author of 39 scientific publications in international journals and a member of the editorial board of *Pathogens*.

pathogens

MDPI

Editorial

# Editorial Comments to the Special Issue: "*Legionella* Contamination in Water Environment"

**Silvia Bonetta [1,*] and Sara Bonetta [2,*]**

1 Department of Life Sciences and Systems Biology, University of Torino, Via Accademia Albertina 13, 10123 Torino, Italy

2 Department of Public Health and Pediatrics, University of Torino, Piazza Polonia 94, 10126 Torino, Italy

* Correspondence: silvia.bonetta@unito.it (S.B.); sara.bonetta@unito.it (S.B.)

Received: 1 December 2020; Accepted: 1 December 2020; Published: 2 December 2020

## 1. Introduction

*Legionella* spp. are ubiquitous microorganisms that are widely distributed in aquatic environments. From these natural reservoirs, this opportunistic pathogen can spread to and colonize artificial aquatic environments [1]. Water systems of large buildings, such as hospitals, thermal baths, hotels, and dental units are often contaminated by legionellae [2,3] and various parameters such as physical, chemical, and microbial building water system characteristics can influence *Legionella* occurrence [4]. *Legionella* are intracellular bacteria whose natural hosts are aquatic protozoa in which these bacteria replicate and are protected from harsh environmental conditions [2].

*Legionella pneumophila* is most frequently associated with human disease (Legionnaire's disease-LD or Pontiac fever); however, other species, including *L. bozemanae*, *L. dumoffii*, and *L. longbeachae* also cause human infections. The most common way of contagion is via aerosols inhalation containing infectious *Legionella* from showerheads, certain medical equipment (e.g., respiratory equipment), cooling towers, hydrotherapy equipment, and decorative fountains [5].

A range of physical and chemical disinfection methods have been proposed with the aim of controlling *Legionella* contamination; however, to date, the most effective procedures have not been defined [6,7]. Therefore, alternative disinfection methods that are effective in controlling the proliferation of *Legionella* could be useful tools to reduce the risk of the spread of Legionnaires' disease.

Surveying and monitoring of legionellae in water systems is needed for risk assessment and prevention of legionellosis. However, although the assessment of *L. pneumophila* in water is typically performed by culture isolation on selective media, it has several limits including the long incubation times and the inability to detect the viable but non-culturable bacteria (VBNC). For this reason, in the last decades, alternative tools for rapid, sensitive, and specific detection of *Legionella* in water samples have been proposed [7,8]. For the identification of possible sources of contamination/infection, high-resolution genotyping of new isolates (e.g., Sequence Based Typing, Multilocus Variable number of tandem repeats) is needed to correlate environmental with clinical isolates.

In order to increase the knowledge on different aspects of *Legionella* contamination in the water environment, this Special Issue aims to bring together research studies related to the occurrence of *Legionella* in water systems of different critical environments (hospital, hotel, large buildings); the role of different factors that can influence the *Legionella* contamination (e.g., disinfection treatment, water characteristics, plumbing materials, and protozoa presence) as well as the advantages and disadvantages of different methodological approaches were also addressed.

## 2. Occurrence of *Legionella* in Large Building Water Systems with Different Methodological Approaches

Many studies have demonstrated that the main sources for LD are the drinking water distribution systems (DWDS) in large buildings such as hospitals and hotels. In particular, the *Legionella* contamination of hospital water systems posed a high risk for patients and hospital staff. The study of Zayed and collaborators [9] showed a different distribution of the *L. pneumophila* population in DWDS of eight hospitals throughout the West Bank, highlighting a low concentration of culturable *L. pneumophila* in water, but a higher prevalence in biofilm. The detection method used influenced the results obtained: in fact, PCR analyses showed a higher detection rate in water and biofilm with respect to culture analyses. The study of environmental isolates is needed for the characterization of the *Legionella* population and to identify possible sources of infection. The genotyping with Multilocus Variable number of tandem repeats Analysis using 13 loci (MLVA-8(12)) identified 20 genotypes only described for the West Bank and they were attributed to individual groundwater based water supplies. The comparison between the MLVA genotyping and the standard Sequence Based Typing (SBT) methods showed that MLVA was highly consistent with SBT, but showed a higher resolution. This method provides a good basis for detailed studies of the health and water management relevant traits of *L. pneumophila* in support of a better clinical and DWDS management. The Whole Genome Sequencing (WGS) also provides detailed genetic information about the *Legionella* strain in large building water system, beyond that obtained from SBT alone, enabling potential subspecies identification, refined taxonomic classification, and genetic profiling for virulence properties [10].

Different studies showed variable rates of contamination and species diversity of *Legionella* in water systems of other large buildings such as hotels, in non-outbreak situation. However, only little information was available on the molecular diversity of *Legionella* spp. in hotel settings. In the study of Yakunin and collaborators [11], the results obtained in Israeli Hotels highlighted a relevant *Legionella* contamination of the DWDS. In 37% of the investigated hotels, *Legionella* spp. counts exceeded the regulatory threshold (1000 CFU/L). The most frequently contaminated water sources were cooling towers followed by faucet, hot tubes, water lines, and storage tanks. In the same study, as also reported by Zayed et al. [9], several *Legionella* strains were found to be related to specific geographical regions. This finding could be associated with the water differences between the regions, i.e., physical and chemical properties. The results obtained highlighted the importance of investigating the prevalence and diversity of *Legionella* strains in hotel buildings in different geographical regions in order to facilitate the risk assessment, surveillance, and control measures of travel-associated Legionnaire's disease.

## 3. Prevention and Control of *Legionella* Contamination

The new revision of the European Drinking Water Directive, such as the WHO guidelines for Drinking Water Quality, suggests the approach of the Water Safety Plan to evaluate the risk associated to the main pathogens involved in waterborne diseases including *Legionella*. The Italian Guidelines support the development of a risk assessment plan and emphasize the need for an adequate environmental surveillance plan. The study of Mazzotta et al. [12] highlighted, during a *Legionella* environmental Surveillance Program in different hospitals, the critical role of Surgical and Washing Outlets (SHWO) with Thermostatic Mixer Valves (TMV) in bacterial growth and Health Care-Associated Infections (HAIs) risk. A non significant difference of *Legionella* contamination between hot and cold samples demonstrated a continuous mixing between two pipelines that create an environment capable of supporting *Legionella* growth. The characteristics of the mixed water produced are also able to influence the distribution of isolates (*L. pneumophila* percentage > in hot water). The results obtained underlined the importance of the implementation of environmental surveillance programs with the aim to deepen the critical points.

A wide variety of disinfection techniques (e.g., chemical disinfection, UV, high temperature) can be used for the prevention and control of *Legionella* contamination in the water network. Girolamini et al. [13] have evaluated that the long term $H_2O_2/Ag^+$ treatment, a low cost disinfectant easy to dose and not very aggressive on the pipelines, is a good strategy to decrease risk in the hospital, reducing the *Legionella* contamination level. However, to guarantee the efficiency of the *Legionella* reduction, it is also necessary to consider the building characteristics, apply an adequate risk assessment plan, increase the monitoring samples size and regulate the dosage in relation to the *Legionella* loads. These infection prevention strategies can be applied to reduce the risk that is also coming from other contaminated healthcare facilities, such as Dental Unit Water Lines (DUWLs), as reported by Tuvo et al. [14]. In this work, the authors highlighted that an implemented risk management plan, that include filters installation and shock disinfection with a solution of 4% hydrogen peroxide and surfactants, appears to be a promising alternative for decreasing *Legionella* colonization in DUWLs of Hospital Clinics. In this context, it is important to highlight that, in the dental unit investigated, a water safety plan, a maintenance plan, and a control program were constantly applied, but there was a low adherence to good practices in DU management. A low adherence to the best practice guidance had probably contributed to biofilm proliferation, making necessary measures that are more restrictive.

## 4. Parameters Influencing the *Legionella* Occurrence in Building Water Systems

Various parameters such as physical (temperature, pH range, hardness), chemical (disinfectant, pipe materials), microbial (free-living amoeba, protozoa), and characteristics of building water systems can influence *Legionella* occurrence. In their paper in the present Special Issue, Cullom and co-workers [15] systematically reviewed the literature to critically examine the varied effects of common metallic (copper, iron) and plastic (PVC, PEX) pipe materials on factors influencing opportunistic pathogens such as *Legionella* growth in drinking water, including the nutrient availability, disinfectant levels, and the composition of the broader microbiome. Plastic pipes demonstrate a lower disinfectant demand while iron pipes exhibit a high disinfectant demand and they can favor the biofilm colonization. Although copper pipes are known for their antimicrobial properties, under some circumstances, copper's interactions with premise plumbing water chemistry and resident microbes can encourage growth of opportunistic pathogens. Plumbing design, configuration, and operation can be manipulated to control such interactions and health outcomes. The influences of pipe materials on opportunistic pathogen physiology should also be considered, including the possibility of influencing virulence and antibiotic resistance. Moreover, the study of Martin and co-workers [16] demonstrated, under controlled laboratory conditions, the importance of considering interactive effects with flow and pipe materials, particularly with respect to relative water corrosivity and influence on residual chlorine levels, in keeping *Legionella* levels low. The complex interaction between the various chemical, physical, and microbiological parameters and *Legionella* contamination is also highlighted by Buse et al. [10]. In their study, negative and positive correlations between *Legionella* and some water characteristics (pH, temperature, turbidity, chlorine, Heterotrophic plate count, and *Vermamoeba vermiformis* contamination) were observed and they varied between location and sample types. The authors concluded that future studies would help elucidate ways to effectively manage the risks associated with *Legionella* exposure within the drinking water distribution systems.

The relationship between *L. pneumophila* contamination and environmental drivers (e.g., temperature, pH, conductivity, iron, nitrate, nitrite, ammonia, copper, phosphate, zinc, hardness, magnesium, calcium of bulk water) was also investigated by Zayed et al. [17]. Statistical analyses with physico-chemical parameters revealed a decrease of *L. pneumophila* abundance in water and biofilms with increasing magnesium concentrations. MLVA-genotype analysis of the *L. pneumophila* isolates and their spatial distribution indicated three niches characterized by distinct physico-chemical parameters and inhabited by specific consortia of genotypes. This study provides novel insights into mechanisms shaping *L. pneumophila* populations and triggering their abundance leading to an understanding of their genotype-specific niches and ecology in support of improved prevention measures.

Some water quality measurements have been suggested as alternative approaches to predict the *Legionella* risk for building's water system instead of directly culturing analysis. However, as reported in the study of Pierre and collaborators [18], a poor correlation and a low positive predictive value between the hot water return line and distal outlet positivity in different buildings was revealed. Moreover, no correlation between *Legionella* distal site positivity and total bacteria, pH, free chlorine, calcium, magnesium, zinc, manganese, copper, temperature, total organic carbon, or incoming cold-water chlorine concentration was observed. These data confirm that these water quality parameters should not be used alone to determine the building's *Legionella* colonization rate and effectiveness of water management programs.

## 5. Sensitivity and Selectivity of Different Culture Media for *Legionella* Detection

The plate culture method using specific media usually supplemented with different combinations of antimicrobial selective substances is considered the gold standard for the detection and enumeration of *Legionella* in water samples. The culture method is generally performed according to standards, such as the International Standard Organization (ISO); in particular, the ISO 11731 is the most used and it has recently been updated. These updates introduced the utilization of different media: (I) the buffered charcoal yeast extract (BCYE) agar, (II) the BCYE with selective supplements (BCYE+AB) containing polymixin B, sodium cefazolin and pimaricin, (III) the highly selective Modified Wadowsky Yee (MWY) agar or, as an alternative, the glycine, vancomycin, polymyxin B, cycloheximide (GVPC) agar. The MWY is the best medium for isolating *L. pneumophila* from potable water samples, while GVPC was proposed in water samples characterized by high interfering microbial flora. Despite these premises, the study conducted by Scaturro et al. [19] in potable water samples with low interfering microorganisms observed that GVPC was more efficient in detecting *Legionella* contamination than the BCYE medium. Moreover, no significant difference of *Legionella* loads (CFU/L) was found between BCYE and GVPC agar plates. Furthermore, the possibility of improving the isolation of *Legionella* non-*pneumophila* species on BCYE was not confirmed. These results make questionable the need to utilized BCYE agar plates to analyze potable water samples.

Given that the recovery of *Legionella* spp. strictly depends on the type of agar being used, quality-assured culture media for water testing are key to consumer safety. In this context, Ditommaso and collaborators [20] reported a comparative assessment of the sensitivity and selectivity of MWY and BCYEα media supplied by two different manufacturers (Xebios Diagnostics and Oxoid) in water samples. Even though the analysis showed an excellent agreement between the recovery rates of the four media tested, the quantitative recovery of *Legionella* spp. colonies using Xebios media was significantly greater than that achieved by Oxoid media. Furthermore, the sensitivity of detection was significantly higher when samples were plated on MWY Xebios agar, while the selectivity of MWY appeared to be the same regardless of the manufacturer. Finally, the MWY Xebios medium enhanced the recovery of non-*pneumophila Legionella* species. The results obtained confirmed that culture protocol standardization, as well as quality control of the culture media, are essential to achieve intra- and interlaboratory reproducibility and accuracy.

Although plate culture methods are the gold standard for *Legionella* detection in water samples, they have high variability in the enumeration, are time consuming, and require significant experience in recognizing *Legionella* colonies. A promising alternative method is the Legiolert test, a liquid culture method based on bacterial enzyme detection technology, which determines the most probable number (MPN) of *L. pneumophila* species in water samples. Scaturro et al. [21] highlighted that the plate culture method (MWY) and Legiolert method were comparable and concluded that Legiolert may be considered as a valuable test for the detection and enumeration of *L. pneumophila* in potable water samples and it can be used as a valid alternative to the traditional plate culture methods.

## 6. The Role of Protozoa in the *Legionella* Contamination in Water Distribution System

The implication of interaction with a protozoan host for the control of *L. pneumophila* as well as the efficacy of potable water disinfection protocols on *L. pneumophila* and host protozoan are essential and they were reviewed by Nisar and collaborators [22]. The systematic review highlights that protozoan hosts facilitate the intracellular replication and packaging of viable *L. pneumophila* in infectious vesicles, while cyst-forming protozoans provide protection from prolonged environmental stress. Moreover, the data collected underline the failure of common disinfection procedures to achieve long-term elimination of *L. pneumophila* and protozoan hosts from potable water. This overview report that the disinfection procedures and protozoan hosts also facilitate biogenesis of viable but non-culturable

(VBNC) *L. pneumophila*, which have been shown to be highly resistant to many water disinfection protocols. However, other studies have demonstrated that all free-living amoebae (FLA) do not exhibit the same behavior when they are exposed to *L. pneumophila* strains. In fact, the *Villaertia magna* strain C2c Maki has been demonstrated to eliminate the *L. pneumophila* serogroup 1 strain Paris. The results obtained in the study of Hasni et al. [23] confirmed that none of the *Legionella* strains tested (Paris, Philadelphia, and Lens) exhibit intracellular growth and that the *V. magna* strain C2c Maki decreases the number of internalized *L. pneumophila*. Thus, these results support the idea that the *V. magna* strain C2c Maki has a unique behavior in regard to *L. pneumophila* strains. The non-permissiveness of *V. magna* C2c Maki toward *L. pneumophila* strains was confirmed by Mameri et al. [24]. This study demonstrated that *V. magna* C2c Maki did not increase the expression of different virulence genes (htpX, icmE, lirR, ccmF, gacA, tatB, and lvrE) of *L. pneumophila* strains in contrast to *Acanthamoeba castellanii*.

**Funding:** This research received no external funding.

**Conflicts of Interest:** The authors declare no conflict of interest.

## References

1. Cunha, C.B.; Cunha, B.A. Legionnaire's Disease since Philadelphia: Lessons Learned and Continued Progress. *Infect. Dis. Clin. N. Am.* **2017**, *31*, 1–5. [CrossRef]
2. Fields, B.S.; Benson, R.F.; Besser, R.E. *Legionella* and Legionnaires' Disease: 25 Years of Investigation. *Clin. Microbiol. Rev.* **2002**, *15*, 506–526. [CrossRef] [PubMed]
3. Bonetta, S.; Ferretti, E.; Balocco, F.; Carraro, E. Evaluation of *Legionella pneumophila* contamination in Italian hotel water systems by quantitative real-time PCR and culture methods. *J. Appl. Microbiol.* **2010**, *108*, 1576–1583. [CrossRef] [PubMed]
4. Bargellini, A.; Marchesi, I.; Righi, E.; Ferrari, A.; Cencetti, S.; Borella, P.; Rovesti, S. Parameters predictive of *Legionella* contamination in hot water systems: Association with trace elements and heterotrophic plate counts. *Water Res.* **2011**, *45*, 2315–2321. [CrossRef] [PubMed]
5. Cunha, B.A.; Burillo, A.; Bouza, E. Legionnaires' disease. *Lancet* **2016**, *387*, 376–385. [CrossRef]
6. Lin, Y.E.; Stout, J.E.; Yu, V.L. Controlling *Legionella* in hospital drinking water: An evidence-based review of disinfection methods. *Infect. Control Hosp. Epidemiol.* **2011**, *32*, 166–173. [CrossRef] [PubMed]
7. Bonetta, S.; Pignata, C.; Bonetta, S.; Meucci, L.; Giacosa, D.; Marino, E.; Gorrasi, I.; Gilli, G.; Carraro, E. Effectiveness of a neutral electrolysed oxidising water (NEOW) device in reducing *Legionella pneumophila* in a water distribution system: A comparison between culture, qPCR and PMA-qPCR detection methods. *Chemosphere* **2018**, *210*, 550–556. [CrossRef]
8. Bonetta, S.; Pignata, C.; Bonetta, S.; Meucci, L.; Giacosa, D.; Marino, E.; Gilli, G.; Carraro, E. Viability of *Legionella pneumophila* in Water Samples: A Comparison of Propidium Monoazide (PMA) Treatment on Membrane Filters and in Liquid. *Int. J. Environ. Res. Public Health* **2017**, *14*, 467. [CrossRef]
9. Zayed, A.R.; Pecellin, M.; Salah, A.; Alalam, H.; Butmeh, S.; Steinert, M.; Lesnik, R.; Brettar, I.; Höfle, M.G.; Bitar, D.M. Characterization of *Legionella pneumophila* Populations by Multilocus Variable Number of Tandem Repeats (MLVA) Genotyping from Drinking Water and Biofilm in Hospitals from Different Regions of the West Bank. *Pathogens* **2020**, *9*, 862. [CrossRef]
10. Buse, H.Y.; Morris, B.J.; Gomez-Alvarez, V.; Szabo, J.G.; Hall, J. *Legionella* Diversity and Spatiotemporal Variation in The Occurrence of Opportunistic Pathogens within a Large Building Water System. *Pathogens* **2020**, *9*, 567. [CrossRef]
11. Yakunin, E.; Kostyal, E.; Agmon, V.; Grotto, I.; Valinsky, L.; Moran-Gilad, J. A Snapshot of the Prevalence and Molecular Diversity of *Legionella pneumophila* in the Water Systems of Israeli Hotels. *Pathogens* **2020**, *9*, 414. [CrossRef] [PubMed]
12. Mazzotta, M.; Girolamini, L.; Pascale, M.R.; Lizzadro, J.; Salaris, S.; Dormi, A.; Cristino, S. The Role of Sensor-Activated Faucets in Surgical Handwashing Environment as a Reservoir of *Legionella*. *Pathogens* **2020**, *9*, 446. [CrossRef] [PubMed]
13. Girolamini, L.; Dormi, A.; Pellati, T.; Somaroli, P.; Montanari, D.; Costa, A.; Savelli, F.; Martelli, A.; Grottola, A.; Serpini, G.F.; et al. Advances in *Legionella* Control by a New Formulation of Hydrogen Peroxide and Silver Salts in a Hospital Hot Water Network. *Pathogens* **2019**, *8*, 209. [CrossRef] [PubMed]

14. Tuvo, B.; Totaro, M.; Cristina, M.L.; Spagnolo, A.M.; Di Cave, D.; Profeti, S.; Baggiani, A.; Privitera, G.; Casini, B. Prevention and Control of *Legionella* and *Pseudomonas* spp. Colonization in Dental Units. *Pathogens* **2020**, *9*, 305. [CrossRef]
15. Cullom, A.C.; Martin, R.L.; Song, Y.; Williams, K.; Williams, A.; Pruden, A.; Edwards, M.A. Critical Review: Propensity of Premise Plumbing Pipe Materials to Enhance or Diminish Growth of *Legionella* and Other Opportunistic Pathogens. *Pathogens* **2020**, *9*, 957. [CrossRef]
16. Martin, R.L.; Strom, O.R.; Pruden, A.; Edwards, M.A. Interactive Effects of Copper Pipe, Stagnation, Corrosion Control, and Disinfectant Residual Influenced Reduction of *Legionella pneumophila* during Simulations of the Flint Water Crisis. *Pathogens* **2020**, *9*, 730. [CrossRef]
17. Zayed, A.R.; Butmeh, S.; Pecellin, M.; Salah, A.; Alalam, H.; Steinert, M.; Höfle, M.G.; Bitar, D.M.; Brettar, I. Biogeography and Environmental Drivers of *Legionella pneumophila* Abundance and Genotype Composition Across the West Bank: Relevance of a Genotype-Based Ecology for Understanding *Legionella* Occurrence. *Pathogens* **2020**, *9*, 1012. [CrossRef]
18. Pierre, D.; Baron, J.L.; Ma, X.; Sidari, F.P., 3rd; Wagener, M.M.; Stout, J.E. Water Quality as a Predictor of *Legionella* Positivity of Building Water Systems. *Pathogens* **2019**, *8*, 295. [CrossRef]
19. Scaturro, M.; Poznanski, E.; Mupo, M.; Blasior, P.; Seeber, M.; Prast, A.M.; Romanin, E.; Girolamo, A.; Rota, M.C.; Bella, A.; et al. Evaluation of GVPC and BCYE Media for *Legionella* Detection and Enumeration in Water Samples by ISO 11731: Does Plating on BCYE Medium Really Improve Yield? *Pathogens* **2020**, *9*, 757. [CrossRef]
20. Ditommaso, S.; Giacomuzzi, M.; Memoli, G.; Garlasco, J.; Zotti, C.M. Sensitivity and Selectivity of Two Commercially Available Media for *Legionella* spp. Recovery from Environmental Water Samples. *Pathogens* **2020**, *9*, 523. [CrossRef]
21. Scaturro, M.; Buffoni, M.; Girolamo, A.; Cristino, S.; Girolamini, L.; Mazzotta, M.; Sabattini, M.A.B.; Zaccaro, C.M.; Chetti, L.; Laboratory, M.A.N.; et al. Performance of Legiolert Test vs. ISO 11731 to Confirm *Legionella pneumophila* Contamination in Potable Water Samples. *Pathogens* **2020**, *9*, 690. [CrossRef]
22. Nisar, M.A.; Ross, K.; Brown, M.H.; Bentham, R.; Edwards, J. *Legionella pneumophila* and Protozoan Hosts: Implications for the Control of Hospital and Potable Water Systems. *Pathogens* **2020**, *9*, 286. [CrossRef] [PubMed]
23. Hasni, I.; Jarry, A.; Quelard, B.; Carlino, A.; Eberst, J.-B.; Abbe, O.; Demanèche, S. Intracellular Behaviour of Three *Legionella pneumophila* Strains within Three Amoeba Strains, Including *Willaertia magna* C2c Maky. *Pathogens* **2020**, *9*, 105. [CrossRef] [PubMed]
24. Mameri, R.M.; Bodennec, J.; Bezin, L.; Demanèche, S. Mitigation of Expression of Virulence Genes in *Legionella pneumophila* Internalized in the Free-Living Amoeba *Willaertia magna* C2c Maky. *Pathogens* **2020**, *9*, 447. [CrossRef] [PubMed]

pathogens

MDPI

Article

# Characterization of *Legionella pneumophila* Populations by Multilocus Variable Number of Tandem Repeats (MLVA) Genotyping from Drinking Water and Biofilm in Hospitals from Different Regions of the West Bank

**Ashraf R. Zayed [1,2], Marina Pecellin [1], Alaa Salah [2], Hanna Alalam [2], Suha Butmeh [2], Michael Steinert [3], Rene Lesnik [1], Ingrid Brettar [1], Manfred G. Höfle [1,*] and Dina M. Bitar [2]**

1 Department of Vaccinology and Applied Microbiology, Helmholtz Centre for Infection Research (HZI), Inhoffenstrasse 7, 38124 Braunschweig, Germany; ashraf_manfan@me.com (A.R.Z.); marinapecellin@gmail.com (M.P.); rene.lesnik@helmholtz-hzi.de (R.L.); ibrettar@web.de (I.B.)

2 Department of Microbiology and Immunology, Al-Quds University, Abu-Dies, East Jerusalem 19356, Palestine; alaa.hisham@uks.eu (A.S.); hanna.alalam@gu.se (H.A.); salbutmeh@staff.alquds.edu (S.B.); dbitar@staff.alquds.edu (D.M.B.)

3 Department of Life Sciences, Institute of Microbiology, Technical University of Braunschweig, Universitätsplatz 2, 38106 Braunschweig, Germany; m.steinert@tu-bs.de

* Correspondence: manfred.hoefle@helmholtz-hzi.de

Received: 30 September 2020; Accepted: 19 October 2020; Published: 22 October 2020

**Abstract:** The West Bank can be considered a high-risk area for Legionnaires' disease (LD) due to its hot climate, intermittent water supply and roof storage of drinking water. *Legionella*, mostly *L. pneumophila*, are responsible for LD, a severe, community-acquired and nosocomial pneumonia. To date, no extensive assessment of *Legionella* spp and *L. pneumophila* using cultivation in combination with molecular approaches in the West Bank has been published. Two years of environmental surveillance of *Legionella* in water and biofilms in the drinking water distribution systems (DWDS) of eight hospitals was carried out; 180 *L. pneumophila* strains were isolated, mostly from biofilms in DWDS. Most of the isolates were identified as serogroup (Sg) 1 (60%) and 6 (30%), while a minor fraction comprised Sg 8 and 10. Multilocus Variable number of tandem repeats Analysis using 13 loci (MLVA-8(12)) was applied as a high-resolution genotyping method and compared to the standard Sequence Based Typing (SBT). The isolates were genotyped in 27 MLVA-8(12) genotypes (Gt), comprising four MLVA clonal complexes (VACC 1; 2; 5; 11). The major fraction of isolates constituted Sequence Type (ST)1 and ST461. Most of the MLVA-genotypes were highly diverse and often unique. The MLVA-genotype composition showed substantial regional variability. In general, the applied MLVA-method made it possible to reproducibly genotype the isolates, and was consistent with SBT but showed a higher resolution. The advantage of the higher resolution was most evident for the subdivision of the large strain sets of ST1 and ST461; these STs were shown to be highly pneumonia-relevant in a former study. This shows that the resolution by MLVA is advantageous for back-tracking risk sites and for the avoidance of outbreaks of *L. pneumophila*. Overall, our results provide important insights into the detailed population structure of *L. pneumophila*, allowing for better risk assessment for DWDS.

**Keywords:** MLVA-genotypes; clonal complex; hospital water; West Bank; *Legionella pneumophila*

---

## 1. Introduction

*Legionella* are opportunistic pathogens with a widespread distribution in freshwater environments. This bacterial genus is well known to cause legionellosis. The term "legionellosis" describes both Legionnaires' disease (LD), a severe form of atypical pneumonia, and a nonpneumonic febrile illness called Pontiac Fever. Due to the low concentrations of nutrients in their aquatic habitats, *Legionella* have adapted to live in biofilms, where they can obtain amino acids and carbon sources that they need for survival, replication and protection from temperature changes and biocide treatment [1]. In biofilms, *Legionella* are part of complex microbial communities [2] where they are subjected to predation by protozoa [3]. The transmission of bacteria from the environment to humans occurs via inhalation or aspiration of *Legionella*-containing aerosols [4,5]. Among the more than 60 species of the genus *Legionella*, *L. pneumophila* is responsible for approximately 90% of all globally reported community- and hospital-acquired cases of legionellosis [6–8]. *L. pneumophila* has 15 serogroups (Sgs); Sg1 is the most common, causing LD. Sg6 comes second, and is also a causative agent of LD [9,10].

Many studies have demonstrated that the main sources for LD are the drinking water distribution systems (DWDS) in large buildings like hospitals and hotels [11–13]. The contamination of hospital water systems with *Legionella* is considered to pose a high risk for patients, especially for those with severe diseases. To this end, it is well known that LD is an important cause of hospital-acquired pneumonia [8]. The presence of *Legionella* in DWDS is a serious health risk to hospital staff and patients, but the magnitude of the problem is often unrecognized [8,14,15]. A high seroprevalence of *L. pneumophila* has been observed among health care workers [16]. The problem is compounded in the West Bank because awareness about the prevalence of *L. pneumophila* or LD is lacking, and few data are available for such arid regions. Furthermore, there are no specific guidelines for *L. pneumophila* surveillance or protection from exposure in hospitals or public buildings.

Cultivation Dependent Analysis (CDA) on specific agar plates is the standard and recommended technique used for environmental surveillance of *L. pneumophila* [17]. One major obstacle in the isolation and quantification of *L. pneumophila* by culture is often its Viable But Non Culturable (VBNC) state, and overgrowth by competing bacteria [18–20]. On the other hand, Cultivation Independent Analyses (CIA) using PCR-based molecular approaches are rapid, sensitive and widely applied for the detection and identification of *L. pneumophila* [21]. Because of the widespread occurrence of *L. pneumophila* in large, man-made, freshwater systems, study of environmental isolates is needed for the implementation of prevention measures, and to identify possible sources of infection [22,23].

For the identification of possible sources of contamination/infection, high resolution genotyping of new isolates is needed to correlate environmental isolates with clinical isolates. This is currently done by two molecular approaches: Sequence Based Typing (SBT) and Multilocus Variable Number of Tandem Repeats (VNTR) Analysis (MLVA). SBT of *L. pneumophila* is done by sequencing a set of seven reference genes per isolate, providing a specific Sequence Type (ST) that can be matched with an International database [24]. MLVA has been widely used to identify different pathogens [25–27]. VNTRs consist of relatively short DNA fragments repeated in tandem that can vary in copy number among strains. For *L. pneumophila*, MLVA has exhibited an excellent ability to distinguish among strains if 8 to 12 different loci are used. It can be performed in a multiplexed manner, followed by capillary electrophoresis, enabling automated sample analysis and data acquisition. These advantages reduce typing time and costs. Several studies have used MLVA for the genotyping of *L. pneumophila* strains [28,29]. They showed the high correspondence between MLVA genotypes and STs with an important increase in resolution when applying MLVA, which is relevant for understanding clonal populations. Recent publications showed that the majority of clinically relevant strains were distributed into a limited number of Clonal Complexes (CCs) defined by MLVA, called VNTR analysis CC (VACC) and characterized by epidemic reference strains such as Paris (VACC1) and Philadelphia-1 (VACC2) [29]. Due to its advantages, MLVA could complement SBT for large sets of isolates and enable insights into the clonal structure of *L. pneumophila* populations, as well as help selecting strains for further whole genome sequencing.

The principal objective of this study was to assess the current distribution of *L. pneumophila* populations from DWDS of hospitals throughout the West Bank and determine their clonal structure and genetic diversity. To this end, *Legionella* abundances were determined for two years in bulk water and biofilms in the hospitals by applying both cultivation-dependent and -independent analyses. By the cultivation analyses, 180 *L. pneumophila* isolates were obtained from water and biofilm. These isolates were subjected to MLVA using 13 loci to reveal their clonal structure and genetic diversity. In comparison to MLVA databases, the uniqueness of the MLVA-genotypes of the West Bank could be assessed. For the West Bank, the clonal structure of *L. pneumophila* was related to the different locations and habitats and to situ-SBT analysis from pneumonia patients.

## 2. Materials and Methods

### 2.1. Study Sites, Water and Biofilm Sampling

Drinking water in the West Bank is derived from groundwater, mainly well water, with some being provided by springs. Water was provided to most of the sampled sites by the Palestinian Water Authority, except for Ramallah (sampling site D), with Mekorot as the provider. Except for site D, water treatment consisted of chlorination in storage sites before provision to the end user. All hospitals had drinking water reservoirs for water storage.

Water samples and biofilm swabs were sampled six times during the period from October 2012 to December 2014 from eight hospitals across the West Bank (Figure S1). The hospitals had the following coordinates: hospital A (coordinates: 32°27′ N, 35°17′ E), hospital B (32°13′ N, 35°14′ E) and hospital C (32°13′ N, 35°15′ E) in northern West Bank, hospital D (31°53′ N, 35°12′ E) and hospital E (31°46′ N, 35°14′ E) in central West Bank, and hospital F (31° 42′ N, 35° 11′ E), hospital G (31°33′ N, 35°4′ E) and hospital H (31°31′ N, 35°5′ E) in southern West Bank. Also, samples were taken from Al-Quds University (AQU) main campus, Abu Dies, East Jerusalem (31°45′ 18.07″ N, 35°15′ 37.614″ E). The six samplings twice covered the main seasons, i.e., spring (March–May), summer (June–August), and autumn (October–December). It should be noted that site D could only be sampled once for spring, summer and autumn, while all other sites were sampled twice for these seasons [30].

Cold and hot water (if available) was collected from a faucet close to the hospital's drinking water reservoir and biofilm swabs were taken from faucets, showerheads, and hoses. This study is representative for Jenin, Nablus, Ramallah, Jerusalem, Bethlehem, and Hebron, going from north to south in the West Bank. The temperature, chlorine, pH, hardness, and conductivity of the water samples were determined upon collection using probes and quantofix sticks (Macherey-Nagel GmbH, Düren, Germany). Further details on sampling and individual results of the physico-chemical parameters are given in Zayed [30].

### 2.2. Cultivation-Dependent Analysis

A total of 72 water samples were collected in sterile 1L plastic bottles after a brief flow time (2–3 min). One liter each of cold and hot water was collected for Heterotrophic Plate Counts (HPC) and again for *Legionella* counts from the hospitals. To neutralize residual free chlorine, 0.5 mL of 0.1 N sodium thiosulphate was added to the sterile bottles for *Legionella* plate counts [31].

For HPC, yeast agar plates (Ant. Er. CP63.1, Carl Roth, Karlsruhe, Germany) were used according to the manufacturer's instruction for each type of water in two sets of triplicates. First, 0.1 mL of the water sample was spread on each agar plate using a sterile glass spreader. The plates were inverted and incubated; three plates each were incubated at 37 °C for 48 h and at 25 °C for 72 h.

Concerning *Legionella* plate counts, 100 mL of water sample was filtered onto a membrane filter (membrane solutions, pore size 0.45 µm, diameter 47 mm, Whatman, England) using sterile filtration unit (Nalgene, Germany). A vacuum of 200 mbar was applied. After filtration, 30 mL of acid buffer (3.9 mL of 0.2 mol/L HCl and 25 mL of sterile 0.2 mol/L KCl were mixed, pH 2.2 ± 0.2) was placed on top of the membrane filter and left for 5 min. The filter was rinsed with 20 mL Page's saline (1.20 g NaCl,

0.04 g $MgSO_4 \cdot 7H_2O$, 0.04 g $CaCl_2 \cdot 2H_2O$), and 1.42 g $Na_2HPO_4$ and 1.36 g $KH_2PO_4$ were dissolved in ten liters of distilled water and autoclaved. The membrane filter was removed from the filtration unit with sterile forceps and placed onto the relevant agar plate. Duplicates of BCYE and GVPC (M809, Himedia, India) agar plates were used according to the manufacturer's instruction. The plates were incubated inverted at 37 °C for 10 days. Plates were checked for growth twice (after three and ten days). Final counts of the triplicates were done after ten days with descriptions of the colonies.

Also, a total of 1136 biofilm swabs from the anterior surfaces of faucets, showerheads or shower hoses in all hospital wards, mainly in areas occupied by high-risk patients (intensive care unit, operating theater, oncology and surgery wards), was obtained using transport medium (Copan, Culture swab transport system, Italy). Swabs for *Legionella* identification were processed immediately by culturing on GVPC agar (medium M809, Himedia, India) based on ISO 11731:2004 [17].

### 2.3. Cultivation-Independent Analysis (16S rDNA PCR)

A total of 72 samples (five liters) each of cold and hot water was collected from the main water source from each site for DNA extraction. Water samples were filtered onto sandwich membrane filters composed of nucleopore-filter (Nuclepore Track-Etch Membrane, MB 90 mm, 0.2 µm, Whatman, UK) and glass fiber-microfilter (GF/F) (GFF, 90 mm, Whatman, UK). Also, a total of 225 biofilm swabs from the anterior surfaces of faucets, showerheads or shower hoses was obtained for DNA extraction using sterile cotton swabs (Cotton Tipped Applicator, Beijing, China).

For the extraction of DNA from the filter sandwiches and the swabs, a modified DNeasy protocol (Qiagen kit No. 69506, Hilden, Germany) was used. Briefly, sandwich filters were cut into small pieces and incubated with enzymatic lysis buffer (20 mM Tris-HCl, 2 mM EDTA, 1.2% Triton X-100 [pH 8.0]) containing 10 mg/mL lysozyme for 60 min in a 37 °C water bath. After the addition of AL buffer from the kit, the samples were incubated at 78 °C in a shaking water bath for 20 min. After filtration through a cell strainer, i.e., 100 µm (DB falcon 352360, Corning, Glendale, AZ, USA), absolute ethanol was added to the filtrate (ratio of filtrate to ethanol is [2:1]) and the mixture was applied to the spin column of the kit. After this step, the protocol was followed according to the manufacturer's instructions.

Three different PCRs were carried out as follows: (i) for the detection of any bacteria, the bacterial common 16S rRNA gene primers (Com), (ii) for *Legionella* genus-specific primers (Lgsp) and (iii) for *L. pneumophila* species-specific primers (Lp1) were applied [32]. Each PCR reaction was carried out using 3 µL (1 ng/µL) of DNA template in a final volume of 25 µL. Amplification was achieved using PCR-ready Master Mix (GoTaq, Green Master Mix, Promega, Madison, WI, USA).

To test the specificity of *L. pneumophila* primers and confirm species identity, six isolates were identified by amplifying and sequencing an internal fragment of the 16S rRNA gene according to Senderovich et al. [33]. The obtained sequences were compared using the NCBI service to certain closest relatives. The sequences were submitted to the GeneBank database (KX778102-KX778107). Sequencing of the 16S rRNA gene of the six isolates confirmed the presence of *L. pneumophila* (≥99.8% 16S rRNA gene similarities).

### 2.4. Sero-Grouping of Legionella Isolates

The serogroups of the 180 *L. pneumophila* isolates were identified by an agglutination test using *Legionella* Latex (Oxoid DR0800, Basingstoke, UK). Using this test, the isolates were sero-grouped as Sg1 and Sg 2–14. Moreover, 47 isolates were sent to the National Reference laboratory for *Legionella* infections in Dresden for analysis by monoclonal antibody subgrouping [34].

### 2.5. Genotyping of L. pneumophila Isolates

For molecular typing of *L. pneumophila* at the strain level, MLVA-13 (MLVA-12 plus 1 additional locus from MLVA-8) designated as (MLVA-8(12)) analysis was performed for 180 isolates. DNA extraction was done either directly from living biomass using (Qiagen kit No. 69504, Hilden, Germany) according to the manufacturer protocol, or from biomass on FTA cards (Whatman, Sigma-Aldrich, Darmstadt, Germany).

For DNA extraction from the FTA cards, the area of the card containing the biomass was punched into 3 mm circular pieces. The pieces were transferred to 0.5 mL sterile water (Roth, Karlsruhe, Germany), incubated for 3 min at room temperature and vortexed three times (after water addition, after 1 min and after 3 min incubation). The FTA punch was removed and 1× Tris-EDTA buffer (Sigma-Aldrich, Darmstadt, Germany) was added to the water to preserve the DNA from degradation. More details on the FTA technology are given by Rajendram et al. [35]. DNA was finally quantified by Nanodrop spectrophotometer (NanoDrop, Thermo Scientific, Dreieich, Germany). MLVA-8 and MLVA-12 molecular genotyping assays by multiplex PCR and capillary electrophoresis were carried out for all isolates, as detailed by Pourcel et al., Sobral el al., Visca et al. and Pecellin [28,29,36,37].

For comparison, a subset of strains representing all MLVA-genotypes was characterized by sequence-based typing (SBT) [38]. In addition, several *L. pneumophila* (Lpn) reference strains were used to generate MLVA-8(12) profiles for comparison and interpretation of the results; these reference strains were Lpn str. Philadelphila-1 (ATCC 33152$^{T}$) [39], Lpn str. Paris (CIP 107629) [40], Lpn str. Bloomington-2 (ATCC 3315) and Lpn str. Corby (NC_009494) [41].

### 2.6. Statistical Analysis

Statistical analysis was performed using the GraphPad Prism software v7.0 (Graph-Pad, San Diego, CA, USA), and cluster analysis and a phylogenetic tree were constructed using PRIMER software v7.0.7 (Primer-e, Auckland, New Zealand). Non-normalized data were normalized. Data are presented as means ± standard deviation (SD). An agglomerative clustering dendrogram was created using the PRIMER software in order to study the similarities between the genotyping characteristics of *L. pneumophila* strains belonging to different VNTR markers (Lpms). The resemblance matrix was calculated using the Bray-Curtis index of association on the VNTR marker.

Capillary electrophoresis data analysis and calculation of the number of repeats for each VNTR marker were performed as described in Pourcel et al. [28]. The numerical code used to designate the MLVA-8 and MLVA-12 genotypes, as well as the joint code for the MLVA-8(12) genotypes, were continued for the isolates. Null alleles ("0") were assigned when no amplicon was detected. Cluster analysis was performed in Bionumerics (version 5.0, Applied Maths, Gent, Belgium). The UPGMA (Unweighted Pair Group Method with Arithmetic Mean) method using a categorical coefficient was applied to define the clusters. The MLVA-8 profiles obtained in this study were compared to those from the *Legionella* MLVA-database, and clusters were defined applying a cut-off of 60% similarity, as done previously [29]. Minimum spanning trees were performed using the categorical coefficient. Simpson's Index of Diversity coefficient was calculated using the online tool provided in https://www.easycalculation.com/statistics/simpson-diversity-index.php. To measure the variation of the number of repeats at each VNTR locus, the Hunter-Gaston Discrimination Index (HGDI), which is a modification of the Simpson's Index of Diversity, was calculated according to Pecellin [37].

## 3. Results

### 3.1. Biological and Physico-Chemical Characteristics of Drinking Water Distribution Systems (DWDS)

The sampled drinking water of the eight hospitals was mainly groundwater based and characterized by a high hardness (on average 230–300 mg/L $CaCO_3$ equivalents) and high conductivity (on average 650–900 µS) (Table 1). The average temperature of the cold water ranged between 21.1 °C and 24.3 °C. The average temperature of hot water ranged between 38.6 °C and 51.9 °C. The average pH of the cold and hot water was 7.6 and 8.0. The conductivity of the hot water was higher than that of the cold water in all hospitals. Chlorine varied on average between 0.2 and 0.7 mg/L. Heterotrophic plate counts at 37 °C ranged from $1.7 \times 10^4$ to $1.6 \times 10^5$ CFU/L.

**Table 1.** Average of biological and physico-chemical parameters from the drinking water systems (DWDS) of the eight hospitals in the West Bank.

| Hospital | Water Type | HPC (CFU/L) at 37 °C ± SD | HPC (CFU/L) at 22 °C ± SD | *Leg.* Count (CFU/L) ± SD | Tempe-Rature °C ± SD | pH ± SD | Conductivity µS/cm ± SD | Chlorine mg/L ± SD | Hardness (mg/L) |
|---|---|---|---|---|---|---|---|---|---|
| A | Cold | $2.1 \times 10^4 \pm 1.6 \times 10^2$ | $7.4 \times 10^3 \pm 7.1 \times 10^1$ | $4.3 \times 10^1 \pm 1.06 \times 10^2$ | 23.1 ± 2.4 | 7.7 ± 0.2 | 780 ± 158 | 0.5 ± 0.3 | 290 ± 36 |
| | Hot | NA | NA | NA | NA | NA | NA | NA | NA |
| B | Cold | $2.0 \times 10^4 \pm 2.2 \times 10^2$ | $1.4 \times 10^3 \pm 1.4 \times 10^2$ | BD | 24.3 ± 2 | 8.0 ± 0.2 | 639 ± 86 | 0.4 ± 0.2 | 299 ± 16 |
| | Hot | $2.2 \times 10^2 \pm 2.9 \times 10^1$ | $1.1 \times 10^2 \pm 1.4 \times 10^1$ | BD | 51.9 ± 16.3 | 7.8 ± 0.2 | 700 ± 30 | NA | NA |
| C | Cold | $1.5 \times 10^4 \pm 2.0 \times 10^2$ | $1.3 \times 10^3 \pm 2.0 \times 10^2$ | BD | 23.1 ± 2.5 | 8.0 ± 0 | 707 ± 60 | 0.5 ± 0 | 299 ± 16 |
| | Hot | NA | NA | NA | NA | NA | NA | NA | NA |
| D | Cold | $1.6 \times 10^5 \pm 8.7 \times 10^3$ | $1.2 \times 10^4 \pm 8.9 \times 10^3$ | BD | 22.5 ± 1.6 | 7.6 ± 0.1 | 509 ± 118 | 0.7 ± 0.35 | 229 ± 26 |
| | Hot | NA | NA | NA | NA | NA | NA | NA | NA |
| E | Cold | $1.7 \times 10^4 \pm 2.0 \times 10^2$ | $1.5 \times 10^3 \pm 2.0 \times 10^1$ | BD | 25.2 ± 5.2 | 7.8 ± 0.3 | 699 ± 82 | 0.2 ± 0.05 | 269 ± 11 |
| | Hot | $1.6 \times 10^4 \pm 3.5 \times 10^2$ | $4.1 \times 10^2 \pm 4.2 \times 10^1$ | BD | 47.9 ± 5.1 | 7.8 ± 0.2 | 761 ± 30 | NA | NA |
| F | Cold | $1.1 \times 10^5 \pm 1.5 \times 10^2$ | $6.3 \times 10^4 \pm 7.1 \times 10^2$ | $1.5 \times 10^2 \pm 2.3 \times 10^2$ | 24.7 ± 2.9 | 7.8 ± 0.2 | 556 ± 106 | 0.4 ± 0.2 | 261 ± 3 |
| | Hot | $2.5 \times 10^4 \pm 2.6 \times 10^2$ | $3.2 \times 10^3 \pm 3.3 \times 10^1$ | $9.1 \times 10^1 \pm 1.3 \times 10^2$ | 43.3 ± 5.7 | 7.8 ± 0.2 | 583 ± 93 | NA | NA |
| G | Cold | $2.9 \times 10^4 \pm 2.0 \times 10^2$ | $2.2 \times 10^3 \pm 2.2 \times 10^2$ | $8.3 \times 10^0 \pm 2.0 \times 10^1$ | 21.6 ± 2.7 | 8.0 ± 0.3 | 648 ± 179 | 0.4 ± 0.2 | 266 ± 19 |
| | Hot | $1.8 \times 10^4 \pm 1.5 \times 10^2$ | $2.7 \times 10^3 \pm 1.8 \times 10^1$ | BD | 38.6 ± 6.8 | 8.0 ± 0.3 | 596 ± 190 | NA | NA |
| H | Cold | $2.7 \times 10^4 \pm 2.5 \times 10^2$ | $1.4 \times 10^3 \pm 1.3 \times 10^2$ | BD | 20.8 ± 1.3 | 7.8 ± 0.1 | 707 ± 191 | 0.4 ± 0.2 | 266 ± 19 |
| | Hot | NA | NA | NA | NA | NA | NA | NA | NA |

NA: Not Available. BD: Below detection limit (<5 CFU/L).

During the study period, *L. pneumophila* was detected in the DWDS of all hospitals and at Al-Quds University. The sampling comprised six campaigns, twice covering the main seasonal changes. Sampling comprised water and biofilms, with comparable numbers of samples taken from each hospital.

From water samples, five *L. pneumophila* strains were isolated from 72 samples. *L. pneumophila* was isolated only from the drinking water of three hospitals (A, F and G). For the collection period, the *Legionella* counts per hospital ranged from 0 to 150 CFU/L (Table 1). In hospital A, on average, 43 CFU/L of *Legionella* spp from cold water were detected; in hospital F, 150 CFU/L of *Legionella* spp were detected from cold water and 91 CFU/L from hot water; finally, in hospital G, on average, 8 CFU/L *Legionella* spp were detected from cold water. Hot water was not continuously available from all hospitals, but *Legionella* spp counts were comparable to cold water [42]. *Legionella* spp, mainly *L. pneumophila*, was isolated from 191 out of 1136 biofilm swab samples (16.8%) (Table 2). The majority of *Legionella* positive samples were detected in 2012 (23.5%), while the fewest samples tested positive in 2013 (8.7%). The highest frequency of *L. pneumophila* in biofilm swabs was detected in hospital F (26.3%), where *Legionella* spp were also detected in the DWDS during 2012–2014. Meanwhile, the lowest frequency was detected in hospital C (3.3%). Finally, a high frequency of *L. pneumophila* was detected in Al-Quds University (36.4%) during the only collection in 2012.

**Table 2.** Occurrence frequencies of *Legionella* in water and biofilm samples obtained from eight hospitals and Al-Quds University in the West Bank during the study period (2012–2014).

| | CDA [1] | | | CIA [2] | | | |
|---|---|---|---|---|---|---|---|
| **Year** | ***L. pneumophila* Isolates/Total Number [3]** | **% of Isolates** | ***Leg.* Counts (Mean of Cold Water) (CFU/l)±SD** | ***Legionella* spp (Lgsp) Positive Samples/Total Number** | **% of Positive Samples** | ***L. pneumophila* (Lpn) Positive Samples/Total Number** | **% of Positive Samples** |
| 2012 | 96/409 | 23.5 | NA | 43/53 | 81.1 | 36/53 | 67.9 |
| 2013 | 30/346 | 8.7 | NA | 64/106 | 60.4 | 55/106 | 51.9 |
| 2014 | 71/453 | 15.7 | NA | 102/138 | 73.9 | 79/138 | 57.2 |
| **Hospital** | | | | | | | |
| A | 30/150 | 20 | $4.3 \times 10^1 \pm 1.1 \times 10^2$ | 29/36 | 80.6 | 23/36 | 63.9 |
| B | 35/156 | 22.4 | BD | 28/42 | 66.7 | 20/42 | 47.6 |
| C | 5/150 | 3.3 | BD | 19/36 | 52.8 | 11/36 | 30.6 |
| D | 18/90 | 20 | BD | 16/21 | 76.2 | 13/21 | 61.9 |
| E | 11/156 | 7.1 | BD | 17/42 | 40.5 | 13/42 | 31 |
| F | 41/156 | 26.3 | $1.5 \times 10^2 \pm 2.3 \times 10^2$ | 37/42 | 88.1 | 34/42 | 81 |
| G | 35/156 | 22.4 | $8.3 \times 10^0 \pm 2.0 \times 10^2$ | 39/42 | 92.9 | 38/42 | 90.5 |
| H | 6/150 | 4 | BD | 24/36 | 66.7 | 18/36 | 50 |
| AQU | 16/44 | 36.4 | NA | NA | NA | NA | NA |
| **Sample type** | | | | | | | |
| Water | 6/72 | 8.3 | | 42/72 | 58.3 | 36/72 | 50 |
| Biofilm | 191/1136 | 16.8 | | 167/225 | 74.2 | 134/225 | 59.5 |

NA: Not Available; BD: Below detection limit (<5 CFU/L); AQU: Al-Quds University; [1] CDA: Cultivation Dependent Analysis; [2] CIA: Cultivation Independent Analysis; [3] The number of isolates corresponds to the number *L. pneumophila* culture-positive water and biofilm samples.

### 3.2. Cultivation Dependent Analysis (CDA) versus Cultivation Independent Analysis (CIA)

Although CDA is the standard and recommended technique for environmental surveillance of *L. pneumophila*, CIA provides higher sensitivity and overcomes the problems of CDA for *Legionella* because of the VBNC state and its overgrowth by competing bacteria. In this study, both methods were used to detect *Legionella* spp in hospital DWDS (Table 2).

A total of 72 water samples and 225 biofilm swabs from the eight hospitals were tested by conventional PCR using three different primers (com, Lgsp, Lpn). Almost all of the samples were positive using com primers (n = 71, 98.6% and n = 225, 100%) for water samples and biofilm swabs respectively. *Legionella* spp were detected in biofilm swabs more than in water samples (n = 167, 74.2% and n = 42, 58.3%), respectively (Table 2). Similar results were obtained using *L. pneumophila*-specific primers: 60% of the biofilm swabs and 50% of the water samples were positive. As expected, the PCR-based CIA showed higher sensitivity than CDA. CIA analysis increased the detection of *L. pneumophila* from 8.3% (CDA) to 50% (CIA) for water samples, and from 16.8% (CDA) to 59.5% (CIA) for biofilm samples.

### *3.3. MLVA-8(12) Genotypes of L. pneumophila Isolates*

MLVA genotyping was carried out for the 180 isolates. MLVA-8, as well as MLVA-12 and the joint scheme MLVA-8(12), were analyzed for the study of the population of *L. pneumophila* isolates. The 180 isolates were categorized as 16 MLVA-8 genotypes (Index of Diversity ID = 0.771, 95% Confidence Interval CI 0.721–0.822), 25 MLVA-12 genotypes (ID = 0.790, 95% CI, 0.739–0.841) and 27 MLVA-8(12) genotypes (ID = 0.790, 95% CI 0.739–0.841). This indicates a lower genotypic resolution for MLVA-8 using eight loci compared to MLVA-8(12) using a total of 13 loci. For details on the comparison based on a larger set of *L. pneumophila* isolates, see [37].

The use of the MLVA-8(12)-genotype nomenclature made it possible to directly compare strains genotyped from MLVA-8 and MLVA-12: the first number reflects the MLVA-8-classification, while the number in brackets reflects the 12 loci-classification, e.g., Gt 4(17) is a Gt 4 according to the MLVA-8, and a Gt 17 according to MLVA-12. Fourteen MLVA-8(12) genotypes were represented by 2 to 74 strains, whereas 13 genotypes were represented by just a single strain from the West Bank isolates (Figure 1). The MLVA-8(12) genotypes comprising the most strains were Gt 4(17), Gt 6(18) and Gt 10(93).

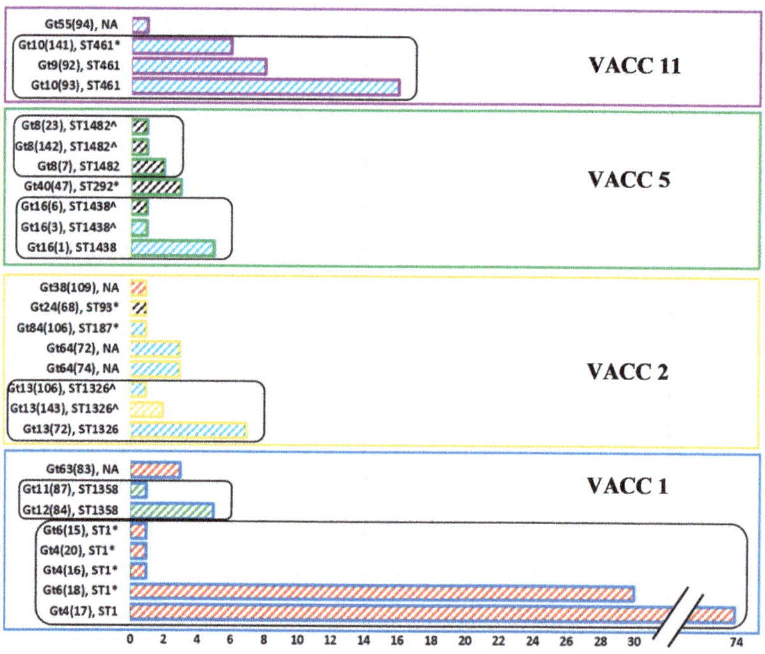


**Figure 1.** Comparison of the number of strains per MLVA8(12)-Genotype (Gt) of the isolated 180 *L. pneumophila* strains and their respective sequence types (ST). Each group of bars, outline color and colored frame indicate VACCs as follows: blue outline represents VACC1, yellow outline represents VACC2, green outline represents VACC5 and purple outline represents VACC11. The black round-edged frames indicate the group of genotypes from the same ST. The wide upward diagonal hatches inside the bars indicate Sg as follows: Sg1—red, Sg 6—sky blue, Sg8—green, Sg10—yellow and Sg2 to Sg14—black. NA—not available ST; *, ST was assessed for strains of the same MLVA-8(12) genotype, and not directly for the West Bank strains; ^, ST was estimated from the MLVA-8 pattern.

### 3.4. VNTR Clonal Complexes (VACC) and Relationship among the Genotypes

An analysis of the relationship among the genotypes was achieved by UPGMA-based cluster analysis of the MLVA-8(12) profiles of the 180 *L. pneumophila* strains (Figure S2). The MLVA clonal complexes (VACC) were defined by a cutoff level of 60% similarity. In addition, the genetic relationship among genotypes was estimated by a minimum-spanning tree based on the MLVA-8(12) profiles (Figure 2).

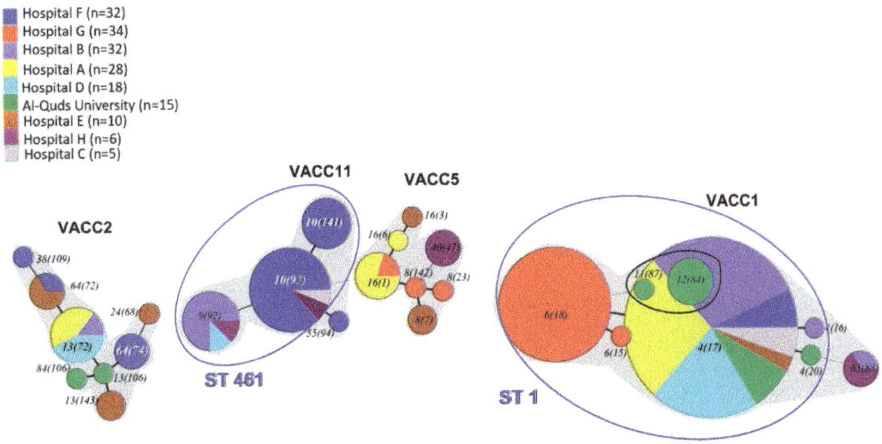


**Figure 2.** Minimum-spanning tree based on MLVA-8(12) profiles of 180 *L. pneumophila* strains isolated from the West Bank. Each circle in the tree represents a different MLVA-8(12) genotype. The genotype number is indicated within or near the circle, whose size is proportional to genotype frequency. Different colors in the pie charts refer to the eight sampling locations (see legend). The thickness of the branches represents the number of different loci. MLVA clonal complexes (VACC) are shaded in grey. The circles representing the Sg8-singletons from Al-Quds University, i.e., Gt 11(87) and Gt 12(84), overlap visually within the circle that represents Gt4(17) due to the high abundance of this genotype. Blue ellipses indicate genotypes that could be candidates for LD. These genotypes belong to ST1 and ST461 (indicated in blue letters) and were assessed by in situ-SBT and present in half of the LD cases in a West Bank study [43].

All MLVA8(12) genotypes were clustered into four MLVA clonal complexes or VACCs (VACC1, VACC2, VACC5 and VACC11) (Figure 2). VACC1, VACC2 and VACC5 were clonal complexes previously defined in the MLVA *Legionella* database. VACC11 is described for the first time in this study (Figure 2 and Figure S2). VACC1 was the largest cluster, including 110 isolates (61.6%). VACC11, VACC2 and VACC5 were, in comparison, smaller clusters, comprising 31, 19 and 14 isolates, respectively (Figure 1).

Not all strains could be included in VACCs. A small group of six isolates that belonged to two different genotypes, i.e., Gt 11(87) and Gt 12(84), separated from the large VACC1 and remained as singletons, i.e., they could not be directly included in a VACC. They differed from the rest of the isolates contained in VACC1 in the number of repeats observed for VNTR markers Lpms31, i.e., 17, in comparison to 4 or 0 in the rest of the profiles of VACC1, and VNTR Lpms33 and Lpms34, which both presented only one repeat in contrast to the 4 and 2 repeats found, respectively, in VACC1. In total, 96.6% of isolates (n = 174) were clustered into the four VACCs, and only 3.6% (n = 6) were found as singletons.

In addition, single linkage clustering using the Bray-Curtis-Index of association was performed with a cutoff value of 60% corresponding to 13 VNTRs. The MLVA8(12) profiles of four additional reference strains were added to the dataset to show their relationship with the genotypes in this study (Figure 3). This clustering showed the same four groups corresponding to the VACCs observed with the UPGMA-clustering and the minimum-spanning tree (Figure 2 and Figure S2). In addition, it revealed a very close relationship of the *L. pneumophila* strain Paris with Gt 4(17) in VACC1, and of Gt 64(74) with reference strain *L. pneumophila* strain Philadelphia1 in VACC2. The dendrogram showed a high discriminatory power and subclustering between *L. pneumophila* isolates with similarity of (>95%). Gt63(83) was shown as an outlier due to the three "Null" alleles in Lpms31, Lpms34 and Lpms35, i.e., no PCR product was obtained for these VNTR-markers (Figure 3). Overall, the Bray-Curtis-grouping of the genotypes confirmed the four VACC cluster. Also, the aforementioned singleton genotypes Gt11(87) and Gt12(84) were grouped separately from VACC1.

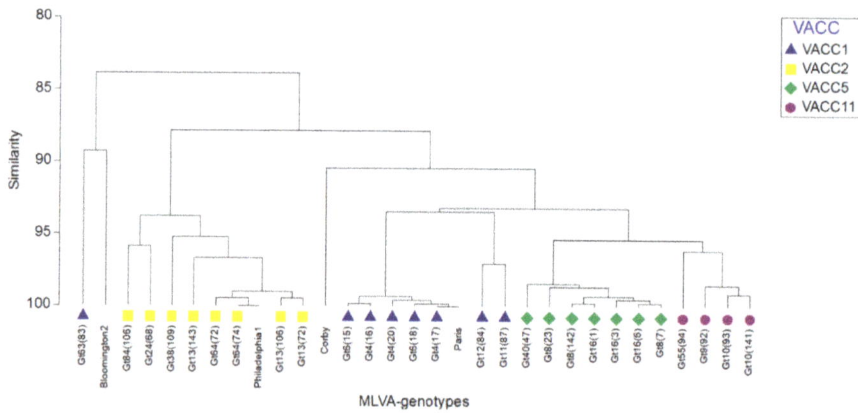


**Figure 3.** Single-linkage cluster dendrogram representing the percentage of similarity between MLVA-8(12) profiles of the genotypes retrieved from the West Bank. For a comparison, MLVA-8(12) profiles of *L. pneumophila* reference strains (Bloomington2, Philadelphia-1, Corby, Paris) were added. The resemblance matrix was calculated using the Bray-Curtis index of association. Strains of different VNTR clonal complexes (VACC) are indicated in different colors (see insert).

Overall, a more detailed analysis of the population structure at the level of the 13 VNTR markers showed a balanced variability in the number of repeats for most VNTR markers among the isolates (Table 3). This could be due to the homogenized habitat and location where the isolates were obtained. Some VNTR markers appeared to be less variable and showed a reduced number of repeats, e.g., Lpms3, Lpms17 and Lpms19, while others showed a greater variability, e.g., Lpms31 and Lpms35. In general, the same repeats prevailed independently of the area from which the isolates had been isolated. Null alleles were present at different frequencies in distinct VNTR markers. Especially high were the frequencies of null alleles in Lpms38. Remarkably, a new allele of Lpms34 was described during this study. The new allele had a size of 634 base pairs and was formed by four repeats. A total of 31 isolates (17.2%) contained this allele. This allele has not previously been described in MLVA studies for *L. pneumophila*.

**Table 3.** VNTR characteristics of the *L. pneumophila* strains isolated from the West Bank.

| VNTR | No. of Repeats | West Bank HGDI [1] (CI 95%) | Null Alleles (%) |
|---|---|---|---|
| Lpms1 | 4 | 0.528 (0.459–0.596) | 0 |
| Lpms3 | 2 | 0.461 (0.420–0.502) | 0 |
| Lpms13 | 5 | 0.579 (0.506–0.652) | 0 |
| Lpms17 | 2 | 0.115 (0.053–0.178) | 0 |
| Lpms19 | 2 | 0.022 (1.000–0.053) | 1.11 |
| Lpms31 | 6 | 0.576 (0.513–0.639) | 1.67 |
| Lpms33 | 4 | 0.575 (0.506–0.643) | 0 |
| Lpms34 | 4 | 0.503 (0.429–0.577) | 2.22 |
| Lpms35 | 6 | 0.687 (0.641–0.733) | 1.67 |
| Lpms38 | 3 | 0.249 (0.168–0.330) | 4.44 |
| Lpms39 | 3 | 0.509 (0.445–0.574) | 0 |
| Lpms40 | 3 | 0.493 (0.444–0.541) | 3.33 |
| Lpms44 | 3 | 0.498 (0.463–0.533) | 0 |

[1] HGDI: Hunter-Gaston Discrimination Index.

### *3.5. Diversity within the Clonal Complexes (VACC)*

The minimum-spanning tree (Figure 2) make it possible to provide an overview of the diversity and the genetic relationship among the MLVA genotypes. While the grouping by VACCs gives a first estimate of the relationship among the total of the genotypes, the relationship of the genotypes within the VACCs is also indicative.

In VACC1, all genotypes pertaining to ST1 were closely related to Gt 4(17). The replicate number of only one locus had changed compared to Gt 4(17). This is comparable to VACC11, where Gt 10(93) has this central position, with only one locus being different compared to Gt 10(141), Gt 9(92) and Gt 55(94). By contrast, in VACC2 and VACC5, a set of more distantly related strains with a broader set of different changes in the VNTR-loci was observed.

The relationship of the genotypes within the cluster as reflected by the minimum spanning tree were consistent with the results of the UPGMA-based cluster analysis (Figure S2) and Bray-Curtis-based analysis (Figure 3).

### *3.6. Comparison of MLVA-8(12) Genotypes and Clonal Complexes with Sequence Types (ST)*

In our comparison, 22 of 27 MLVA8(12) genotypes could be assigned to nine sequence types (STs). Most STs with a larger set of strains could be divided in two to five genotypes (Figure 1 and Figure S2). ST1 was split into five MLVA-8(12) genotypes all adhering to VACC1. ST1 comprised the largest fraction of strains, i.e., 111 strains of the total of 180 strains. ST1 comprised the MLVA-8(12) genotypes with the most isolates, i.e., Gt 4(17) and Gt 6(18) (Figures 1 and 2). ST461 comprised 30 strains and three genotypes of VACC11. It was the second largest ST. VACC11 strains were not present in any other ST. Three genotypes were also present in each of ST1326, ST1438, and ST1482. Two MLVA genotypes were present in ST1358. The remaining three STs (9, 93, 187) comprised one to three strains and constituted a single MLVA-8(12) genotype.

In summary, all STs with a larger set of strains were split up in several MLVA-genotypes, i.e., MLVA-8(12)-genotyping showed a substantially higher resolution than SBT. All STs comprised only strains of the same VNTR clonal complex (VACC). Our dataset indicates a high level of consistency between SBT and MLVA8(12)-genotyping. This observation was confirmed by the analysis of a larger and more diverse set of *L. pneumophila* isolates [37].

### 3.7. Serotype Distribution of L. pneumophila Isolates and Relationship with MLVA Genotypes and Clonal Complexes

The 180 PCR-confirmed *L. pneumophila* environmental isolates were tested for serogroups (Table 4 and Figure 1). Most of the isolates were characterized as serogroup 1 (Sg1) (n = 111); the remaining 69 isolates were non Sg1. A subset of ten Sg1 isolates was subgrouped according to the monoclonal antibody; all belonged to the MAb 3/1 negative OLDA subtype, which is considered to lack the virulence- associated epitope. The 69 non-Sg1 isolates were analyzed by monoclonal subgrouping; 54 of them were serotyped as Sg6, followed by Sg8 (n = 6) and Sg10 (n = 2). The rest of the non-Sg.1 were characterized as serogroups 2–14, as determined using an agglutination kit.

**Table 4.** Serogroup and monoclonal antibody subtyping of 180 environmental *L. pneumophila* isolates from the West Bank.

| Serogroup | mAb [1] Subgroup | *L. pneumophila* Isolates | |
|---|---|---|---|
| | | No. | Frequency (%) |
| 1 | OLDA | 10 | 5.6 |
| 1 | NA [2] | 101 | 56.1 |
| *Total Sg1* | | *111* | *61.6* |
| 6 | Dresden | 54 | 30.0 |
| 8 | NA [2] | 6 | 3.3 |
| 10 | NA [2] | 2 | 1.1 |
| (2–14) | NA [2] | 7 | 3.9 |
| *Total non-Sg1* | | *69* | *38.3* |
| Total | | 180 | 100 |

[1] mAb: monoclonal Antibody; [2] NA: Not analyzed.

In terms of the number of isolates, the *L. pneumophila* population showed a dominance of Sg1, followed by Sg6 (Figure 1). In terms of MLVA-genotypes, 11 could be attributed to Sg6 and seven to Sg1, while Sg8 was represented by only two and Sg10 by only one genotype. In terms of clonal complexes, VACC11 comprised only strains of Sg6. VACC1 comprised mainly Sg1 strains, except for Sg8 for genotypes Gt11(87) and Gt12(84). VACC5 comprised strains of either Sg6 or Sg2–14. By contrast, VACC2 comprised a serogroup-divers set of genotypes, i.e., Sg1, Sg6, Sg10 and Sg2–14. The six singleton isolates were all Sg8-strains isolated from Al-Quds University.

### 3.8. Prevalence and Abundance of L. pneumophila MLVA-Genotypes and Clonal Complexes (VACCs)

An overview of the association of the MLVA-8(12) genotypes and VACCs with the different sampling sites is reflected in the minimum-spanning tree (Figure 2). Details on the strains retrieved and their characteristics for each sampling site are listed in Table 5.

**Table 5.** MLVA-8(12) genotype abundances at the sampling sites in the West Bank.

| Location | MLVA-8 (12)-Genotype | No of Strains per Genotype (%) | Sg-MAb | MLVA-Clonal Complex (VACC) | No. of Strains per VACC (%) |
|---|---|---|---|---|---|
| Hospital A | Gt4(17) | 20 (71) | 1 | VACC1 | 20(71) |
| | Gt16(1) | 4 (14) | 6 Dresden | VACC5 | 5(18) |
| | Gt13(72) | 3 (11) | 6 Dresden | VACC2 | 3(11) |
| | Gt16(6) | 1 (4) | (2–14) | 0 | 0 |
| | *Total* | *28(100)* | *0* | *0* | *28(100)* |
| Hospital B | Gt4(17) | 21(66) | 1 | VACC1 | 24(75) |
| | Gt9(92) | 7(22) | 6 Dresden | VACC11 | 7(22) |
| | Gt63(83) | 2(6) | 1 | VACC2 | 1(3) |
| | Gt4(16) | 1(3) | 1 | 0 | 0 |
| | Gt13(72) | 1(3) | 6 Dresden | 0 | 0 |
| | *Total* | *32(100)* | *0* | *0* | *32(100)* |
| Hospital C | Gt4(17) | 4(80) | 1 | VACC1 | 4(80) |
| | Gt10(93) | 1(20) | 6 Dresden | VACC11 | 1(20) |
| | *Total* | *5(100)* | *0* | *0* | *5(100)* |
| Hospital D | Gt4(17) | 15(83) | 1 | VACC1 | 15(83) |
| | Gt13(72) | 3(17) | 6 Dresden | VACC2 | 3(17) |
| | *Total* | *18(100)* | *0* | *0* | *18(100)* |
| Hospital E | Gt13(143) | 2(20) | 10 | VACC2 | 5(50) |
| | Gt64(72) | 2(20) | 6 Dresden | VACC5 | 3(30) |
| | Gt8(7) | 2(20) | (2–14) | VACC1 | 2(20) |
| | Gt4(17) | 2(20) | 1 | 0 | 0 |
| | Gt24(68) | 1(10) | (2–14) | 0 | 0 |
| | Gt16(3) | 1(10) | (2–14) | 0 | 0 |
| | *Total* | *10(100)* | *0* | *0* | *10(100)* |
| Hospital F | Gt10(93) | 14(44) | 6 Dresden | VACC11 | 21(66) |
| | Gt10(141) | 6(19) | 6 Dresden | VACC1 | 6(19) |
| | Gt4(17) | 6(19) | 1 | VACC2 | 5(16) |
| | Gt64(74) | 3(9) | 6 Dresden | 0 | 0 |
| | Gt55(94) | 1(3) | 6 Dresden | 0 | 0 |
| | Gt64(72) | 1(3) | 6 Dresden | 0 | 0 |
| | Gt38(109) | 1(3) | 1 | 0 | 0 |
| | *Total* | *32(100)* | *0* | *0* | *32(100)* |
| Hospital G | Gt6(18) | 30(88) | 1 | VACC1 | 31(91) |
| | Gt6(15) | 1(3) | 1 | VACC5 | 3(9) |
| | Gt16(1) | 1(3) | 6 Dresden | 0 | 0 |
| | Gt8(142) | 1(3) | (2–14) | 0 | 0 |
| | Gt8(23) | 1(3) | (2–14) | 0 | 0 |
| | *Total* | *34(100)* | *0* | *0* | *34(100)* |
| Hospital H | Gt40(47) | 3(50) | 6 Dresden | VACC5 | 3(50) |
| | Gt10(93) | 1(17) | 6 Dresden | VACC11 | 2(33) |
| | Gt9(92) | 1(17) | 6 | VACC1 | 1(17) |
| | Gt63(83) | 1(17) | 1 | 0 | 0 |
| | *Total* | *6(100)* | *0* | *0* | *6(100)* |
| AQU * | Gt4(17) | 6(40) | 1 OLDA | VACC1 | 13(87) |
| | Gt12(84) | 5(33) | 8 | VACC2 | 2(13) |
| | Gt4(20) | 1(7) | 1 OLDA | 0 | 0 |
| | Gt11(87) | 1(7) | 8 | 0 | 0 |
| | Gt13(106) | 1(7) | 6 | 0 | 0 |
| | Gt84(106) | 1(7) | 6 | 0 | 0 |
| | *Total* | *15(100)* | *0* | *0* | *15(100)* |

*: Al-Quds University.

In terms of VACC prevalence, Figure 2 shows that at least two distinct clonal complexes were present at each hospital as well as at the Al-Quds University. VACC1, the largest clonal complex, was present across the West Bank. Genotypes belonging to it were isolated from all eight hospitals and from AQU. VACC2 isolates were isolated from four hospitals distributed throughout the West Bank and AQU. Although VACC11 was present at five hospitals, it was the major clonal complex at hospital F (n = 22, 66.7%). Isolates grouped into VACC5, the smallest clonal complex, were, however, found at four hospitals located throughout the West Bank (Table 5).

At the genotype level, only nine out of the 27 MLVA-8(12) genotypes were isolated in more than one location. The remaining 18 MLVA-8(12) genotypes were isolated exclusively in one specific site. Gt 4(17), the main VACC1-genotype that also comprises *L. pneumophila* strain Paris, was the only genotype present in all hospitals except for hospital G. Furthermore, it represented a high fraction of the isolates in several hospitals. Gt 4(17) was the most abundant genotype in hospital B (68.7%), A (71.4%), D (83.4%), and C (80%). At Al-Quds University, Gt 4(17) accounted for 40% of the isolates. Gt 6(18) was the second most abundant VACC1-genotype; it is closely related to GT 4(17), differing by just one additional repeat in the VNTR Lpms35. Gt 6(18) was endemic in the West Bank and found exclusively in hospital G, where it was the most abundant genotype (90.9%). Genotype Gt 10(93), a member of the newly-described VACC11, was found in hospitals C and F; it was isolated only once in hospital C, but was the most abundant genotype in hospital F (n = 15, 45.5%). The remaining genotypes that were found in more than one location had a rather restricted distribution, i.e., they were observed only in one or two additional sampling sites (Figure 2 and Table 5).

In general, the most frequent genotypes in each hospital were isolated repeatedly during samplings performed in following years. Gt 4(17) was recurrently isolated in hospitals A, B, C, D and F between 2012 and 2014. The endemic Gt 6(18) was isolated in hospital G in 2013 and 2014. Genotypes Gt 10(141), Gt 10(93), and Gt 9(92) were isolated in their respective sites (hospital F, B) from 2012 to 2014. The MLVA-8(12) genotypes were shared among the north, central and southern West Bank. According to the geographical distribution of the West Bank, 64 (35.6%), 43 (23.9%) and 73 (40.6%) isolates were isolated from northern, central and southern areas, respectively. Nevertheless, the most abundant and broadly distributed genotype was common to the whole West Bank, i.e., Gt 4(17), which comprised 68.8%, 53.5% and 9.6% of total isolates in northern, central and southern West Bank, respectively. Surprisingly, four genotypes, i.e., Gt 9(92), Gt 10(93), Gt 16(1) and Gt 63(83), were shared between more distant sites, i.e., northern and southern West Bank. As a tendency, the diversity of the genotypes observed decreased from the Southern to the Northern West Bank.

Interestingly, Gt 4(17) was never obtained from water samples, but only from biofilm. Genotypes of rather limited distribution were the only genotypes retrieved from water, i.e., Gt 10(93) and the endemic Gt 10(141), were obtained from hospital F, the endemic Gt 6(18) from G, and Gt 16(1) from A, respectively.

In summary, two thirds of the MLVA-8(12) genotypes were endemic, i.e., they were found exclusively in one hospital or the Al-Quds University. Only one third of the MLVA-8(12) genotypes were isolated in more than a single location, and these common genotypes were usually much more frequent in one of the locations. An exception was Gt 4(17), that was present in most locations and occurred often in high abundance.

## 4. Discussion

### 4.1. *Legionella Abundance in Hospital Water and Biofilm of the West Bank*

This study is based on the first extensive sampling campaign examining the prevalence of *Legionella* spp in DWDS of hospitals in the West Bank. The analysis of water and biofilm samples was done using cultivation-dependent and -independent methods targeting *Legionella* from the genus to the clone level for *L. pneumophila* by molecular techniques including MLVA-8(12) genotyping.

In general, water samples had a far lower prevalence of *Legionella* compared to biofilms. Water samples tested positive for the presence of *Legionella* with a prevalence of 8.3% by cultivation dependent analysis and 50% by cultivation independent analysis. Biofilms had a higher prevalence, with 16.8% positive by cultivation dependent analysis and 61.3% by cultivation independent analysis. The findings of increased PCR-based detection in water and biofilms are consistent with other studies [44] and were analyzed in detail by Zayed et al. [42].

The MLVA-genotypes of the five water isolates were always present in the biofilm of the respective sites, usually making up a high fraction of the local biofilm isolates. In more detail, only from hospital A (one isolate), hospital F (three isolates) and hospital G (one isolate) were isolates from water obtained. In hospital A, the most abundant genotype from biofilm was Gt 4(17) (71% of the isolates), whereas Gt 16(1) obtained from water had a lower abundance (11%) of the biofilm isolates of this site. The three water isolates from hospital F belonged to Gt 10(93) or Gt 10(141). The water isolate from hospital G belonged to Gt 6(18). The water isolates from hospital F and G were the most abundant biofilm genotypes from these hospitals, i.e., Gt 10(93), Gt 10(141) and Gt 6(18) (Table 5 and Table S1).

The low prevalence of *Legionella* in culture-based studies is in accordance with studies in Israel and Greece [45–48]. However, many studies showed a much higher culturable *L. pneumophila* prevalence in water, e.g., 21.6%, 22% and 40% in Kuwait, Tunisia [49,50] and Jordan [51], respectively. The prevalence of *L. pneumophila* was even higher (68.5%) in a study from northern Israel [52]. The low prevalence of *L. pneumophila* in the West Bank was, at least to some extent, attributed to the high magnesium content of the drinking water [42].

Most of the *L. pneumophila* isolates from the West Bank (n = 175, 97%) were obtained from biofilm samples (Table 2). This is consistent with the results by Douterelo et al. [53], showing that more than 95% of the microbial biomass in a DWDS is found in the biofilms attached to the pipe lines due to the multiple advantages that biofilms represent for microorganisms, such as providing protection from external factors and beneficial interactions with other microorganisms [54]. Additionally, from the point of view of public health, biofilm sampling has a great importance, since it has been observed that *L. pneumophila* strains derived from biofilm replicate significantly more in murine macrophages than plankton-derived strains [55].

### 4.2. General Health Relevant Aspects of the Isolated Strains

Cultivation is still considered the gold standard for the detection of *Legionella* in the environment, even though other, nonculture methods are available, such as serology or nucleic acid-based detection methods [56]. Cultivation can be inaccurate as a result of overgrowth by other microorganisms on the agar plates, and can be ineffective due to the presence of viable but nonculturable (VNBC) *Legionella* cells [57]. However, cultivation makes it possible to obtain isolates that can be identified and characterized phenotypically and genetically, which is essential for epidemiological studies.

According to current epidemiological data available from around the world, different *L. pneumophila* serogroups cause legionellosis with a distinct estimated risk. Overall, the great majority of strains isolated from the area under study were characterized as Sg1 (62.3%). This fact followed the tendency already reported by other studies that have described Sg1 as the most frequently detected Sg of environmental isolates in different geographic regions [58–60]. Besides the high prevalence of Sg1, other serogroups were isolated, where the fraction of non-Sg1 isolates went up to 37.7%. In our study, Sg6 was particularly abundant (30%). Sg8 and Sg10 were also isolated, although in smaller proportions (3.3% and 1.1%, respectively) (Table 3). The results obtained here were consistent with those of two studies on the distribution of *L. pneumophila* serogroups not related to human disease in man-made water systems [60,61], and were comparable, climate-wise to the area of study, i.e., Greece [48,62] *L. pneumophila* Sg1 was the most frequently isolated serogroup, followed by Sg6 in France and the UK, where Sg10 was also found. Sg6 is the serogroup which is second most responsible for cases of LD after Sg1, according to European surveillance data [63]. Furthermore, these specific serogroups (Sg1 and Sg6) are the most frequent and virulent among clinical cases [34,64–66].

ST1 was the most prevalent sequence type in the West Bank, and is the most dominant ST worldwide [22,60,67,68]. The high abundance of ST1 in the environment has been reported in several studies. In Japan, as well as in South Korea, the majority of environmental isolates comprised ST1 [69]; for the latter, ST1 was distributed across all sampled facilities and regions and accounted for 48.1% of the isolates [70]. ST1 was the most abundant sequence type among environmental isolates in Canada, and was found ubiquitously across the country [22]. In a study conducted across the United States, ST1 was the most frequent sequence type between both clinical sporadic and environmental isolates, accounting for 25% and 49% of the total number of isolates, respectively [68]. In Europe, ST1 has also been reported as the most predominant sequence type among environmental isolates in Germany [71], England and Wales [60], Portugal [72], Spain [73], France [74] and Italy [75].

Typically, the climate in the West Bank is Mediterranean, slightly cool to cold in winter and dry to humid and warm to hot in summer. Previous studies have suggested that the incidence of LD may increase under warm and wet meteorological conditions, which could be exacerbated by global climate change [76]. Therefore, surveillance of environmental sources and proper maintenance of man-made freshwater systems is key in the prevention of legionellosis. Surveillance of *Legionella* in the environment is also essential to validate the efficacy of decontamination procedures, and for risk assessment when evaluating potential transmission or amplification sources.

### 4.3. Genotyping Using MLVA—What Resolution Is Needed for Ecological and Clinical Issues?

SBT is considered as the gold-standard for *L. pneumophila* genotyping, primarily due to a large International database created by the "*L. pneumophila* community". SBT has high typeability, interlaboratory reproducibility and generally a high index of discrimination [77]. However, the resolution is not as high as is often needed, e.g., for ST1 and its many health-relevant strains occurring world-wide [78,79]. It is suggested that the number of sequenced genes be increased to about fifty, guided by genome analyses.

MLVA is a rather well-established genotyping method currently used for 32 pathogenic bacterial species [27]. To date, it has been applied mostly for clinical strains. MLVA-8(12) for *L. pneumophila* was developed and its resolution analyzed by Sobral et al. Visca et al. and Pecellin [29,36,37]. All these studies demonstrated that MLVA-8, and even more, MLVA-8(12), have a higher resolution than SBT and are rather consistent with SBT. This was also shown in this study: larger strain sets adhering to a specific ST could always be distinguished into different MLVA-genotypes. MLVA-genotyping was always consistent with SBT, i.e., strains of the same genotype were not assigned to different STs. In this study, nine STs were split into 22 MLVA-8(12) genotypes, with large STs comprising several MLVA-genotypes.

The required resolution for the genotyping of strains is dependent on the tasks to be performed. For clinical reasons, genotyping should allow for a distinction with respect to virulence traits and antibiotic resistance [80]. Sharaby et al. [81,82] showed that *L. pneumophila* isolates from Israel had MLVA-8 genotype-specific virulence traits. Even their resistance to antibiotics showed a strong correlation with the MLVA-genotypes. Interestingly, environmental strains were more resistant to antibiotics than clinical ones. Moreover, there were major differences between different MLVA-genotypes associated with ST1 that could not have been distinguished by SBT.

For the management of *L. pneumophila* abundance in DWDS, the ecology of *L. pneumophila* has to be studied. As first shown by Rodriguez-Martinez et al. [52], the MLVA-genotyping level makes it possible to distinguish among specific ecotypes, i.e., by assessing the strains' environmental preferences. It was demonstrated that the preference for temperature and the respective growth speed could be well differentiated using MLVA-genotyping [83]. Furthermore, in these studies, there were relevant differences between different MLVA-genotypes affiliated with ST1. A more detailed analysis on environmental preferences for the West Bank strains was provided by Zayed et al. [42]. For the West Bank strains, specific environmental traits could be assigned to all MLVA-8(12) genotypes comprising a larger set of strains.

Microbial source tracking is of high relevance for *L. pneumophila* due to its occurrence in environmental freshwater and its transfer, specifically during outbreaks, from the environment to humans. MLVA is rather economical and can be run in a fully automated manner using capillary sequencing. Due to the need for rapid analyses of large sets of strains in an outbreak scenario, MLVA can be a cost-efficient and fast option. The successful source tracking by MLVA of an *L. pneumophila* outbreak was successfully demonstrated by Sobral et al. [29] in the French city of Rennes.

Another point of interest is intraspecies evolution. Garcia et al. [84] successfully followed the microevolution of *Vibrio parahaemolyticus* using MLVA technology in an experimental setting. Based on the experimentally-derived mutation rates, they estimated the worldwide evolution and time scale for *V. parahaemolyticus* populations. In general, clustering in VACCs gives the basis for good estimates among MLVA-genotypes. Links between the strains indicate their lines of evolution. For this study, relationships of the occurrence of genotypes over time and in neighboring sampling sites could indicate source strains and evolutionary tendencies [37].

### 4.4. Comparison of L. pneumophila MLVA-8(12) Genotypes from the West Bank with the International Data Base and A Study on A Larger Set of Strains from Germany and Israel

To address the distribution of the 27 MLVA-8(12) genotypes outside of the West Bank, the genotypes were compared to the International MLVA database (http://microbesgenotyping.i2bc.paris-saclay.fr/) and a larger strain analysis was performed by Pecellín [37]. The study by Pecellín [37] described a set of 610 clinical and environmental *L. pneumophila* strains retrieved from Germany, Israel and Palestine. The set of 180 strains from the West Bank described in this study were included in the study by Pecellín [37].

A comparison with the international database and the study by Pecellín [37] showed that there were a few highly ubiquitous genotypes. First of all, the MLVA-genotype comprised *L. pneumophila* Paris, i.e., Gt 4(17), that was associated with VACC1 and ST1. This genotype has a high relevance worldwide as a clinical and environmental genotype [68,74]. Another prominent member is Gt64(74), that comprises *L. pneumophila* strain Philadelphia-1 (ST 440). Two more genotypes occurred in the study of the three countries, i.e., Gt 4(16), a ST1-genotype closely related to Gt 4(17), and Gt 24(68), which is related to ST93. Gt 4(17) has an eminent role in Israel and the West Bank. This genotype occurs at high abundances in both regions. In Israel, it plays a relevant role as a clinical isolate [81]. At Oranim campus close to Haifa, concentrations of Gt 4(17) were high in the water and adjacent biofilms. In comparison with Germany, one more genotype occurred, i.e., Gt 40(74), which was associated with ST292. In comparison with Israel, two genotypes of environmental and clinical origin were observed, i.e., Gt 6(18) and Gt 6(15), both of which were affiliated with ST1 [81].

In summary, this means that 20 of the 27 MLVA-genotypes were unique for the West Bank. In addition, the VNTR clonal cluster VACC11 was described in the present study for the first time for the West Bank. Both the high percentage of new genotypes and the new VACC indicate the uniqueness of the West Bank strains. One possible line of ecological reasoning could be that groundwater is mostly the source water for DWDS in the West Bank [85], with an individualistic supply due to specific wells and springs as local water sources. This high diversity among sampling sites and the uniqueness of most of the genotypes may therefore be due to the very diverse water sources (Figure 2 and Table 5).

### 4.5. Conceivable Health Relevance of the L. pneumophila MLVA-8(12) Genotypes in the DWDS

PCR analysis of sputum and Broncho-Alveolar-Lavage (BAL) samples from pneumonia patients by Jaber et al. [43] in the West Bank revealed a rather high fraction of *L. pneumophila* contamination, i.e., 15% and 35%, respectively. By in situ SBT, they identified 29% of the detected *L. pneumophila* contaminations as ST1, and 21% as ST461. This is rather consistent with the MLVA-8(12) genotypes retrieved from the DWDS of the West Bank: the largest fraction of strains retrieved were affiliated with ST1 and ST461 (Figure 2). Unfortunately, there were no isolates obtained from patients due to previous antibiotic treatment. Therefore, no clinical strains could have been submitted to MLVA-8(12) analysis [43].

In the study by Sharaby et al. [81], clinical strains associated with ST1 were mostly Gt 4(17) followed by Gt 6(18). Therefore, it can be assumed that these MLVA-genotypes may have been responsible for infections in patients where ST1 was detected. For patients with ST461, all strains affiliated with this ST could be responsible for pneumonia, i.e., all genotypes (Gt10(141), Gt 10(93), Gt 9(92)) of the newly-described VACC11. The role of the remaining genotypes from the West Bank (non-ST1 and non-ST461) is unknown and remains to be elucidated in further studies. Hints may come from some publications [29] where strains of VACC2, including the genotype of *L. pneumophila* Philadelphia-1, were often observed as sources of LD in local outbreaks.

Another relevant point for public health and infection is the concentration of *L. pneumophila* in drinking water. For a site with comparable climate, Sharaby et al. [86] showed the risk at high concentrations of culturable *L. pneumophila* in drinking water of the Oranim campus (Haifa). From this perspective, it can be regarded as good news that the level of culturable *L. pneumophila* in the West Bank hospitals was, on average, rather low or undetectable. However, the noncontinuous water supply in the West Bank may cause disruption of biofilms in the DWDS, leading to short-term increases of the levels of *L. pneumophila* in drinking water that remain undetected during measurement campaigns.

Another health aspect is DWDS in private homes that has not yet been assessed. Since *L. pneumophila* STs found in drinking water were observed for at least half of the investigated pneumonia patients, the drinking water in private homes of the West Bank should be considered as a potential health risk with respect to LD. Overall, the situations in private homes, i.e., discontinuous water supply and roof storage of water, are still underassessed; future studies should investigate the risks associated with the water quality and supply aspects.

In summary, our study provided the first comprehensive and long-term overview of the prevalence of *L. pneumophila* in DWDS with respect to water and biofilm, achieved by cultivation and PCR-based methods. Genotyping of the isolated *L. pneumophila* strains by high-resolution genotyping methods allowed us to group the isolates on a subspecies level and make international comparisons. In combination with genotype abundance and regional distribution, this provides better insights into potential health risks and may indicate where and which prevention measures might be needed.

## 5. Conclusions

A two-year study of *L. pneumophila* populations in water and biofilms in the drinking water distribution systems (DWDS) of eight hospitals across the West Bank demonstrated low and rare abundance of culturable *L. pneumophila* in water, but substantially higher prevalence in biofilm. PCR-based analyses consistently showed a higher detection rate in water and biofilm. Based on high resolution MLVA-8(12) genotyping, the 180 isolates retrieved in the West Bank could be characterized as a rather diverse population, with four clonal complexes (VACC). Most of the genotypes (20 out of 27) were unique, and so far, have only been described for the West Bank, including those forming a new clonal complex (VACC11). In addition, seven genotypes were also observed outside of the West Bank, including two genotypes of worldwide abundance, i.e., Gt 4(17) comprising *L. pneumophila* strain Paris, and Gt 64(74) comprising strain Philadelphia-1. The observed uniqueness of the genotypes and the variability from site to site were attributed to individual groundwater-based water supplies. In addition, the isolated strains seemed to be of high health relevance, especially strains of VACC1 and VACC11. MLVA-genotyping was shown to be highly consistent with SBT but showed a higher resolution. Since the most health relevant ST1 (VACC1) and ST461 (VACC11) strains could be further distinguished into several MLVA-genotypes, MLVA-genotyping could provide an excellent basis for future source tracking in the West Bank. MLVA-genotyping provides an adequate resolution, and thus, a good basis for detailed studies of the health- and water-management-relevant traits of *L. pneumophila* [42,81,83,86] in support of a better clinical and DWDS management in the West Bank.

**Supplementary Materials:** The following are available online at http://www.mdpi.com/2076-0817/9/11/862/s1, Table S1. List of *L. pneumophila* strains (n = 180) isolated from the West Bank analyzed in this study, MLVA-8(12) Figure S1: Sampling map of the eight hospitals and Al-Quds University in the West Bank. Figure S2: UPGMA based clustering analysis of the MLVA-8(12) profiles of 180 *L. pneumophila* strains isolated from water and biofilm samples of the Al-Quds University campus and eight hospitals of the West Bank.

**Author Contributions:** A.R.Z. performed sample acquisition and analysis, strain isolation, data analysis and contributed to drafting of the manuscript. I.B., M.G.H., M.S. and D.M.B. designed the work, contributed to data evaluation and wrote the manuscript. M.P., A.S., H.A., S.B., and R.L. contributed to sampling, and sample and strain analysis. All authors have read and agreed to the published version of the manuscript.

**Funding:** This work was supported by DFG grant (HO 930/5-1&2) and the Hildegard-Elisabeth Foundation.

**Acknowledgments:** We thank the administration of the hospitals for allowing us to collect the samples. We thank the Palestinian Water Authority (PWA) for helpful support, information and data. We are grateful to Dalia Abu Hilal, Verena Maiberg and Josefin Koch for technical assistance. Christian Lück of the TU Dresden kindly provided the Sequence Types of the isolates.

**Conflicts of Interest:** The authors declare no conflict of interest.

## References

1. Murga, R.; Pruckler, J.M.; Forster, T.S.; Donlan, R.M.; Brown, E.; Fields, B.S. Role of biofilms in the survival of *Legionella pneumophila* in a model potable-water system. *Microbiology* **2001**, *147*, 3121–3126. [CrossRef] [PubMed]
2. Declerck, P. Biofilms: The environmental playground of *Legionella pneumophila*. *Environ. Microbiol.* **2010**, *12*, 557–566. [CrossRef] [PubMed]
3. Rowbotham, T.J. Preliminary report on the pathogenicity of *Legionella pneumophila* for freshwater and soil amoebae. *J. Clin. Pathol.* **1980**, *33*, 1179–1183. [CrossRef]
4. Blatt, S.P.; Parkinson, M.D.; Pace, E.; Hoffman, P.; Dolan, D.; Lauderdale, P.; Zajac, R.A.; Melcher, G.P. Nosocomial legionnaires' disease: Aspiration as a primary mode of disease acquisition. *Am. J. Med.* **1993**, *95*, 16–22. [CrossRef]
5. Breiman, R.F.; Cozen, W.; Fields, B.S.; Mastro, T.D.; Carr, S.J.; Spika, J.S.; Mascola, L. Role of air sampling in investigation of an outbreak of Legionnaires' disease associated with exposure to aerosols from an evaporative condenser. *J. Infect. Dis.* **1990**, *161*, 1257–1261. [CrossRef]
6. Adeleke, A.A.; Fields, B.S.; Benson, R.F.; Daneshvar, M.I.; Pruckler, J.M.; Ratcliff, R.M.; Harrison, T.G.; Weyant, R.S.; Birtles, R.J.; Raoult, D.; et al. *Legionella drozanskii* sp. nov., *Legionella rowbothamii* sp. nov. and *Legionella fallonii* sp. nov.: Three unusual new *Legionella* species. *Int. J. Syst. Evol. Microbiol.* **2001**, *51*, 1151–1160. [CrossRef]
7. Benson, R.F.; Fields, B.S. Classification of the genus *Legionella*. *Semin. Respir. Infect.* **1998**, *13*, 90–99.
8. Yu, V.L.; Stout, J.E. Hospital characteristics associated with colonization of water systems by *Legionella* and risk of nosocomial Legionnaires' disease: A cohort study of 15 hospitals. *Infect. Control. Hosp. Epidemiol.* **2000**, *21*, 434–435. [CrossRef]
9. Amemura-Maekawa, J.; Kura, F.; Helbig, J.H.; Chang, B.; Kaneko, A.; Watanabe, Y.; Isobe, J.; Nukina, M.; Nakajima, H.; Kawano, K.; et al. Characterization of *Legionella pneumophila* isolates from patients in Japan according to serogroups, monoclonal antibody subgroups and sequence types. *J. Med. Microbiol.* **2010**, *59*, 653–659. [CrossRef]
10. Messi, P.; Annalisa, B.; Immacolata, A.; Isabella, M.; De Niederhäusern, S.; Moreno, B. Protozoa and human macrophages infection by *Legionella pneumophila* environmental strains belonging to different serogroups. *Arch. Microbiol.* **2012**, *195*, 89–96. [CrossRef]
11. Mavridou, A.; Smeti, E.; Mandilara, G.; Pappa, O.; Plakadonaki, S.; Grispou, E.; Polemis, M. Prevalence study of *Legionella* spp. contamination in Greek hospitals. *Int. J. Environ. Heal. Res.* **2008**, *18*, 295–304. [CrossRef]
12. Fraser, D.W.; Tsai, T.R.; Orenstein, W.; Parkin, W.E.; Beecham, H.J.; Sharrar, R.G.; Harris, J.; Mallison, G.F.; Martin, S.M.; McDade, J.E.; et al. Legionnaires' disease. *N. Engl. J. Med.* **1977**, *297*, 1189–1197. [CrossRef] [PubMed]
13. Buse, H.Y.; Morris, B.J.; Gomez-Alvarez, V.; Szabo, J.G.; Hall, J.S. *Legionella* diversity and spatiotemporal variation in the occurrence of opportunistic pathogens within a large building water system. *Pathogens* **2020**, *9*, 567. [CrossRef]
14. Schijven, J.F.; Husman, A.D.R. Effect of climate changes on waterborne disease in The Netherlands. *Water Sci. Technol.* **2005**, *51*, 79–87. [CrossRef]

15. Fiore, A.E.; Nuorti, J.P.; Levine, O.S.; Marx, A.; Weltman, A.C.; Yeager, S.; Benson, R.F.; Pruckler, J.; Edelstein, P.H.; Greer, P.; et al. Epidemic Legionnaires' disease two decades later: Old sources, new diagnostic methods. *Clin. Infect. Dis.* **1998**, *26*, 426–433. [CrossRef]
16. Graham, F.F.; Hales, S.; White, P.S.; Baker, M.G. Review Global seroprevalence of legionellosis-A systematic review and meta-analysis. *Sci. Rep.* **2020**, *10*, 1–11. [CrossRef]
17. International Standards Organization. *Water Quality-Detection and Enumeration of Legionella*; ISO 11731; International Standards Organization: Geneva, Switzerland, 1998; Available online: https://www.iso.org/standard/19653.html (accessed on 15 July 2016).
18. Steinert, M.; Hentschel, U.; Hacker, J. *Legionella pneumophila*: An aquatic microbe goes astray. *FEMS Microbiol. Rev.* **2002**, *26*, 149–162. [CrossRef]
19. Steinert, M.; Emödy, L.; Amann, R.; Hacker, J. Resuscitation of viable but nonculturable *Legionella pneumophila* Philadelphia JR32 by *Acanthamoeba castellanii*. *Appl. Environ. Microbiol.* **1997**, *63*, 2047–2053. [CrossRef] [PubMed]
20. Nederbragt, A.J.; Balasingham, A.; Sirevåg, R.; Utkilen, H.; Jakobsen, K.S.; Anderson-Glenna, M.J. Multiple-locus variable-number tandem repeat analysis of *Legionella pneumophila* using multi-colored capillary electrophoresis. *J. Microbiol. Methods* **2008**, *73*, 111–117. [CrossRef]
21. Villari, P.; Motti, E.; Farullo, C.; Torre, I. Comparison of conventional culture and PCR methods for the detection of *Legionella pneumophila* in water. *Lett. Appl. Microbiol.* **1998**, *27*, 106–110. [CrossRef]
22. Reimer, A.R.; Au, S.; Schindle, S.; Bernard, K.A. *Legionella pneumophila* monoclonal antibody subgroups and DNA sequence types isolated in Canada between 1981 and 2009: Laboratory Component of National Surveillance. *Eur. J. Clin. Microbiol. Infect. Dis.* **2009**, *29*, 191–205. [CrossRef] [PubMed]
23. Tijet, N.; Tang, P.; Romilowych, M.; Duncan, C.; Ng, V.; Fisman, D.N.; Jamieson, F.B.; Low, D.E.; Guyard, C. New endemic *Legionella pneumophila* serogroup I clones, Ontario, Canada. *Emerg. Infect. Dis.* **2010**, *16*, 447–454. [CrossRef]
24. Mercante, J.W.; Winchell, J.M. Current and emerging *Legionella* diagnostics for laboratory and outbreak investigations. *Clin. Microbiol. Rev.* **2015**, *28*, 95–133. [CrossRef] [PubMed]
25. Chenal-Francisque, V.; Diancourt, L.; Cantinelli, T.; Passet, V.; Tran-Hykes, C.; Bracq-Dieye, H.; Leclercq, A.; Pourcel, C.; Lecuit, M.; Brisse, S. Optimized multilocus variable-number tandem-repeat analysis assay and its complementarity with pulsed-field gel electrophoresis and multilocus sequence typing for Listeria monocytogenes clone identification and surveillance. *J. Clin. Microbiol.* **2013**, *51*, 1868–1880. [CrossRef] [PubMed]
26. Xue, G.; Wang, Q.; Yan, C.; Jeoffreys, N.; Wang, L.; Li, S.; Gilbert, G.L.; Sun, H. Molecular characterizations of PCR-positive *Mycoplasma pneumoniae* specimens collected from Australia and China. *J. Clin. Microbiol.* **2014**, *52*, 1478–1482. [CrossRef]
27. Lindstedt, B.-A.; Torpdahl, M.; Vergnaud, G.; Le Hello, S.; Weill, F.-X.; Tietze, E.; Malorny, B.; Prendergast, D.M.; Ghallchoir, E.N.; Lista, R.F.; et al. Use of multilocus variable-number tandem repeat analysis (MLVA) in eight European countries, 2012. *Eurosurveillance* **2013**, *18*, 20385. [CrossRef]
28. Pourcel, C.; Visca, P.; Afshar, B.; D'Arezzo, S.; Vergnaud, G.; Fry, N. Identification of variable-number tandem-repeat (VNTR) sequences in *Legionella pneumophila* and development of an optimized multiple-locus VNTR analysis typing scheme. *J. Clin. Microbiol.* **2007**, *45*, 1190–1199. [CrossRef]
29. Sobral, D.; Le Cann, P.; Gerard, A.; Jarraud, S.; Lebeau, B.; Loisy-Hamon, F.; Vergnaud, G.; Pourcel, C. High-throughput typing method to identify a non-outbreak-involved *Legionella pneumophila* strain colonizing the entire water supply system in the town of Rennes, France. *Appl. Environ. Microbiol.* **2011**, *77*, 6899–6907. [CrossRef]
30. Zayed, A.R. Ecology, virulence and genomics of *Legionella pneumophila* Isolates from the West Bank, Palestine. Ph.D. Thesis, Technical University of Braunschweig, Braunschweig, Germany, 2018; p. 107ff. Available online: https://publikationsserver.tu-braunschweig.de/receive/dbbs_mods_00066061 (accessed on 13 October 2020).
31. Centers for Disease Control and Prevention CDC. Procedures for the Recovery of Legionella from the Environment Centers for Disease Control and Prevention. 2005; pp. 1–13. Available online: https://www.cdc.gov/legionella/labs/procedures-manual.html (accessed on 21 October 2020).
32. Kahlisch, L.; Henne, K.; Groebe, L.; Draheim, J.; Höfle, M.G.; Brettar, I. Molecular analysis of the bacterial drinking water community with respect to live/dead status. *Water Sci. Technol.* **2010**, *61*, 9–14. [CrossRef]

33. Senderovich, Y.; Gershtein, Y.; Halewa, E.; Halpern, M. *Vibrio cholerae* and Aeromonas: Do they share a mutual host? *ISME J.* **2008**, *2*, 276–283. [CrossRef]
34. Helbig, J.H.; Bernander, S.; Castellani-Pastoris, M.; Etienne, J.; Gaia, V.; Lauwers, S.; Lindsay, D.; Luck, T.; Marques, S.; Mentula, S.; et al. Pan-European study on culture-proven Legionnaires' disease: Distribution of *Legionella pneumophila* serogroups and monoclonal subgroups. *Eur. J. Clin. Microbiol. Infect. Dis.* **2002**, *21*, 710–716. [CrossRef] [PubMed]
35. Rajendram, D.; Ayenza, R.; Holder, F.; Moran, B.; Long, T.; Shah, H. Long-term storage and safe retrieval of DNA from microorganisms for molecular analysis using FTA matrix cards. *J. Microbiol. Methods* **2006**, *67*, 582–592. [CrossRef] [PubMed]
36. Visca, P.; D'Arezzo, S.; Ramisse, F.; Gelfand, Y.; Benson, G.; Vergnaud, G.; Fry, N.; Pourcel, C. Investigation of the population structure of *Legionella pneumophila* by analysis of tandem repeat copy number and internal sequence variation. *Microbiology* **2011**, *157*, 2582–2594. [CrossRef] [PubMed]
37. Pecellín, M. Structure and Virulence of *Legionella pneumophila* Populations from Freshwater Systems in Germany and Middle East. Ph.D. Thesis, Technische Universität Braunschweig, Braunschweig, Germany, 2016. Available online: https://publikationsserver.tu-braunschweig.de/servlets/solr/mods_nameIdentifier?q=mods.name:%22Pecell%C3%ADn%20Monjo%20Marina%22&owner=createdby:guest (accessed on 13 October 2020).
38. Lück, P.C.; Fry, N.K.; Helbig, J.H.; Jarraud, S.; Harrison, T.G. Typing methods for *Legionella*. *Recent Results Cancer Res.* **2012**, *954*, 119–148.
39. Chien, M.; Morozova, I.; Shi, S.; Sheng, H.; Chen, J.; Gomez, S.M.; Asamani, G.; Hill, K.; Nuara, J.; Feder, M.; et al. The Genomic Sequence of the Accidental Pathogen *Legionella pneumophila*. *Science* **2004**, *305*, 1966–1968. [CrossRef] [PubMed]
40. Cazalet, C.; Rusniok, C.; Brüggemann, H.; Zidane, N.; Magnier, A.; Ma, L.; Tichit, M.; Jarraud, S.; Bouchier, C.; Vandenesch, F.; et al. Evidence in the *Legionella pneumophila* genome for exploitation of host cell functions and high genome plasticity. *Nat. Genet.* **2004**, *36*, 1165–1173. [CrossRef]
41. Glöckner, G.; Albert-Weissenberger, C.; Weinmann, E.; Jacobi, S.; Schunder, E.; Steinert, M.; Hacker, J.; Heuner, K. Identification and characterization of a new conjugation/type IVA secretion system (trb/tra) of *Legionella pneumophila* Corby localized on two mobile genomic islands. *Int. J. Med. Microbiol.* **2008**, *298*, 411–428. [CrossRef]
42. Zayed, A.R.; Butmeh, S.; Pecellin, M.; Salah, A.; Alalam, H.; Höfle, M.G.; Bitar, D.M.; Brettar, I. Biogeography and environmental drivers of *Legionella pneumophila* abundance and genotype composition across the West Bank: Relevance of a genotype-based ecology for understanding *Legionella* occurrence. *Pathogens* **2020**. under review.
43. Jaber, L.; Amro, M.; Abu Tair, H.; Bahader, S.A.; AlAlam, H.; Butmeh, S.; Abu Hilal, D.; Brettar, I.; Höfle, M.G.; Bitar, D.M. Comparison of in situ sequence type analysis of *Legionella pneumophila* in respiratory tract secretions and environmental samples of a hospital in East Jerusalem. *Epidemiol. Infect.* **2018**, *146*, 2116–2121. [CrossRef]
44. Bonetta, S.; Ferretti, E.; Balocco, F.; Carraro, E.; Bonetta, S. Evaluation of *Legionella pneumophila* contamination in Italian hotel water systems by quantitative real-time PCR and culture methods. *J. Appl. Microbiol.* **2010**, *108*, 1576–1583. [CrossRef]
45. Beauté, J.; Robesyn, E.; De Jong, B. Legionnaires' disease in Europe: All quiet on the eastern front? *Eur. Respir. J.* **2013**, *42*, 1454–1458. [CrossRef] [PubMed]
46. Yarom, R.; Sheinman, R.; Armon, R. *Legionella pneumophila* serogroup 3 prevalence in drinking water survey in Israel (2003–2007). *Water Supply* **2010**, *10*, 746–752. [CrossRef]
47. Huson, D.H.; Bryant, D. Application of phylogenetic networks in evolutionary studies. *Mol. Biol. Evol.* **2005**, *23*, 254–267. [CrossRef] [PubMed]
48. Katsiaflaka, A.; Pournaras, S.; Kristo, I.; Mouchtouri, V.A.; Kyritsi, M.; Velonakis, E.; Vatopoulos, A.C.; Hadjichristodoulou, C. Epidemiological investigation of *Legionella pneumophila* serogroup 2 to 14 isolates from water samples by amplified fragment length polymorphism and sequence-based typing and detection of virulence traits. *Appl. Environ. Microbiol.* **2016**, *82*, 6102–6108. [CrossRef]
49. Ghrairi, T.; Chaftar, N.; Jarraud, S.; Berjeaud, J.M.; Hani, K.; Frère, J. Diversity of legionellae strains from Tunisian hot spring water. *Res. Microbiol.* **2013**, *164*, 342–350. [CrossRef] [PubMed]

50. Al-Matawah, Q.A.; Al-Zenki, S.F.; Qasem, J.A.; Al-Waalan, T.E.; Ben Heji, A.H. Detection and quantification of *Legionella pneumophila* from water systems in Kuwait residential facilities. *J. Pathog.* **2012**, *13*, 8389. [CrossRef]
51. Ma'Ayeh, S.Y.; Al-Hiyasat, A.S.; Hindiyeh, M.; Khader, Y.; Hindiyeh, M.Y. *Legionella pneumophila* contamination of a dental unit water line system in a dental teaching centre. *Int. J. Dent. Hyg.* **2008**, *6*, 48–55. [CrossRef]
52. Rodríguez-Martínez, S.; Sharaby, Y.; Pecellín, M.; Brettar, I.; Höfle, M.; Halpern, M. Spatial distribution of *Legionella pneumophila* MLVA-genotypes in a drinking water system. *Water Res.* **2015**, *77*, 119–132. [CrossRef]
53. Douterelo, I.; Boxall, J.B.; Deines, P.; Sekar, R.; Fish, K.E.; Biggs, C.A. Methodological approaches for studying the microbial ecology of drinking water distribution systems. *Water Res.* **2014**, *65*, 134–156. [CrossRef]
54. Berry, D.; Xi, C.; Raskin, L. Microbial ecology of drinking water distribution systems. *Curr. Opin. Biotechnol.* **2006**, *17*, 297–302. [CrossRef]
55. Reuter, S.; Harrison, T.G.; Köser, C.U.; Ellington, M.J.; Smith, G.P.; Parkhill, J.; Peacock, S.J.; Bentley, S.D.; Török, M.E. A pilot study of rapid whole-genome sequencing for the investigation of a *Legionella* outbreak. *BMJ Open* **2013**, *3*, e002175. [CrossRef] [PubMed]
56. Whiley, H.; Keegan, A.; Fallowfield, H.; Ross, K. Uncertainties associated with assessing the public health risk from *Legionella*. *Front. Microbiol.* **2014**, *5*, 501. [CrossRef]
57. Oliver, J.D. The viable but nonculturable state in bacteria. *J. Microbiol.* **2005**, 93–100.
58. Lee, H.K.; Kang, Y.H.; Yu, J.Y. Genomic diversity of *Legionella pneumophila* serogroup 1 from environmental water sources and clinical specimens using pulsed-field gel electrophoresis (PFGE) from 1985 to 2007, Korea. *J. Microbiol.* **2010**, *48*, 547–553. [CrossRef] [PubMed]
59. Uzel, A.; Uçar, F.; Hameş-Kocabaş, E.E. Prevalence of *Legionella pneumophila* serogroup 1 in water distribution systems in İzmir province of Turkey. *APMIS* **2005**, *113*, 664–669. [CrossRef]
60. Harrison, T.G.; Afshar, B.; Doshi, N.; Fry, N.K.; Lee, J.V. Distribution of *Legionella pneumophila* serogroups, monoclonal antibody subgroups and DNA sequence types in recent clinical and environmental isolates from England and Wales (2000–2008). *Eur. J. Clin. Microbiol. Infect. Dis.* **2009**, *28*, 781–791. [CrossRef]
61. Doleans, A.; Aurell, H.; Reyrolle, M.; Lina, G.; Freney, J.; Vandenesch, F.; Etienne, J.; Jarraud, S. Clinical and environmental distributions of *Legionella* strains in France are different. *J. Clin. Microbiol.* **2004**, *42*, 458–460. [CrossRef]
62. Coscolla, M.; González-Candelas, F. Comparison of clinical and environmental samples of *Legionella pneumophila* at the nucleotide sequence level. *Infect. Genet. Evol.* **2009**, *9*, 882–888. [CrossRef]
63. Martin-Loeches, I.; Torres, A.; Rinaudo, M.; Terraneo, S.; De Rosa, F.; Ramirez, P.; Díaz, E.; Fernandez-Barat, L.; Bassi, G.L.L.; Ferrer, M. Resistance patterns and outcomes in intensive care unit (ICU)-acquired pneumonia. Validation of European Centre for Disease Prevention and Control (ECDC) and the Centers for Disease Control and Prevention (CDC) classification of multidrug resistant organisms. *J. Infect.* **2015**, *70*, 213–222. [CrossRef]
64. Yu, V.L.; Plouffe, J.F.; Pastoris, M.C.; Stout, J.E.; Schousboe, M.; Widmer, A.; Summersgill, J.; File, T.; Heath, C.M.; Paterson, D.L.; et al. Distribution of *Legionella* species and serogroups isolated by culture in patients with sporadic community-acquired Legionellosis: An international collaborative survey. *J. Infect. Dis.* **2002**, *186*, 127–128. [CrossRef]
65. Fendukly, F.; Bernander, S.; Hanson, H.-S. Nosocomial Legionnaires' disease caused by *Legionella pneumophila* serogroup 6: Implication of the sequence-based typing method (SBT). *Scand. J. Infect. Dis.* **2007**, *39*, 213–216. [CrossRef] [PubMed]
66. Zhang, Q.; Zhou, H.; Chen, R.; Qin, T.; Ren, H.; Liu, B.; Ding, X.; Sha, D.; Zhou, W. Legionnaires' Disease caused by *Legionella pneumophila* serogroups 5 and 10, China. *Emerg. Infect. Dis.* **2014**, *20*, 1242–1243. [CrossRef] [PubMed]
67. Cazalet, C.; Jarraud, S.; Ghavi-Helm, Y.; Kunst, F.; Glaser, P.; Etienne, J.; Buchrieser, C. Multigenome analysis identifies a worldwide distributed epidemic *Legionella pneumophila* clone that emerged within a highly diverse species. *Genome Res.* **2008**, *18*, 431–441. [CrossRef] [PubMed]
68. Kozak-Muiznieks, N.A.; Lucas, C.E.; Brown, E.; Pondo, T.; Taylor, T.H.; Frace, M.; Miskowski, D.; Winchell, J.M. Prevalence of sequence types among clinical and environmental isolates of *Legionella pneumophila* serogroup 1 in the United States from 1982 to 2012. *J. Clin. Microbiol.* **2013**, *52*, 201–211. [CrossRef] [PubMed]

69. Amemura-Maekawa, J.; Kikukawa, K.; Helbig, J.H.; Kaneko, S.; Suzuki-Hashimoto, A.; Furuhata, K.; Chang, B.; Murai, M.; Ichinose, M.; Ohnishi, M.; et al. Distribution of monoclonal antibody subgroups and sequence-based types among *Legionella pneumophila* serogroup 1 isolates derived from cooling tower water, bathwater, and soil in Japan. *Appl. Environ. Microbiol.* **2012**, *78*, 4263–4270. [CrossRef] [PubMed]
70. Lee, H.K.; Shim, J.I.; Kim, H.E.; Yu, J.Y.; Kang, Y.H. Distribution of *Legionella* species from environmental water sources of public facilities and genetic diversity of *L. pneumophila* serogroup 1 in South Korea. *Appl. Environ. Microbiol.* **2010**, *76*, 6547–6554. [CrossRef] [PubMed]
71. Borchardt, J.; Helbig, J.H.; Lück, P.C. Occurrence and distribution of sequence types among *Legionella pneumophila* strains isolated from patients in Germany: Common features and differences to other regions of the world. *Eur. J. Clin. Microbiol. Infect. Dis.* **2007**, *27*, 29–36. [CrossRef]
72. Chasqueira, M.J.; Rodrigues, L.; Nascimento, M.; Ramos, M.; Marques, T. Genetic diversity and evolutionary relationships among *Legionella pneumophila* clinical isolates, Portugal, 1987 to 2012. *Eurosurveillance* **2014**, *19*, 20965. [CrossRef]
73. Sánchez-Busó, L.; Coscollá, M.; Pinto-Carbó, M.; Catalán, V.; González-Candelas, F. Genetic characterization of *Legionella pneumophila* isolated from a common watershed in Comunidad Valenciana, Spain. *PLoS ONE* **2013**, *8*, e61564. [CrossRef]
74. David, S.; Afshar, B.; Mentasti, M.; Ginevra, C.; Podglajen, I.; Harris, S.R.; Chalker, V.J.; Jarraud, S.; Harrison, T.G.; Parkhill, J. Seeding and establishment of *Legionella pneumophila* in hospitals: Implications for genomic investigations of nosocomial Legionnaires' disease. *Clin. Infect. Dis.* **2017**, *64*, 1251–1259. [CrossRef]
75. Fontana, S.; Scaturro, M.; Rota, M.C.; Caporali, M.G.; Ricci, M.L. Molecular typing of *Legionella pneumophila* serogroup 1 clinical strains isolated in Italy. *Int. J. Med. Microbiol.* **2014**, *304*, 597–602. [CrossRef] [PubMed]
76. García-Vidal, C.; Labori, M.; Viasus, D.; Simonetti, A.; Garcia-Somoza, D.; Dorca, J.; Gudiol, F.; Carratalà, J. Rainfall is a risk factor for sporadic cases of *Legionella pneumophila* pneumonia. *PLoS ONE* **2013**, *8*, e61036. [CrossRef] [PubMed]
77. Gaia, V.; Fry, N.K.; Harrison, T.G.; Peduzzi, R. Sequence-based typing of *Legionella pneumophila* serogroup 1 offers the potential for true portability in *Legionellosis* outbreak investigation. *J. Clin. Microbiol.* **2003**, *41*, 2932–2939. [CrossRef] [PubMed]
78. Lawrence, C.; Reyrolle, M.; Dubrou, S.; Forey, F.; Decludt, B.; Goulvestre, C.; Matsiota-Bernard, P.; Etienne, J.; Nauciel, C. Single clonal origin of a high proportion of *Legionella pneumophila* serogroup 1 isolates from patients and the environment in the area of Paris, France, over a 10-year period. *J. Clin. Microbiol.* **1999**, *37*, 2652–2655. [CrossRef] [PubMed]
79. David, S.; Rusniok, C.; Mentasti, M.; Gomez-Valero, L.; Harris, S.R.; Lechat, P.; Lees, J.; Ginevra, C.; Glaser, P.; Ma, L.; et al. Multiple major disease-associated clones of *Legionella pneumophila* have emerged recently and independently. *Genome Res.* **2016**, *26*, 1555–1564. [CrossRef]
80. David, S.; Mentasti, M.; Tewolde, R.; Aslett, M.; Harris, S.R.; Afshar, B.; Underwood, A.; Fry, N.K.; Parkhill, J.; Harrison, T.G. Evaluation of an optimal epidemiological typing scheme for *Legionella pneumophila* with whole-genome sequence data using validation guidelines. *J. Clin. Microbiol.* **2016**, *54*, 2135–2148. [CrossRef]
81. Sharaby, Y.; Rodríguez-Martínez, S.; Pecellin, M.; Sela, R.; Peretz, A.; Höfle, M.G.; Halpern, M.; Brettar, I. Virulence traits of environmental and clinical *Legionella pneumophila* Multilocus Variable-Number Tandem-Repeat Analysis (MLVA) Genotypes. *Appl. Environ. Microbiol.* **2018**, *84*. [CrossRef] [PubMed]
82. Sharaby, Y.; Nitzan, O.; Brettar, I.; Höfle, M.G.; Peretz, A.; Halpern, M. Antimicrobial agent susceptibilities of *Legionella pneumophila* MLVA-8 genotypes. *Sci. Rep.* **2019**, *9*, 6138. [CrossRef]
83. Sharaby, Y.; Rodríguez-Martínez, S.; Oks, O.; Pecellin, M.; Mizrahi, H.; Peretz, A.; Brettar, I.; Höfle, M.G.; Halpern, M. Temperature-dependent growth modeling of environmental and clinical *Legionella pneumophila* Multilocus Variable-Number Tandem-Repeat Analysis (MLVA) genotypes. *Appl. Environ. Microbiol.* **2017**, *83*, e03295-16. [CrossRef]
84. García, K.; Gavilán, R.G.; Höfle, M.G.; Martínez-Urtaza, J.; Espejo, R.T. Microevolution of pandemic *Vibrio parahaemolyticus* assessed by the number of repeat units in short sequence tandem repeat regions. *PLoS ONE* **2012**, *7*, e30823. [CrossRef] [PubMed]

85. Husseini, H. Palestinian Water Authority: Developments and Challenges-Legal Framework and Capacity. In *Water Resources in the Middle East*; Springer Science and Business Media LLC: New York, NY, USA, 2007; pp. 301–308.
86. Sharaby, Y.; Rodríguez-Martínez, S.; Höfle, M.; Brettar, I.; Halpern, M. Quantitative microbial risk assessment of *Legionella pneumophila* in a drinking water supply system in Israel. *Sci. Total. Environ.* **2019**, *671*, 404–410. [CrossRef] [PubMed]


**Publisher's Note:** MDPI stays neutral with regard to jurisdictional claims in published maps and institutional affiliations.

pathogens

MDPI

Article

# *Legionella* Diversity and Spatiotemporal Variation in the Occurrence of Opportunistic Pathogens within a Large Building Water System

**Helen Y. Buse [1,*], Brian J. Morris [2], Vicente Gomez-Alvarez [3], Jeffrey G. Szabo [1] and John S. Hall [1]**

1 Homeland Security and Materials Management Division, Center for Environmental Solutions & Emergency Response (CESER), Office of Research and Development (ORD), US Environmental Protection Agency (USEPA), Cincinnati, OH 45268, USA; Szabo.Jeff@epa.gov (J.G.S.); Hall.John@epa.gov (J.S.H.)

2 Pegasus Technical Services, Inc c/o US EPA, Cincinnati, OH 45268, USA; morris.brian@epa.gov

3 Water Infrastructure Division, Center for Environmental Solutions & Emergency Response (CESER), US Environmental Protection Agency (USEPA), Office of Research and Development (ORD), Cincinnati, OH 45268, USA; Gomez-Alvarez.Vicente@epa.gov

* Correspondence: buse.helen@epa.gov; Tel.: +1-513-569-7930

Received: 8 June 2020; Accepted: 7 July 2020; Published: 13 July 2020

**Abstract:** Understanding *Legionella* survival mechanisms within building water systems (BWSs) is challenging due to varying engineering, operational, and water quality characteristics unique to each system. This study aimed to evaluate *Legionella*, mycobacteria, and free-living amoebae occurrence within a BWS over 18–28 months at six locations differing in plumbing material and potable water age, quality, and usage. A total of 114 bulk water and 57 biofilm samples were analyzed. *Legionella* culturability fluctuated seasonally with most culture-positive samples being collected during the winter compared to the spring, summer, and fall months. Positive and negative correlations between *Legionella* and *L. pneumophila* occurrence and other physiochemical and microbial water quality parameters varied between location and sample types. Whole genome sequencing of 19 presumptive *Legionella* isolates, from four locations across three time points, identified nine isolates as *L. pneumophila* serogroup (sg) 1 sequence-type (ST) 1; three as *L. pneumophila* sg5 ST1950 and ST2037; six as *L. feeleii*; and one as *Ochrobactrum*. Results showed the presence of a diverse *Legionella* population with consistent and sporadic occurrence at four and two locations, respectively. Viewed collectively with similar studies, this information will enable a better understanding of the engineering, operational, and water quality parameters supporting *Legionella* growth within BWSs.

**Keywords:** potable water; first draw; second draw; biofilm; whole genome sequencing; environmental monitoring; premise plumbing systems

## 1. Introduction

Various *Legionella* species, such as *L. anisa*, *L. feeleii*, *L. longbeachae*, *L. pneumophila*, and *L. micdadei*, can cause legionellosis, which are bacterial infections resulting in either a mild flu-like illness (Pontiac Fever) or a potentially fatal form of pneumonia (Legionnaires' Disease (LD)) [1]. These infections are primarily caused by the inhalation of *Legionella*-contaminated aerosols generated from engineered water systems [2,3]. Of the 74 drinking water-associated outbreaks reported in the US between 2011 and 2014, *Legionella* was responsible for 61% of those outbreaks causing 17% (241/1437) of the illness cases, 88% (200/226) of hospitalizations, and 100% of the outbreak deaths (27/27) [4,5]. *Legionella* presence in the building water systems (BWSs) was cited as the main deficiency leading to those outbreaks underscoring the need to control and prevent *Legionella* growth within these BWSs.

Over the past 40 years, *Legionella* occurrence in drinking water distribution systems (DWDSs) has been well-studied since the first speculation that exposure to aerosols, derived from contaminated water chillers, caused the deadly 1976 American Legion convention outbreak in Philadelphia [6]. *Legionella* has been shown to occur at low levels in drinking water treatment plants and distribution systems and at higher levels in both cold and hot water within BWSs [7–9]. *Legionella* occurrence has been documented in both antiquated (>100 years old) and newly constructed buildings [10,11]. Moreover, numerous studies have reported stable colonization of, and continuing infections caused by, *L. pneumophila* within BWSs over a ≥ 15 year period, despite repeated cycles of shock chlorination, superheating and thermal inactivation, and flushing [12–16].

*Legionella* occurrence in diverse environments and conditions (e.g., in sea-, fresh-, rain-, and treated water; in soil; at temperatures between 4 and 60 °C and a pH range of 2.7–8.3) [1] suggests the presence of heterogeneous survival mechanisms acquired through effective evolutionary processes. Indeed, *L. pneumophila* is genetically well-equipped due to their (1) high recombination rate and DNA exchange among strains and different *Legionella* species [17] and (2) large repertoire of effector proteins allowing exploitation of conserved cellular pathways in various eukaryotic hosts [18,19].

Due to their stringent nutrient requirements, it is hypothesized that *Legionella* growth within DWDSs is largely dependent on their parasitization of drinking water-associated eukaryotic hosts, specifically free-living amoebae (FLA). FLA, such as *Acanthamoeba* spp. and *Vermamoeba vermiformis*, are problematic in drinking waters systems as they can cause diseases like keratitis and can also amplify other human pathogens such as *L. pneumophila* and mycobacteria, including *Mycobacterium intracellulare* [20,21]. *M. intracellulare*, a member of the non-tuberculous mycobacteria (NTM) group, is a significant cause of pulmonary NTM infections [22] and has been isolated in higher frequencies and concentrations in both bulk water and biofilms from drinking water distribution systems compared to *Mycobacterium avium* [23].

Numerous studies have demonstrated extracellular growth of *L. pneumophila* in drinking water. Specifically, four different types of non-*Legionella* drinking water bacteria [24,25]; heat-inactivated cooling tower biofilms, *Escherichia coli*, and *Pseudomonas putida* [26]; extracellular cyanobacterial components [27]; and filtered sterilized drinking water [28] were capable of supporting *L. pneumophila* growth, with the latter also supporting their colonization and growth within biofilms for prolonged periods. Collectively, *Legionella* associations with, and dependencies on, other microbes can explain their vast environmental distribution, ability to survive in DWDSs, and human pathogenicity, especially in engineered environments where confined, close proximity to humans have increased disease risk from this pathogen.

Thus, current approaches to *Legionella* exposure prevention, e.g., through environmental monitoring [29] and/or implementation of building water management plans [30,31], require a thorough understanding of their persistence and transmission mechanisms in premise plumbing systems. In this study, to better understand BWS conditions supporting *Legionella* survival, microbial water quality parameters, such as the occurrence of *Legionella* spp., *L. pneumophila*, and *M. intracellulare*; and the FLA hosts, *Acanthamoeba* spp. and *Vermamoeba vermiformis*, and heterotrophic plate count levels, were monitored, along with chemical water quality, engineering, and operational parameters. Determining how various physical, chemical, and microbial BWS characteristics influence *Legionella* occurrence may elucidate ways to minimize and eliminate their growth within these systems.

## 2. Results

### 2.1. Description of Sampling Locations and Water Quality Characteristics

First draw and second draw (post-flushing) cold bulk water samples and biofilm swab samples were collected every three months (fall, F; winter, W; spring, Sp; and summer, Su) at six locations within a large commercial building (Table 1 and Section 4.1). This building water system (BWS) contained a variety of plumbing materials, varying water flow/rates, and usage patterns at each of the sampled locations (Table 1) with previous detection of *Legionella* in the cold water (data not shown). This building was also chosen because it contains both office and production facilities where large volumes of water are used in the latter portions (e.g., for production processes and cooling) and smaller volumes in the office spaces for employees. Three locations had polyvinyl chloride (PVC) schedule 80 valves that supplied monochloramine-treated water (PVC-MA) or chlorine-treated water with and without passage through the building's chiller/refrigeration system (PVC-R and PVC-FC, respectively); and the other three sampling locations supplied chlorine-treated water from a cast brass spigot (Spigot), a chrome-plated, forged brass faucet (Faucet), and a drinking water fountain with a stainless steel and ethylene propylene diene bubbler head (Fountain) (Figure S1, Table 1).

The water usage ranged from 5 L to 1.4 million L per season with an average of 278,030 L ± a standard deviation (SD) of 598,785 L per season for each sampling location. The PVC-R location is a high water usage site due to the presence of multiple online and remote water quality sensors that require constant water flow; thus, when this site is excluded, the water usage of the other five locations had an average of 35,194 ± 76,876 L per season. The pipe material from the outlet to the supply feed were either all copper (Spigot, Faucet, and Fountain), all PVC (PVC-MA), a mixture of both (PVC-FC), or a mixture of fiberglass ductile iron and PVC reinforced plastic tubing (PVC-R).

In this study, a total of 114 bulk water and 57 biofilm samples were collected across the six locations. Bulk water samples were analyzed for pH, turbidity, temperature, and free and total chlorine (Table 2). There were no statistical differences between the pH of the first and second draw samples for each location and sampling time points, but there were differences in pH between the PVC-MA samples and the other five locations ($P < 0.001$). There were no statistical differences between the turbidity of all water samples at each location most likely due to the large range in Nephelometric Turbidity Units (NTUs) observed for each sample (Table 2, Figure S2).

Temperatures between first and second draw samples within each location and time point were also not significant, except for Spigot-first draw versus Spigot-second draw samples ($P < 0.05$). Free and total chlorine levels for first and second draw samples between, and within, each location were not significant, except for Faucet-first draw versus Faucet-second draw and PVC-R-first vs Faucet-first draw samples ($P < 0.01$, Table 2, Figure S3). There were no statistical differences between monochloramine and total chlorine levels in the bulk water from location PVC-MA ($P > 0.05$). There were statistical differences between the heterotrophic plate count (HPC) levels of the first vs second draw samples at only the PVC-R, Spigot, and Faucet locations ($P < 0.01$, Table 2). For biofilm samples, the average HPC level observed at the Faucet location was the highest compared to other locations but was statistically different only from the PVC-R biofilm samples ($P < 0.01$, Table 2).

**Table 1.** Description of each sampling location.

| | Sampling Locations | | | | | | | | | | | |
|---|---|---|---|---|---|---|---|---|---|---|---|---|
| **Site Name** | **PVC-MA** | | **PVC-R** | | **PVC-FC** | | **Spigot** | | **Faucet** | | **Fountain** | |
| **Disinfectant type** | monochloramine | | chlorine | | chlorine | | chlorine | | chlorine | | chlorine | |
| **Water type** | potable water | | chiller water | | potable water | | potable water | | potable water | | potable water | |
| **Approx. usage per season** | 172,709 L | | 1,492,209 L | | 5 L | | 1703 L | | 864 L | | 689 L | |
| **Outlet type** | valve | | valve | | valve | | spigot | | faucet | | bubbler | |
| **Outlet material** | PVC80; PharMed BPT[1] | | PVC80 | | PVC80 | | cast brass | | chrome plated, forged brass | | SS/EDPM rubber | |
| **Pipe material** | PVC80 | | FDI and PPT[2] | | PVC80 and copper[2] | | copper | | copper | | copper | |
| **Total distance to feed water** | 14 m | 46 ft | 396 m | 1299 ft | 21 m | 70 ft | 102 m | 334 ft | 20 m | 65 ft | 9.7 m | 32 ft |
| Section 1 | 14 m | 46 ft | 366 m | 1201 ft | 9 m | 30 ft | 5 m | 16 ft | 8 m | 26 ft | 9.1 m | 30 ft |
| **pipe diameter** | 15 cm | 6 in | 7.6 cm | 3 in | 2.5 cm | 1 in | 1.9 cm | 0.75 in | 1.9 cm | 0.75 in | 1.3 cm | 0.5 in |
| Section 2 | - | | 30 m | 98 ft | 12 m | 40 ft | 97 m | 318 ft | 12 m | 39 ft | 0.6 m | 2 ft |
| **pipe diameter** | - | | 1.3 cm | 0.5 in | 2.5 cm | 1 in | 5.1 cm | 2 in | 5.1 cm | 2 in | 0.6 cm | 0.25 in |

Abbreviations: -, not applicable; cm, centimeter; EDPM, ethylene propylene diene; FDI, fiberglass ductile iron; ft, feet; in, inch; m, meter; PPT, PVC-reinforced plastic tubing; PVC80, polyvinyl chloride schedule 80; SS, stainless steel. [1] Masterflex PharMed BPT tubing approximately 1 m (3.5 ft) in length and inner diameter 4.8 mm (0.189 in). [2] Sections 1 and 2, respectively.

**Table 2.** Summary of water quality parameters for each sampling location.

| Location Name & Sample Type | | HPC | | pH | | Turbidity | | Temp | | Free $Cl_2$ | | Total $Cl_2$ | |
|---|---|---|---|---|---|---|---|---|---|---|---|---|---|
| | | Mean | SEM | Mean | SEM | Mean | SEM | Mean | SEM | Mean | SEM | Mean | SEM |
| PVC-MA [a] | First Draw | 5.5 | 0.1 | 8.21 | 0.08 | 0.23 | 0.03 | 20.2 | 0.3 | 0.06 | 0.01 | 1.21 | 0.11 |
| | Second Draw | 4.9 | 0.2 | 8.19 | 0.07 | 0.26 | 0.04 | 20.2 | 0.3 | 0.06 | 0.01 | 1.22 | 0.11 |
| | Biofilm | 3.4 | 0.2 | - | - | - | - | - | - | - | - | - | - |
| PVC-R | First Draw | 5.4 | 0.2 | 8.53 | 0.07 | 6.94 | 3.23 | 24.8 | 1.4 | 1.04 | 0.10 | 1.13 | 0.09 |
| | Second Draw | 4.5 | 0.2 | 8.68 | 0.05 | 2.13 | 0.96 | 24.8 | 1.4 | 0.90 | 0.06 | 1.02 | 0.06 |
| | Biofilm | 2.6 | 0.1 | - | - | - | - | - | - | - | - | - | - |
| PVC-FC | First Draw | 4.2 | 0.6 | 8.47 | 0.15 | 0.42 | 0.15 | 20.7 | 2.5 | 0.83 | 0.07 | 0.92 | 0.08 |
| | Second Draw | 4.0 | 0.1 | 8.60 | 0.17 | 0.22 | 0.02 | 18.6 | 3.0 | 0.99 | 0.04 | 1.10 | 0.05 |
| | Biofilm | 2.0 | 0.6 | - | - | - | - | - | - | - | - | - | - |
| Spigot | First Draw | 5.4 | 0.1 | 8.67 | 0.06 | 0.81 | 0.35 | 24.8 | 1.2 | 0.53 | 0.12 | 0.59 | 0.12 |
| | Second Draw | 2.8 | 0.2 | 8.73 | 0.06 | 0.23 | 0.08 | 18.4 | 2.3 | 1.00 | 0.06 | 1.12 | 0.04 |
| | Biofilm | 2.5 | 0.2 | - | - | - | - | - | - | - | - | - | - |
| Faucet | First Draw | 5.5 | 0.2 | 8.67 | 0.06 | 0.17 | 0.02 | 24.0 | 0.7 | 0.35 | 0.10 | 0.40 | 0.11 |
| | Second Draw | 4.2 | 0.2 | 8.72 | 0.06 | 0.18 | 0.02 | 20.0 | 1.7 | 1.05 | 0.04 | 1.13 | 0.04 |
| | Biofilm | 4.2 | 0.2 | - | - | - | - | - | - | - | - | - | - |
| Fountain | First Draw | 4.0 | 0.2 | 8.75 | 0.06 | 0.25 | 0.06 | 13.5 | 0.4 | 0.59 | 0.07 | 0.69 | 0.07 |
| | Second Draw | 3.8 | 0.3 | 8.78 | 0.07 | 0.18 | 0.02 | 15.4 | 1.0 | 0.89 | 0.03 | 0.98 | 0.03 |
| | Biofilm | 4.0 | 0.5 | - | - | - | - | - | - | - | - | - | - |

Abbreviations: -, no data; $Cl_2$, chlorine, mg $L^{-1}$; HPC, heterotrophic plate count, $log_{10}$ CFU per 100 mL or $cm^2$; LOD, limit of detection; NTU, Nephelometric Turbidity Unit; SEM, standard error mean; Temp, temperature in degrees Celsius. [a] see Materials and Methods, Section 4.1, for average monochloramine and free ammonia levels during the entire sampling period.

## 2.2. *Legionella* Culture Results

Bulk water and biofilm samples were processed and enumerated for *Legionella* colony forming units (CFU) as described in Sections 4.2 and 4.4. No culturable *Legionella* was detected in samples from the PVC-MA, Spigot, or Fountain location; from the second draw and biofilm samples from PVC-R; and biofilm samples from PVC-FC. However, culturable *Legionella* was detected at various time points for the first and second draw samples from PVC-FC; all sample types from the Faucet; and only the W2018 first draw sample from PVC-R, which contained *Legionella* non-*pneumophila* bacteria as confirmed by colony lysate PCR and latex agglutination (Table 3).

**Table 3.** Summary of *Legionella* spp. culture results for positive sites.

| Time Point | PVC-R | | | PVC-FC | | | | | | Faucet | | | | | | | | |
|---|---|---|---|---|---|---|---|---|---|---|---|---|---|---|---|---|---|---|
| | First Draw | | | First Draw | | | Second Draw | | | First Draw | | | Second Draw | | | Biofilm | | |
| | CFU | sg | PCR | CFU | sg | PCR | CFU | sg | PCR | CFU | sg | PCR | CFU | sg | PCR | CFU | sg | PCR |
| F2016 | - | - | - | nd | nd | nd | nd | nd | nd | - | - | - | - | - | - | 1.9 | 1 & 2-14 | Leg/Lp |
| W2017 | - | - | - | nd | nd | nd | nd | nd | nd | - | - | - | - | - | - | - | - | - |
| Sp2017 | - | - | - | nd | nd | nd | nd | nd | nd | 1.2 | 2-14 | Leg/Lp | - | - | - | - | - | - |
| Su2017 | - | - | - | - | - | - | - | - | - | - | - | - | - | - | - | - | - | - |
| F2017 | - | - | - | - | - | - | - | - | - | 1.2 | neg | Leg | - | - | - | - | - | - |
| W2018 | 2.6 | neg | Leg | 4.5 | neg | Leg | 2.6 | neg | Leg | 2.1 | 1 & 2-14 | Leg/Lp | - | - | - | - | - | - |
| Sp2018 | - | - | - | - | - | | - | - | - | - | - | - | - | - | - | - | - | - |
| Su2018 | - | - | - | 3.8 | neg & 2-14 | Leg & Leg/Lp | 2.8 | neg & 2-14 | Leg & Leg/Lp | 2.7 | 1 | Leg/Lp | - | - | - | - | - | - |
| F2018 | - | - | - | - | - | | 3.0 | neg | Leg | - | - | - | - | - | - | - | - | - |
| W2019 | - | - | - | 2.1 | neg | Leg | 2.6 | neg | Leg | 2.8 | 1 | Leg/Lp | 2.1 | 1 | Leg/Lp | - | - | - |

No culturable *Legionella* was detected in the PVC-MA; Spigot; Fountain; PVC-R Distal and Biofilm; and PVC-FC Biofilm samples. Abbreviations: -, below LOD; B, biofilm; LOD, limit of detection; nd, no data; qPCR, quantitative PCR; sg, serogroup. CFU is expressed as $\log_{10}$ CFU 100 $mL^{-1}$ for bulk water samples (LOD 1.0 $\log_{10}$ CFU 100 $mL^{-1}$) and $\log_{10}$ CFU $cm^{-2}$ for biofilm samples (LOD 0.7 $\log_{10}$ CFU $cm^{-2}$). sg results: 1, sg 1 positive; 2-14, sg 2-14 positive; 1 & 2-14, different colonies on the plate were sg 1 and 2-14 positive; neg, agglutination negative. PCR results: Leg, *Legionella* spp. assay positive only; Leg/Lp, *Legionella* spp. and *L. pneumophila* assay positive.

For the PVC-FC bulk water samples, *Legionella* non-*pneumophila* colonies were detected in both the first and second draw samples during W2018 (4.5 and 2.6 $\log_{10}$ CFU 100 $mL^{-1}$, respectively) and W2019 (2.1 and 2.6 $\log_{10}$ CFU 100 $mL^{-1}$, respectively) and only the second draw (3.0 $\log_{10}$ CFU 100 $mL^{-1}$) during F2018 via colony confirmation PCR (Table 3). The F2018 isolate was identified as *L. feeleii* via indirect immunofluorescent antibody assay as described in Section 4.4. *L. pneumophila* and non-*pneumophila* colonies were detected in the first and second draw samples (3.8 and 2.8 $\log_{10}$ CFU 100 $mL^{-1}$, respectively) at this location during Su2018 via colony confirmation PCR. The *L. pneumophila* PCR-positive Su2018 colonies either gave an inconclusive latex agglutination result or were identified as belonging to serogroup (sg) 2–14. Three Su2018 isolates (one from the first draw and two from the second draw samples), one F2018 isolate (from the second draw sample), and four W2019 isolates (two from the first and two from the second draw samples) were processed for whole genome sequencing. One of the two Su2018 second draw isolates was identified as *L. pneumophila* sg5 via indirect immunofluorescent antibody assay.

Throughout the study period, only one second draw (W2019) and one biofilm (F2016) sample, at the Faucet location had culturable *Legionella* bacteria at concentrations of 2.1 $\log_{10}$ CFU 100 $mL^{-1}$ and 1.9 $\log_{10}$ CFU $cm^{-2}$, respectively (Table 3). The Faucet second draw sample was confirmed as *L. pneumophila* sg1, and the biofilm sample contained a mixture of *L. pneumophila* sg1 and 2–14 colonies as confirmed by PCR and latex agglutination. Fifty percent of the Faucet first draw samples were *Legionella* culture positive: Sp2017, F2017, W2018, Su2018, and W2019 with an average ± SD concentration of 2.0 ± 0.8 $\log_{10}$ CFU 100 $mL^{-1}$ (Table 3). *Legionella* identified within these Faucet first draw samples were diverse with only *L. pneumophila* sg1 being identified in the Su2018 and W2019 samples; only *L. pneumophila* sg2–14 in the Sp2017 sample; a mixture of *L. pneumophila* sg1 and 2–14 in the W2018 sample; and *Legionella* non-*pneumophila* identified in the F2017 sample. Five isolates obtained from the Su2018 and one from the W2019 Faucet first draw samples were processed for whole genome sequencing.

For only the Su2018, F2018, and W2019 time points, *L. pneumophila* most probable number (MPN) in bulk water and biofilm samples were enumerated using Legiolert® as described in Section 4.4. All samples at each location had no detectable *L. pneumophila* MPN except for the Faucet first draw (Su2018 1.8 $\log_{10}$ MPN 100 $mL^{-1}$; W2019 3.8 $\log_{10}$ MPN 100 $mL^{-1}$) and second draw (Su2018 1.4 $\log_{10}$ MPN 100 $mL^{-1}$; W2019 2.5 $\log_{10}$ MPN 100 $mL^{-1}$) samples. Four wells from the first draw and three wells from the second draw Legiolert® tray, read as *L. pneumophila* positive, were sampled and pure colonies were obtained as described in Section 4.4. Isolates were *Legionella* and *L. pneumophila* PCR-positive but gave an inconclusive latex agglutination result. One of the W2019 Faucet second draw Legiolert® isolates was processed for whole genome sequencing and identified as sg5 via indirect immunofluorescent antibody assay.

### 2.3. *Detection of Water-based Pathogens and Free-living Amoebae (FLA) via Quantitative PCR Analyses*

#### 2.3.1. Legionella spp. and L. pneumophila Occurrence

In addition to culture methods, bulk water and biofilm samples were also analyzed for the presence of *Legionella* spp. and *L. pneumophila* via qPCR as described in Section 4.6. In agreement with the specificity and sensitivity of the *Legionella* genus and *L. pneumophila* species-specific qPCR assays used in this study, the total *Legionella* levels detected in the bulk water and biofilm samples were higher than the *L. pneumophila* levels observed in the corresponding samples (Figure 1a–f and g–l). The presence of *Legionella* and *L. pneumophila* was detected at all sampling locations and sample types and within each location, detection generally occurred more frequently, and at higher levels, in the bulk water compared to the biofilm samples (Figure 1, first draw, circles, and second draw, squares, compared to biofilm, triangles).

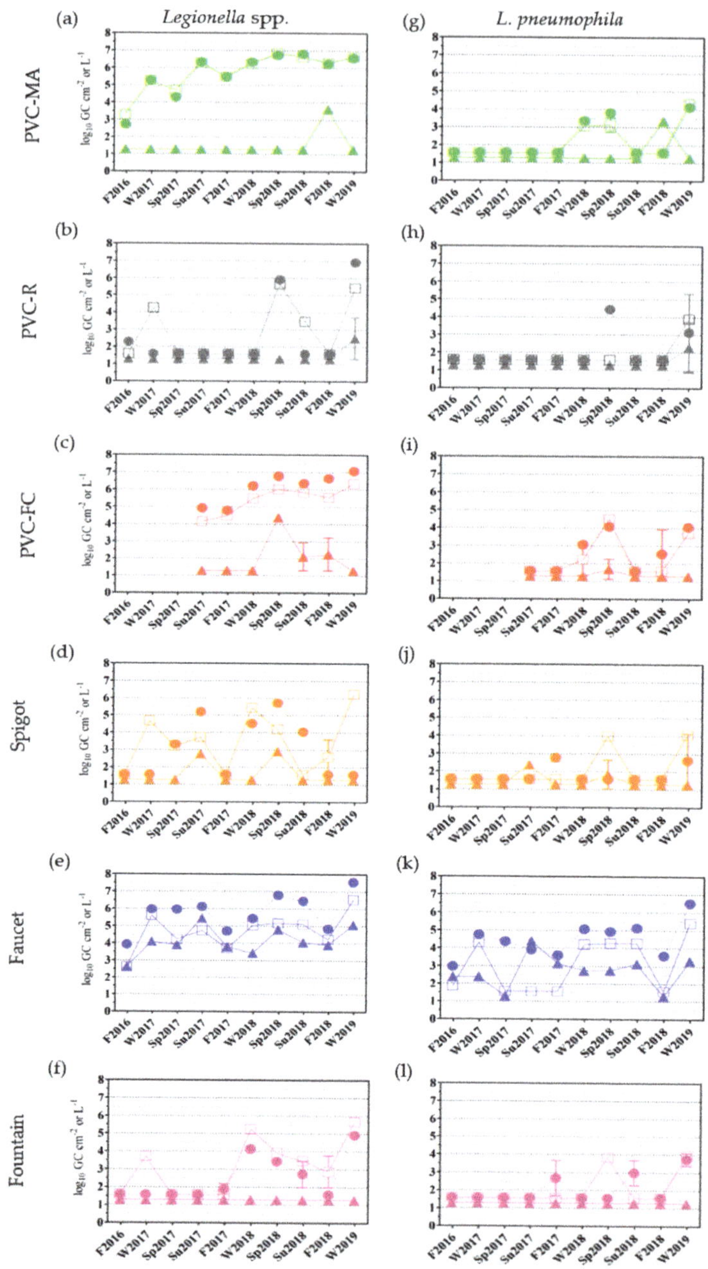


**Figure 1.** qPCR detection of *Legionella* at six locations within a building water system. Bulk water (first draw, filled circles •; second draw, open squares □) and biofilm (filled triangles ▲) samples were analyzed by *Legionella* spp. (**a–f**) and *L. pneumophila* (**g–l**) 16S rRNA qPCR as described in Section 4.6 from each of the six locations listed on the left. Each data point is the mean of duplicate wells with standard deviation. No sampling occurred in F2016–Sp2017 for location PVC-FC. The limit of detection for bulk water and biofilm samples were 1.6 $\log_{10}$ genomic copies (GC) $L^{-1}$, and 1.3 $\log_{10}$ GC $cm^{-2}$, respectively.

*Legionella* was detected in all bulk water samples at locations PVC-MA, PVC-FC, and Faucet (Figure 1a, c, and e). In contrast, only 30% (3/10 first draw; 3/10 second draw) of PVC-MA, 50% (4/7 first draw; 3/7 second draw) of PVC-FC, and 80% (10/10 first draw; 6/10 second draw) of Faucet samples were *L. pneumophila* positive (Figure 1g, i, and k). *Legionella* was detected sporadically in the bulk water samples at locations PVC-R (35%, 3/10 first draw; 4/10 second draw), Spigot (60%, 5/10 first draw; 7/10 second draw), and Fountain (55%, 5/10 first draw; 6/10 second draw) (Figure 1b, d, and f). *L. pneumophila* was detected in 15% (2/10 first draw; 1/10 second draw) of the PVC-R, in 20% (2/10 first draw; 2/10 second draw) of the Spigot, and in 25% (3/10 first draw; 2/10 second draw) of the Fountain bulk water samples (Figure 1h, j, l).

Biofilm samples from all locations had sporadic detection of *Legionella* and *L. pneumophila* except for the Faucet and Fountain locations (Figure 1, triangles). At the Faucet location, both were detected in all biofilm samples except for the Sp2017 and F2018 time points for *L. pneumophila* (Figure 1e and f, blue triangles). At the Fountain location, there was no detectable *Legionella* or *L. pneumophila* at all time points (Figure 1f and l, pink triangles). Additionally, for the biofilm samples, *Legionella* was detected in only one PVC-MA (F2018) and PVC-R (W2019); three PVC-FC (Sp2018, Su2018, and F2018); and two Spigot (Su2017 and Sp2018) biofilm samples (Figure 1a–d, triangles). Similarly, *L. pneumophila* was detected in one PVC-MA (F2018), PVC-R (W2019) and only one of the three *Legionella* positive PVC-FC (Sp2018) samples; and two Spigot (Su2017 and Sp2018) biofilm samples (Figure 1g–j, triangles).

### 2.3.2. Free-living Amoebae and *Mycobacterium intracellulare* Occurrence

To determine if *Legionella* occurrence correlated with the presence of their eukaryotic hosts and another water-based human opportunistic pathogen, bulk water and biofilm samples at each location were analyzed for two different free-living amoebae, *Vermamoeba vermiformis* and *Acanthamoeba* spp., and *Mycobacterium intracellulare* (Figure 2). *V. vermiformis* was detected in the bulk water and biofilm samples at various time points for locations PVC-MA, PVC-FC, Spigot, and Faucet; in only the bulk water and not biofilms for the Fountain location; and was not detected in any samples from location PVC-R (Figure 2a–f). For *Acanthamoeba* spp., only five time points were analyzed for each location (F2016, Sp2018 to W2019), except for PVC-FC where only four were analyzed (Sp2018 to W2019) (Figure 2g–l). Notably, only the second draw sample at location PVC-MA had detectable *Acanthamoeba* spp. (Figure 2g, green square mean ± SD of 2.4 ± 0.0 $\log_{10}$ CE $L^{-1}$).

At location PVC-MA, *V. vermiformis* was detected in 80% of the bulk water samples (9/10 first draw; 7/10 second draw) and 40% of the biofilm samples (4/10). *V. vermiformis* was detected in 43% of the bulk water samples at the PVC-FC (1/7 first draw; 5/7 second draw); 30% at the Spigot (6/10 first draw; 0/10 second draw); and 45% at the Faucet (7/10 first draw; 2/10 second draw) locations (Figure 2c–e). Only one biofilm sample was positive for *V. vermiformis* at the PVC-FC (Sp2018), Spigot (W2018), and Faucet (Su2017) locations (Figure 2c–e). For the Fountain location, *V. vermiformis* was detected in 15% (2/10 first draw; 1/10 second draw) of the bulk water samples (Figure 2f). All samples were negative for *M. intracellulare* at locations PVC-MA, PVC-FC, Faucet, and Fountain (Figure 2m, o, e–f). At location PVC-R, only 15% (2/10 first draw; 1/10 second draw) of water samples were positive (Figure 2n). *M. intracellulare* was more frequently detected in the Spigot bulk water (65%, 8/10 first draw; 5/10 second draw) and biofilm (40%, 4/10) samples.

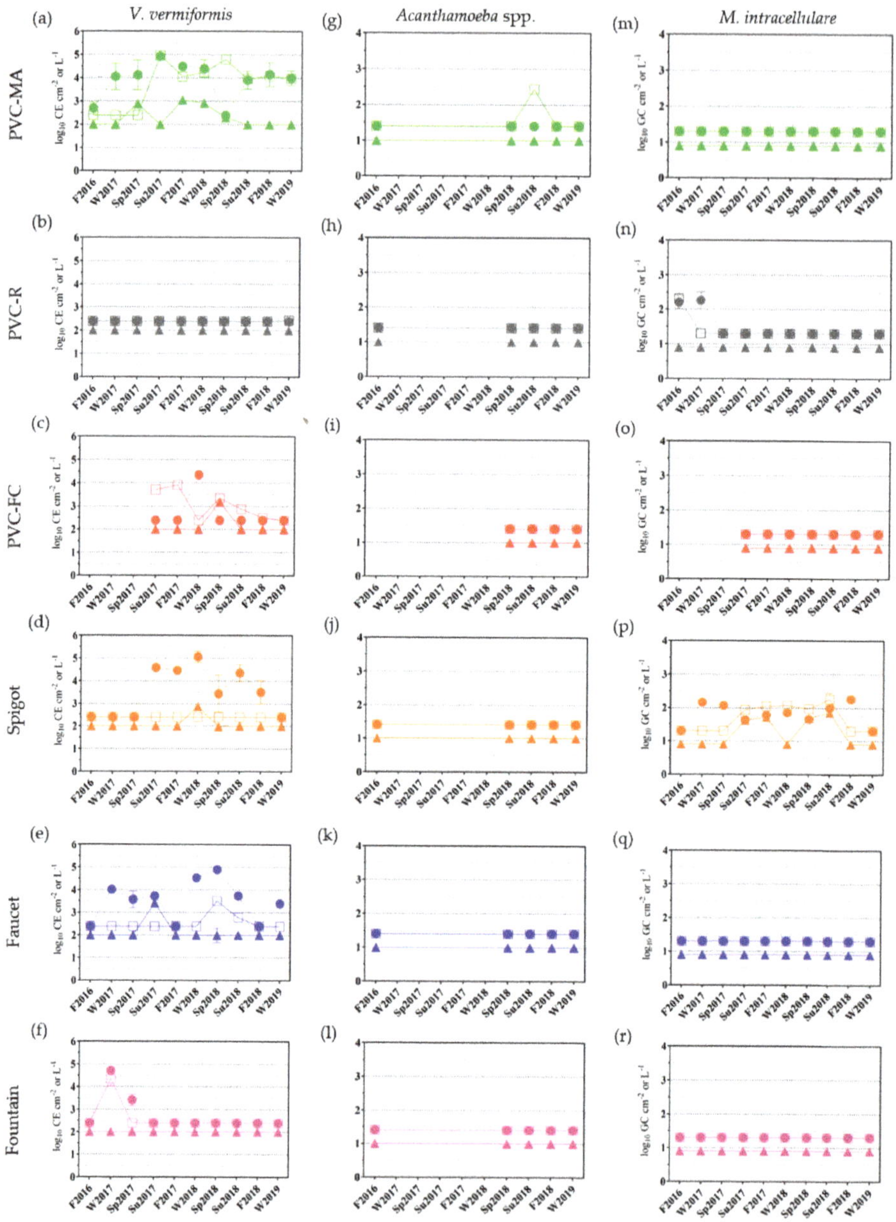


**Figure 2.** qPCR detection of free-living amoeba and *Mycobacterium intracellulare*. Bulk water (first draw, filled circles •; second draw, open squares □) and biofilm (filled triangles ▲) samples were analyzed for *Vermamoeba vermiformis* (**a–f**), *Acanthamoeba* spp. (**g–l**), and *M. intracellulare* (**m–r**) by qPCR as described in Section 4.6 from each of the six locations listed on the left. Each data point is the mean of duplicate wells with standard deviation. No sampling occurred from fall (F) 2016 to spring (Sp) 2017 for location PVC-FC. The limit of detection for bulk water and biofilm samples were 2.4 $\log_{10}$ cell equivalents (CE) $L^{-1}$ and 2.0 $\log_{10}$ CE $cm^{-2}$ for the *V. vermiformis*; 1.4 $\log_{10}$ CE $L^{-1}$ and 1.0 $\log_{10}$ CE $cm^{-2}$ for the *Acanthamoeba* spp.; and 1.3 $\log_{10}$ GC $L^{-1}$ and 0.9 $\log_{10}$ GC $cm^{-2}$ for *M. intracellulare* assays, respectively.

### 2.4. Additional Sampling Sites

During the W2019 time point, 1 L of first draw samples were collected at two additional locations within the large BWS: an outlet off of the incoming main water line located in the boiler room (BWS Supply Line) and a recirculating pipe loop (PVC-Loop) previously described [32]. The bulk water sample from the BWS Supply Line had a turbidity of 0.34 NTU; 0.00/0.05 mg $L^{-1}$ free/total chlorine; pH of 8.53; temperature of 39.5 °C; and HPC levels of 1.6 $\log_{10}$ CFU 100 $mL^{-1}$. For the PVC-Loop, the bulk water sample had a turbidity of 0.21 NTU; 0.01/0.01 mg $L^{-1}$ free/total chlorine; pH of 8.31; temperature of 20.6 °C; and HPC levels of 4.7 $\log_{10}$ CFU 100 $mL^{-1}$.

The BWS Supply Line contained 3.0 $\log_{10}$ CFU and MPN 100 $mL^{-1}$ of *L. pneumophila* sg1 and 2–14 as confirmed by colony lysate PCR and latex agglutination. Two colonies were *Legionella* and *L. pneumophila* PCR positive, identified as sg1, and processed for whole genome sequencing. The BWS Supply Line sample was negative for *Acanthamoeba* spp. and *M. intracellulare* but contained 3.9 $\log_{10}$ CE 100 $mL^{-1}$ of *V. vermiformis*, 7.3 $\log_{10}$ GC 100 $mL^{-1}$ of *Legionella* spp., and 6.9 $\log_{10}$ GC 100 $mL^{-1}$ of *L. pneumophila*. The PVC-Loop bulk water sample was negative for *V. vermiformis* and *M. intracellulare* but contained 3.0 $\log_{10}$ CE 100 $mL^{-1}$ of *Acanthamoeba* spp., 7.4 $\log_{10}$ GC 100 $mL^{-1}$ of *Legionella* spp., and 7.3 $\log_{10}$ GC 100 $mL^{-1}$ of *L. pneumophila*.

The PVC-Loop had a high level of non-*Legionella* background that negatively impacted the enumeration of presumptive *Legionella* colonies. Although *Legionella* CFU could not be determined, 3.4 $\log_{10}$ MPN 100 $mL^{-1}$ of *L. pneumophila* was detected using Legiolert®. Four wells of the PVC-Loop Legiolert® tray, read as *L. pneumophila* positive following manufacturer's protocols, were sampled, and pure colonies were obtained as described in Section 4.4. Legiolert® isolates were analyzed by PCR and latex agglutination. One of the four Legiolert® isolates was *Legionella* and *L. pneumophila* PCR negative and processed for whole genome sequencing. Three of the four Legiolert® isolates were *Legionella* and *L. pneumophila* PCR positive with 2/3 giving an inconclusive agglutination result and 1/3 identified as sg5. This PCR positive, *L. pneumophila* sg5 isolate was processed for whole genome sequencing.

### 2.5. Statistical Correlations between Water Quality Characteristics

Correlation analysis was performed, as described in Section 4.8, to determine if there were negative or positive associations between the observed water quality characteristics within each sampling location. As expected, there was a strong positive correlation between free and total chlorine in both the first and second draw bulk water samples at all locations supplied with chlorinated water, (r = 0.7–1.0, $P < 0.05$). For the other pairwise comparisons within each location, only the statistically significant ($P < 0.05$) correlations between water quality characteristics are shown in Figure 3.

*Legionella* and/or *L. pneumophila* occurrence was negatively correlated with disinfectant residual at three of the six sampling locations. *Legionella* spp. was negatively correlated with $NH_2Cl$ (r = −0.8) and TCl (r = −0.9) in the first draw, but only with TCl in the second draw samples (r = −0.9), at the PVC-MA location. At the PVC-FC location, *Legionella* spp. and *L. pneumophila* were negatively correlated with TCl and free chlorine (FCl), respectively, only in the second draw samples (r = −0.8, $P < 0.05$). Similarly, *L. pneumophila* was negatively correlated to both FCl and TCl in the Spigot second draw samples (r = −0.7). Notably, for the Faucet location that had consistent levels of *Legionella* spp. and *L. pneumophila* (Figure 1e and k), as well as culturable *Legionella* (Table 3), no statistical correlations were made between *Legionella* and disinfectant residual (Figure 3, Faucet). Moreover, occurrence of *V. vermiformis* was also negatively associated with FCl and TC (r = −0.8) at only the Faucet location.

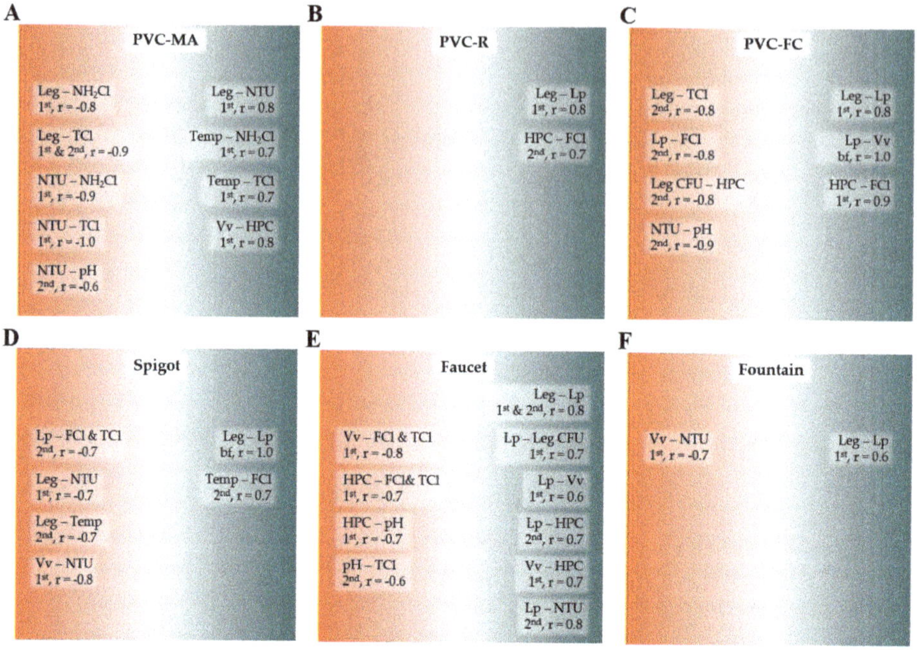


**Figure 3.** Negative and positive correlations between water quality characteristics. Pairwise comparisons with statistically significant and strong correlations are shown for each location (**A**: PVC-MA; **B**: PVC-R; **C**: PVC-FC; **D**: Spigot; **E**: Faucet; **F**: Fountain) and sample type. Negative and positive correlations are shown on the left and right half of each square, respectively. Abbreviations: $1^{st}$, first draw; $2^{nd}$, second draw; bf, biofilm; FCl, free chlorine; HPC, heterotrophic plate count; Leg, *Legionella* spp.; Leg CFU, culturable *Legionella*; Lp, *L. pneumophila*; $NH_2Cl$, monochloramine; NTU, turbidity; pH, potential of hydrogen; TCl, total chlorine; Vv, *V. vermiformis*.

At all sampling locations except for PVC-MA, *Legionella* spp. was positively correlated with *L. pneumophila* in the first draw, second draw, and biofilm samples depending on the location (Fountain, r = 0.6; PVC-R, PVC-FC, and Faucet, r = 0.8; and Spigot, r = 1.0). Additionally, at the Faucet location, culturable *Legionella* was positively correlated with *L. pneumophila* detection in the first draw water samples (r = 0.7). For *L. pneumophila* and *V. vermiformis*, a positive correlation was found at locations PVC-FC (biofilm, r = 1.0) and Faucet (first draw, r = 0.6). Both positive and negative correlations between HPCs and chlorine residuals were observed at three of the six locations. There were positive correlations between HPC and FCl in the first draw (PVC-FC, r = 0.9) and second draw (PVC-R, r = 0.7) samples and negative correlations between HPC and both FCl and TCl in the first draw samples at the Faucet location (r = −0.7). At the Faucet location, HPCs were negatively correlated with pH (r = −0.7) in the first draw samples. HPCs were also negatively associated with culturable *Legionella* in the PVC-FC second draw samples (r = −0.8), but positively associated with *L. pneumophila* molecular detection in the Faucet second draw samples (r = 0.7). Only a positive correlation was found between HPCs and *V. vermiformis* detection in the first draw samples at two locations, PVC-MA (r = 0.8) and Faucet (r = 0.7).

There were strong correlations between *Legionella* detection and various physiochemical water quality parameters. *Legionella* spp. was negatively correlated to temperature at only one location (Spigot, second draw, r = −0.7). Turbidity (NTU) was positively correlated with *L. pneumophila* detection in the Faucet second draw samples (r = 0.8); however, both positive and negative correlations were observed between NTU and *Legionella* spp. detection at two separate locations PVC-MA (first draw, r = 0.8) and Spigot (first draw, r = −0.7), respectively. Only a negative correlation was found between NTU and *V. vermiformis* detection in the first draw samples at two locations, Spigot (r = −0.8) and Fountain (r = −0.7).

Correlations between the physiochemical water quality parameters included a positive association between temperature and both $NH_2Cl$ and FCl at the PVC-MA (first draw, r = 0.7) and between temperature and FCl at the Spigot (second draw, r = 0.7) locations, respectively. There was a negative correlation between NTU and pH in the second draw samples at the PVC-MA (r = −0.6) and PVC-FC (r = −0.9) locations; and negative correlation between NTU and $NH_2Cl$ (r = −0.9) and TCl (−1.0) in the first draw samples at the PVC-MA location.

### 2.6. Whole Genome Sequencing of Drinking Water Isolates

Nineteen representative *L. pneumophila* and non-*pneumophila* isolates from the Su2018, F2018, and W2019 time points and PVC-FC, Faucet, BWS Supply Line, and PVC-Loop locations were submitted for whole genome sequencing as described in Section 4.7 (Table 4). All seven Faucet isolates from the Su2018 (Faucet 1–5) and W2019 (Faucet Legiolert® 1 and Faucet 2) time points, and the W2019 BWS Supply Line 1 and 2 isolates, were identified as *L. pneumophila* Sequence Type (ST) 1 with an average genome size of 3.6 million base pairs (Mbp), a guanine–cytosine (G + C) content of 38%, and approximately 3200 predicted genes. The average nucleotide identity (ANI) between these seven Faucet and two BWS Supply Line isolates was between 99.97% and 100% indicating that these isolates were the same *L. pneumophila* strain (Figure 4). These isolates were also identified as sg1 via latex agglutination as described in Section 2.2 and 2.4.

Of the eight PVC-FC isolates, two (Su2018 PVC-FC 1–2) were identified as *L. pneumophila* ST2037, with a comparable genome size (~3.5 Mbp), 38% G + C content, and 100% ANI (Table 4, Figure 4), with Su2018 PVC-FC 2 identified as *L. pneumophila* sg5 via indirect immunofluorescent antibody assay as described in Section 2.2. The remaining six PVC-FC isolates (Su2018 PVC-FC 3, F2018 PVC-FC, and W2019 PVC-FC 1–4) were identified as *Legionella* with a genome size of 3.0–3.3 Mbp, 41% G + C content, and 99.99–100% ANI (Table 4, Figure 4), with F2018 PVC-FC identified as *L. feeleii* via indirect immunofluorescent antibody assay. The W2019 PVC-Loop Legiolert® 1 isolate, derived from a *L. pneumophila* positive well, was identified as *Ochrobactrum* with a higher genome size of 4.7 Mbp, higher G + C content of 67% compared to the W2019 PVC-Loop Legiolert® 2 isolate identified as *L. pneumophila* ST1950, with a genome size of 3.4 Mbp, 38% G + C content, and 3057 predicted genes (Table 4).

Between the *L. pneumophila* sg5 ST1950 isolate and sg5 ST2037 isolates, the ANI was 92%, suggesting that they belong to the same species, but are different strains (Figure 4). Notably, the nine *L. pneumophila* sg1 ST1 isolates had a higher ANI of 96% with the *L. pneumophila* sg5 ST1950 isolate compared to the 92% ANI with the *L. pneumophila* sg5 ST2037 isolates. This higher nucleotide similarity between the *L. pneumophila* sg1 ST1 and *L. pneumophila* sg5 ST1950 can be visualized with their clustering in the phylogenetic tree (Figure 5). The *L. pneumophila* sg1 ST1 strains (W2019 BWS Supply Line, Su2018 and W2019 Faucet) were in the same branch as the *L. pneumophila* sg5 ST1950 strain, while the *L. pneumophila* sg5 ST2037 strain (Su2018 PVC-FC) was in a separate branch. The PVC-FC isolates were identified as *L. feeleii* via indirect immunofluorescent antibody assay and sequence similarities to two *L. feeleii* reference genomes (strain WO-44C and NCTC11978) which displayed >99.9% similarity based on 16S rRNA and ANI of 98.5%. Figure 5 shows the *L. feeleii* isolates in their own cluster with the reference genomes of *L. massiliensis* and *L. nautarum* in the next closest branch. The two *L. feeleii* reference genomes were not included in the sequence database at the time of this analysis and thus are not represented in the phylogenetic tree (Figure 5).

**Table 4.** Summary statistics of whole-genome assemblies for the drinking water isolates

| Isolate | Lineage | Genome Size (bp) | No. of Contigs | Contig $N_{50}$ (bp) | G+C Content (%) | No. Predicted Genes | MLST |
|---|---|---|---|---|---|---|---|
| Su2018 Faucet 1 | *L. pneumophila* | 3,589,286 | 72 | 160,018 | 38 | 3240 | 1 |
| Su2018 Faucet 2 | *L. pneumophila* | 3,589,059 | 75 | 160,010 | 38 | 3239 | 1 |
| Su2018 Faucet 3 | *L. pneumophila* | 3,589,228 | 71 | 160,018 | 38 | 3236 | 1 |
| Su2018 Faucet 4 | *L. pneumophila* | 3,574,346 | 99 | 88,934 | 38 | 3228 | 1 |
| Su2018 Faucet 5 | *L. pneumophila* | 3,590,303 | 69 | 160,010 | 38 | 3240 | 1 |
| W2019 Faucet Legiolert®1 | *L. pneumophila* | 3,564,531 | 124 | 59,538 | 38 | 3229 | 1 |
| W2019 Faucet 2 | *L. pneumophila* | 3,562,418 | 77 | 123,108 | 38 | 3224 | 1 |
| Su2018 PVC-FC 1 | *L. pneumophila* | 3,510,698 | 39 | 413,452 | 38 | 3151 | 2037 |
| Su2018 PVC-FC 2 | *L. pneumophila* | 3,498,273 | 40 | 223,695 | 38 | 3138 | 2037 |
| Su2018 PVC-FC 3 | *Legionella* | 3,285,021 | 57 | 260,915 | 41 | 3028 | - |
| F2018 PVC-FC | *Legionella* | 3,287,868 | 49 | 348,799 | 41 | 3028 | - |
| W2019 PVC-FC 1 | *Legionella* | 3,063,339 | 43 | 348,799 | 41 | 2812 | - |
| W2019 PVC-FC 2 | *Legionella* | 3,288,042 | 53 | 321,108 | 41 | 3033 | - |
| W2019 PVC-FC 3 | *Legionella* | 3,286,784 | 53 | 283,321 | 41 | 3031 | - |
| W2019 PVC-FC 4 | *Legionella* | 3,285,778 | 45 | 466,387 | 41 | 3022 | - |
| W2019 PVC-Loop Legiolert®1 | *Ochrobactrum* | 4,764,477 | 55 | 478,728 | 58 | 4621 | - |
| W2019 PVC-Loop Legiolert®2 | *L. pneumophila* | 3,388,353 | 42 | 255,126 | 38 | 3057 | 1950 |
| W2019 BWS Supply Line 1 | *L. pneumophila* | 3,589,200 | 67 | 176,930 | 38 | 3237 | 1 |
| W2019 BWS Supply Line 2 | *L. pneumophila* | 3,588,658 | 70 | 160,010 | 38 | 3242 | 1 |

Abbreviations: -, not applicable; bp, base pair; BWS, building water system; LST, multi-locus sequence typing; No., number.

| Sequenced Isolates | Su2018 Faucet 1 | Su2018 Faucet 2 | Su2018 Faucet 3 | Su2018 Faucet 4 | Su2018 Faucet 5 | W2019 Faucet Legiolert® 1 | W2019 Faucet 2 | Su2018 PVC-FC 1 | Su2018 PVC-FC 2 | Su2018 PVC-FC 3 | F2018 PVC-FC | W2019 PVC-FC 1 | W2019 PVC-FC 2 | W2019 PVC-FC 3 | W2019 PVC-FC 4 | W2019 PVC-Loop Legiolert® 1 | W2019 PVC-Loop Legiolert® 2 | W2019 BWS Supply Line 1 | W2019 Bldg Supply Line 2 |
|---|---|---|---|---|---|---|---|---|---|---|---|---|---|---|---|---|---|---|---|
| Su2018 Faucet 1 | | | | | | | | | | | | | | | | | | | |
| Su2018 Faucet 2 | 100 | | | | | | | | | | | | | | | | | | |
| Su2018 Faucet 3 | 100 | 100 | | | | | | | | | | | | | | | | | |
| Su2018 Faucet 4 | 99.99 | 99.99 | 99.99 | | | | | | | | | | | | | | | | |
| Su2018 Faucet 5 | 100 | 100 | 100 | 99.99 | | | | | | | | | | | | | | | |
| W2019 Faucet Legiolert® 1 | 99.98 | 99.98 | 99.98 | 99.99 | 99.99 | | | | | | | | | | | | | | |
| W2019 Faucet 2 | 100 | 100 | 100 | 99.99 | 100 | 99.98 | | | | | | | | | | | | | |
| Su2018 PVC-FC 1 | 91.82 | 91.82 | 91.83 | 91.83 | 91.8 | 91.8 | 91.8 | | | | | | | | | | | | |
| Su2018 PVC-FC 2 | 91.84 | 91.83 | 91.84 | 91.82 | 91.83 | 91.83 | 91.8 | 100 | | | | | | | | | | | |
| Su2018 PVC-FC 3 | <80 | <80 | <80 | <80 | <80 | <80 | <80 | <80 | <80 | | | | | | | | | | |
| F2018 PVC-FC | <80 | <80 | <80 | <80 | <80 | <80 | <80 | <80 | <80 | 100 | | | | | | | | | |
| W2019 PVC-FC 1 | <80 | <80 | <80 | <80 | <80 | <80 | <80 | <80 | <80 | 100 | 100 | | | | | | | | |
| W2019 PVC-FC 2 | <80 | <80 | <80 | <80 | <80 | <80 | <80 | <80 | <80 | 100 | 100 | 100 | | | | | | | |
| W2019 PVC-FC 3 | <80 | <80 | <80 | <80 | <80 | <80 | <80 | <80 | <80 | 100 | 100 | 100 | 100 | | | | | | |
| W2019 PVC-FC 4 | <80 | <80 | <80 | <80 | <80 | <80 | <80 | <80 | <80 | 100 | 100 | 100 | 100 | 100 | | | | | |
| W2019 PVC-Loop Legiolert® 1 | <80 | <80 | <80 | <80 | <80 | <80 | <80 | <80 | <80 | <80 | <80 | <80 | <80 | <80 | <80 | | | | |
| W2019 PVC-Loop Legiolert® 2 | 96.55 | 96.54 | 96.56 | 96.57 | 96.56 | 96.58 | 96.54 | 92.16 | 92.17 | <80 | <80 | <80 | <80 | <80 | <80 | <80 | | | |
| W2019 BWS Supply Line 1 | 100 | 100 | 100 | 99.99 | 99.99 | 99.97 | 99.99 | 91.81 | 91.82 | <80 | <80 | <80 | <80 | <80 | <80 | <80 | 96.54 | | |
| W2019 Bldg Supply Line 2 | 100 | 100 | 100 | 99.99 | 100 | 99.98 | 100 | 91.81 | 91.83 | <80 | <80 | <80 | <80 | <80 | <80 | <80 | 96.56 | 100 | |

**Figure 4.** Average nucleotide identity (ANI) matrix for the whole-genome sequenced isolates

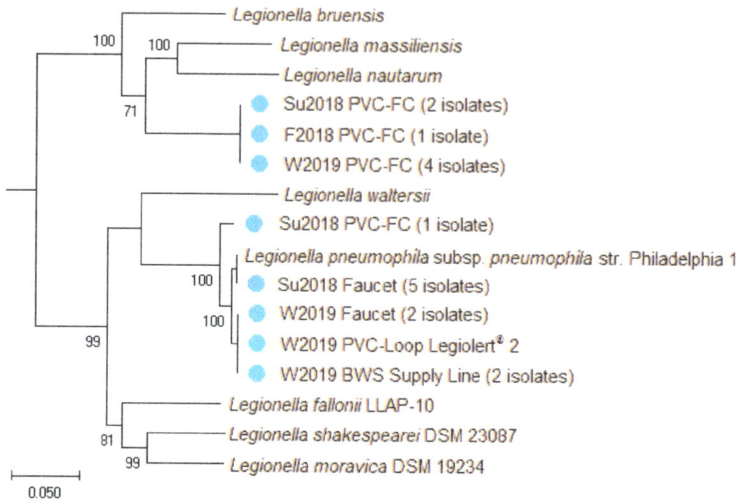


**Figure 5.** Phylogenetic tree illustrating isolate relatedness to reference genomes. Representative strains (**blue dots**) from each sampling location and time were chosen for construction of this phylogenetic tree. Numbers presented are confidence values (**bootstrapping**) used by FastTree 2 to estimate maximum likelihood. The scale bar represents 0.05 nucleotide substitutions per site.

## 3. Discussion

The main objective of this study was to gain a better understanding of water quality parameters that correlated with *Legionella* occurrence within a large complex building water system (BWS) over an extended period of time. Cold bulk water and biofilm samples were collected to monitor the occurrence and distribution of *Legionella*, other water-based pathogens, and eukaryotic hosts, while considering seasonal fluctuations, and varying engineering, operational, and water quality characteristics. Cold water was analyzed in this study given (1) the high prevalence of *Legionella* contamination previously identified in cold water samples within public building, private residences, healthcare facilities, and water storage tanks; (2) *Legionella* transmission and infections epidemiologically linked to cold water exposure; and (3) the current recommendations to monitor both hot and cold water in BWS to control for *Legionella* [33]. In this study, both culture and molecular methods were used for *Legionella* detection and culture isolates were submitted for whole genome sequencing for further genetic characterization.

Two percent of biofilm samples (1/57) and 12% of bulk water samples (14/114) were culture-positive for *Legionella* of which 57% (8/14) were isolated during the winter; 21% (3/14) during the summer; 14% (2/14) during the fall; and 7% (1/14) during the spring. Legionnaires' disease (LD) cases typically peak during the summer and fall seasons [34,35]. However, other epidemiological studies reported a winter peak for non-travel related cases [36] and no monthly or seasonal correlations for nosocomial-acquired [37] and community-acquired *Legionella* pneumonia cases [38]. Higher incidences of legionellosis have been associated with various meteorological factors (e.g., humidity, temperature, rainfall, atmospheric pressure); geographic location; and properties of the local watershed and source water (e.g., proximity and water temperature, levels, and flow) [39–42]. Thus, the exact environmental mechanisms and triggers of *Legionella* transmission and subsequent disease incidences are yet to be clearly defined.

Culture and qPCR are frequently used methods for *Legionella* environmental detection [43–45]. However, due to discrepancies between and within each of these methods, interpretations of *Legionella* occurrence across various studies and correlations to legionellosis health risks within BWSs are challenging [46]. Moreover, false positivity rates for the Legiolert® method have been reported to be between 0% and 3.3% for potable water and 4.9% and 11% for non-potable water [47–49]. Such as with *Ochrobactrum* identified in the PVC-Loop location, stable colonization of a Legiolert® false positive causing strain at an environmental monitoring site may continually overestimate *L. pneumophila* levels and confound interpretations of their occurrence at those sites. Culture-based methods are a reliable indicator of pathogen viability and potential health risks associated with *Legionella* detection; however, periodic validation of culture-based results can be performed using molecular methods.

In this study, the PVC-R W2018 first draw sample was *Legionella* culture positive, but *Legionella* qPCR negative (Table 3, Figure 1b), while the PVC-FC Su2018 second draw sample was *Legionella* and *L. pneumophila* culture positive, but only *L. pneumophila* qPCR negative (Table 3, Figure 1c and i). Culture-positivity/qPCR-negativity has been observed previously and was associated more with drinking water compared to cooling tower water samples [50–52]. Furthermore, *L. pneumophila* qPCR levels were higher than those of *Legionella* spp. in four first draw samples (Spigot F2017, W2019 and Fountain F2017, Su2018) (Figure 1d, j and f, l) and *L. pneumophila* was also observed in hot water, but not cooling tower water samples [53]. These discrepancies have been attributed to the presence of PCR inhibitors and competing non-target organisms, and varying culture conditions (e.g., growth temperature, agar type) and sample processing steps that may select for, or inhibit growth of, different strains of *Legionella* [46].

For *Legionella* environmental monitoring, samples that can be collected for analysis include (1) biofilm materials, which contain a concentration of surface attached microorganisms; (2) first draw, stagnant water samples, which represent water quality at the outlet; and (3) second draw, post-flushed samples, which represent water quality supplied to the outlet from within the building water system [54,55]. Of the *Legionella pneumophila* and non-*pneumophila* culture-positive bulk water samples, 64% (9/14) were first draw and 36% (5/14) were second draw samples, while only 2% (1/57) of biofilm samples were culture-positive (Table 3). From the qPCR analyses, *Legionella* was detected in 30% (17/57) of biofilm samples and 74% (84/114) of bulk water samples of which 48% (40/84) were first draw and 52% (44/84) were second draw samples. *L. pneumophila* was detected in 23% (13/57) of biofilm samples and 36% (41/114) of bulk water samples of which 59% (24/41) were first draw and 41% (17/41) were second draw samples (Figure 1).

Differences in physiochemical and microbial water quality parameters between first and second draw samples were previously reported with microbial loads generally higher in the first draw depending on the sampling location and volume collected [56,57]. As stated in Section 2.1, there were no statistical differences in pH, temperature, free chlorine, monochloramine, and total chlorine between the first and second draw samples at all locations with the exceptions of temperature for the Spigot samples and free and total chlorine for the Faucet samples. Three of the six locations had statistical differences between the first and second draw samples for HPCs: PVC-R, Spigot, and Faucet. For *Legionella* levels detected by qPCR, there were strong positive correlations between the first and second draw samples from all locations: PVC-MA ($r = 0.9$, $P < 0.001$), PVC-R ($r = 0.6$, $P < 0.05$), PVC-FC ($r = 0.9$, $P < 0.01$), Faucet ($r = 0.8$, $P < 0.01$), and Fountain ($r = 0.8$, $P < 0.01$), except for the Spigot location ($r = 0.2$, $P = 0.575$). For *L. pneumophila* levels as detected by qPCR, there were only strong positive correlations between the first and second draw samples at three of the six locations: PVC-MA ($r = 1.0$, $P < 0.001$), PVC-FC ($r = 0.9$, $P < 0.01$), and Faucet ($r = 0.7$, $P < 0.05$).

The results between the bulk water samples suggested that an analysis of either the first or second draw samples was able to indicate the presence of *Legionella*. Moreover, it is unclear why only the Faucet location showed statistical differences of both free and total chlorine between the first and second draw samples as water usage at this site was higher than two other locations and the distance of the Faucet outlet was greater than and almost equal to three other sampling locations (Table 1). The only difference between the Faucet location and all others was the ability to draw hot water from this outlet, but only cold water was analyzed in this study; thus, hot water may be contributing to bacterial contamination in the Faucet.

After conversion to monochloramine disinfection within BWSs, there were reductions in the number of distal sites testing positive for *Legionella* (39–100% to 0–18% positivity); however, there was a large range in the log reduction of *Legionella* levels (0.2 to 3 $\log_{10}$ CFU $L^{-1}$) with one study reporting no changes in levels post-conversion during the one- to three-year monitoring period [58–60]. Control of biofilm-associated *Legionella* was also observed [58,61], most likely due to the better penetration of monochloramine into biofilms compared to chlorine [62]; however, control of biofilm-associated *L. pneumophila* (Lp) was previously reported to be pipe material specific during chlorine and monochloramine treatment [63]. The PVC-MA location used in this study has been operating with a monochloramine residual for approximately 10 years with a two-month chlorine conversion from December 2013 to February 2014. During the sampling period, no culturable *Legionella* was detected, but molecular analyses indicated a consistent and high level of *Legionella* and sporadic detection of *L. pneumophila* in the bulk water and biofilm samples. Specifically, 100% (20/20) of the bulk water samples and 10% (1/10) of biofilm samples had detectable *Legionella*; while *L. pneumophila* was detected in 30% (6/20) of bulk water samples (three first draw and three second draw samples) and 10% (1/10) of biofilm samples (Figure 1a and g). Using *E. coli*, monochloramine was shown to disrupt protein-mediated metabolic processes with no damage to the cell envelope or nucleic acids [64]. This finding is supported by other studies demonstrating no significant impacts on *Legionella* 16S rRNA gene transcript levels during monochloramine versus chlorine treatment [63] and *Pseudomonas aeruginosa* extracellular polymeric substances material limiting and delaying monochloramine access to the cell surface [65]. Thus, it is unclear what the exact mechanisms are for monochloramine control of *Legionella* bacteria and whether different surface properties (e.g., lipopolysaccharides used for *L. pneumophila* serogroup identification) would result in varying degrees of inactivation with intermediate stages allowing for *Legionella* regrowth.

Numerous studies have reported more frequent detection of *V. vermiformis* within BWSs compared to *Acanthamoeba* spp. [56,66,67]. Concordantly, *Acanthamoeba* spp. were undetectable in the bulk water and biofilm samples during F2016, thus the following five sampling time points were excluded; however, the analysis was resumed in Sp2018 to confirm the low frequency and/or undetectable observation for *Acanthamoeba* spp. in BWS samples (Figure 2g–l). *Acanthamoeba* spp. was detected in only one second draw sample at the PVC-MA location (Figure 2g). Similarly, *M. intracellulare* was infrequently detected in this study (Figure 2n and p) as previously observed for water samples from chlorinated BWSs [23,68]. *V. vermiformis* was detected at all locations except PVC-R (Figure 2a–f) with positive correlations between *L. pneumophila* and *V. vermiformis* in PVC-FC biofilms and Faucet first draw samples. Given that FLA detection has been correlated to *Legionella* and *Mycobacterium* in drinking water systems [69], there is utility in monitoring for FLA to better understand the conditions and FLA members contributing to *Legionella* and mycobacterial survival within BWSs.

Although diverse populations of *Legionella* bacteria have been described in drinking water [70,71], environmental monitoring of BWSs focuses primarily on *L. pneumophila* detection since this species makes up the majority of clinical isolates, with > 80% of those isolates belonging to serogroup (sg) 1 [72,73]. Serotyping allows for the differentiation of *L. pneumophila* isolates based on their reactivity to the Dresden Panel of antibodies that recognize distinct structures on the bacterial lipopolysaccharide molecule [74,75]. *L. pneumophila* contains 17 serogroups and 10 subgroups within sg1 [76,77]. The latex agglutination serotyping method used in this study has been shown to produce false negatives as *Legionella*-like colonies isolated from drinking water samples, confirmed as *L. pneumophila* via 16S rRNA sequence analyses, were agglutination negative (this study; [78]). Thus, molecular-based methods, such as sequence-based typing (SBT) and whole genome sequencing (WGS), are increasingly being used due to their reliability, better resolution, and discriminatory power for describing genetic diversity, environmental distribution, evolution, population structure, clonal expansion, and virulence properties of *Legionella* isolates [77,79].

In this study, *L. feeleii*, *L. pneumophila* sg1 sequence-type (ST) 1, and *L. pneumophila* sg5 ST 1950 and ST2037 were identified via 16S rRNA-based analyses, serotyping (latex agglutination and indirect immunofluorescent antibody assays), WGS, and SBT analyses. These unique *Legionella* strains were isolated from the Faucet, PVC-FC, PVC-Loop, and BWS Supply Line locations during the Su2018, F2018, and W2019 time points (Table 4 andFigure 4). ST1 is the most commonly identified and globally distributed strain isolated from both environmental and clinical samples ([79,80]; this study). Identification of previously unknown STs (e.g., ST1950 and ST2037 from this study) supports the continued SBT of clinical and environmental isolates. This will help elucidate which STs may be more globally distributed or geographically confined, and which are more associated with disease cases, such that when those are environmentally identified, preventative measures can be implemented to limit public health and exposure risks to these pathogens. WGS analyses provide more detailed genetic information about the *Legionella* strain, beyond those obtained from SBT alone, enabling potential subspecies identification, refined taxonomic classification, and genetic profiling for virulence properties [77]. *Legionella* diversity and distribution data may also reveal environmental parameters that influence *Legionella* occurrence and survival within specific environments such as BWSs.

Due to lengthy incubation periods required for *Legionella* culture, the tendency for qPCR to overestimate their levels, and the discrepancies associated with these methods, as described above, use of other microbial and/or physiochemical water quality parameters as potential indicators for *Legionella* presence in BWSs has been investigated [70]. As described in Section 2.5, there were strong statistical correlations observed between various water quality parameters and *Legionella* occurrence such as *V. vermiformis*, HPC, chlorine residual, temperature, and turbidity (Figure 3). However, these correlations were location and sample type specific with conflicting positive and negative correlations for turbidity and HPC. Conflicting correlations between the latter and *Legionella* levels have been reported previously with either strong correlations [81], no correlations [82–85], or possible seasonal dependencies for these correlations [7]. Other conflicting correlations were reported between *Legionella* occurrence and pH, temperature, various minerals and metals (Ca, Cu, Fe, Mg, Mn, and Zn), total organic carbon, conductivity, and free chlorine at the sampling site and building supply feed [85–87].

Thus, further analyses of existing data and correlations to *Legionella* occurrence, as well as more in-depth studies on identifying these correlations, need to be performed. This will enable determinations as to whether correlations are dependent on complex factors, either individually or in concert, such as disinfectant type; source water quality fluctuations; genetic background of detected *Legionella* populations; presence of other drinking water microorganisms; or certain engineering and operational water system aspects specific to sampled locations. The notable observations from this study were (1) detection and culture of *Legionella* from outlets that neither supply, nor are connected to plumbing for, hot water; (2) isolation of diverse *L. pneumophila* and non-*pneumophila* strains from different locations; (3) utility of whole genome sequencing and sequence-based typing for enhanced isolate description and characterization of their distribution; (4) *Legionella* levels detected during monitoring can significantly differ between the first and second draw sample; and (5) negative and positive correlations between *Legionella* and various water quality parameters were location and sample type specific.

Confoundingly, occurrence alone is not the most important factor for legionellosis risk. Environmental, bacterial, and host specific factors such as aerosolization into respirable droplets and their potential for human exposure; virulence of the environmental *Legionella* strain; and host immune status and susceptibility to infection, collectively play an important role in exposure risks and disease outcome. The main objective of this study was to gain a better understanding of *Legionella* occurrence and water quality parameters supporting their growth within a large, complex building water system. Information from this, and future studies, will help elucidate ways to effectively manage the risks associated with *Legionella* exposure within these drinking water distribution systems.

## 4. Materials and Methods

### 4.1. Sampling Locations

A 40-year old, 33,000 square ft. building with an average potable water usage of 3.6 million gallons (13.6 million L) per year was used in this study. Water usage at any given location varies widely depending upon the activity (floor washing, water storage tank cleanout, etc.) being conducted and the facility cooling demand during warmer months. The building's potable water supply is derived from river water treated by coagulation, flocculation, and sedimentation; followed by sand, gravel, and granular activated carbon filtration; and then chlorination. Cold potable water samples were collected seasonally every three months from six locations throughout the building (Table 1, Figure S1). The total number of samples for each site was 30 (10 first and second draw bulk water and 10 biofilm samples) collected over a 28-month period, October 2016 to February 2019; except for site PVC-FC, where the total number of samples was 21 (seven first and second draw bulk water and biofilm samples) collected over an 18-month period, August 2017 to February 2019. Sampling time points are denoted F, for fall; W, for winter; Sp, for spring; and Su, for summer followed by the corresponding year. F, W, Sp, and Su samples were collected during the months of October, November, February, May, and August, respectively.

Within this building, a semi-closed pipe loop distribution system simulator was fed with the chlorinated municipal drinking water, described above, and amended with ammonium hydroxide and sodium hypochlorite (Sigma Aldrich, St. Louis, MO, USA) to yield a 2 mg $L^{-1}$ monochloramine residual as previously described [88]. Average monochloramine and ammonia levels (± SD) during this sampling period were 1.25 ± 0.37 and 0.16 ± 0.07 ppm, respectively.

### 4.2. Sample Collection and Processing

For each sampling location, the first draw sample was taken immediately after turning the tap on, while the second draw was collected after 10 s of flushing (approximately 4 L), except for Fountain where the second draw sample was collected after 30 s of flushing (approximately 2 L). The 10–30-second flush time was used to ensure collection of non-stagnant water that was still representative of water quality conditions within the BWS. Sampling took place early in the morning after an overnight stagnation period. Water samples were collected in sterile 1 L plastic bottles and 1 mL of 10% w/v sodium thiosulfate was added to neutralize any disinfectant residual. An additional 100 mL was also collected for water quality analysis as described below. Approximately 1 L of each bulk water sample was filtered through a 0.2 μm polyethersulfone membrane (Supor® Membrane, PALL Life Sciences, Nassau, NY, USA). Filters were placed into 11 mL of UV-light dechlorinated, 0.22 μm filtered drinking water ($dfH_2O$), and vortexed at maximum speed for 1 min to resuspend the concentrated bulk water material. For biofilm collection, a sterile polyester tipped applicator was used to swab an approximate area of 2 $cm^2$ inside the tap. The applicator was then placed in a 14-mL round bottom tube containing 2 mL of $dfH_2O$ and vortexed vigorously for 1 min to resuspend the collected biofilm material.

Approximately 1 mL of the concentrated bulk water and biofilm suspension was analyzed for CFU, as described in Section 4.4, and the remaining volume was centrifuged at high speed (13,000 rcf, room temperature, 10 min; Eppendorf, Foster City, CA, USA). Pellets were resuspended in 200 μL of $dfH_2O$ and placed in a Lysing Matrix A tube (MP Biomedicals, Solon, OH, USA) along with the washed filter or biofilm swab for nucleic acid extraction as described below.

### 4.3. Water Quality Analysis

Bulk water samples were analyzed for pH, turbidity, temperature, disinfectant residual, and heterotrophic plate count (HPC). Free chlorine and total chlorine measurements were performed using the DPD colorimetric method (Powder Pillows; Hach USA) and monochloramine and free ammonia measurements were performed using the indophenol method (method 10200, Powder Pillows, free ammonia chlorinating solution; Hach USA). HPCs were enumerated by the spread plate method on Reasoner's 2A agar (R2A, Difco Laboratories, Detroit, MI, USA) following incubation at 28 °C for 7 d. The limit of detection (LOD) for bulk water samples was 1.0 $\log_{10}$ CFU 100 $mL^{-1}$ and 0.7 $\log_{10}$ CFU $cm^{-2}$ for biofilm samples.

### 4.4. *Legionella* Enumeration and Presumptive Colony Analysis

For colony forming unit (CFU) enumeration, undiluted and serially diluted suspensions were spread plated on buffered charcoal yeast extract (BCYE) agar plates (BD Diagnostics, Franklin Lakes, NJ, USA) and incubated for 4–6 days at 37 °C [44]. Presumptive *Legionella* colonies were counted; and a subset was isolated and confirmed as *Legionella* spp. or *L. pneumophila* via polymerase chain reaction (PCR) using the 16S rRNA gene assays described in Section 4.6. An aliquot of the processed bulk water and biofilm samples was also heat-treated (incubation in a 55 °C water bath for 30 min) before plating on BCYE agar plates to evaluate potential differences in *Legionella* recovery from this pretreatment method [44]. Although growth of non-*Legionella* bacteria was inhibited by heat treatment, there were no significant differences between *Legionella* CFU observed between unheated and heated samples (data not shown).

For most probable number (MPN) enumeration, Legiolert® (Idexx Laboratories, Westbrook, ME, USA) was used to analyze 10 mL of the unconcentrated bulk water samples and 0.5 mL of the resuspended biofilm samples for only the Su2018, F2018, and W2019 time points. To obtain pure isolates from the Legiolert® tray, positive wells were punctured using a 26-gauge needle and 50–1000 µL of the well contents was collected. A 20 µL aliquot of the sampled well was streaked onto a BCYE agar plate and incubated for 4–6 days at 37 °C.

Those identified as *L. pneumophila* by PCR were serotyped using the Oxoid™ *Legionella* Latex Agglutination Kit (ThermoFisher, Waltham, MA, USA), which allows for the separate identification of *L. pneumophila* serogroup 1 and serogroups 2–14 and detection of seven other *Legionella* species (*L. anisa*; *L. bozemanii* 1 and 2; *L. dumoffii*; *L. gormanii*; *L. jordanis*; *L. longbeachae* 1 and 2; and *L. micdadei*). Two *L. pneumophila* isolates identified as belonging to serogroups 2–14 via latex agglutination (Su2018 PVC-FC 1 and W2019 PVC-Loop Legiolert® 2) and one *Legionella* spp. PCR positive isolate (F2018 PVC-FC) were sent to an external laboratory (EMSL Analytical Inc., Cinnaminson, NJ, USA) for further identification via indirect immunofluorescent antibody assay [44].

To account for zero values, 1 was added to all data points before conversion to the log10 scale (e.g., log10 (CFU + 1)). Calculations from CFU and molecular analyses were adjusted and expressed as units per mL or $cm^2$ for bulk water samples and biofilms, respectively. The LOD for bulk water samples was 1.0 $\log_{10}$ CFU 100 $mL^{-1}$ and 0.7 $\log_{10}$ CFU $cm^{-2}$ for biofilm samples.

### 4.5. Isolation and Preparation of Total DNA

DNA was extracted from bacterial cells using the MasterPure™ Complete DNA purification kit (Epicentre Biotechnologies Inc., Madison, WI, USA) according to manufacturer's protocol and the Mini-Beadbeater–16 (Biospec Products, Bartlesville, OK, USA) where samples were processed twice for 30 s at 3450 oscillations $min^{-1}$. The DNA pellet was resuspended in 100 µL of molecular grade water.

### 4.6. Quantitative Polymerase Chain Reaction (qPCR)

Biofilm and bulk water DNA samples were analyzed in duplicate using the Applied Biosystems QuantStudio 6 Flex Fast Real-Time PCR system (ThermoFisher, Waltham, MA, USA). A 10-fold dilution of each sample was also analyzed in duplicate to test for presence of environmental qPCR inhibitors. The TaqMan qPCR assay for *Legionella* spp., *L. pneumophila*, *Mycobacterium intracellulare* detection, targeting the 16S rRNA gene, was performed as previously described [63,89,90]. The TaqMan qPCR assay for *Acanthamoeba* spp. and SYBR green qPCR assay for *Vermamoeba vermiformis* detection, targeting the 18S rRNA gene, was performed as previously described [91,92].

The forward and reverse primers and probe sequences (5' to 3') and cycling parameters used in this study for the *Legionella* spp. qPCR assay, respectively, are 16S-LegF1c: TAG TGG AAT TTC CGG

TGT A; 16S-LegR1c: CCA ACA GCT AGT TGA CAT C; 16S-LegP1: CGG CTA CCT GGC CTA ATA CTG A; and 50 °C for 2 min, 95 °C for 10 min, 40 cycles of 95 °C for 10 s and 50 °C for 30 s, and at 70 °C for 30 s [90]. The forward and reverse primers and probe sequences (5′ to 3′) and cycling parameters used in this study for the *L. pneumophila* qPCR assay, respectively, are LpneuF1: CGG AAT TAC TGG GCG TAA AGG-3; LpneuR1: GAG TCA ACC AGT ATT ATC TGA CCG T; LpneuP1: AAG CCC AGG AAT TTC ACA GAT AAC TTA ATC AAC CA; and 95 °C for 10 min, 40 cycles of 95 °C for 10 s, and at 60 °C for 1 min [63]. The forward and reverse primers and probe sequences (5′ to 3′) and cycling parameters used in this study for the *M. intracellulare* qPCR assay, respectively, are F: GGG TGA GTA ACA CGT GTG CAA; R: CCA CCT AAA GAC ATG CGA CTA AA; P: TGC ACT TCG GGA TAA GCC TGG GAA A; and 50 °C for 2 min, 95 °C for 10 min, 40 cycles of 95 °C for 15 s, and 60 °C for 1 min [89]. The forward and reverse primers and probe sequences (5′ to 3′) and cycling parameters used in this study for the *Acanthamoeba* spp. qPCR assay, respectively, are TaqAcF1: CGA CCA GCG ATT AGG AGA CG; TaqAcR1: CCG ACG CCA AGG ACG AC; TaqAcP1: TGA ATA CAA AAC ACC ACC ATC GGC GC; and 50 °C for 2 min, 95 °C for 10 min, followed by 40 cycles at 95 °C for 15 s and 60 °C for 1 min, respectively [92]. The forward and reverse primer sequences (5′ to 3′) and cycling parameters used in this study for the *V. vermiformis* spp. qPCR assay, respectively, are Hv1227F: TTA CGA GGT CAG GAC ACT GT; Hv1728R: GAC CAT CCG GAG TTC TCG; and 95 °C for 3 min, followed by 40 cycles at 95 °C for 20 s, 56 °C for 30 s, and 72 °C for 40 s, and then 72 °C for 10 min [91].

For *Legionella* spp. and *L. pneumophila* qPCR assays, standard curves were generated, on each plate, using a plasmid vector (pUCIDT-AMP; Integrated DNA Technologies, Inc., Coralville, IA, USA) containing a cloned 189-bp region of the *L. pneumophila* Philadelphia-1 16S rRNA gene (NCBI reference sequence NC_002942.5, positions 609325 to 609513) that contains the targets for each of these qPCR assays. *M. intracellulare* standard curves were generated from serially diluted purified genomic DNA. Cell-based calibration curves were constructed for *Acanthamoeba* spp. and *V. vermiformis* by preparing 10-fold serial dilutions of DNA extracted from amoeba cell cultures of known densities.

Standards ranging from 1 to $10^7$ gene copy (GC) for *Legionella* spp. and *L. pneumophila* qPCR assays; 4 to $10^4$ GC for *M. intracellulare* qPCR assays; and 1 to $10^5$ cell equivalents (CE) for the amoeba qPCR assays were generated and analyzed in triplicate along with duplicate no-template control for each 96-well plate. Data were expressed as $\log_{10}$ gene copy or CE or GU per mL or $cm^2$. The limits of detection for bulk water and biofilm samples were 1.6 $\log_{10}$ GC $L^{-1}$ and 1.3 $\log_{10}$ GC $cm^{-2}$ for the *Legionella* spp. and *L. pneumophila* assays; 1.3 $\log_{10}$ GC $L^{-1}$ and 0.9 $\log_{10}$ GC $cm^{-2}$ for the *M. intracellulare* assay; 1.4 $\log_{10}$ CE $L^{-1}$ and 1.0 $\log_{10}$ CE $cm^{-2}$ for the *Acanthamoeba* spp. assay; and 2.4 $\log_{10}$ CE $L^{-1}$ and 2.0 $\log_{10}$ CE $cm^{-2}$ for the *V. vermiformis* assay, respectively.

### 4.7. Whole Genome Sequencing and Sequence Analyses

Twenty-one bulk water isolates were chosen for whole genome sequencing. Total DNA from each strain was isolated as described in Section 4.5. DNA concentrations were estimated using the Nanodrop ND–1000 Spectrophotometer (NanoDrop Technologies, Inc., Wilmington, DE, USA). Total DNA was submitted for whole genome sequencing (Wright Labs LLC, Huntingdon, PA, USA) where genomic libraries were prepared using the Nextera XT Index Kit v2 Set A and sequenced on the HiSeq 4000 platform (Illumina Inc., San Diego, CA, USA) with a HiSeq 3000/4000 PE Cluster kit (2 × 150 bp). Prior to assembly, libraries were (i) cleaned from contaminants (adapters, phiX, artifacts, and human), (ii) error corrected, (iii) normalized to ≤ 100 ×, (iv) removed of low (<6×) coverage reads, and (v) filtered to a minimum length read of 100 nt. Reads were processed using the software package BBMap v37.90 (http://sourceforge.net/projects/bbmap) and de novo assembly using the software Unicycler v0.4.4 [93]. The Illumina reads are deposited in the National Center for Biotechnology Information (NCBI) Sequence Read Archive database under the BioProject accession number PRJNA558750.

Sequence-based typing (SBT) analysis was performed in silico with legsta and multi-locus sequence typing (mlst) as described previously [94]. The phylogenetic tree was constructed combining the sequenced genomes from this study and a set of closely related genomes. Relatedness is determined by alignment similarity to a select subset of COG (Clusters of Orthologous Groups) domains. The phylogenetic tree is reconstructed using FastTree 2 [95] to determine maximum likelihood phylogeny. Average nucleotide identity (ANI), an index of similarity between two genomes [96], was calculated using FastANI v1.3 (https://github.com/ParBLiSS/FastANI) [97]. ANI is defined as mean nucleotide identity of orthologous gene pairs shared between two microbial genomes. No ANI output is reported for a genome pair if the ANI value is below 80%.

### 4.8. Statistical Analysis

For each water quality parameter, a Shapiro–Wilk normality test was conducted for each site to determine distribution of the data throughout the sampling period. A one-way analysis of variance (ANOVA) using the Tukey multiple comparisons test was conducted between each site and sample type. $P$-values of $< 0.05$ were considered statistically significant. Analyses were performed using Prism 8 (GraphPad Software, San Diego, CA, USA). The R functions cor() and cor.test() were used, with a Spearman correction, to determine the direction (positive or negative) and significance of correlation between pairs of water quality characteristics within each sample location [98]. The correlation between the pairs was denoted by the number, r, which varies between −1 and +1, with 0 meaning no correlation, +1 a complete positive correlation, and −1 a complete negative correlation.

**Supplementary Materials:** The following are available online at http://www.mdpi.com/2076-0817/9/7/567/s1; Figure S1: Images of outlets at each sampling location; Figure S2: Scatterplot of turbidity measurements for bulk water samples; Figure S3: Scatterplot of free (A) and total (B) chlorine measurements for bulk water samples.

**Author Contributions:** Conceptualization and methodology, H.Y.B. and J.S.H.; validation, H.B.; formal analysis, H.Y.B. and V.G.-A.; investigation, H.Y.B., B.J.M., J.G.S., and J.S.H.; data curation, H.Y.B.; writing—original draft preparation, H.Y.B.; writing—review and editing, H.Y.B., B.J.M., V.G.-A., J.G.S., and J.S.H.; visualization, H.Y.B. All authors have read and agreed to the published version of the manuscript.

**Funding:** The U.S. Environmental Protection Agency (EPA) through its Office of Research and Development funded and managed the research described herein under contract #EP-C−11-006 to Pegasus Technical Services, Inc.

**Acknowledgments:** We would like to thank our EPA colleagues, James Goodrich and Hodon Ryu, for their technical review of this manuscript and Chelsea Hintz, for her assistance with the correlation analyses using R.

**Disclaimer:** This manuscript has been subjected to the Agency's review and has been approved for publication. The views expressed in this article are those of the authors and do not necessarily represent the views or policies of the Agency. Mention of trade names, commercial products, and/or services does not imply an endorsement or recommendation for use by the U.S. Government or EPA.

**Conflicts of Interest:** The authors declare no conflict of interest. With EPA funding, researchers designed the study; collected, analyzed, and interpreted the data; and wrote the manuscript intended for publication.

## References

1. WHO. *Legionella and the Prevention of Legionellosis*; World Health Organization Press: Geneva, Switzerland, 2007.
2. Davis, G.S.; Winn, W.C., Jr.; Gump, D.W.; Craighead, J.E.; Beaty, H.N. Legionnaires' pneumonia after aerosol exposure in guinea pigs and rats. *Am. Rev. Respir. Dis.* **1982**, *126*, 1050–1057.
3. Kaufmann, A.F.; McDade, J.E.; Patton, C.M.; Bennett, J.V.; Skaliy, P.; Feeley, J.C.; Anderson, D.C.; Potter, M.E.; Newhouse, V.F.; Gregg, M.B.; et al. Pontiac fever: Isolation of the etiologic agent (*Legionella pneumophilia*) and demonstration of its smode of transmission. *Am. J. Epidemiol.* **1981**, *114*, 337–347. [CrossRef]
4. Beer, K.B.; Gargano, J.W.; Roberts, V.A.; Reses, H.E.; Hill, V.R.; Garrison, L.E.; Kutty, P.K.; Hilborn, E.D.; Wade, T.J.; Fullerton, K.E.; et al. Outbreaks associated with environmental and undetermined water exposures—United States, 2011–2012. *MMWR Morb. Mortal. Wkly. Rep.* **2015**, *64*, 849–851. [CrossRef]
5. Bededict, K.M.; Reses, H.E.; Vigar, M.; Roth, D.M.; Roberts, V.A.; Mattioli, M.; Cooley, L.A.; Hilborn, E.D.; Wade, T.J.; Fullerton, K.E.; et al. Surveillance for waterborne disease outbreaks associated with drinking water—United States, 2013–2014. *MMWR Morb Mortal. Wkly. Rep.* **2017**, *66*, 1216–1221. [CrossRef]
6. Fraser, D.W.; Tsai, T.R.; Orenstein, W.; Parkin, W.E.; Beecham, H.J.; Sharrar, R.G. Legionnaires' disease: Description of an epidemic of pneumonia. *N. Engl. J. Med.* **1977**, *297*, 1189–1197. [CrossRef]
7. Rodríguez-Martínez, S.; Sharaby, Y.; Pecellín, M.; Brettar, I.; Höfle, M.; Halpern, M. Spatial distribution of *Legionella pneumophila* MLVA-genotypes in a drinking water system. *Water Res.* **2015**, *77*, 119–132. [CrossRef]

8. King, D.N.; Donohue, M.J.; Vesper, S.J.; Villegas, E.N.; Ware, M.W.; Vogel, M.E.; Furlong, E.F.; Kolpin, D.W.; Glassmeyer, S.T.; Pfaller, S. Microbial pathogens in source and treated waters from drinking water treatment plants in the United States and implications for human health. *Sci. Total Environ.* **2016**, *562*, 987–995. [CrossRef]
9. Donohue, M.J.; King, D.; Pfaller, S.; Mistry, J.H. The sporadic nature of *Legionella pneumophila, Legionella pneumophila* Sg1 and *Mycobacterium avium* occurrence within residences and office buildings across 36 states in the United States. *J. Appl. Microbiol.* **2019**, *126*, 1568–1579. [CrossRef]
10. Stout, J.E.; Brennen, C.; Muder, R.R. Legionnaires' disease in a newly constructed long-term care facility. *J. Am. Geriatr. Soc.* **2000**, *48*, 1589–1592. [CrossRef]
11. Orsi, G.B.; Vitali, M.; Marinelli, L.; Ciorba, V.; Tufi, D.; Del Cimmuto, A.; Ursillo, P.; Fabiani, M.; De Santis, S.; Protano, C.; et al. *Legionella* control in the water system of antiquated hospital buildings by shock and continuous hyperchlorination: 5 years experience. *BMC Infect. Dis.* **2014**, *14*, 394. [CrossRef] [PubMed]
12. Perola, O.; Kauppinen, J.; Kusnetsov, J.; Kärkkäinen, U.-M.; Lück, P.C.; Katila, M.-L. Persistent *Legionella pneumophila* colonization of a hospital water supply: Efficacy of control methods and a molecular epidemiological analysis. *APMIS* **2005**, *113*, 45–53. [CrossRef]
13. Scaturro, M.; Dell'eva, I.; Helfer, F.; Ricci, M.L. Persistence of the same strain of *Legionella pneumophila* in the water system of an Italian hospital for 15 years. *Infect. Control. Hosp. Epidemiol.* **2007**, *28*, 1089–1092. [CrossRef]
14. Cooper, I.R.; White, J.; Mahenthiralingam, E.; Hanlon, G.W. Long-term persistence of a single *Legionella pneumophila* strain possessing the mip gene in a municipal shower despite repeated cycles of chlorination. *J. Hosp. Infect.* **2008**, *70*, 154–159. [CrossRef] [PubMed]
15. Garcia-Nuñez, M.; Sopena, N.; Ragull, S.; Pedro-Botet, M.L.; Morera, J.; Sabria, M. Persistence of *Legionella* in hospital water supplies and nosocomial Legionnaires' disease. *FEMS Immunol. Med. Microbiol.* **2008**, *52*, 202–206. [CrossRef]
16. Bédard, E.; Boppe, I.; Kouamé, S.; Martin, P.; Pinsonneault, L.; Valiquette, L.; Racine, J.; Prévost, M. Combination of heat shock and enhanced thermal regime to control the growth of a persistent *Legionella pneumophila* strain. *Pathogens* **2016**, *5*, 35. [CrossRef]
17. Gomez-Valero, L.; Rusniok, C.; Jarraud, S.; Vacherie, B.; Rouy, Z.; Barbe, V.; Medigue, C.; Etienne, J.; Buchrieser, C. Extensive recombination events and horizontal gene transfer shaped the *Legionella pneumophila* genomes. *BMC Genom.* **2011**, *12*, 536. [CrossRef]
18. Segal, G.; Shuman, H.A. *Legionella pneumophila* utilizes the same genes to multiply within *Acanthamoeba castellanii* and human macrophages. *Infect. Immun.* **1999**, *67*, 2117–2124. [CrossRef]
19. Ensminger, A.W. *Legionella pneumophila*, armed to the hilt: Justifying the largest arsenal of effectors in the bacterial world. *Curr. Opin. Microbiol.* **2016**, *29*, 74–80. [CrossRef]
20. Ashbolt, N.J. Environmental (saprozoic) pathogens of engineered water systems: Understanding their ecology for risk assessment and management. *Pathogens* **2015**, *4*, 390–405. [CrossRef]
21. Ben Salah, I.; Drancourt, M. Surviving within the amoebal exocyst: The *Mycobacterium avium* complex paradigm. *BMC Microbiol.* **2010**, *10*, 99. [CrossRef]
22. Griffith, D.E.; Aksamit, T.; Brown-Elliott, B.A.; Catanzaro, A.; Daley, C.; Gordin, F.; Holland, S.M.; Horsburgh, R.; Huitt, G.; Iademarco, M.F.; et al. An official ATS/IDSA statement: Diagnosis, treatment, and prevention of nontuberculous mycobacterial diseases. *Am. J. Respir. Crit. Care Med.* **2007**, *175*, 367–416. [CrossRef] [PubMed]
23. Falkinham, J.O., 3rd; Norton, C.D.; LeChevallier, M.W. Factors influencing numbers of *Mycobacterium avium, Mycobacterium intracellulare*, and other Mycobacteria in drinking water distribution systems. *Appl. Environ. Microbiol.* **2001**, *67*, 1225–1231. [CrossRef] [PubMed]
24. Wadowsky, R.M.; Yee, R.B. Satellite growth of *Legionella pneumophila* with an environmental isolate of *Flavobacterium breve*. *Appl. Environ. Microbiol.* **1983**, *46*, 1447–1449. [CrossRef]
25. Wadowsky, R.M.; Yee, R.B. Effect of non-*Legionellaceae* bacteria on the multiplication of *Legionella pneumophila* in potable water. *Appl. Environ. Microb.* **1985**, *49*, 1206–1210. [CrossRef]
26. Temmerman, R.; Vervaeren, H.; Noseda, B.; Boon, N.; Verstraete, W. Necrotrophic growth of *Legionella pneumophila*. *Appl. Environ. Microbiol.* **2006**, *72*, 4323–4328. [CrossRef]
27. Tison, D.L.; Pope, D.H.; Cherry, W.B.; Fliermans, C.B. Growth of *Legionella pneumophila* in association with blue-green algae (cyanobacteria). *Appl. Environ. Microb.* **1980**, *39*, 456–459. [CrossRef]

28. Rogers, J.; Dowsett, A.B.; Dennis, P.J.; Lee, J.V.; Keevil, C.W. Influence of temperature and plumbing material selection on biofilm formation and growth of *Legionella pneumophila* in a model potable water system containing complex microbial flora. *Appl. Environ. Microbiol.* **1994**, *60*, 1585–1592. [CrossRef]
29. Cristino, S.; Legnani, P.P.; Leoni, E. Plan for the control of *Legionella* infections in long-term care facilities: Role of environmental monitoring. *Int. J. Hyg. Environ. Health* **2012**, *215*, 279–285. [CrossRef]
30. CDC. *Developing a Water Management Program to Reduce Legionella Growth & Spread in Buildings: A Practical Guide to Implementing Industry Standards*; USDHHS, Ed.; CDC: Atlanta, GA, USA, 2017; p. 36.
31. ANSI/ASHRAE. *Standard 188-2018 Legionellosis: Risk Management for Building Water Systems*; ANSI/ASHRAE: Atlanta, GA, USA, 2018.
32. Szabo, J.G.; Meiners, G.; Heckman, L.; Rice, E.W.; Hall, J. Decontamination of *Bacillus* spores adhered to iron and cement-mortar drinking water infrastructure in a model system using disinfectants. *J. Environ. Manag.* **2017**, *187*, 1–7. [CrossRef]
33. National Academies of Science, Engineering, and Medicine. *Management of Legionella in Water Systems*; The National Academies Press: Washington, DC, USA, 2020.
34. Neil, K.; Berkelman, R. Increasing incidence of legionellosis in the United States, 1990–2005: Changing epidemiologic trends. *Clin. Infect. Dis.* **2008**, *47*, 591–599. [CrossRef]
35. Alarcon Falconi, T.M.; Cruz, M.S.; Naumova, E.N. The shift in seasonality of legionellosis in the USA. *Epidemiol. Infect.* **2018**, *146*, 1824–1833. [CrossRef] [PubMed]
36. Bhopal, R.S.; Fallon, R.J. Seasonal variation of Legionnaires' disease in Scotland. *J. Infect.* **1991**, *22*, 153–160. [CrossRef]
37. Joseph, C.A.; Watson, J.M.; Harrison, T.G.; Bartlett, C.L.R. Nosocomial Legionnaires' disease in England and Wales, 1980–92. *Epidemiol. Infect.* **1994**, *112*, 329–346. [CrossRef]
38. Von Baum, H.; Ewig, S.; Marre, R.; Suttorp, N.; Gonschior, S.; Welte, T.; Lück, C.; Group CNfCAPS. Community-acquired *Legionella* pneumonia: New insights from the German Competence Network for Community Acquired Pneumonia. *Clin. Infect. Dis.* **2008**, *46*, 1356–1364. [CrossRef]
39. Ng, V.; Tang, P.; Jamieson, F.; Drews, S.J.; Brown, S.; Low, D.E.; Johnson, C.C.; Fisman, D.N. Going with the flow: Legionellosis risk in Toronto, Canada is strongly associated with local watershed hydrology. *Ecohealth* **2008**, *5*, 482–490. [CrossRef]
40. Simmering, J.E.; Polgreen, L.A.; Hornick, D.B.; Sewell, D.K.; Polgreen, P.M. Weather-dependent risk for Legionnaires' Disease, United States. *Emerg. Infect. Dis.* **2017**, *23*, 1843–1851. [CrossRef]
41. Cassell, K.; Gacek, P.; Warren, J.L.; Raymond, P.A.; Cartter, M.; Weinberger, D.M. Association between sporadic legionellosis and river systems in Connecticut. *J. Infect. Dis.* **2018**, *217*, 179–187. [CrossRef]
42. Barskey, A.; Lackraj, D.; Tripathi, P.S.; Cooley, L.A.; Lee, S.; Smith, J.; Edens, C. *Legionnaires' Disease Surveillance Summary Report, United States—2016 and 2017*; Centers for Disease Control and Prevention: Atlanta, GA, USA, 2020.
43. ISO. *ISO 12869:2012 Water Quality—Detection and Quantification of Legionella spp. and/or Legionella Pneumophila by Concentration and Genic Amplification by quantitative Polymerase Chain Reaction (qPCR)*; ISO/TR 13843; International Organization for Standardization: Geneva, Switzerland, 2012.
44. ISO. *ISO 11731:2017 Water Quality—Enumeration of Legionella*; International Organization for Standardization: Geneva, Switzerland, 2017.
45. Kozak, N.A.; Lucas, C.E.; Winchell, J.M. Identification of *Legionella* in the environment. In *Legionella Methods and Protocols*; Buchrieser, C., Hilbi, H., Eds.; Springer: New York, NY, USA; Humana Press: London, UK, 2013.
46. Whiley, H.; Taylor, M. *Legionella* detection by culture and qPCR: Comparing apples and oranges. *Crit. Rev. Microbiol.* **2016**, *42*, 65–74. [CrossRef]
47. Sartory, D.P.; Spies, K.; Lange, B.; Schneider, S.; Langer, B. Evaluation of a most probable number method for the enumeration of *Legionella pneumophila* from potable and related water samples. *Lett. Appl. Microbiol.* **2017**, *64*, 271–275. [CrossRef]
48. Petrisek, R.; Hall, J. Evaluation of a most probable number method for the enumeration of *Legionella pneumophila* from North American potable and nonpotable water samples. *J. Water Health* **2018**, *16*, 25–33. [CrossRef] [PubMed]
49. Rech, M.M.; Swalla, B.M.; Dobranic, J.K. Evaluation of Legiolert for quantification of *Legionella pneumophila* from non-potable water. *Curr. Microbiol.* **2018**, *75*, 1282–1289. [CrossRef] [PubMed]

50. Levi, K.; Smedley, J.; Towner, K.J. Evaluation of a real-time PCR hybridization assay for rapid detection of *Legionella pneumophila* in hospital and environmental water samples. *Clin. Microbiol Infect.* **2003**, *9*, 754–758. [CrossRef] [PubMed]
51. Lee, J.V.; Lai, S.; Exner, M.; Lenz, J.; Gaia, V.; Casati, S.; Hartemann, P.; Luck, C.; Pangon, B.; Ricci, M.L.; et al. An international trial of quantitative PCR for monitoring *Legionella* in artificial water systems. *J. Appl. Microbiol.* **2011**, *110*, 1032–1044. [CrossRef] [PubMed]
52. Toplitsch, D.; Platzer, S.; Pfeifer, B.; Hautz, J.; Mascher, F.; Kittinger, C. *Legionella* detection in environmental samples as an example for successful implementation of qPCR. *Water* **2018**, *10*, 1012. [CrossRef]
53. Joly, P.; Falconnet, P.A.; André, J.; Weill, N.; Reyrolle, M.; Vandenesch, F.; Maurin, M.; Etienne, J.; Jarraud, S. Quantitative real-time *Legionella* PCR for environmental water samples: Data interpretation. *Appl. Environ. Microbiol.* **2006**, *72*, 2801–2808. [CrossRef]
54. ISO. *ISO 19458:2006 Water Quality—Sampling for Microbiological Analysis*; International Organization for Standardization: Geneva, Switzerland, 2006.
55. The European Guidelines Working Group. *Infections EWGfL: European Technical Guidelines 2017: Minimising the Risk from Legionella Infections in Building Water Systems*; The European Guidelines Working Group: London, UK, 2017.
56. Lu, J.; Buse, H.; Struewing, I.; Zhao, A.; Lytle, D.; Ashbolt, N. Annual variations and effects of temperature on *Legionella* spp. and other potential opportunistic pathogens in a bathroom. *Environ. Sci. Pollut. Res. Int.* **2017**, *24*, 2326–2336. [CrossRef]
57. Bédard, E.; Laferrière, C.; Déziel, E.; Prévost, M. Impact of stagnation and sampling volume on water microbial quality monitoring in large buildings. *PLoS ONE* **2018**, *13*, e0199429. [CrossRef]
58. Flannery, B.; Gelling, L.B.; Vugia, D.J.; Weintraub, J.M.; Salerno, J.J.; Conroy, M.J.; Stevens, V.A.; Rose, C.E.; Moore, M.R.; Fields, B.S.; et al. Reducing *Legionella* colonization in water systems with monochloramine. *Emerg. Infect. Dis.* **2006**, *12*, 588–596. [CrossRef]
59. Duda, S.; Kandiah, S.; Stout, J.E.; Baron, J.L.; Yassin, M.; Fabrizio, M.; Ferrelli, J.; Hariri, R.; Wagener, M.M.; Goepfert, J.; et al. Evaluation of a new monochloramine generation system for controlling *Legionella* in building hot water systems. *Infect. Control. Hosp. Epidemiol.* **2014**, *35*, 1356–1363. [CrossRef]
60. Mancini, B.; Scurti, M.; Dormi, A.; Grottola, A.; Zanotti, A.; Cristino, S. Effect of monochloramine treatment on colonization of a hospital water distribution system by *Legionella* spp.: A 1 year experience study. *Environ. Sci. Technol.* **2015**, *49*, 4551–4558. [CrossRef]
61. Waak, M.B.; LaPara, T.M.; Hallé, C.; Hozalski, R.M. Occurrence of *Legionella* spp. in water-main biofilms from two drinking water distribution systems. *Environ. Sci. Technol.* **2018**, *52*, 7630–7639. [CrossRef] [PubMed]
62. Lee, W.H.; Pressman, J.G.; Wahman, D.G. Three-dimensional free chlorine and monochloramine biofilm penetration: Correlating penetration with biofilm activity and viability. *Environ. Sci. Technol.* **2018**, *52*, 1889–1898. [CrossRef] [PubMed]
63. Buse, H.Y.; Morris, B.J.; Struewing, I.T.; Szabo, J.G. Chlorine and monochloramine disinfection of *Legionella pneumophila* colonizing copper and polyvinyl chloride drinking water biofilms. *Appl. Environ. Microbiol.* **2019**, *85*, e02956-18. [CrossRef] [PubMed]
64. Jacangelo, J.G.; Olivieri, V.P.; Kawata, K. Investigating the mechanism of inactivation of *Escherichia coli* B by monochloramine. *J. AWWA* **1991**, *83*, 80–87. [CrossRef]
65. Xue, Z.; Hessler, C.M.; Panmanee, W.; Hassett, D.J.; Seo, Y. *Pseudomonas aeruginosa* inactivation mechanism is affected by capsular extracellular polymeric substances reactivity with chlorine and monochloramine. *FEMS Microbiol. Ecol.* **2013**, *83*, 101–111. [CrossRef] [PubMed]
66. Thomas, V.; Herrera-Rimann, K.; Blanc, D.S.; Greub, G. Biodiversity of amoebae and amoeba-resisting bacteria in a hospital water network. *Appl. Environ. Microbiol.* **2006**, *72*, 2428–2438. [CrossRef]
67. Pagnier, I.; Valles, C.; Raoult, D.; Scola, B. Isolation of *Vermamoeba vermiformis* and associated bacteria in hospital water. *Microb. Pathog.* **2015**, *80*, 14–20. [CrossRef]
68. Le Dantec, C.; Duguet, J.P.; Montiel, A.; Dumoutier, N.; Dubrou, S.; Vincent, V. Occurrence of mycobacteria in water treatment lines and in water distribution systems. *Appl. Environ. Microbiol.* **2002**, *68*, 5318–5325. [CrossRef]
69. Garcia, A.; Goni, P.; Cieloszyk, J.; Fernandez, M.T.; Calvo-Begueria, L.; Rubio, E.; Fillat, M.F.; Peleato, M.L.; Clavel, A. Identification of free-living amoebae and amoeba-associated bacteria from reservoirs and water treatment plants by molecular techniques. *Environ. Sci. Technol.* **2013**, *47*, 3132–3140. [CrossRef] [PubMed]

70. Buse, H.Y.; Schoen, M.E.; Ashbolt, N.J. Legionellae in engineered systems and use of quantitative microbial risk assessment to predict exposure. *Water Res.* **2012**, *46*, 921–933. [CrossRef]
71. Lesnik, R.; Brettar, I.; Höfle, M.G. *Legionella* species diversity and dynamics from surface reservoir to tap water: From cold adaptation to thermophily. *ISME J.* **2016**, *10*, 1064. [CrossRef] [PubMed]
72. Yu, V.L.; Plouffe, J.F.; Pastoris, M.C.; Stout, J.E.; Schousboe, M.; Widmer, A.; Summersgill, J.; File, T.; Heath, C.M.; Paterson, D.L.; et al. Distribution of *Legionella* species and serogroups isolated by culture in patients with sporadic community-acquired legionellosis: An international collaborative survey. *J. Infect. Dis.* **2002**, *186*, 127–128. [CrossRef]
73. Harrison, T.G.; Doshi, N.; Fry, N.K.; Joseph, C.A. Comparison of clinical and environmental isolates of *Legionella pneumophila* obtained in the UK over 19 years. *Clin. Microbiol. Infect.* **2007**, *13*, 78–85. [CrossRef] [PubMed]
74. Ciesielski, C.A.; Blaser, M.J.; Wang, W.L. Serogroup specificity of *Legionella pneumophila* is related to lipopolysaccharide characteristics. *Infect. Immun.* **1986**, *51*, 397–404. [CrossRef]
75. Otten, S.; Iyer, S.; Johnson, W.; Montgomery, R. Serospecific antigens of *Legionella pneumophila*. *J. Bacteriol.* **1986**, *167*, 893–904. [CrossRef]
76. Helbig, J.H.; Amemura-Maekawa, J. Serotyping of *Legionella pneumophila* and epidemiological investigations. *Nowa Medycyna* **2009**, *16*, 69–75.
77. Kozak-Muiznieks, N.A.; Morrison, S.S.; Mercante, J.W.; Ishaq, M.K.; Johnson, T.; Caravas, J.; Lucas, C.E.; Brown, E.; Raphael, B.H.; Winchell, J.M. Comparative genome analysis reveals a complex population structure of *Legionella pneumophila* subspecies. *Infect. Genet. Evol.* **2018**, *59*, 172–185. [CrossRef]
78. Carvalho, F.R.; Vazoller, R.F.; Foronda, A.S.; Pellizari, V.H. Phylogenetic study of *legionella* species in pristine and polluted aquatic samples from a tropical Atlantic forest ecosystem. *Curr. Microbiol.* **2007**, *55*, 288–293. [CrossRef]
79. Kozak-Muiznieks, N.A.; Lucas, C.E.; Brown, E.; Pondo, T.; Taylor, T.H., Jr.; Frace, M.; Miskowski, D.; Winchell, J.M. Prevalence of sequence types among clinical and environmental isolates of *Legionella pneumophila* serogroup 1 in the United States from 1982 to 2012. *J. Clin. Microbiol.* **2014**, *52*, 201–211. [CrossRef] [PubMed]
80. Quero, S.; Párraga-Niño, N.; Barrabeig, I.; Sala, M.R.; Pedro-Botet, M.L.; Monsó, E.; Jané, M.; Sabrià, M.; Garcia-Núñez, M. Population structure of environmental and clinical *Legionella pneumophila* isolates in Catalonia. *Sci. Rep.* **2018**, *8*, 6241. [CrossRef]
81. Edagawa, A.; Kimura, A.; Doi, H.; Tanaka, H.; Tomioka, K.; Sakabe, K.; Nakajima, C.; Suzuki, Y. Detection of culturable and nonculturable *Legionella* species from hot water systems of public buildings in Japan. *J. Appl. Microbiol.* **2008**, *105*, 2104–2114. [CrossRef] [PubMed]
82. Leoni, E.; Legnani, P.P. Comparison of selective procedures for isolation and enumeration of *Legionella* species from hot water systems. *J. Appl. Microbiol.* **2001**, *90*, 27–33. [CrossRef] [PubMed]
83. Asghari, F.B.; Nikaeen, M.; Hatamzadeh, M.; Hassanzadeh, A. Surveillance of *Legionella* species in hospital water systems: The significance of detection method for environmental surveillance data. *J. Water Health* **2013**, *11*, 713–719. [CrossRef] [PubMed]
84. Duda, S.; Baron, J.L.; Wagener, M.M.; Vidic, R.D.; Stout, J.E. Lack of correlation between *Legionella* colonization and microbial population quantification using heterotrophic plate count and adenosine triphosphate bioluminescence measurement. *Environ. Monit. Assess.* **2015**, *187*, 393. [CrossRef]
85. Pierre, D.; Baron, J.L.; Ma, X.; Sidari, F.P., 3rd; Wagener, M.M.; Stout, J.E. Water quality as a predictor of *Legionella* positivity of building water systems. *Pathogens* **2019**, *8*, 295. [CrossRef]
86. Rakić, A.; Perić, J.; Foglar, L. Influence of temperature, chlorine residual and heavy metals on the presence of *Legionella pneumophila* in hot water distribution systems. *Ann. Agric. Environ. Med.* **2012**, *19*, 431–436.
87. Rakić, A.; Štambuk-Giljanović, N.; Foglar, L. Monitoring *Legionella pneumophila* in drinking water distribution systems in southern Croatia. *Fresenius Environ. Bull.* **2013**, *22*, 3390–3397.
88. Revetta, R.P.; Gomez-Alvarez, V.; Gerke, T.L.; Curioso, C.; Santo Domingo, J.W.; Ashbolt, N.J. Establishment and early succession of bacterial communities in monochloramine-treated drinking water biofilms. *FEMS Microbiol. Ecol.* **2013**, *86*, 404–414. [CrossRef]
89. Chern, E.C.; King, D.; Haugland, R.; Pfaller, S. Evaluation of quantitative polymerase chain reaction assays targeting *Mycobacterium avium*, *M. intracellulare*, and *M. avium* subspecies *paratuberculosis* in drinking water biofilms. *J. Water Health* **2015**, *13*, 131–139. [CrossRef]

90. Lu, J.; Struewing, I.; Yelton, S.; Ashbolt, N. Molecular survey of occurrence and quantity of *Legionella* spp., *Mycobacterium* spp., *Pseudomonas aeruginosa* and amoeba hosts in municipal drinking water storage tank sediments. *J. Appl. Microbiol.* **2015**, *119*, 278–288. [CrossRef]
91. Kuiper, M.W.; Valster, R.M.; Wullings, B.A.; Boonstra, H.; Smidt, H.; van der Kooij, D. Quantitative detection of the free-living amoeba *Hartmannella vermiformis* in surface water by using real-time PCR. *Appl. Environ. Microbiol.* **2006**, *72*, 5750–5756. [CrossRef]
92. Riviére, D.; Szczebara, F.M.; Berjeaud, J.M.; Frére, J.; Héchard, Y. Development of a real-time PCR assay for quantification of *Acanthamoeba* trophozoites and cysts. *J. Microbiol. Methods* **2006**, *64*, 78–83. [CrossRef]
93. Wick, R.R.; Judd, L.M.; Gorrie, C.L.; Holt, K.E. Unicycler: Resolving bacterial genome assemblies from short and long sequencing reads. *PLoS Comput. Biol.* **2017**, *13*, e1005595. [CrossRef] [PubMed]
94. Gomez-Alvarez, V.; Boczek, L.; King, D.; Pemberton, A.; Pfaller, S.; Rodgers, M.; SantoDomingo, J.; Revetta, R.P. Draft genome sequences of seven *Legionella pneumophila* isolates from a hot water system of a large building. *Microbiol. Resour. Announc.* **2019**, *8*, e00384-19. [CrossRef] [PubMed]
95. Price, M.N.; Dehal, P.S.; Arkin, A.P. FastTree 2—Approximately maximum-likelihood trees for large alignments. *PLoS ONE* **2010**, *5*, e9490. [CrossRef] [PubMed]
96. Goris, J.; Konstantinidis, K.T.; Klappenbach, J.A.; Coenye, T.; Vandamme, P.; Tiedje, J.M. DNA-DNA hybridization values and their relationship to whole-genome sequence similarities. *Int. J. Syst. Evol. Microbiol.* **2007**, *57*, 81–91. [CrossRef]
97. Jain, C.; Rodriguez-R, L.M.; Phillippy, A.M.; Konstantinidis, K.T.; Aluru, S. High throughput ANI analysis of 90K prokaryotic genomes reveals clear species boundaries. *Nat. Commun.* **2018**, *9*, 5114. [CrossRef]
98. Team RC. *R: A Language and Environment for Statistical Computing*; R Foundation for Statistical Computing: Vienna, Austria, 2013; Available online: http://www.R-project.org/ (accessed on 1 April 2020).

*pathogens*

MDPI

*Article*

# A Snapshot of the Prevalence and Molecular Diversity of *Legionella pneumophila* in the Water Systems of Israeli Hotels

**Eugenia Yakunin [1], Eszter Kostyal [2], Vered Agmon [1], Itamar Grotto [1,3], Lea Valinsky [1] and Jacob Moran-Gilad [1,3,*]**

1 Central Laboratories and Public Health Services, Ministry of Health, Jerusalem 9134302, Israel; eugenia.yakunin@moh.gov.il (E.Y.); vered.agmon@moh.gov.il (V.A.); itamar.grotto@moh.gov.il (I.G.); lea.valinsky@moh.gov.il (L.V.)

2 Department of Water Microbiology, Biolab Ltd., Jerusalem 9134001, Israel; ester@biolab-chemicals.com

3 Department of Health Systems Management, School of Public Health, Faculty of Health Sciences, Ben-Gurion University of the Negev, Beer-Sheva 8410501, Israel

* Correspondence: giladko@post.bgu.ac.il

Received: 20 April 2020; Accepted: 23 May 2020; Published: 27 May 2020

**Abstract:** Exposure to *Legionella* spp. contaminated aerosols in hotel settings confers risk for travel-associated Legionnaire's disease (TALD). In this study, we investigated the prevalence of *Legionella* contamination and its molecular diversity in hotels and resorts across Israel. The study was comprised of a convenience sample of water systems from 168 hotels and resorts countrywide, routinely inspected between March 2015 and February 2017. Isolation and quantitation of *Legionella* were performed in a water laboratory using the ISO 11731 method. The distribution of *Legionella* isolates was analyzed according to geography and source. The genetic diversity of a subset of isolates was analyzed by sequence-based typing (SBT) at the National Reference Laboratory for *Legionella* and compared to the national database. Out of 2830 samples tested, 470 (17%) obtained from 102 different premises (60% of hotels) were positive for *Legionella* spp. In 230 samples (49% of all positive, 8% of total samples), accounting for 37% of hotels, *Legionella* spp. counts exceeded the regulatory threshold of 1000 CFU/L. The most frequently contaminated water sources were cooling towers (38%), followed by faucets, hot tubs, water lines, and storage tanks (14–17% each). Furthermore, 32% and 17% of samples obtained from cooling towers and hot tubs, respectively, exceeded the regulatory thresholds. SBT was performed on 78 strains and revealed 27 different sequence types (STs), including two novel STs. The most prevalent STs found were ST1 (26%), ST87 (10%), ST93 (6%), and ST461 and ST1516 (5% each). Several *L. pneumophila* STs were found to be limited to certain geographical regions. This is the first study to investigate the prevalence and diversity of *Legionella* in hotels and resorts in Israel during non-outbreak environmental inspections. These findings will inform risk assessment, surveillance, and control measures of TALD.

**Keywords:** *Legionella pneumophila*; SBT; molecular diversity; hotels

## 1. Introduction

*Legionella* is a Gram-negative bacterium found ubiquitously in aqueous environments, which can multiply quickly in man-made water systems [1]. *Legionella* spp. have a complex life cycle, and exist in the environment as free-living bacteria in microbial consortia of environmental organisms or as intracellular pathogens. *L. pneumophila* has plenty of virulence factors, which it uses effectively to infect aquatic protozoa or human lung alveolar macrophages [2].

*L. pneumophila* is the major causative agent of Legionnaires' disease (LD), a severe pneumonia with a fatality rate of up to 15%, and a flu-like illness called Pontiac fever [3,4]. Humans can contract the disease during exposure to contaminated water aerosols generated by hot and cold water systems, cooling towers, showering facilities, and spa pools [5]. *Legionella* bacteria is an opportunistic pathogen [2]. The risk factors include old age, underlying diseases, and smoking [6]. Although many *Legionella* spp. are considered potentially pathogenic for humans, *Legionella pneumophila* (Lp) causes the vast majority of LD cases, and of the 16 known Lp serogroups (sg), sg1 accounts for over 80% of LD cases [7,8].

Legionellosis is often associated with staying in hotel accommodations, and LD is recognized as a major form of travel-associated pneumonia (TALD) [9]. Since 2010, TALD cases have accounted for 20% of all reported LD cases in Europe each year. The number of cases reported to the European TALD surveillance scheme continues to rise annually, with a 20% increase observed between 2014 and 2015 [10]. Moreover, *Legionella pneumophila* has significant outbreak potential. Since its first fatal outbreak in a hotel in Philadelphia, United States, in 1976, many clusters and outbreaks linked to hotel settings have been investigated globally [11–14]. Factors shown to contribute to the *Legionella* spp. spread and colonization are the complexity, old age, and poor maintenance of a distribution system, warm water temperature, and the presence of amoebae [1,15–17]. Several recent studies have focused on the prevalence and distribution of *Legionella* in water systems of hotels in non-outbreak situations. These studies revealed variable rates of contamination and species diversity [18–21], but limited data is published on the molecular diversity of *Legionella* spp. in hotel settings [22–24].

In Israel, where international and domestic tourism is an important branch of the national economy, TALD has accounted for 15% of all LD cases between 2006 and 2011 [25]. According to recent national epidemiology surveillance data of the Ministry of Health, the majority of TALD cases in Israel are sporadic or imported from abroad, and no major change in trends was observed during the last decade. While isolates from TALD cases undergo molecular typing, a few of them have been linked to a specific accommodation sites. It is likely that a great proportion of cases go unnoticed, due to the mild symptoms and underdiagnosis, the long incubation period of *Legionella* spp., and the short-term nature of domestic tourism. Of note is that no comprehensive data are available concerning the abundance of *Legionella* spp. in Israeli hotel water systems. In this study, we investigated, for the first time, the prevalence and characteristics of environmental *Legionella* spp. in the Israeli hotel setting as part of routine inspections.

## 2. Results

### 2.1. *Legionella* Contamination Rates

During the study period, 2830 water specimens were collected routinely from the water systems of 168 hotels and resorts in six districts across Israel. *Legionella* spp. were isolated from 470 samples (17%) originating from 102 (60%) hotels. The percentage of *Legionella*-positive samples was lower in the Southern, Jerusalem, and Tel Aviv districts (13%, 15%, and 14%, respectively), of which the largest number of samples was submitted (1139, 794, and 447 samples). A higher level of contamination was found in the North (40%), but only 42 samples were collected (Table 1 and Figure 1). In 230 samples (49% of all positive, 8% of total samples), accounting for 37% of hotels, *Legionella* spp. concentrations exceeded the national regulatory thresholds. The percentage of exceeding samples per district ranged from 6% to 33% (Table 1).

Analysis of *Legionella* spp. prevalence according to sample source showed that both cold and hot water distribution systems were colonized. The leading contaminated water sources were cooling towers (38%), followed by hot tubs, outlets, and main water lines (14–17% each). Of 277 *Legionella*-positive water samples from the outlets representing hotel rooms, 166 (59.9%) and 111 (40.1%) were from hot and cold water systems, respectively. The respective positivity rates were 15.3% and 17.1%. Levels exceeding regulatory *Legionella* counts were detected in 84 samples from the outlets, of which 67 samples (79.7%) originated from hot and 17 (20.2%) from cold water systems. The respective exceedance rates above regulatory thresholds were 6.2% and 2.6%. The majority of contaminated main water lines represented hot water distribution systems and accounted for 63 (92.6%) of 68 positive samples from water lines. Only five (7.3%) positive samples were from cold water lines. Of 35 samples with levels of contamination exceeding regulatory thresholds, 33 (94.2%) were from hot water lines at a rate of 8.6%. In 74 (32%) of 232 samples from cooling towers, and 36 (17%) of 218 samples from hot tubs, levels of *Legionella* spp. exceeded the regulatory thresholds of 1000 CFU/L established for potable water and the 1 CFU/100 mL threshold for the hot tubs (Table 2).

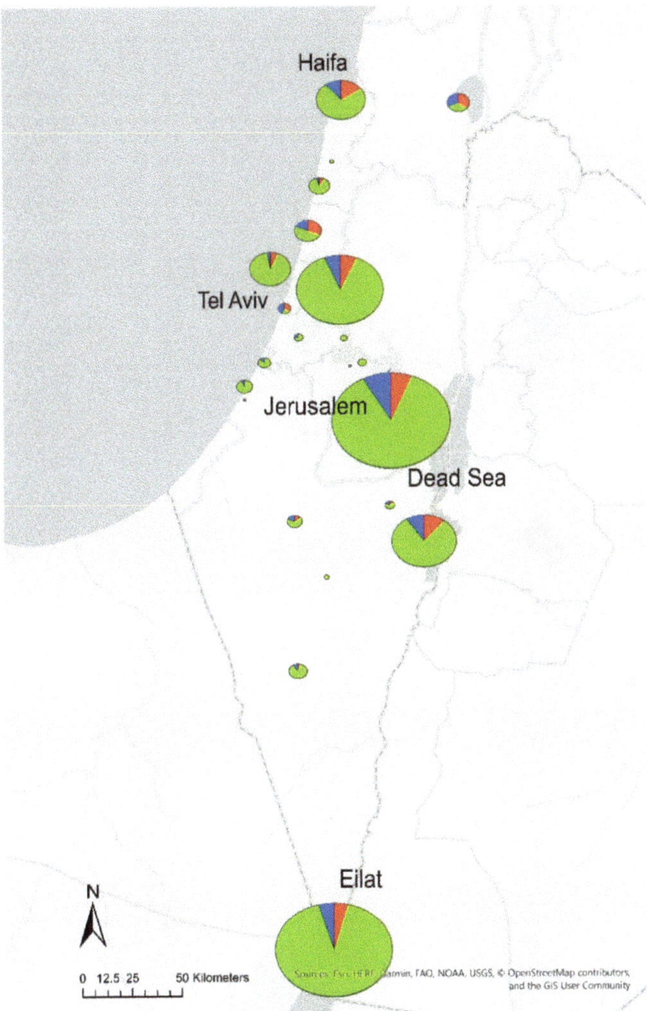


**Figure 1.** Geographic distribution and number of samples obtained from the 168 hotels and resorts included in the study. Samples are represented as pie charts at different locations; the size is proportional to the number of samples obtained from a specific location. Negative samples are shown in green, and positive and exceeding samples are shown in blue and red, respectively. Four major tourist sites with the largest number of samples (negative/positive/exceeding the regulatory thresholds) are Eilat (695/38/34), Jerusalem (662/71/52), the Tel-Aviv region (384/31/32) and the Dead Sea region (210/28/32).

**Table 1.** Distribution of premises and samples, according to administrative region.

| District | Total Tested | | Positive Samples | | Exceeding Samples [1] | |
|---|---|---|---|---|---|---|
| | No. of Hotels | No. of Samples | No. of Hotels | No. of Samples (% Per District) | No. of Hotels | No. of Samples (% Per District) |
| North | 3 | 42 | 2 | 27 (64) | 1 | 14 (33) |
| Center | 9 | 207 | 6 | 54 (26) | 4 | 30 (14) |
| South | 78 | 1139 | 44 | 151 (13) | 28 | 72 (6) |
| Haifa | 9 | 201 | 7 | 52 (26) | 4 | 30 (15) |
| Tel Aviv | 20 | 447 | 12 | 63 (14) | 8 | 32 (7) |
| Jerusalem | 49 | 794 | 30 | 123 (15) | 17 | 52 (7) |
| **Total** | **168 [a]** | **2830** | **101 [b]** | **470 (17)** | **62 [c]** | **230 (8)** |

[1] Cut-off values of 1000 CFU/L for all categories and 1 CFU/100 mL for hot tubs, according to the national regulations; [a] Out of 168 hotels and resorts tested for *Legionella* spp., 119 were tested more than once (range of 1–25 sampling days, median 2); [b] Of 101 hotels with positive samples under the regulatory cutoffs, 56 were sampled at least twice (range of 1–17 sampling days, median 2); [c] Of 62 hotels with *Legionella* concentrations above the national regulatory cutoffs, 29 were sampled at least twice (range of 1–12 sampling days, median 1).

**Table 2.** Prevalence of *L. pneumophila*, according to source type.

| Source Type | No. of Samples | | | | Positive Samples | | Exceeding Samples [1] | |
|---|---|---|---|---|---|---|---|---|
| | Cold Water | Hot Water | Mixed Water [2] | Total | No. of Samples | Per Category (%) | No. of Samples | Per Category (%) |
| Outlet | 649 | 1084 | | 1733 | 277 | 16% | 84 | 5% |
| Main water line | 96 | 383 | | 479 | 68 | 14% | 35 | 7% |
| Cooling tower | 232 | | | 232 | 87 | 38% | 74 | 32% |
| Hot tub | 5 | 9 | 204 | 218 | 36 | 17% | 36 | 17% |
| Fountain | 24 | 4 | | 28 | 1 | 4% | 0 | 0% |
| Pool | 11 | | | 11 | 0 | 0% | 0 | 0% |
| Air conditioning | 4 | | | 4 | 1 | 25% | 1 | 25% |
| Not available [3] | 38 | 87 | | 125 | 0 | 0% | 0 | 0% |
| **Total:** | **1059** | **1567** | **204** | **2830** | **470** | **17%** | **230** | **8%** |

[1] Cut off values of 1000 CFU/L for all categories and 1 CFU/100 mL for hot tubs, according to the national regulations; [2] According to the national regulations, the water temperature range of 37–39 °C for hot tubs can be achieved by mixing hot and cold water. [3] Source type not indicated in the laboratory requisition form accompanying the samples.

### 2.2. Distribution of Serotypes

Serotyping of a convenience sample comprised of 162 isolates revealed that 104 isolates (64%) from 44 hotels belonged to sg2–14, and 53 (33%) isolates from 22 hotels were sg1. Both sg1 and sg2–14 were found in the water systems of nine hotels. Two percent of isolates (4/162) for which serogroup identification failed and one recognized by the kit as *Legionella* spp. were subjected to *mip* sequencing, and subsequently identified as *L. pneumophila* and *L. bozemanii*, respectively. Overall, Lp accounted for the vast majority of the 162 serotyped isolates (99%).

Furthermore, we analyzed serogroup prevalence according to sample type. The majority of sg2–14 isolates were recovered from hot and mixed water samples, and accounted for 81 (77.8%) of the 104 isolates. Lp sg1 isolates were predominant in cold water systems, and 40 (75.4%) of the 53 isolates originated from cold water.

### 2.3. Phylogenetic Analysis

Phylogenetic analysis using SBT of a subset of 78 isolates revealed 27 different STs, including two novel STs (ST2169, ST2284), with the index of diversity being 0.912. Twelve STs were associated with more than one isolate, and 16 STs were identified with one single isolate. The most prevalent STs found were ST1 (26%), ST87 (10%), ST93 (6%), and ST461 and ST1516 (5% each). Of all Lp sg1 isolates, ST1 accounted for 63% (20/32), while the leading Lp non-sg1 subtype, ST87, comprised 17% of all sg2–14 isolates (8/46). Amplification failure of the *flaA* fragment occurred in two allelic profiles (0,4,16,1,7,13,206 and 0,14,16,1,7,13,206), and therefore no STs could be obtained for them. The clustering analysis of the 27 strains is shown in Figure 2.

**Similarity, percentage**

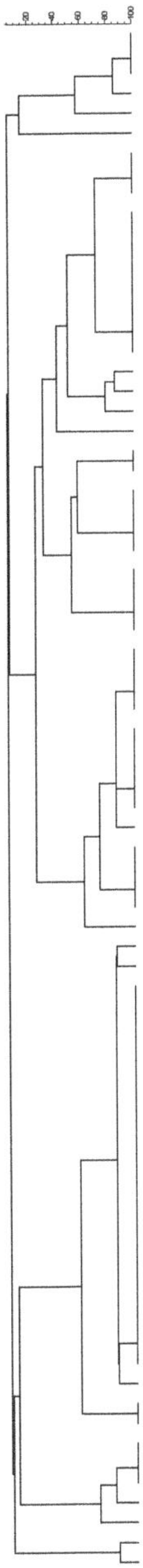

| flaA | pilE | asd | mip | mompS | proA | neuA | ST | Sg | Source type | District |
|---|---|---|---|---|---|---|---|---|---|---|
| 7 | 6 | 17 | 3 | 13 | 11 | 11 | 59 | 1 | Faucet | Jerusalem |
| 7 | 6 | 17 | 3 | 13 | 11 | 11 | 59 | 1 | Faucet | Jerusalem |
| 7 | 6 | 17 | 3 | 13 | 11 | 11 | 59 | 1 | Hot tub | Jerusalem |
| 7 | 6 | 17 | 3 | 13 | 11 | 56 | 1979 | 2-14 | Faucet | Haifa |
| 7 | 6 | 3 | 8 | 13 | 11 | 3 | 80 | 2-14 | Faucet | Center |
| 8 | 6 | 34 | 9 | 53 | 8 | 209 | 1413 | 2-14 | Storage tank | Haifa |
| 7 | 10 | 3 | 28 | 9 | 4 | 3 | 1641 | 2-14 | Storage tank | Jerusalem |
| 7 | 10 | 3 | 28 | 9 | 4 | 3 | 1641 | 2-14 | Faucet | Jerusalem |
| 7 | 10 | 3 | 28 | 9 | 4 | 3 | 1641 | 2-14 | Water line | Jerusalem |
| 2 | 10 | 3 | 28 | 9 | 4 | 13 | 87 | 2-14 | Faucet | South |
| 2 | 10 | 3 | 28 | 9 | 4 | 13 | 87 | 2-14 | Faucet | South |
| 2 | 10 | 3 | 28 | 9 | 4 | 13 | 87 | 2-14 | Faucet | South |
| 2 | 10 | 3 | 28 | 9 | 4 | 13 | 87 | 2-14 | Faucet | Tel Aviv |
| 2 | 10 | 3 | 28 | 9 | 4 | 13 | 87 | 2-14 | Faucet | South |
| 2 | 10 | 3 | 28 | 9 | 4 | 13 | 87 | 2-14 | Water line | Center |
| 2 | 10 | 3 | 28 | 9 | 4 | 13 | 87 | 2-14 | Shower | Center |
| 2 | 10 | 3 | 28 | 9 | 4 | 13 | 87 | 2-14 | Tap | South |
| 6 | 10 | 3 | 28 | 19 | 4 | 3 | 2180 | 2-14 | Hot tub | South |
| 6 | 10 | 19 | 28 | 19 | 4 | 3 | 292 | 2-14 | Hot tub | North |
| 6 | 10 | 19 | 28 | 19 | 4 | 207 | 1409 | 2-14 | Hot tub | Jerusalem |
| 2 | 10 | 14 | 10 | 21 | 4 | 3 | 1119 | 1 | Hot tub | South |
| 6 | 10 | 15 | 12 | 9 | 14 | 11 | 356 | 1 | Cooling tower | Center |
| 6 | 10 | 15 | 12 | 9 | 14 | 11 | 356 | 1 | Cooling tower | Center |
| 6 | 10 | 15 | 28 | 4 | 14 | 207 | 1516 | 2-14 | Hot tub | South |
| 6 | 10 | 15 | 28 | 4 | 14 | 207 | 1516 | 2-14 | Hot tub | South |
| 6 | 10 | 15 | 28 | 4 | 14 | 207 | 1516 | 2-14 | Hot tub | South |
| 6 | 10 | 15 | 28 | 4 | 14 | 207 | 1516 | 2-14 | Hot tub | South |
| 6 | 10 | 14 | 28 | 21 | 14 | 9 | 461 | 2-14 | Storage tank | South |
| 6 | 10 | 14 | 28 | 21 | 14 | 9 | 461 | 2-14 | Storage tank | South |
| 6 | 10 | 14 | 28 | 21 | 14 | 9 | 461 | 2-14 | Water line | South |
| 6 | 10 | 14 | 28 | 21 | 14 | 9 | 461 | 2-14 | Water line | Jerusalem |
| 3 | 10 | 1 | 28 | 14 | 9 | 207 | 1326 | 2-14 | Faucet | Jerusalem |
| 3 | 10 | 1 | 28 | 14 | 9 | 207 | 1326 | 2-14 | Water line | Jerusalem |
| 3 | 10 | 1 | 28 | 14 | 9 | 207 | 1326 | 2-14 | Faucet | Jerusalem |
| 3 | 10 | 1 | 28 | 14 | 9 | 207 | 1326 | 2-14 | Shower | Jerusalem |
| 3 | 10 | 1 | 28 | 14 | 9 | 13 | 93 | 2-14 | Shower | South |
| 3 | 10 | 1 | 28 | 14 | 9 | 13 | 93 | 2-14 | Storage tank | Jerusalem |
| 3 | 10 | 1 | 28 | 14 | 9 | 13 | 93 | 2-14 | Storage tank | Jerusalem |
| 3 | 10 | 1 | 28 | 14 | 9 | 13 | 93 | 2-14 | Hot tub | South |
| 3 | 10 | 1 | 28 | 14 | 9 | 13 | 93 | 2-14 | Hot tub | South |
| 3 | 10 | 1 | 28 | 14 | 9 | 3 | 187 | 2-14 | Cooling tower | Jerusalem |
| 3 | 6 | 1 | 28 | 14 | 9 | 3 | 1642 | 2-14 | Shower | Jerusalem |
| 3 | 6 | 1 | 28 | 14 | 9 | 3 | 1642 | 2-14 | Faucet | North |
| 3 | 6 | 1 | 28 | 14 | 9 | 3 | 1642 | 2-14 | Faucet | South |
| 3 | 6 | 1 | 28 | 14 | 9 | 3 | 1642 | 2-14 | Tap | Tel Aviv |
| 3 | 6 | 1 | 14 | 14 | 9 | 11 | 40 | 1 | Faucet | Tel Aviv |
| 1 | 4 | 3 | 1 | 1 | 1 | 60 | 2284 | 1 | Cooling tower | Tel Aviv |
| 1 | 4 | 3 | 1 | 1 | 1 | 11 | 296 | 1 | Cooling tower | Tel Aviv |
| 1 | 4 | 3 | 1 | 1 | 1 | 1 | 1 | 1 | Cooling tower | Tel Aviv |
| 1 | 4 | 3 | 1 | 1 | 1 | 1 | 1 | 1 | Cooling tower | Tel Aviv |
| 1 | 4 | 3 | 1 | 1 | 1 | 1 | 1 | 1 | Cooling tower | Haifa |
| 1 | 4 | 3 | 1 | 1 | 1 | 1 | 1 | 1 | Faucet | Center |
| 1 | 4 | 3 | 1 | 1 | 1 | 1 | 1 | 1 | Cooling tower | South |
| 1 | 4 | 3 | 1 | 1 | 1 | 1 | 1 | 1 | Cooling tower | Tel Aviv |
| 1 | 4 | 3 | 1 | 1 | 1 | 1 | 1 | 1 | Faucet | North |
| 1 | 4 | 3 | 1 | 1 | 1 | 1 | 1 | 1 | Faucet | Tel Aviv |
| 1 | 4 | 3 | 1 | 1 | 1 | 1 | 1 | 1 | Hot tub | South |
| 1 | 4 | 3 | 1 | 1 | 1 | 1 | 1 | 1 | Cooling tower | Tel Aviv |
| 1 | 4 | 3 | 1 | 1 | 1 | 1 | 1 | 1 | Cooling tower | Tel Aviv |
| 1 | 4 | 3 | 1 | 1 | 1 | 1 | 1 | 1 | Cooling tower | Tel Aviv |
| 1 | 4 | 3 | 1 | 1 | 1 | 1 | 1 | 1 | Cooling tower | Haifa |
| 1 | 4 | 3 | 1 | 1 | 1 | 1 | 1 | 1 | Cooling tower | South |
| 1 | 4 | 3 | 1 | 1 | 1 | 1 | 1 | 1 | Cooling tower | Jerusalem |
| 1 | 4 | 3 | 1 | 1 | 1 | 1 | 1 | 1 | Cooling tower | Jerusalem |
| 1 | 4 | 3 | 1 | 1 | 1 | 1 | 1 | 1 | Cooling tower | Jerusalem |
| 1 | 4 | 3 | 1 | 1 | 1 | 1 | 1 | 1 | Cooling tower | South |
| 1 | 4 | 3 | 1 | 1 | 1 | 1 | 1 | 1 | Water line | Jerusalem |
| 1 | 4 | 3 | 1 | 1 | 1 | 1 | 1 | 1 | Hot tub | North |
| 1 | 4 | 3 | 1 | 14 | 1 | 1 | 286 | 1 | Cooling tower | Haifa |
| 1 | 6 | 3 | 10 | 1 | 1 | 11 | 284 | 1 | Cooling tower | Haifa |
| 1 | 6 | 3 | 10 | 1 | 1 | 11 | 284 | 1 | Cooling tower | Haifa |
| 11 | 14 | 16 | 25 | 7 | 13 | 206 | 1334 | 2-14 | Cooling tower | South |
| 11 | 14 | 16 | 25 | 7 | 13 | 206 | 1334 | 2-14 | Cooling tower | South |
| 11 | 14 | 16 | 25 | 7 | 13 | 206 | 1334 | 2-14 | Water line | North |
|  | 14 | 16 | 1 | 7 | 13 | 206 | * | 2-14 | Cooling tower | Tel Aviv |
|  | 4 | 16 | 1 | 7 | 13 | 206 | * | 2-14 | Cooling tower | Jerusalem |
| 5 | 1 | 22 | 30 | 6 | 10 | 213 | 2169 | 2-14 | Cooling tower | South |
| 5 | 1 | 22 | 30 | 6 | 10 | 1 | 856 | 2-14 | Cooling tower | South |

**Figure 2.** Similarity dendrogram of 78 *L. pneumophila* strains isolated from hotel and resort water systems in Israel. A phylogenetic tree was constructed using unweighted pair group method with arithmetic averages (UPGMA) clustering. The sequence type (ST), serogroup, source by category type, and region of isolation are indicated. The asterisks indicate untypeable isolates (failed *flaA* gene amplification).

While several *L. pneumophila* sequence types were distributed widely throughout the country (ST1, ST1642, and ST461), a number of strains have been found to be limited to certain geographical regions. Specifically, ST59, ST1326, and ST1641 were unique to the Jerusalem district, and ST1516 was only found in the Southern region. Moreover, the Southern and Jerusalem districts displayed the most diverse *L. pneumophila* population, with 11 and 10 different STs, respectively (Figure 3).

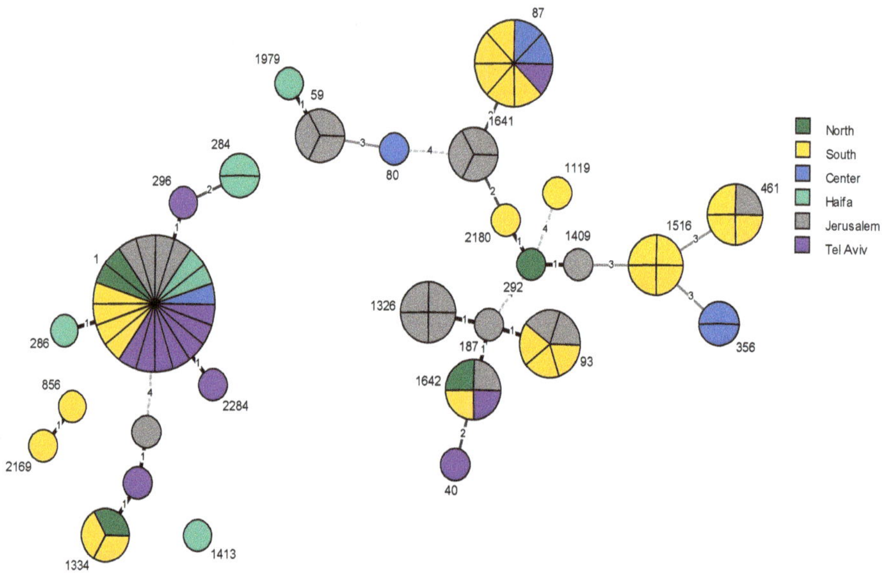


**Figure 3.** Minimum spanning tree (MST) based on sequence-based typing (SBT) profiles of 78 Lp isolates from hotel and resort water systems. Sequence type (ST) is indicated next to the circles, circle size is proportional to the number of isolates sharing the same ST, and each isolate is shown as a segment of the relevant circle. Branches connect the STs and show the genetic distance between them. STs differing by five or more alleles are not connected. Color-coding of the circles denotes geographic region. Nodes without an ST number represent strains closely related to ST1334 with failed *flaA* typing.

## 3. Discussion

The abundance of Lp in the tourism sector is a continuous focus of attention in *Legionella* research, due to its possible implications to public health. A summary of earlier publications reporting national surveys of tourist accommodations in different countries is presented in Table 3.

This study shows, for the first time, the distribution and prevalence of *Legionella* spp. in the Israeli hotel sector. By analyzing 2830 water specimens, taken from 168 hotels over the two-year period between 2015 and 2017, we demonstrate that 60% of the examined hotels were colonized with *L. pneumophila*, and in 37% of them, the concentrations of *Legionella* in water exceeded the national regulatory thresholds. Of all 2830 specimens collected, 17% were *Legionella*-positive, with half of those exceeding threshold levels of *Legionella*.

We analyzed the results of *Legionella* quantitation, according to the category of water source, which included cooling towers, hot tubs, waterlines, showering facilities, storage tanks, and room tap water. The most affected source type was cooling towers (38%), while specimens from other sources showed lower rates of *Legionella* colonization, at around 15%. Furthermore, 32% of the samples from cooling towers exceeded the 1000 CFU/L regulatory threshold for *Legionella* concentrations, making this the water source with the highest proportion of exceeding samples.

**Table 3.** Summary of earlier publications reporting national surveys of tourist accommodations in different countries.

| Author | Country of Study | Year of Publication | Geography | Sample Selection | Sample Size | Time of Study | Laboratory Methods |
|---|---|---|---|---|---|---|---|
| Borella et al. | Italy | 2005 | Five representative cities, northern, central, and southern Italian regions | The hotels were selected based on the water distribution systems in the cities, the characteristics of the buildings, and hotel cooperation. | 119 water samples from 40 hotels (3–5 samples from each hotel) | September 2003–July 2004 | *Legionella* isolation, enumeration and serotyping; PFGE analysis; physical and chemical water analyses |
| Lee et al. | South Korea | 2010 | Seven geographic regions throughout South Korea | The number of samples and isolates depended on the number of facilities located in each region | 4938 water samples from water systems of different settings, including hotels. | June–September 2008 | *Legionella* isolation, enumeration and serotyping; molecular identification of *L.* spp (16S rRNA, *mip*, or *rpoB*); SBT |
| Napoli et al. | Italy | 2010 | Southeastern Italy | Representative samples from different building types and water systems.<br>Re-inspection samples excluded. | 13,286 water samples, including 5009 samples from 305 hotels | January 2000–December 2009 | *Legionella* isolation, enumeration, and serotyping |
| Bonetta et al. | Italy | 2010 | Northern, central, and southern Italy | Samples representative of 18 towns and types of water systems. | 76 water samples from 19 hotels | October 2006–February 2007 | *Legionella* isolation, enumeration, and serotyping; real-time PCR; physical and chemical analyses |
| Chochlakis et al. | Greece | 2013 | Four regions of Crete island | Eight to 15 representative samples from each hotel, depending on hotel size and water system type. | 1494 water samples from 124 hotels | 2004–2011 | *Legionella* isolation, enumeration, and serotyping; molecular identification of *Legionella* spp (16S rRNA, *mip*); MALDI-TOF mass spectrometry; SBT; |
| Sepin Özen et al. | Turkey | 2017 | Antalya region | Samples from different water systems | 1403 water samples from 54 hotels | January–December 2010 | *Legionella* isolation, enumeration, and serotyping |

Cooling towers are the most frequently reported water source of LD outbreaks worldwide [26–28], and can involve a large number of cases [29,30]. The role of cooling towers in the urban spread of Lp has also been demonstrated recently in a genomic analysis of isolates over time in Switzerland [31]. The high proportion of Lp-contaminated cooling towers reported here is a public health concern that should prompt further investigation, due to the high population density in urban areas. However, in contrast to the reports from other countries, LD cases in Israel have not been linked to cooling towers. Since not all *Legionella* spp. and Lp strains are suggested to have the capacity to cause LD [32,33], this might be a reason for the discrepancy. It would be interesting, therefore, to look specifically at the population structure of *Legionella* spp. in cooling towers nationwide.

Serotyping of a subset of 162 presumed *Legionella* isolates revealed that 33% belonged to Lp sg1, while 64% belonged to Lp sg2–14. Several studies have explored the distribution of Lp sg1 in the environment. A study from South Korea demonstrated the significant predominance of Lp sg1 in manmade water systems, including hotels, with prevalence rates up to 55% [22]. In Italy, the Lp sg1 distribution rates in the hotel setting differed greatly between two studies, at 27.7% and 55%, including mixed cultures [15,34]. On the other hand, findings from Italy [19], Greece [23], and Turkey [21] have shown that the most frequent colonizers of the hotel water systems in these studies were Lp sg2–14.

A growing body of evidence shows the *Legionella* strains' ability for long-term persistence in manmade water systems, without a significant fluctuation of population diversity [12,35,36]. Based on this hypothesis, we assume that our findings reflect the rates of Lp sg1 distribution in Israeli hotels, though more investigation is needed to extend our knowledge on the persistence of local Lp strains in water systems associated with different settings.

Using SBT applied on a convenience sample of 78 Lp isolates, we have identified 27 STs, including two novel STs. Nine STs belonged to sg1. Lp sg1 ST1 was the prevalent type, accounting for 26% (20/78) of the sequence-typed isolates.

ST1 has been described by numerous studies, amongst a few other STs, as a main causative agent of LD globally, supporting its high pathogenicity [37,38]. Moreover, in contrast to other highly pathogenic clinical strains rarely isolated from the environment, ST1 has been shown to be among the predominant environmental Lp sg1 strains [22,39–43].

In our study, ST1 comprised 63% (20/32) of all sg1 isolates from hotel water systems. The high rate of the environmental predominance of Lp sg1 ST1 corresponds with our national surveillance data, where ST1 is by far the most common cause of LD in Israel [44]. This abundance of ST1 in the environment poses a challenge for public health services, limiting their capability to ultimately identify a source of infection during investigations of ST1-associated outbreaks using traditional SBT.

Amongst the non-sg1 isolates (46/78), two isolates failed to generate a full seven-allele profile, due to no amplification of a *flaA* PCR product. Lp strains with mutations at the SBT *flaA* primer-binding site have been described elsewhere [45], including an Lp subtype from Israel (0,14,16,25,7,13,206) that had been further identified by the whole genome sequencing (WGS) approach as having the *flaA11* allele, and which has been assigned to ST1334 [46]. However, the two strains found in this study differ from ST1334 (0,4,16,1,7,13,206 and 0,14,16,1,7,13,206), supporting the idea of an ongoing dissemination of the mutation.

Concerning the geographical distribution of the Lp population in this study, we observed a relative abundance of a number of strains in some districts in Israel. For example, sequence type ST1 was identified in each of the six Israeli districts, and ST1642 was found in four of the districts. On the contrary, other subtypes were associated with only one geographical region. For example, ST59, ST1326, and ST1641 were found in the Jerusalem district, and ST1516 was limited to the Southern region. Even though these strains are not unique to Israel, apart from ST1641, we observe their strong association with these two regions from our surveillance programs and during epidemiological investigations [47]. An explanation might be water-related differences between the regions (i.e., physical and chemical properties) caused by the climatic and topographical characteristics of the geographic regions. However, more data is needed to verify this assumption.

In our study, we have found an unexpectedly low rate of non-Lp spp. in the hotel water systems. In fact, we identified only one *L. bozemanii* strain from isolates subjected for serotyping. Other studies that have explored the environmental distribution of *Legionella* spp. have detected Lp and non-Lp co-existence in water. A recent study from the United States has demonstrated that 72 of culture-positive environmental samples collected during summer 2016, where Lp sg1 was recovered, also contained at least one other Lp or non-Lp *Legionella* spp. [30]. In another study from Crete, Greece [23], carried out from 2004 to 2011, about 50 non-Lp *Legionella* spp. were identified in the water systems of the hotel setting. Variability in the prevalence of non-Lp spp. between the studies can possibly be explained by the differences in the isolate selection procedure for subsequent analysis. In our study, the initial identification of *Legionella* spp. was carried out by serotyping, followed by the *mip* sequencing of non-groupable strains. In contrast, the application of PCR-based techniques for *Legionella* spp. screening on all samples would probably have yielded results that are more diverse. Moreover, the isolation processing methods used in our study may have reduced the detection of non-*pneumophila Legionella* spp., due to the overgrowth of other bacteria, and may have underestimated their overall abundance in samples. The membrane plating method is subject to some issues of overgrowth (especially in non-potable water), has been reinforced in a few studies [48,49], and is discussed in the new ISO 11731:2017 [50]. Thus, *Legionella* spp. distribution may not be fully represented here.

This study has several limitations. First, despite the considerable overall number of 2830 water specimens analyzed, several geographical regions, such as the Northern district, were underrepresented in our study. Second, the study was based on a convenience sample, and thus may not accurately represent the entire tourism sector in Israel, or in certain districts.

In this survey, a large set of water samples was examined routinely without any sampling efforts, due to sporadic travel-associated LD cases or outbreaks. Therefore, our findings on *Legionella* prevalence in hotel settings in Israel are fully representative of non-outbreak-related surveillance. Regarding the molecular structure of *L. pneumophila* population, this study demonstrates, for the first time, the molecular profile of Lp strains in the water systems of Israeli hotels and resorts.

Altogether, our findings contribute to the existing knowledge concerning the understanding of the environmental distribution of *Legionella* spp. in our region, and may facilitate international activities, such as TALD surveillance. The peculiar geographic distribution of different strains should be further investigated.

## 4. Materials and Methods

In total, 2830 convenience samples from 168 hotels and resorts were collected via routine surveillance, according to the regulations of the Israeli Ministry of Health for the prevention of *Legionella* growth in water distribution systems and hot tubs [51]. The study took place between March 2015 and the end of February 2017 across six Israeli districts (Northern, Center, Southern, Haifa, Tel Aviv, and Jerusalem). Hotels and other tourist accommodation are obliged by the regulations to monitor their water systems for the presence of *Legionella* spp. The minimum mandatory testing routine schedule depends on the hotel's size: once every two years for sites containing <50 rooms, once a year for those with 50–300 rooms, and twice a year for those with >300 rooms. Both hot and cold water distribution systems should be tested as part of this procedure. For hot tubs, the minimum sampling routine is quarterly. Selection of the sampling points depends on a hotel water system maintenance plan and is comprised of hot and cold water from outlets representing the rooms (faucets, showers) and mains (hot water return lines, hot and cold water supply, and storage tanks): cold water from cooling towers, decorative fountains, pools, air conditioning systems, and cold/hot/mixed water from hot tubs. Samples were taken after flushing for 2 minutes and the disinfection of the outlet, as per the requirements of the Israeli Public Health guidelines for routine monitoring of water distribution systems [51]. In addition, following regulatory requirements, water systems with *Legionella* concentrations above the thresholds of 1000 CFU/L for potable water and 1 CFU/100 mL for hot tubs were re-tested after the appropriate treatment [51]. Overall, of the 168 hotel water systems included in this study, 119 were probed at least twice. At each sampling point, 1–2 water samples (hot and/or cold water) were collected in 1 L sterile plastic bottles containing sodium thiosulfate, in order to neutralize the residual-free chlorine. All water samples were stored at 4 °C and processed within 24 h of their collection.

The detection and quantitation of *Legionella* spp. were performed in a certified water testing laboratory per the ISO 11731-2:2004 method [52]. Potable and non-potable water samples were filtered with 0.45 μm sterile gray membrane filter paper, treated with 30 mL of acid buffer containing 0.2 M KCL and 0.2 M HCL for 5 min, and washed with 20 mL of PAGE's saline. Water samples originated from cooling towers and fountains were processed in four dilutions (1:10000, 1:1000, 1:100, and 1:10), in order to avoid the overgrowth of microbial flora. Membranes were transferred to Glycine Vancomycin Polymyxin Cycloheximide (GVPC) medium (cat. no. 257007, BD, Heidelberg, Germany) and after incubation at 35 ± 0.5 °C for 10 days, colonies suggestive of *Legionella* spp. were subcultured to Buffered Charcoal Yeast Extract (BCYE) and 5% sheep blood agar media (P073 and P049, HyLabs, Rehovot, Israel). Subsets of representative isolates identified as *Legionella* spp. were regularly referred to the National Reference Laboratory for *Legionella* at the Ministry of Health, according to regulations [51].

The total amount of 164 *Legionella* isolates from hotel water systems was obtained during the two-year study. Serotyping was performed with the Legionella Latex Test kit (Cat. No. DR0800, Oxoid, Basingstoke, UK).

Strains not readily confirmed by serotyping as *L. pneumophila* were identified to species level by sequencing the *mip* gene, as described by Ratcliff et al. [7], and comparing the sequence to the *mip* database [53]. The molecular characterization of *L. pneumophila* strains was conducted according to the European Society of Clinical Microbiology and Infectious Diseases (ESCMID) Study Group for *Legionella* Infections (ESGLI) sequence-based typing (SBT) scheme [54,55]. The choice of the isolates subjected to SBT monthly was based on the data provided by the referring laboratory and guided by epidemiological and risk assessment criteria, such as high *Legionella* CFU counts, a source type with high public health risk potential, or a new sampling site. After the exclusion of duplicate isolates arising from the same sampling points, 78 isolates were examined in this study. Sequences obtained by Sanger sequencing were analyzed with the BioNumerics software (Version 7.6, Applied Maths) and compared to the ESGLI database for assigning the ST. New allelic profiles were submitted to the ESGLI SBT database [56]. The strain diversity index was calculated according to the modified method of Hunter and Gaston [57].

BioNumerics software (Version 7.6, Applied Maths) was used for phylogenetic analysis. Clustering was created using the unweighted pair group method with arithmetic averages (UPGMA) [58]. The minimum spanning tree (MST) was created using a predefined MST for the categorical data template, with single- and double-locus variance priority rules. Geomap was created using ArcGIS Pro 2.5 (Esri, Redlands, CA, USA).

**Author Contributions:** Conceptualization, J.M.-G.; methodology, E.Y. and J.M.-G.; formal analysis, L.V. and E.K.; investigation, E.K.; resources, V.A. and I.G.; data curation, E.Y. and E.K.; writing—original draft preparation, E.Y.; writing—review and editing, J.M.-G. and L.V.; supervision, L.V.; funding acquisition, V.A. and I.G. All authors have read and agreed to the published version of the manuscript.

**Funding:** This research received no external funding.

**Conflicts of Interest:** The authors declare no conflict of interest.

## References

1. Fields, B.S.; Benson, R.F.; Besser, R.E. Legionella and Legionnaires' disease: 25 years of investigation. *Clin. Microbiol. Rev.* **2002**, *15*, 506–526. [CrossRef] [PubMed]
2. Oliva, G.; Sahr, T.; Buchrieser, C. The Life Cycle of L. pneumophila: Cellular Differentiation Is Linked to Virulence and Metabolism. *Front. Cell. Infect. Microbiol.* **2018**, *8*, 3. [CrossRef] [PubMed]

3. McDade, J.E.; Shepard, C.C.; Fraser, D.W.; Tsai, T.R.; Redus, M.A.; Dowdle, W.R. Legionnaires' disease: Isolation of a bacterium and demonstration of its role in other respiratory disease. *N. Engl. J. Med.* **1977**, *297*, 1197–1203. [CrossRef] [PubMed]
4. Phin, N.; Parry-Ford, F.; Harrison, T.; Stagg, H.R.; Zhang, N.; Kumar, K.; Lortholary, O.; Zumla, A.; Abubakar, I. Epidemiology and clinical management of Legionnaires' disease. *Lancet Infect. Dis.* **2014**, *14*, 1011–1021. [CrossRef]
5. Khodr, A.; Kay, E.; Gomez-Valero, L.; Ginevra, C.; Doublet, P.; Buchrieser, C.; Jarraud, S. Molecular epidemiology, phylogeny and evolution of Legionella. *Infect. Genet. Evol.* **2016**, *43*, 108–122. [CrossRef]
6. Burillo, A.; Pedro-Botet, M.L.; Bouza, E. Microbiology and Epidemiology of Legionnaire's Disease. *Infect. Dis. Clin. N. Am.* **2017**, *31*, 7–27. [CrossRef]
7. Ratcliff, R.M.; Lanser, J.A.; Manning, P.A.; Heuzenroeder, M.W. Sequence-based classification scheme for the genus Legionella targeting the mip gene. *J. Clin. Microbiol.* **1998**, *36*, 1560–1567. [CrossRef]
8. Yu, V.L.; Plouffe, J.F.; Pastoris, M.C.; Stout, J.E.; Schousboe, M.; Widmer, A.; Summersgill, J.; File, T.; Heath, C.M.; Paterson, D.L.; et al. Distribution of Legionella species and serogroups isolated by culture in patients with sporadic community-acquired legionellosis: An international collaborative survey. *J. Infect. Dis.* **2002**, *186*, 127–128. [CrossRef]
9. Bartram, J.; Chartier, Y.; Lee, J.V.; Pond, K.; Surman-Lee, S. *Legionella and the Prevention of Legionellosis*; World Health Organization: Geneva, Switzerland, 2007.
10. ECDC. Legionnaires' Disease—Annual Epidemiological Report for 2015. Available online: https://www.ecdc.europa.eu/sites/portal/files/documents/AER_for_2015-legionnaires-disease_0.pdf (accessed on 21 November 2017).
11. Smith, S.S.; Ritger, K.; Samala, U.; Black, S.R.; Okodua, M.; Miller, L.; Kozak-Muiznieks, N.A.; Hicks, L.A.; Steinheimer, C.; Ewaidah, S.; et al. Legionellosis Outbreak Associated With a Hotel Fountain. *Open Forum Infect. Dis.* **2015**, *2*, ofv164. [CrossRef]
12. Ahlen, C.; Aas, M.; Krusnell, J.; Iversen, O.J. A single Legionella pneumophila genotype in the freshwater system in a ship experiencing three separate outbreaks of legionellosis in 6 years. *Microb. Ecol. Health Dis.* **2016**, *27*, 31148.
13. Sánchez-Busó, L.; Guiral, S.; Crespi, S.; Moya, V.; Camaró, M.L.; Olmos, M.P.; Adrián, F.; Morera, V.; González-Morán, F.; Vanaclocha, H.; et al. Genomic Investigation of a Legionellosis Outbreak in a Persistently Colonized Hotel. *Front. Microbiol.* **2016**, *6*, 1556. [CrossRef] [PubMed]
14. Chochlakis, D.; Sandalakis, V.; Keramarou, M.; Tselentis, Y.; Psaroulaki, A. Legionellosis: A Walk-through to Identification of the Source of Infection. *Cent. Eur. J. Public Health* **2017**, *25*, 235–239. [CrossRef] [PubMed]
15. Borella, P.; Montagna, M.T.; Stampi, S.; Stancanelli, G.; Romano-Spica, V.; Triassi, M.; Marchesi, I.; Bargellini, A.; Tato, D.; Napoli, C.; et al. Legionella Contamination in Hot Water of Italian Hotels. *Appl. Environ. Microbiol.* **2005**, *71*, 5805–5813. [CrossRef] [PubMed]
16. Leoni, E.; De Luca, G.; Legnani, P.P.; Sacchetti, R.; Stampi, S.; Zanetti, F. Legionella waterline colonization: Detection of Legionella species in domestic, hotel and hospital hot water systems. *J. Appl. Microbiol.* **2005**, *98*, 373–379. [CrossRef]
17. Mouchtouri, V.A.; Rudge, J.W. Legionnaires' Disease in Hotels and Passenger Ships: A Systematic Review of Evidence, Sources, and Contributing Factors. *J. Travel Med.* **2015**, *22*, 325–337. [CrossRef]
18. Mouchtouri, V.; Velonakis, E.; Tsakalof, A.; Kapoula, C.; Goutziana, G.; Vatopoulos, A.; Kremastinou, J.; Hadjichristodoulou, C. Risk Factors for Contamination of Hotel Water Distribution Systems by Legionella Species. *Appl. Environ. Microbiol.* **2007**, *73*, 1489–1492. [CrossRef]
19. Bonetta, S.; Bonetta, S.; Ferretti, E.; Balocco, F.; Carraro, E. Evaluation ofLegionella pneumophila contamination in Italian hotel water systems by quantitative real-time PCR and culture methods. *J. Appl. Microbiol.* **2010**, *108*, 1576–1583. [CrossRef]
20. Fragou, K.; Kokkinos, P.; Gogos, C.; Alamanos, Y.; Vantarakis, A. Prevalence of Legionella spp. in water systems of hospitals and hotels in South Western Greece. *Int. J. Environ. Health Res.* **2012**, *22*, 340–354. [CrossRef]
21. Sepin Özen, N.; Tuğlu Ataman, Ş.; Emek, M. Exploring the Legionella pneumophila positivity rate in hotel water samples from Antalya, Turkey. *Environ. Sci. Pollut. Res.* **2017**, *24*, 12238–12242. [CrossRef]

22. Lee, H.K.; Shim, J.I.; Kim, H.E.; Yu, J.Y.; Kang, Y.H. Distribution of Legionella Species from Environmental Water Sources of Public Facilities and Genetic Diversity of L. pneumophila Serogroup 1 in South Korea. *Appl. Environ. Microbiol.* **2010**, *76*, 6547–6554. [CrossRef]
23. Chochlakis, D.; Sandalakis, V.; Panoulis, C.; Goniotakis, I.; Makridaki, E.; Tselentis, Y.; Psaroulaki, A. Typing ofLegionella strains isolated from environmental samples in Crete, Greece, during the period 2004–2011. *J. Water Health* **2013**, *11*, 762. [CrossRef] [PubMed]
24. Katsiaflaka, A.; Pournaras, S.; Kristo, I.; Mouchtouri, V.A.; Kyritsi, M.; Velonakis, E.; Vatopoulos, A.C.; Hadjichristodoulou, C. Epidemiological Investigation of Legionella pneumophila Serogroup 2 to 14 Isolates from Water Samples by Amplified Fragment Length Polymorphism and Sequence-Based Typing and Detection of Virulence Traits. *Appl. Environ. Microbiol.* **2016**, *82*, 6102–6108. [CrossRef] [PubMed]
25. Moran-Gilad, J.; Mentasti, M.; Lazarovitch, T.; Huberman, Z.; Stocki, T.; Sadik, C.; Shahar, T.; Anis, E.; Valinsky, L.; Harrison, T.G.; et al. Molecular epidemiology of Legionnaires' disease in Israel. *Clin. Microbiol. Infect.* **2014**, *20*, 690–696. [CrossRef] [PubMed]
26. Weiss, D.; Boyd, C.; Rakeman, J.L.; Greene, S.K.; Fitzhenry, R.; McProud, T.; Musser, K.; Huang, L.; Kornblum, J.; Nazarian, E.J.; et al. A Large Community Outbreak of Legionnaires' Disease Associated With a Cooling Tower in New York City, 2015. *Public Health Rep.* **2017**, *132*, 241–250. [CrossRef]
27. Fitzhenry, R.; Weiss, D.; Cimini, D.; Balter, S.; Boyd, C.; Alleyne, L.; Stewart, R.; McIntosh, N.; Econome, A.; Lin, Y.; et al. Legionnaires' Disease Outbreaks and Cooling Towers, New York City, New York, USA. *Emerg. Infect. Dis.* **2017**, *23*, 1769. [CrossRef]
28. Petzold, M.; Ehricht, R.; Slickers, P.; Pleischl, S.; Brockmann, A.; Exner, M.; Monecke, S.; Luck, C. Rapid genotyping of Legionella pneumophila serogroup 1 strains by a novel DNA microarray-based assay during the outbreak investigation in Warstein, Germany 2013. *Int. J. Hyg. Environ. Health* **2017**, *220*, 673–678. [CrossRef]
29. Bassett, M.T.; Balter, S. Regulating Cooling Towers to Prevent Outbreaks of Legionnaires' Disease. *Public Health Rep.* **2017**, *132*, 133–135. [CrossRef]
30. Llewellyn, A.C.; Lin, B.; Lucas, C.E.; Roberts, S.E.; Brown, E.W.; Nayak, B.S.; Raphael, B.H.; Winchell, J.M. Distribution of Legionella and bacterial community composition among regionally diverse US cooling towers. *PLoS ONE* **2017**, *12*, e0189937. [CrossRef]
31. Wuthrich, D.; Gautsch, S.; Spieler-Denz, R.; Dubuis, O.; Gaia, V.; Moran-Gilad, J.; Hinic, V.; Seth-Smith, H.M.; Nickel, C.H.; Tschudin-Sutter, S.; et al. Air-conditioner cooling towers as complex reservoirs and continuous source of Legionella pneumophila infection evidenced by a genomic analysis study in 2017, Switzerland. *Eur. Commun. Dis. Bull.* **2019**, *24*, 1800192. [CrossRef]
32. Amemura-Maekawa, J.; Kikukawa, K.; Helbig, J.H.; Kaneko, S.; Suzuki-Hashimoto, A.; Furuhata, K.; Chang, B.; Murai, M.; Ichinose, M.; Ohnishi, M.; et al. Distribution of monoclonal antibody subgroups and sequence-based types among Legionella pneumophila serogroup 1 isolates derived from cooling tower water, bathwater, and soil in Japan. *Appl. Environ. Microbiol.* **2012**, *78*, 4263–4270. [CrossRef]
33. Lawrence, A.; Eglezos, S.; Huston, W. Environmental Legionella spp. collected in urban test sites of South East Queensland, Australia, are virulent to human macrophages in vitro. *Res. Microbiol.* **2016**, *167*, 149–153. [CrossRef] [PubMed]
34. Napoli, C.; Fasano, F.; Iatta, R.; Barbuti, G.; Cuna, T.; Montagna, M.T. Legionella spp. and legionellosis in southeastern Italy: Disease epidemiology and environmental surveillance in community and health care facilities. *BMC Public Health* **2010**, *10*, 660. [CrossRef] [PubMed]
35. Perola, O.; Kauppinen, J.; Kusnetsov, J.; Karkkainen, U.M.; Luck, P.C.; Katila, M.L. Persistent Legionella pneumophila colonization of a hospital water supply: Efficacy of control methods and a molecular epidemiological analysis. *APMIS* **2005**, *113*, 45–53. [CrossRef] [PubMed]
36. Cooper, I.R.; White, J.; Mahenthiralingam, E.; Hanlon, G.W. Long-term persistence of a single Legionella pneumophila strain possessing the mip gene in a municipal shower despite repeated cycles of chlorination. *J. Hosp. Infect.* **2008**, *70*, 154–159. [CrossRef]
37. Vekens, E.; Soetens, O.; De Mendonca, R.; Echahidi, F.; Roisin, S.; Deplano, A.; Eeckhout, L.; Achtergael, W.; Pierard, D.; Denis, O.; et al. Sequence-based typing of Legionella pneumophila serogroup 1 clinical isolates from Belgium between 2000 and 2010. *Eurosurveillance* **2012**, *17*, 20302.

38. David, S.; Rusniok, C.; Mentasti, M.; Gomez-Valero, L.; Harris, S.R.; Lechat, P.; Lees, J.; Ginevra, C.; Glaser, P.; Ma, L.; et al. Multiple major disease-associated clones of Legionella pneumophila have emerged recently and independently. *Genome Res.* **2016**, *26*, 1555–1564. [CrossRef]
39. Qin, T.; Zhou, H.; Ren, H.; Guan, H.; Li, M.; Zhu, B.; Shao, Z. Distribution of sequence-based types of legionella pneumophila serogroup 1 strains isolated from cooling towers, hot springs, and potable water systems in China. *Appl. Environ. Microbiol.* **2014**, *80*, 2150–2157. [CrossRef]
40. Lévesque, S.; Lalancette, C.; Bernard, K.; Pacheco, A.L.; Dion, R.; Longtin, J.; Tremblay, C. Molecular Typing of Legionella pneumophila Isolates in the Province of Quebec from 2005 to 2015. *PLoS ONE* **2016**, *11*, e0163818. [CrossRef]
41. Kozak-Muiznieks, N.A.; Lucas, C.E.; Brown, E.; Pondo, T.; Taylor, T.H., Jr.; Frace, M.; Miskowski, D.; Winchell, J.M. Prevalence of sequence types among clinical and environmental isolates of Legionella pneumophila serogroup 1 in the United States from 1982 to 2012. *J. Clin. Microbiol.* **2014**, *52*, 201–211. [CrossRef]
42. Sanchez-Buso, L.; Coscolla, M.; Palero, F.; Camaro, M.L.; Gimeno, A.; Moreno, P.; Escribano, I.; Lopez Perezagua, M.M.; Colomina, J.; Vanaclocha, H.; et al. Geographical and Temporal Structures of Legionella pneumophila Sequence Types in Comunitat Valenciana (Spain), 1998 to 2013. *Appl. Environ. Microbiol.* **2015**, *81*, 7106–7113. [CrossRef]
43. Quero, S.; Parraga-Nino, N.; Barrabeig, I.; Sala, M.R.; Pedro-Botet, M.L.; Monso, E.; Jane, M.; Sabria, M.; Garcia-Nunez, M. Population structure of Environmental and Clinical Legionella pneumophila isolates in Catalonia. *Sci. Rep.* **2018**, *8*, 6241. [CrossRef] [PubMed]
44. Ministry of Health Central Laboratories Annual Report, pp. 66–67. Available online: https://www.health.gov.il/PublicationsFiles/LAB_JER2017.pdf (accessed on 25 May 2020).
45. Harrison, T.G.; Moran-Gilad, J.; Mentasti, M.; David, S.; Afshar, B.; Andersen, P.S.; Stegger, M.; Valinsky, L.; Yakunin, E.; Uldum, S.A. Use of whole genome sequence (WGS) data to investigate anomalies in the ESGLI Legionella pneumophila DNA-sequence based typing (SBT) method. In Proceedings of the 2nd ESCMID Study Group for Legionella Infections Meeting (ESGLI), Barcelona, Spain, 17–19 September 2014.
46. Qadreyah Al-Matawah, S.A.-Z. Søren Uldum Sequence-Based Typing for Legionella Pneumophila Isolated from Water Systems of Residential Facilities in Kuwait. *J. Environ. Sci.* **2016**, *2*, 016.
47. Yakunin, E.; Ohayon, S.; Schnaidman, B.; Marva, E.; Agmon, V.; Eizenkraft, A.; Wagnert, L.; Grotto, I.; Valinsky, L.; Moran-Gilad, J. Prevalence and Diversity of Legionella pneumophila in the Defence Setting. In Proceedings of the Israeli Society for Microbiology Annual Meeting, Beit Dagan, Israel, 13–14 September 2017.
48. Leoni, E.; Legnani, P.P. Comparison of selective procedures for isolation and enumeration of Legionella species from hot water systems. *J. Appl. Microbiol.* **2001**, *90*, 27–33. [CrossRef] [PubMed]
49. Kirschner, A.K.T. Determination of viable legionellae in engineered water systems: Do we find what we are looking for? *Water Res.* **2016**, *93*, 276–288. [CrossRef]
50. International Organization for Standardization. *Water Quality—Enumeration of Legionella*; ISO 11731:2017; ISO: Geneva, Switzerland, 2017.
51. Guidelines on Preventing the Proliferation of the Legionella Bacteria in Water Systems. Available online: https://www.health.gov.il/Subjects/Environmental_Health/drinking_water/Pages/Legionella.aspx (accessed on 25 May 2020).
52. International Organization for Standardization. *Water Quality-Detection and Enumeration of Legionella-Part 2: Direct Membrane Filtration method for Waters with Low Bacterial Counts*; ISO 11731-2:2004; ISO: Geneva, Switzerland, 2004.
53. Public Health England. mip Sequencing Database. Available online: https://www.gov.uk/guidance/phe-data-and-analysis-tools (accessed on 19 November 2018).
54. Gaia, V.; Fry, N.K.; Afshar, B.; Luck, P.C.; Meugnier, H.; Etienne, J.; Peduzzi, R.; Harrison, T.G. Consensus sequence-based scheme for epidemiological typing of clinical and environmental isolates of Legionella pneumophila. *J. Clin. Microbiol.* **2005**, *43*, 2047–2052. [CrossRef]
55. Ratzow, S.; Gaia, V.; Helbig, J.H.; Fry, N.K.; Luck, P.C. Addition of neuA, the gene encoding N-acylneuraminate cytidylyl transferase, increases the discriminatory ability of the consensus sequence-based scheme for typing Legionella pneumophila serogroup 1 strains. *J. Clin. Microbiol.* **2007**, *45*, 1965–1968. [CrossRef]

56. Public Health England. *L. pneumophila* Sequenced-Based Typing. Available online: https://webarchive.nationalarchives.gov.uk/20190501130700/http://bioinformatics.phe.org.uk/legionella/legionella_sbt/php/sbt_homepage.php (accessed on 20 November 2019).
57. Hunter, P.R.; Gaston, M.A. Numerical index of the discriminatory ability of typing systems: An application of Simpson's index of diversity. *J. Clin. Microbiol.* **1988**, *26*, 2465–2466. [CrossRef]
58. Sokal, R.R.; Michener, C.D. A statistical method for evaluating systematic relationships. *Univ. Kans. Sci. Bull.* **1958**, *38*, 1409–1438.

*pathogens*

MDPI

*Article*

# The Role of Sensor-Activated Faucets in Surgical Handwashing Environment as a Reservoir of *Legionella*

**Marta Mazzotta [1], Luna Girolamini [1], Maria Rosaria Pascale [1], Jessica Lizzadro [1], Silvano Salaris [1], Ada Dormi [2] and Sandra Cristino [1,*]**

1 Department of Biological, Geological, and Environmental Sciences, University of Bologna, via San Giacomo 12, 40126 Bologna, Italy; marta.mazzotta2@unibo.it (M.M.); luna.girolamini2@unibo.it (L.G.); mariarosaria.pascal2@unibo.it (M.R.P.); jessica.lizzadro2@unibo.it (J.L.); silvano.salaris@unibo.it (S.S.)

2 Department of Medical and Surgical Science, University of Bologna, via San Giacomo 12, 40126 Bologna, Italy; ada.dormi@unibo.it

* Correspondence: sandra.cristino@unibo.it; Tel.: +39-051-209-4811; Fax: +39-051-209-4829

Received: 3 May 2020; Accepted: 3 June 2020; Published: 5 June 2020

**Abstract:** Surgical handwashing is a mandatory practice to protect both surgeons and patients in order to control Healthcare-Associated Infections (HAIs). The study is focused on *Legionella* and *Pseudomonas aeruginosa* contamination in Surgical Handwashing Outlets (SHWOs) provided by sensor-activated faucets with Thermostatic Mixer Valves (TMVs), as correlated to temperature, technologies, and disinfection used. Samples were analyzed by standard culture techniques, comparing hot- and cold-water samples. *Legionella* isolates were typed by an agglutination test and by *mip* sequencing. *Legionella* contamination showed the same distribution between hot and cold samples concerning positive samples and mean concentration: 44.5% and 1.94 $Log_{10}$ cfu/L vs. 42.6% and 1.81 $Log_{10}$ cfu/L, respectively. Regarding the distribution of isolates (*Legionella pneumophila* vs. *Legionella* non-*pneumophila* species), significant differences were found between hot- and cold-positive samples. The contamination found in relation to ranges of temperature showed the main positive samples (47.1%) between 45.1–49.6 °C, corresponding to high *Legionella* concentrations (2.17 $Log_{10}$ cfu/L). In contrast, an increase of temperature (>49.6 °C) led to a decrease in positive samples (23.2%) and mean concentration (1.64 $Log_{10}$ cfu/L). A low level of *Pseudomonas aeruginosa* was found. For SHWOs located in critical areas, lack of consideration of technologies used and uncorrected disinfection protocols may lead to the development of a high-risk environment for both patients and surgeons.

**Keywords:** Healthcare-Associated Infections (HAIs); Surgical Handwashing Outlets (SHWOs); sensor-activated faucets; *Legionella* spp.; risk assessment plan

## 1. Introduction

Nosocomial infections, also known as Healthcare-Associated Infections (HAIs), are defined as infections which were absent at the time of hospital admission that a patient acquires during their stay in a hospital or other healthcare facilities [1]. Populations that are at risk for HAIs are immunocompromised patients in Intensive Care Units (ICUs), those in burn units, those undergoing organ transplants, or older patients and neonates. Extensive studies have been carried out by the World Health Organization (WHO) showing that the most frequent nosocomial infections globally include catheter-associated urinary tract infections, central-line associated bloodstream infections, ventilator-associated pneumonia, and surgical site infections [1].

It has been estimated that, in Italy, 5–8% of hospitalized patients contract nosocomial infections every year and 450,000–700,000 HAIs occur in hospitalized patients; these data refer to urinary infections, followed by infections of surgical wounds, pneumonia, and sepsis [2,3].

Risk factors that promote nosocomial infections—other than patient susceptibility, such as immunosuppressed patients in ICUs—include poor hygienic conditions such as improper hand hygiene of Healthcare Staff (HCS) or contaminated air and water [1]. The water supply system in hospitals may constitute a source of HAIs caused by opportunistic pathogens such as *Pseudomonas aeruginosa* (*P. aeruginosa*), *Legionella* spp., *Acinetobacter* species, and fungi [4,5]. These organisms are transmitted by direct or indirect contact with water or by inhalation of aerosol generated by a water source [6–8]. *Legionella* spp. are ubiquitous aquatic organisms associated with community-acquired pneumoniae as well as hospital-acquired pneumonia. Direct inhalation of aerosols from environmental colonization is typically the source of infection. As *Legionella* infection is not spread between humans, environmental monitoring of potable water, cooling towers, and related sources is crucial to control the incidence of disease. *Legionella* is able to survive for long periods in water and even to replicate in the presence of disinfectants and some conditions (e.g., pipeline materials, stagnation and sludge formation, parasitism of amoebas and protozoic cysts, and so on) [9].

In recent years, the increasing incidence of both nosocomial and community-acquired *Legionella* infections has been a major public health concern: in 2018, 2964 cases were notified to the National Surveillance System in Italy, with an incidence of 48.9 cases per million inhabitants with lethality rate for community and healthcare cases of 10.9% and 51.7%, respectively [10].

The risk of illness increases dramatically if the germ is found in certain wards such as ICUs, hematology-oncology units, cardiology units, hemodialysis units, and pulmonology units due to the critical nature of these wards for their hospitalized patients [11]. Nevertheless, the real risk of other sources of infection remains partially underestimated when making a correct *Legionella* risk assessment plan in water systems, such as suggested by the Italian Guidelines as the correct strategy to minimize the risk of colonization [12].

Different guidelines and studies have suggested that water outlets for handwashing in hospitals are frequently contaminated with *P. aeruginosa* and other Gram-negative bacteria, such as *Legionella*, which have been linked to nosocomial infections [8,13]. In particular, the presence of *Legionella* in outlets poses a risk of infection during handwashing practices due to aerosol generation.

The key factors for prevention of HAIs in the surgical area are associated with hand hygiene, surgeon handwashing characteristics, and appropriately timed glove use. Hand hygiene is an extremely important measure implemented to reduce HAIs; the WHO published guidelines in 2006 and in 2009 for routine and surgical hand hygiene protocols directed to control resident flora as well as transient microflora [14,15]. There are two primary methods for hand hygiene: antimicrobial or non-antimicrobial soap and water scrub, called the "scrub method", and Alcohol-Based Hand Rub, called the "rub method" [16]. Concerning the surgeon handwashing station characteristics, they are generally made entirely of stainless steel with a tank made of a single plate to guarantee the continuity of the surfaces and to avoid all possible areas of bacterial proliferation (e.g., spaces or grooves). The front part is slanted by 30° in order to prevent splashing and direct water contact with operators [17].

Moreover, surgical handwashing points are equipped with two main types of faucets: manual faucets, with a long clinical lever that dispenses and mixes water by use of the elbow or foot to avoid direct contact with the hands, or non-touch water taps, provided with photocell-operated water supply as electronically managed by a photocell sensor, some of them provided by Thermostatic Mixer Valves (TMVs) [18].

Non-touch water taps, also called sensor-activated faucets with TMVs, have been gradually introduced into private and public hospital facilities to prevent patients or HCS from risk of acquiring infection or transferring infection during surgical procedures by touching contaminated taps. These taps work only when the hands are put in front of a magnetic/sensor valve which causes water to flow out and, when hands are removed, the water flow to stop. The presence of a TMV permits the flushing of water through a single pipeline at a fixed temperature (generally about 36 °C). The mixing is due to the presence of a cartridge which is able to recall cold water, leading to the desired temperature when mixed with hot water. Hospitals and other healthcare facilities where hygienic measures are very important have started to install this type of touch-free tap system to promote lower water consumption, thus saving costs and preventing HCS from potential recontamination upon hand contact with faucet valves [18]. However, there are no current data that support a decrease in HAIs associated with the use of non-touch water taps [19].

Periodic monitoring of the presence of *Legionella* or other waterborne pathogens in all outlets used for hand hygiene—in particular, during the preoperative phases of hand hygiene in surgeons—represents a preventive measure to avoid handwashing contamination before starting surgical procedures and to control the possible exposure of patients and health professionals.

Our research is presented as the result of a *Legionella* environmental monitoring program, conducted from 2013 to 2019 in 11 hospitals located in different regions of Italy. The analysis of data has identified, as critical points, 52 Surgical Handwashing Outlets (SHWOs) provided by sensor-activated faucets with TMVs with high levels of *Legionella* contamination.

The focus of the study is the analysis of microbiological contamination of SHWOs concerning *Legionella* and *P. aeruginosa*, comparing hot- and cold-water samples supplied by a municipal distribution system. The data obtained are also studied in relation to the SHWO temperatures measured as well as compare the SHWOs technologies—sensor-activated faucets with TMVs versus manual clinical valves without TMVs—to understand the key elements of contamination that could develop a reservoir for *Legionella* and could enhance the risk of infection.

## 2. Results

All results are presented, first of all, by considering the general contamination found in SHWOs and, then later, by dividing *Legionella* contamination between hot- and cold-water samples. The data about *Legionella* concentration are expressed in $Log_{10}$ cfu/L (Log cfu/L).

The same method is used to correlate the microbial contamination found with temperature values measured in SHWOs and their distribution between hot- and cold-water samples.

### 2.1. *Legionella Contamination in SHWOs*

The results of mean *Legionella* concentrations found in 52 SHWOs from 11 hospitals are shown in Figure 1. Seven of the hospitals showed *Legionella* contamination (7/11, 63.6%), where three (3/7, 42.8%) of them showed values over the level of risk indicated by Italian Guidelines, that is, at >100 cfu/L (>2 Log cfu/L) [12]. The contamination was found in hot or cold samples and in both water distribution systems for each hospital.

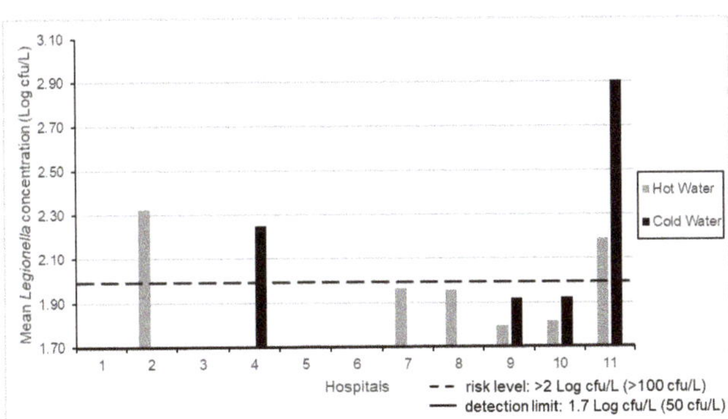


**Figure 1.** Mean *Legionella* concentrations in 11 hospitals.

The results of microbial contamination from 669 SHWO samples show that *Legionella* was detected in 293/669 (43.8%) of samples.

An analysis of *Legionella* contamination was then performed between hot-water ($n$ = 427) and cold-water samples ($n$ = 242). The differences between the numbers of hot- and cold-water samples were linked to a higher concentration of *Legionella* found in hot-water samples which, according to the suggestions of the Italian Guidelines, requires resampling from the same positive outlets [12].

In particular, the analysis of results between hot- and cold-water distribution systems showed 190/427 (44.5%) of positive hot-water samples and 103/242 (42.6%) positive cold-water samples. The positive samples over the *Legionella* level of risk (>2 Log cfu/L) were 140/190 (73.7%) for hot- and 70/103 (68.0%) for cold-water samples.

In Table 1, the data of mean temperature and disinfectant residue with relative minimum (min) and maximum (max) values, the percentage of *Legionella* positive samples, mean concentrations, and the range of contamination (min–max) found in hot and cold-water samples are listed, respectively. Data about temperature, disinfectant residues, and *Legionella* concentration are expressed as mean ± Standard Deviation (SD).

**Table 1.** Surgical Handwashing Outlet (SHWO) microbiological and physical-chemical parameters measured: hot- vs. cold-water samples.

| SHWO Distribution Systems | Temperature Mean ± SD (Min–Max) (°C) | $H_2O_2$ Residue Mean ± SD (Min–Max) (mg/L) | Number of Total SHWO Water Samples | Number of *Legionella*-Positive Samples/Total Samples (%) | *Legionella* Concentration Mean ± SD (Min–Max) (Log cfu/L) | Mean *Legionella* Concentration Comparison Hot vs. Cold Samples Mann–Whitney Test *p*-Value (*p*) |
|---|---|---|---|---|---|---|
| Hot water samples | 47.7 ± 4.95 (21.9–60.1) | 10 ± 6.67 (5–25) | 427 | 190/427 (44.5) | 1.94 ± 1.07 (1.70–5.8) | 0.34 |
| Cold water samples | 19.1 ± 4.38 (9.2–44.7) | 2.5 ± 1.5 (0.5–5) | 242 | 103/242 (42.6) | 1.81 ± 0.88 (1.70–4.7) | |

No significant difference ($p$ = 0.34) is found between hot and cold samples concerning *Legionella* levels.

Regarding the *Legionella* isolates distribution in SHWOs between hot- and cold-positive samples, the results showed samples contaminated only by *Legionella pneumophila (L. pneumophila)*, samples contaminated only by *Legionella* non-*pneumophila* species (other *Legionella* spp.) and others contaminated by both species. Significant differences ($p$ = 0.001), obtained with the statistical $\chi^2$ test, were found concerning the *Legionella* spp. distribution between hot and cold samples as follows: in hot-water samples, the main isolate belonged to *L. pneumophila* 123/190 (64.7%), followed by samples with both species (*L. pneumophila* and other *Legionella* spp.) 41/190 (21.6%) and, finally, by 26/190 (13.7%) showing only the presence of other *Legionella* spp. In cold-water samples, we found the same trend, with 44/103 (42.7%) of samples with *L. pneumophila*, 30/103 (29.1%) contaminated by both species, and finally, 29/103 (28.1%) with only other *Legionella* spp. The isolates of *L. pneumophila* were identified by an agglutination test as belonging to serogroups 1, 3, 4, 6, and 8. The typing of *Legionella* non-*pneumophila* species by *mip* gene sequencing, indicated the presence of *Legionella anisa (L. anisa)*, *Legionella rubrilucens (L. rubrilucens)*, *Legionella tauriniensis (L. tauriniensis)*, *Legionella nautarum (L. nautarum)*, and *Legionella steelei (L. steelei)*.

The study of *Legionella* isolates in terms of mean concentration ± standard deviation (Log cfu/L ± SD) between hot- and cold-positive samples is presented in Table 2. Multiple comparisons were performed between isolates found in hot- and cold-water samples (horizontal lines), while the comparison between hot- and cold-water samples for each type of *Legionella* isolate is shown in the columns. High *L. pneumophila* concentrations were found in hot-water samples (2.92 ± 1.08 Log cfu/L) with significant difference compared to samples colonized by only other *Legionella* spp. ($p$ = 0.03) and with respect to cold-water samples ($p$ = 0.008). In cold-water samples, despite a high other *Legionella* spp. mean concentration (2.47 ± 0.72 Log cfu/L), a significant difference was found only with respect to samples colonized by both species ($p$ = 0.0046).

**Table 2.** *Legionella* isolate mean concentration comparison in SHWOs: hot- vs. cold-water samples.

| *Legionella* Isolate | Samples with Only *L. pneumophila* Mean ± SD (Log cfu/L) | Samples with Only Other *Legionella* spp. Mean ± SD (Log cfu/L) | Samples with *L. pneumophila* and Other *Legionella* spp. Mean ± SD (Log cfu/L) | Legionella Isolate Mean Comparison in Hot and Cold Water Mann–Whitney Test *p*-Value (*p*) | |
|---|---|---|---|---|---|
| Hot water samples | 2.92 ± 1.08 | 2.31 ± 0.66 | 3.13 ± 0.85 | *L. pneumophila* vs. other *Legionella* spp. | 0.03 * |
| | | | | *L. pneumophila* vs. *L. pneumophila* and other *Legionella* spp. | 0.40 |
| | | | | Other *Legionella* spp. vs. *L. pneumophila* and other *Legionella* spp. | 0.00012 * |
| Cold water samples | 2.43 ± 0.83 | 2.47 ± 0.72 | 3.09 ± 0.63 | *L. pneumophila* vs. other *Legionella* spp. | 1.00 |
| | | | | *L. pneumophila* vs. *L. pneumophila* and Other *Legionella* spp. | 0.0012 * |
| | | | | Other *Legionella* spp. vs. *L. pneumophila* and Other *Legionella* spp. | 0.0046 * |
| Legionella Isolate Mean Comparison between hot vs. cold samples Mann–Whitney test *p*-value (*p*) | 0.008 * | 0.4 | 0.7 | | |

* Values are statistically significant at $p \leq 0.05$.

## 2.2. *Legionella* Contamination in Relation to Water Temperature

Regarding the temperature measured between hot and cold samples, we found a range between 21.9–60.1 °C (mean value of 47.7 °C) and a range between 9.2–44.7 °C (mean value of 19.1 °C) for hot and cold samples, respectively.

The *Legionella* contamination found considering all SHWOs samples was distributed in four ranges of temperature, which were linked to relevant considerations about the environment of *Legionella* as follows:

- the first range, called "I", represents the samples collocated at temperature values < 21 °C. This temperature range corresponds to the standard one for drinking water for human consumption;
- the second range, called "II", was 21–45 °C, corresponding to the mixed water produced by outlets provided by TMVs;
- the third range, called "III", corresponds to the range between 45.1–49.6 °C. This range represents the setting temperature generally measured during environmental monitoring on hot-water system producers (e.g., boilers, electric tanks, heater-exchangers, and so on), other than the values suggested to reduce energy costs [20]; and
- the fourth range, called "IV", corresponds to temperature values > 49.6 °C. This is the optimal value suggested by the Italian Guidelines to control *Legionella* proliferation in water-distribution systems.

A multiple comparison was performed between each range by an ANOVA test, showing significant differences, as indicated in Table 3 with the (*) symbol.

**Table 3.** Mean *Legionella* concentration in relation to ranges of temperature measured (I, II, III, and IV).

| Range of Temperature (°C) | | Number of Samples | Number of Positive Samples (%) | Mean *Legionella* Concentration (Log cfu/L) | 95% Confidence Interval (CI) | Range of Temperature Comparison | ANOVA Test *p*-Value (*p*) |
|---|---|---|---|---|---|---|---|
| I | <21 | 168 | 54 (32.1) | 1.78 | 1.65–1.91 | vs. II<br>vs. III<br>vs. IV | 0.464<br>0.002*<br>1.000 |
| II | 21–45 | 157 | 59 (37.6) | 1.98 | 1.81–2.15 | vs. I<br>vs. III<br>vs. IV | 0.464<br>0.474<br>0.012 * |
| III | 45.1–49.6 | 172 | 81 (47.1) | 2.17 | 2.00–2.34 | vs. I<br>vs. II<br>vs. IV | 0.002 *<br>0.474<br>0.001 * |
| IV | >49.6 | 172 | 40 (23.2) | 1.64 | 1.51–1.78 | vs. I<br>vs. II<br>vs. III | 1.000<br>0.012 *<br>0.001 * |

* Values are statistically significant at $p \leq 0.05$.

The contamination of samples in relation to the temperature measured during the sampling reveals that the main *Legionella* positive samples (47.1%) were in the third range (III), which was also the main contaminated source in terms of *Legionella* mean concentration (2.17 Log cfu/L). By contrast, the lowest percentage of positive samples (23.2%) and mean concentration (1.64 Log cfu/L) were found in the fourth range (IV).

In Figure 2, the distribution of mean *Legionella* concentration in relation to temperature values measured is represented, with hot and cold samples separately considered, in ranges between 21.9–60.1 °C (mean value of 47.7 °C) and between 9.2–44.7 °C (mean value of 19.1 °C).

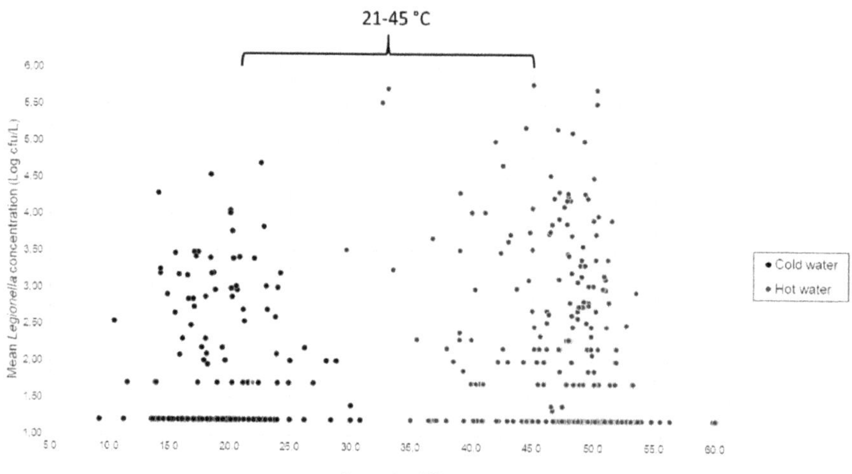


**Figure 2.** Mean *Legionella* concentration distribution in relation to water sample temperatures measured (°C).

An analysis of the results considering only samples in the range of 21–45 °C (e.g., the range for SHWO mixed water) showed 98/427 (23.0%) and 81/242 (33.5%) contaminated hot- and cold-water samples, with mean concentrations of 2.12 ± 1.22 Log cfu/L and 1.87 ± 0.92 Log cfu/L, respectively.

Considering only *Legionella*-positive samples, we found 52/98 (53.0%) in hot water—respectively 39/81 (48.1%) in cold water—with mean *Legionella* concentration higher in hot (2.94 ± 1.17 Log cfu/L) than cold samples (2.60 ± 0.87 Log cfu/L). The nonsignificant difference was found using the Mann–Whitney test ($p = 0.22$).

### 2.3. *Legionella* Contamination before and after the SHWO Replacement

In three hospitals (called 1, 8, and 11), following renovation works, replacement of sensor-activated faucets with TMVs by clinical valves without TMVs was carried out. The reassessment of *Legionella* contamination on the same SHWOs after replacement permitted us to observe changes in the *Legionella* concentration. Analyzing the contamination found in 110 of 669 total samples collected in these hospitals, we compared the contamination before ($n = 55$) and after ($n = 55$) replacement. As shown in Table 4, we observed a significant decrease in terms of *Legionella* contamination ($p = 0.001$) with the same significant trend in each hospital, other than with an increase of hot-water temperature and a consequent decrease of *Legionella* levels.

**Table 4.** Mean *Legionella* concentration in three hospitals before and after the replacement of sensor-activated faucets with Thermostatic Mixer Valves (TMVs).

| ID Hospitals | Number of SHWOs (Total Samples) | Time of Renovation Works of SHWOs (Total Samples) | Mean Temperature Samples (°C) | Number of *Legionella* Positive Samples/Total of Samples (%) | Number of *Legionella* Samples Over Risk Value/Positive Samples (%) | Mean *Legionella* Concentration ± SD (Log cfu/L) | t-Student and Wilcoxon Test *p*-Value (*p*) |
|---|---|---|---|---|---|---|---|
| 1 | 5 (28) | Before (14) | 42.73 | 5/28 (17.9) | 4/5 (80.0) | 1.98 ± 1.34 | 0.046 * |
| | | After (14) | 49.15 | 1/28 (3.6) | 0 | 1.23 ± 0.13 | |
| 8 | 3 (50) | Before (25) | 42.89 | 19/50 (38.0) | 13/19 (68.4) | 2.59 ± 1.34 | 0.001 * |
| | | After (25) | 46.98 | 5/50 (10.0) | 0 | 1.3 ± 0.20 | |
| 11 | 14 (32) | Before (16) | 48.96 | 15/32 (46.9) | 15/15 (100.0) | 3.03 ± 1.04 | 0.001 * |
| | | After (16) | 49.50 | 9/32 (28.1) | 5/9 (55.6) | 1.8 ± 0.86 | |

* Values are statistically significant at $p \leq 0.05$.

### 2.4. *P. aeruginosa* Contamination in SHWOs

The data about *P. aeruginosa* contamination indicated that 27/669 (4.0%) samples were contaminated. Considering the contamination in relation to hot- and cold-water circuits, we found a higher contamination in cold-water samples compared to hot-water samples: 22/242 (9.0%) and 5/427 (1.2%), respectively. However, the low number of positive samples did not permit us to find a statistical correlation between the data analyzed ($p = 0.65$).

### 2.5. Disinfectant Residue Analysis

Concerning the disinfectant residue measured, the mean concentration of hydrogen peroxide ($H_2O_2$) component was about 2.5 mg/L and 10 mg/L in cold- and hot-water samples, respectively. Although only the hot water network is treated with hydrogen peroxide/$Ag^+$ ($H_2O_2/Ag^+$), we found the presence of disinfectant residue in all cold-water samples, with a range between 0.5–5 mg/L.

## 3. Discussion

The prevention of HAIs is an important problem, particularly in high-risk patient care. The risk of infections has been linked to interactions between pathogens and hosts which involves the number of microorganisms, their virulence factors, and the host's immune defenses [21]. To reduce the impact on human health as well as to avoid economic, legal, and political issues, particular attention must be directed to a hospital's hygiene and environment. This aim can arise only through the development of a risk assessment plan which is linked to knowledge of the hospital and patient characteristics, the health-care procedures already in place and to be improved, and the hospital environment where patients and HCS may be in contact with microorganisms through the air, water, and contaminated surfaces [6,22,23].

The new revision of the European Drinking Water Directive, such as the WHO Guidelines for Drinking Water Quality, suggests the approach of the Water Safety Plan to identify the main pathogens involved in waterborne diseases, to understand their pathogenic pathways, and to contain their impact on public health [24–28]. *Legionella* and *P. aeruginosa* are two of the main waterborne pathogens involved in hospital environments associated with nosocomial infections [6,29,30].

This study reports knowledge acquired during a *Legionella* environmental surveillance program performed in hospitals, where high *Legionella* levels were detected in SHWOs with TMVs, some of them with concentrations over the risk level (>2 Log cfu/L), suggesting their critical role in bacterial growth and HAI risk. It has been well documented that temperature is a key factor in microbial growth and that, in particular, the mixing of hot and cold water creates an optimal temperature for bacterial environment, which can occur in SHWOs [8,23,31,32].

To analyze the contamination found in SHWOs, hot- and cold-water data sets were separately studied in terms of percentage of positive samples, level of contamination, and *Legionella* isolates distribution, including temperature as a possible determining factor for data fluctuations in the microbial parameters analyzed.

The results showed a similar percentage of hot- and cold-water samples (44.5% and 42.6%, respectively) contaminated by *Legionella*, with the same trend regarding samples over the *Legionella* risk level (73.7% hot vs. 68.0% cold). A nonsignificant difference in terms of *Legionella* contamination between hot and cold samples ($p = 0.34$) demonstrates how hot- and cold-water circuits are not separate with continuous mixing between two pipelines, creating an environment capable of supporting *Legionella* growth.

These results are supported by the residues of $H_2O_2$ disinfectant found also in cold-water samples. This disinfectant introduced in hospitals is injected only into the return line of the hot-water distribution system, and generally, when the two main distribution systems (e.g., hot and cold) are well separated, the cold water is expected to be free of disinfectant residues. This observation can be attributed to damage on the TMV cartridge because, during cold-water sampling, although the TMVs were deactivated, we found disinfectant residues in all samples. Moreover, damage in the TMV device was supported by the temperatures measured, which revealed a decrease in hot-water values and an increase in cold-water values, as demonstrated by the large ranges of temperature: 21.9–60.1 °C and 9.2–44.7 °C for hot and cold, respectively.

Considering the distribution of *Legionella* isolates, a significant difference was found between hot- and cold-water-positive samples ($p = 0.001$), showing that the characteristics of the mixed water produced are able to influence the distribution of isolates. According to knowledge about *Legionella* ecology and epidemiological data, the main positive samples found in hot and cold water (64.7% vs. 42.7%) belonged to *L. pneumophila* (serogroups 1, 3, 4, 6, and 8). In a low percentage of hot- and cold-water samples, we found isolates belonging to *Legionella* non-*pneumophila* species (*L. anisa*, *L. rubrilucens*, *L. tauriniensis*, *L. nautarum*, and *L. steelei*), with high values in cold water compared to hot water (28.1% vs. 13.7%). The same differences were found between cold and hot samples regarding the percentage of positive samples contaminated by both species (*L. pneumophila* and other *Legionella* spp.).

These data required supplementary analysis regarding the level of contamination found inside each distribution system and between them. In hot-water samples, we found a higher *Legionella* contamination in samples contaminated by both isolates, with a significant difference with respect to the level of contamination found in samples with only other *Legionella* spp. ($p = 0.00012$). A significant difference was, therefore, found in terms of the level of contamination between *L. pneumophila* and other *Legionella* spp. ($p = 0.03$).

In cold-water samples, we observed a different trend, with high samples contaminated by both species showing significant differences with respect to samples having only *L. pneumophila* ($p = 0.0012$) and samples contaminated by only other *Legionella* spp. ($p = 0.0046$).

Considering the comparison of mean concentration found for each isolate between hot versus cold samples, a significant difference was found only for *L. pneumophila* ($p = 0.008$).

Relevant information comes from these results regarding the ecology of isolates in water distribution systems.

*Legionella* lives in a water environment, with optimal growth in warm environments. Therefore, the abundance of *L. pneumophila* in hot-water samples found was in line with data about the high incidence of this species in human disease. In hot-water environments, there is likely a selective pressure of *L. pneumophila* on *Legionella* non-*pneumophila* species, which is suppressed in cold-water distribution systems, as demonstrated by the high number of samples with both species when the water temperature was mixed. Our hypothesis is also based on observations done during *Legionella* culture, where we generally find a lower *Legionella* non-*pneumophila* species isolation rate, due to their slow growth and late detection after 10–15 days of incubation when *L. pneumophila* is more abundant. When the culture technique was conducted up to 10 days, some of these species were missing and, consequently, underestimated; by contrast, an extension of culture timing permits their detection.

The poor awareness of these species and their underestimation is also associated to the low rate of clinical isolation, to their low correlation with human disease, and to the non-detection by diagnostic techniques (e.g., antigenic urinary tests) [33,34].

Another important point that can explain the high presence of *Legionella* non-*pneumophila* species in cold water is related to the disinfection treatment that often, as seen in this study, is performed on the hot-water circuit, leaving the cold-water distribution system without any type of control (monitoring by culture, temperature measures, flushing, and disinfectant residues measures). This represents a reservoir for other *Legionella* species. The absence of disinfectant or low levels of disinfectant residues measured usually require high temperatures for their activation and are unable to control their growth.

These results were also confirmed by our previous data [35] regarding the ability of *Legionella* to colonize and increase its concentration in cold-water distribution systems, inducing a change of cold water microflora; during renovation works, pipeline, TMV, and faucet damage; or when rapid breakdown of hot temperatures occurs. The presence of a high percentage of positive samples with high *Legionella* concentration contaminated by both species in both distribution systems confirms that SHWOs with mixed water develop an environment favorable to *Legionella* growth.

The high contamination of SHWOs are therefore supported by a wide fluctuation of temperatures found in samples: both low and high temperatures are able to favor bacteria growth. The analysis of contamination levels with respect to temperatures was analyzed by dividing the temperature values measured between four ranges, each of them associated to the ecology of *Legionella*.

The possibility to maintain separation between cold- and hot-water pipelines is one of the strategies suggested by National and European directives in order to contain the proliferation of bacteria. Our data demonstrated an inverse correlation between the temperature and bacteria load: at higher temperatures (>49.6 °C), a lower *Legionella* mean concentration (1.64 Log cfu/L) was observed, according with the directive's suggestions about the value of >50 °C being able to perform complete control of the level of *Legionella* [12].

The results obtained inside the II and III ranges of temperature (21–45 °C and 45.1–49.6 °C, respectively) showed approximately the same *Legionella* concentrations with a nonsignificant difference inside these ranges ($p = 0.474$). These data confirm that samples with temperature close to the optimum *Legionella* growth range (25–42 °C) are more contaminated and that an increase of temperature (>49.6 °C) leads to control of the *Legionella* proliferation (II vs. IV, $p = 0.012$; III vs. IV, $p = 0.001$) [36].

The contamination in SHWOs and the wide range of temperatures found can be explained, moreover, by taking into account the SHWO technology provided by hospitals. All of them are characterized by the presence of magnetic valves, which are the principal part of electronic/non-touch/sensor tap systems. Cold and hot water from the junctions of the central water pipeline system are mixed to provide an acceptable and comfortable setting temperature, generally, around 36 °C. The magnetic valves in the cartridge are made of material membranes—for example, made of rubber, plastic, or Polyvinyl Chloride (PVC)—which are very hard to disinfect and easily enhance bacterial growth and biofilm development, which can become a protective envelope against biocides and disinfectants. Furthermore, in these tap systems, flushing procedures are forbidden by the presence of a photocell system, leading to low water pressure and flow [18]. These considerations were supported by data about positive samples located in the range of the TMVs' working temperature (21–45 °C), which is very close to the temperature associated with optimal *Legionella* growth, where we did not find a difference between hot and cold samples.

Our hypothesis is strengthened furthermore by the observation that, in three hospitals which implemented a substitution program from sensor-activated faucets with TMVs to manual clinical valves without TMVs, an increase of mean temperature was measured, corresponding to a significant reduction trend in *Legionella* concentration levels.

As concerning *P. aeruginosa* SHWO contamination, the lower presence of positive samples coming from the eleven hospitals suggests the general good performance of disinfection procedures applied by hospital staff on faucets. The choice of tapware provided by faucet aerators guarantees low pressure without an internal thread, and descaling and disinfection procedures are applied weekly, permitting to avoid bacterial growth on outlets and preventing biofilm development [32].

The data regarding the higher *P. aeruginosa* contamination in cold-water samples can be explained by the same consideration as for sensor-activated faucets in *Legionella* contamination due to the sharing of these bacteria in the same habitat.

These findings led to the following considerations:

- the implementation of environmental monitoring in the cold-water distribution system, where *Legionella* surveillance is often missing, helps to explain the lower hot-water temperature sometimes observed also in hot water, which is often associated to damage in the mixing water system (e.g., in TMVs, levers, or faucets);
- the replacement of broken devices avoids the necessity of use of disinfection treatment in the whole distribution system, which can enhance bacterial resistance according to Berjeaud et al. [37]; and
- the continuous mixing between hot and cold water produced by TMVs leads to a mixture regarding the distribution of *Legionella* isolates in hot- and cold-water systems, as suggested by our data, developing a potential source of infection in cold water.

## 4. Materials and Methods

The eleven hospitals of this study, numbered 1 to 11, were involved in a *Legionella* environmental surveillance program from 2013 to 2019. After the introduction of last version of the Italian Guidelines in 2015, the 11 hospitals developed a risk assessment plan for *Legionella* control, considering the locations of buildings, their types of patients, and the water distribution system characteristics. All hospitals were supplied by municipal water that, after softener treatment, was heated by a heat-exchanger along with a hot-water return line.

All hospitals performed a six-month plan of *Legionella* environmental monitoring and active surveillance to control nosocomial *Legionella* infection by urinary-antigen test. Therefore, a complete program of maintenance procedures by measuring and recording temperatures, flushing outlet points, continuous disinfection of the system by $H_2O_2/Ag^+$, and a fortnightly plan regarding aerator disinfection and/or replacement was undertaken.

During environmental monitoring, we found a higher *Legionella* concentration in SHWOs with respect to other hospital outlets involved in monitoring, indicating the necessity of a supplementary investigation.

In Table 5, the number of SHWOs ($n$ = 52) in each hospital is reported, all of them equipped with sensor-activated faucets with TMVs (Figure 3). The main distribution system supplied hot-water outlets in a temperature range between 40–50 °C, while the cold-water outlets showed a temperature range of 15–20 °C. In SHWOs, the presence of TMVs produced a continuous mixed water at a set temperature around 36 °C.

**Table 5.** Number of SHWOs/hospitals.

| Number of SHWOs ($n$ = 52) | ID Hospitals | | | | | | | | | | |
|---|---|---|---|---|---|---|---|---|---|---|---|
| | 1 | 2 | 3 | 4 | 5 | 6 | 7 | 8 | 9 | 10 | 11 |
| | 5 | 5 | 5 | 1 | 6 | 2 | 2 | 3 | 4 | 5 | 14 |

**Figure 3.** SHWOs with sensor-activated faucets and TMVs (**a**) and a sensor-activated faucet (**b**).

During the study, eight hospitals had not implemented any replacement in SHWOs; however, three hospitals (1, 8, and 11) implemented a substitution program for their surgical hand preparation points regarding the faucet apparatuses: sensor-activated faucets with TMVs were removed and substituted with elbow-operated manual faucets without TMVs (Figure 4).

**Figure 4.** SHWOs with elbow-operated manual faucets without TMVs.

The environmental surveillance program consisted of *Legionella* and *P. aeruginosa* monitoring, according to the risk assessment plans provided by hospital healthcare directives.

The hot-water circuit in all hospitals was treated by $H_2O_2/Ag^+$ disinfectant, which was added by a pump proportionally to the volume of cold water supply at a concentration around 50 mg/L in order to allow a residue at outlets between 10–20 mg/L, following manufacturer's instructions.

To assess the complete monitoring of water microbiological quality supplied by SHWOs and to evaluate differences in terms of contamination between hot- and cold-water distribution systems, both circuits were sampled.

For the three hospitals that implemented a substitution program with manual clinical valves without TMVs, the data of cold-water samples were not available, as the risk assessment plan after replacement involved only hot SHWO samples; therefore, comparison in terms of *Legionella* contamination before and after the substitution program was considered only in hot-water circuits.

### 4.1. SHWO Sampling

According to the Italian Guidelines for the Prevention and Control of Legionellosis [12], analysis of *Legionella* contamination was performed by collecting two liters of cold and hot SHWO samples. In particular, in order to determine the quality in the main distribution system, post-flushing sampling was applied, which consisted of removing the filter or faucets, disinfection of taps with ethanol (70%), open taps, flushing for 2 min, and collection of cold before hot samples [38]. For cold samples, the TMVs were deactivated; by contrast, TMVs were reactivated to collect hot samples at the setting temperature for SHWOs (36 °C).

From the 52 SHWOs, 669 samples were collected (427 hot and 242 cold), where two liters of water were sampled using 1-liter sterile polytetrafluoroethylene (PTFE) bottles containing sodium thiosulphate (20 mg/L) [38,39].

The samples were processed by a membrane-filtration technique using polyethersulfone membrane filters with a porosity of 0.22 µm (Sartorius, Bedford, MA, USA), according to the International Standard Organization (ISO) 11731:2017 procedure [40].

### 4.2. *Legionella* and *P. aeruginosa* Culture and Typing

The *Legionella* culture was performed on Glycine-Polymyxin B-Vancomycin-Cycloheximide (GVPC) plates (Thermo Fisher Diagnostic, Basingstoke, UK) and subsequently incubated at 36 ± 1 °C with 2.5% $CO_2$. *Legionella* growth was evaluated every 2 days for a total of 15 days of culture.

After the incubation period, the colonies with morphologies associated to the *Legionella* genus were enumerated and five suspected colonies for each morphology, as indicated by ISO 11731:2017, were subcultured on Buffered Charcoal Yeast Extract (BCYE) agar with L-cysteine (cys+) and without L-cysteine (cys−) as supplement, which is a selective media used for *Legionella* isolation. The positive *Legionella* colonies were those that grew on *Legionella* BCYE cys+ agar but failed to grow on *Legionella* BCYE cys− agar.

The isolates grown on BCYE cys+ were serologically typed by an agglutination test (*Legionella* latex test kit, Thermo Fisher Diagnostic, Basingstoke, UK). The isolates identified as *L. pneumophila* were then processed for serogroup identification by polyclonal latex reagents (Biolife, Milan, Italy).

Colonies identified by the agglutination test as belonging to *Legionella* non-*pneumophila* species were subsequently analyzed by *mip* gene sequencing and by Polymerase Chain Reaction (PCR) using degenerate primers (as described by Ratcliff et al. [41]) and modified by M13 tailing to avoid noise in the DNA sequence [42]. Gene amplification was carried out in a 50-µL reaction containing DreamTaq Green PCR Master Mix 2× (Thermo Fisher Diagnostic) and 40 picomoles of each primer; 100 nanograms of DNA extracted from the presumptive colonies of *Legionella* was added as a template. The same amounts of DNA from *L. pneumophila* type strain EUL00137, provided by the European Working Group for *Legionella* Infections (EWGLI) [43], and fetal bovine serum were used as positive and negative controls, respectively.

Following purification, DNA was sequenced using BigDye Chemistry and analyzed on an ABI PRISM 3100 Genetic Analyzer (Applied Biosystems, Foster City, CA, USA). Specifically, *mip* amplicons (661–715 base pairs) were sequenced using M13 forward and reverse primers (mip-595R-M13R caggaaacagctatgaccCATATGCAAGACCTGAGGGAAC; mip-74F-M13F tgtaaaacgacggccagtGCTGCAACCGATGCCAC) to obtain complete coverage of the sequenced region of interest. Raw sequencing data were assembled using the CLC Main Workbench 7.6.4 software (QIAGEN, Redwood City, CA, USA). The sequences were compared with sequences deposited in the *Legionella mip* gene sequence database using a similarity analysis tool. Identification on the species level was done based on ≥98% similarity to a sequence in the database [44].

The results regarding *Legionella* contamination in the samples were expressed as colony formant unit (cfu) per liter (cfu/L). According to ISO 11731:2017, a negative result (absence of bacteria growth) was expressed as the lower limit of detection, that is, <50 cfu/L [40].

The same samples ($n$ = 669) were analyzed to quantify the presence of *P. aeruginosa* due to its role in biofilm formation and to its capacity to inhibit *Legionella* growth during isolation culture, producing inaccurate results [45]. The analysis was performed on a volume of 100 mL of hot and cold samples, filtered using a cellulose nitrate membrane filter with a 0.45-μm pore size (Sartorius, Bedford, MA, USA), according to UNI EN ISO 16266:2008 [46,47].

The membrane was seeded on *Pseudomonas*-selective agar plate (PSA, Biolife, Milan, Italy) and incubated for 48 h in 36 °C incubators. Colonies that showed green-blue fluorescence when placed under a Wood's lamp (ultraviolet light at 365 nm) were subcultured on Nutrient agar (NA, Biolife, Milan, Italy) for 18–24 h. Subsequently, the colonies were identified biochemically as *P. aeruginosa* by indole, oxidase reaction tests, and BBL Crystal Enteric/Non Fermenter ID Kit (Becton Dickinson Systems, Cockeysville, MD, USA), according to the manufacturer's instructions.

The results are expressed in terms of cfu/100 mL.

### 4.3. Physical and Chemical Analyses

The physical and chemical parameters—the temperature of the water samples as well as the disinfectant residues at SHWOs—were measured during the collection of samples.

The temperature (°C) (T) was measured by a conductivity meter coupled with a thermistor probe (Temp 6 basic for probe Pt100 RTD from −50 to +199 °C; Thermo Fisher Scientific Inc., Eutech Instruments Pte Ltd., Singapore). An on-site commercial kit for the residual hydrogen peroxide component of $H_2O_2/Ag^+$ (mg/L) was used. The kit uses a colorimetric test based on peroxidase activity to transfer peroxide oxygen to an organic redox indicator, which produces a blue oxidation product.

The hydrogen peroxide concentration was measured semiquantitatively by visual comparison of the result seen on the reaction zone of the test strip with the fields on a color scale in a range of 0.5–25 mg/L $H_2O_2$.

### 4.4. Statistical Analysis

The *Legionella* concentration data were converted into $Log_{10}$ cfu/L (Log cfu/L) to normalize the non normal distributions. According to the Italian Guidelines for *Legionella,* the detection limit corresponding to 50 cfu/L (1.7 Log cfu/L) was used; by contrast, the risk value, >100 cfu/L, was expressed as >2 Log cfu/L.

To compare data of hot- and cold-water *Legionella* concentrations, the Mann–Whitney test was used (Table 1).

The distribution of different *Legionella* isolates between hot and cold positive samples was studied by chi-squared test ($\chi^2$). Therefore, the differences in *Legionella* isolate concentrations in hot- or cold-water samples were studied by the Kruskal–Wallis test: the significant data found were then also analyzed by the Mann–Whitney test (Table 2).

Multiple comparisons between *Legionella* concentrations and the four ranges of temperature measured were performed by using the ANOVA test (Table 3).

Regarding the three hospitals that implemented the replacement program (e.g., with elbow-operated clinical valves without TMVs), the data analysis to compare *Legionella* levels before and after replacement was performed by parametric *t*-Student test when considering a number of values $n > 30$ and by nonparametric Wilcoxon test for $n < 30$ (Table 4).

The *P. aeruginosa* results were converted into Log cfu/100mL. The contamination found was studied by Mann–Whitney test to compare hot- and cold-water samples.

Statistical analyses were performed using the SPSS software for Windows version 23 (IBM SPSS, Inc., Chicago, IL, USA).

The data were considered significant for $p$ values ($p$) ≤ 0.05.

## 5. Conclusions

In conclusion, sensor-activated faucets with TMVs are generally more contaminated than clinical valves without thermostatic mixers. This allows us to conclude that the technologies typically chosen by a hospital do not correspond with the water microbiological environment that can develop in the SHWOs. The microbial interaction with the selected technologies, pipeline and faucet materials, and chemical-physical water characteristics result in an environment that, in semi-critical and critical areas, can lead to serious risks for patients, hospital staff, and stakeholders involved in maintenance procedures. The limit of this study is the lack of data on cold water after the replacement program developed by three hospitals due to there being no cold-water monitoring in the risk assessment plan, to poor knowledge, and to cost-containment demands.

The authors wish to encourage infection control teams to evaluate the use of non-touch fittings in hospitals, especially when installed in high-risk areas, and wish to promote water microbial monitoring in both hot- and cold-water distribution systems according to a water safety plan that can guide the hospital's choices based on epidemiological data, technological knowledge, and applied maintenance procedures.

**Author Contributions:** S.C. and M.M. conceived and designed the experiments and wrote the paper. L.G., M.R.P., and J.L. performed sample collection and the experiments. S.S. and A.D. performed the statistical analyses. All authors have read and agreed to the published version of the manuscript.

**Funding:** This work was partially supported by a research grant (RFO 2018) from the University of Bologna, Italy.

**Acknowledgments:** The authors would like to thank the GVM Engineering staff for technical assistance and Graziella Ciani for microbiological analyses.

**Conflicts of Interest:** The authors declare no conflict of interest.

## References

1. Khan, H.A.; Baig, F.K.; Mehboob, R. Nosocomial infections: Epidemiology, prevention, control and surveillance. *Asian Pac. J. Trop. Biomed.* **2017**, *7*, 478–482. [CrossRef]
2. Antonioli, P.; Bolognesi, N.; Valpiani, G.; Morotti, C.; Bernardini, D.; Bravi, F.; Di Ruscio, E.; Stefanati, A.; Gabutti, G. A 2-year point-prevalence surveillance of healthcare-associated infections and antimicrobial use in Ferrara University Hospital, Italy. *BMC Infect. Dis.* **2020**, *20*, 75. [CrossRef] [PubMed]
3. Istituto Superiore di Sanità, Italia. Infezioni correlate all'assistenza aspetti epidemiologici. Available online: https://www.epicentro.iss.it/infezioni-correlate/epidemiologia (accessed on 15 April 2020).
4. Squier, C.; Yu, V.L.; Stout, J.E. Waterborn nosocomial infections. *Curr. Infect. Dis. Rep.* **2000**, *2*, 490–496. [CrossRef] [PubMed]
5. Spagnolo, A.M.; Cristina, M.L.; Casini, B.; Perdelli, F. Legionella pneumophila in healthcare facilities. *Rev. Med. Microbiol.* **2013**, *24*, 70–80. [CrossRef]
6. Anaissie, E.J.; Penzak, S.R.; Dignani, M.C. The hospital water supply as a source of nosocomial infections: A plea for action. *Arch. Intern. Med.* **2002**, *162*, 1483–1492. [CrossRef] [PubMed]
7. Guidelines for Environmental Infection Control in Health-Care Facilities: Recommendations of CDC and the Healthcare Infection Control Practices Advisory Committee (HICPAC), 2003. MMWR Recomm. Rep. 52(RR-10), 1–42. Available online: https://www.cdc.gov/mmwr/preview/mmwrhtml/rr5210a1.htm (accessed on 12 February 2020).
8. Cristina, M.L.; Spagnolo, A.M.; Casini, B.; Baggiani, A.; Del Giudice, P.; Brusaferro, S.; Poscia, A.; Moscato, U.; Perdelli, F.; Orlando, P. The impact of aerators on water contamination by emerging Gram-negative opportunists in at-risk hospital departments. *Infect. Control. Hosp. Epidemiol.* **2014**, *35*, 122–129. [CrossRef]

9. Fields, B.S.; Benson, R.F.; Besser, R.E. Legionella and Legionnaires' disease: 25 years of investigation. *Clin. Microbiol. Rev.* **2002**, *15*, 506–526. [CrossRef]
10. Rota, M.C.; Caporali, M.G.; Bella, A.; Scaturro, M.; Giannitelli, S.; Ricci, M.L. Rapporto annuale sulla legionellosi in Italia nel 2018. Notiziario Ist. Super Sanità 2019, 32 (11). Available online: http://www.legionellaonline.it/Not%20ISS%20nov%202019.pdf (accessed on 3 February 2020).
11. Laganà, P.; Facciolà, A.; Palermo, R.; Delia, S. Environmental Surveillance of Legionellosis within an Italian University Hospital-Results of 15 Years of Analysis. *Int. J. Environ. Res. Public Health* **2019**, *16*, 1103. [CrossRef]
12. Italian Health Ministry. Guidelines for Prevention and Control of Legionellosis. Approvate in Conferenza Stato-Regioni Seduta Del 7 Maggio 2015. Italy, 2015. Available online: http://www.salute.gov.it/imgs/C_17_pubblicazioni_2362_allegato.pdf (accessed on 2 March 2020).
13. Mitchell, A.; Spencer, M.; Edmiston, C., Jr. Role of healthcare apparel and other healthcare textiles in the transmission of pathogens: A review of the literature. *J. Hosp. Infect.* **2015**, *90*, 285–292. [CrossRef]
14. World Health Organization. Guidelines on Hand Hygiene in Health Care (Advanced Draft). 2006. Available online: https://www.who.int/patientsafety/information_centre/Last_April_versionHH_Guidelines%5B3%5D.pdf (accessed on 12 February 2020).
15. World Health Organization. Guidelines on Hand Hygiene in Health Care: First Global Patient Safety Challenge Clean Care is Safer Care. 2009. Available online: https://apps.who.int/iris/bitstream/handle/10665/44102/9789241597906_eng.pdf?sequence=1 (accessed on 12 February 2020).
16. Ellingson, K.; Haas, J.P.; Aiello, A.E.; Kusek, L.; Maragakis, L.L.; Olmsted, R.N.; Perencevich, E.; Polgreen, P.M.; Schweizer, M.L.; Trexler, P.; et al. Strategies to Prevent Healthcare-Associated Infections through Hand Hygiene. *Infect. Control Hosp. Epidemiol.* **2014**, *35*, 937–960. [CrossRef]
17. International Health Facility Guidelines—Part B, version 5, 2017—Hand Hygiene. Available online: http://www.healthfacilityguidelines.com/ViewPDF/ViewIndexPDF/iHFG_part_d_hand_hygiene (accessed on 2 March 2020).
18. Halabi, M.; Wiesholzer-Pittl, M.; Schöberl, J.; Mittermayer, H. Non-touch fittings in hospitals: A possible source of *Pseudomonas aeruginosa* and *Legionella* spp. *J. Hosp. Infect.* **2001**, *49*, 117–121. [CrossRef] [PubMed]
19. Merrer, J.; Girou, E.; Ducellier, D.; Clavreul, N.; Cizeau, F.; Legrand, P.; Leneveu, M. Should electronic faucets be used in intensive care and hematology units? *Intensive Care Med.* **2005**, *31*, 1715–1718. [CrossRef] [PubMed]
20. Decret 26.08.1993, n. 412. Regolamento Recante Norme per la Progettazione, L'installazione, L'esercizio e la Manutenzione Degli Impianti Termici Degli Edifici ai Fini del Contenimento dei Consumi di Energia, in Attuazione Dell'art. 4, Comma 4, Della Legge 09.01.1991, n. 10. G.U. Serie Generale n.242 14.10.1993—Suppl. Ordinario n. 96. Available online: https://www.bosettiegatti.eu/info/norme/statali/1993_0412.htm (accessed on 12 February 2020).
21. Vonberg, R.P.; Eckmanns, T.; Bruderek, J.; Rüden, H.; Gastmeier, P. Use of terminal tap water filter systems for prevention of nosocomial legionellosis. *J. Hosp. Infect.* **2005**, *60*, 159–162. [CrossRef] [PubMed]
22. Falkinham, J.O.; Pruden, A.; Edwards, M. Opportunistic Premise Plumbing Pathogens: Increasingly Important Pathogens in Drinking Water. *Pathogens* **2015**, *4*, 373–386. [CrossRef]
23. Liu, S.; Gunawan, C.; Barraud, N.; Rice, S.A.; Harry, E.J.; Amal, R. Understanding, Monitoring, and Controlling Biofilm Growth in Drinking Water Distribution Systems. *Environ. Sci. Technol.* **2016**, *50*, 8954–8976. [CrossRef]
24. Council Directive 98/83/EC of 3 November 1998 on the Quality of Water Intended for Human Consumption. Available online: http://data.europa.eu/eli/dir/1998/83/oj (accessed on 12 February 2020).
25. Commission directive (EU) 2015/1787 of 6 October 2015 Amending Annexes II and III to Council Directive 98/83/EC on the Quality of Water Intended for Human Consumption. Available online: http://data.europa.eu/eli/dir/2015/1787/oj (accessed on 12 February 2020).
26. D. Lgs. 02.02.2001, n.31. Directive 98/83/CE Relativa Alla Qualità Delle Acque Destinate al Consumo Umano. G.U. Della Repubblica Italiana n. 52, 03.03.2001. Available online: http://www.camera.it/parlam/leggi/deleghe/01031dl.htm (accessed on 12 February 2020).

27. DM 14.06.2017. Directive (UE) 2015/1787 che Modifica gli Allegati II e III Della Direttiva 98/83/CE Sulla Qualità Delle Acque Destinate al Consumo Umano. Modifica Degli Allegati II e III del Decreto Legislativo 2 Febbraio 2001, n. 31. G.U. Serie Generale n.192, 18.08.2017. Available online: http://www.si-t.eu/wordpress/wp-content/uploads/2017/09/6.pdf (accessed on 12 February 2020).
28. World Health Organization. Guidelines for Drinking-Water Quality. 2011. Available online: http://apps.who.int/iris/bitstream/10665/44584/1/9789241548151_eng.pdf (accessed on 2 March 2020).
29. Exner, M.; Kramer, A.; Lajoie, L.; Gebel, J.; Engelhart, S.; Hartemann, P. Prevention and control of health care-associated waterborne infections in health care facilities. *Am. J. Infect. Control* **2005**, *33*, S26–S40. [CrossRef]
30. Falkinham, J.O.; Hilborn, E.D.; Arduino, M.J.; Pruden, A.; Edwards, M.A. Epidemiology and Ecology of Opportunistic Premise Plumbing Pathogens: *Legionella pneumophila*, *Mycobacterium avium*, and *Pseudomonas aeruginosa*. *Environ. Health Perspect.* **2015**, *123*, 749–758. [CrossRef]
31. Syndor, E.R.M.; Bova, G.; Gimburg, A.; Cosgrove, S.E.; Perl, T.M.; Maragakis, L.L. Electronic-Eye Faucets: *Legionella Species* Contamination in Healthcare Settings. *Infect. Control Hosp. Epidemiol.* **2012**, *33*, 235–239. [CrossRef]
32. Charron, D.; Bédard, E.; Lalancette, C.; Laferrière, C.; Prévost, M. Impact of electronic faucets and water quality on the occurrence of *Pseudomonas aeruginosa* in water: A multi-hospital study. *Infect. Control Hosp. Epidemiol.* **2015**, *36*, 311–319. [CrossRef]
33. Rota, M.C.; Fontana, S.; Montaño-Remacha, C.; Scaturro, M.; Caporali, M.G.; Vullo, V.; Scorzolini, L.; Ercole, A.; Ricci, M.L. Legionnaires' disease pseudoepidemic due to falsely positive urine antigen test results. *J. Clin. Microbiol.* **2014**, *52*, 2279–2280. [CrossRef]
34. Svarrer, C.W.; Lück, C.; Elverdal, P.L.; Uldum, S.A. Immunochromatic kits Xpect *Legionella* and BinaxNOW *Legionella* for detection of *Legionella pneumophila* urinary antigen have low sensitivities for the diagnosis of Legionnaires' disease. *J. Med. Microbiol.* **2012**, *61*, 213–217. [CrossRef] [PubMed]
35. Girolamini, L.; Dormi, A.; Pellati, T.; Somaroli, P.; Montanari, D.; Costa, A.; Savelli, F.; Martelli, A.; Grottola, A.; Fregni Serpini, G.; et al. Advances in *Legionella* Control by a New Formulation of Hydrogen Peroxide and Silver Salts in a Hospital Hot Water Network. *Pathogens* **2019**, *8*, 209. [CrossRef] [PubMed]
36. Fliermans, C.B.; Cherry, W.B.; Orrison, L.H.; Smith, S.J.; Tison, D.L.; Pope, D.H. Ecological distribution of *Legionella pneumophila*. *Appl. Environ. Microbiol.* **1981**, *41*, 9–16. [CrossRef] [PubMed]
37. Berjeaud, J.M.; Chevalier, S.; Schlusselhuber, M.; Portier, E.; Loiseau, C.; Aucher, W.; Lesouhaitier, O.; Verdon, J. *Legionella pneumophila*: The Paradox of a Highly Sensitive Opportunistic Waterborne Pathogen Able to Persist in the Environment. *Front. Microbiol.* **2016**, *7*, 486. [CrossRef]
38. EN ISO 19458:2006—Water Quality—Sampling for Microbiological Analysis. Available online: https://www.iso.org/standard/33845.html (accessed on 2 March 2020).
39. Liguori, G.; Cavallotti, I.; Arnese, A.; Amiranda, C.; Anastasi, D.; Angelillo, I.F. Microbiological quality of drinking water from dispensers in Italy. *BMC Microbiol.* **2010**, *10*, 19. [CrossRef]
40. ISO 11731:2017 Water quality—Enumeration of *Legionella*. International Organization for Standardization, Geneva, Switzerland. Available online: https://www.iso.org/standard/61782.html (accessed on 2 March 2020).
41. Ratcliff, R.M.; Lanser, J.A.; Manning, P.A.; Heuzenroeder, M.W. Sequence-based classification scheme for the genus *Legionella* targeting the *mip* gene. *J. Clin. Microbiol.* **1998**, *36*, 1560–1567. [CrossRef]
42. Mentasti, M.; Fry, N.K.; Afshar, B.; Palepou-Foxley, C.; Naik, F.C.; Harrison, T.G. Application of *Legionella pneumophila* specific quantitative real-time PCR combined with direct amplification and sequence-based typing in the diagnosis and epidemiological investigation of Legionnaires' disease. *Eur. J. Clin. Microbiol.* **2012**, *8*, 2017–2028. [CrossRef]
43. Fry, N.K.; Alexiou-Daniel, S.; Bangsborg, J.M.; Bernander, S.; Castellani Pastoris, M.; Etienne, J.; Forsblom, B.; Gaia, V.; Helbig, J.H.; Lindsay, D.; et al. A multi-center evaluation of genotypic methods for the epidemiological typing of *Legionella pneumophila* serogroup 1: Results of a pan-European study. *Clin. Microbiol. Infec.* **1999**, *5*, 462–477. [CrossRef]
44. Fry, N.K.; Afshar, B.; Bellamy, W.; Underwood, A.P.; Ratcliff, R.M.; Harrison, T.G. Identification of *Legionella* spp. by 19 European reference laboratories: results of the European Working Group for *Legionella* Infections External Quality Assessment Scheme using DNA sequencing of the macrophage infectivity potentiator gene and dedicated online tools. *Clin. Microbiol. Infect.* **2007**, *13*, 1119–1124. [CrossRef]

45. Kimura, S.; Tateda, K.; Ishii, Y.; Horikawa, M.; Miyairi, S.; Gotoh, N. *Pseudomonas aeruginosa* Las quorum sensing autoinducer suppresses growth and biofilm production in Legionella species. *Microbiology* **2009**, *155*, 1934–1939. [CrossRef]
46. UNI EN ISO 16266:2008—Water Quality—Detection and Enumeration of *Pseudomonas aeruginosa*—Method by Membrane Filtration. International Organization for Standardization, Geneva, Switzerland. Available online: https://www.iso.org/standard/39272.html (accessed on 12 February 2020).
47. Wu, W.; Jin, Y.; Bai, F.; Jin, S. *Chapter 41: Pseudomonas aeruginosa. Molecular Medical Microbiology (Second Edition), 2015, 2, 753–767*. Available online: https://doi.org/10.1016/B978-0-12-397169-2.00041-X (accessed on 12 February 2020).

*pathogens*

MDPI

*Article*

# Advances in *Legionella* Control by a New Formulation of Hydrogen Peroxide and Silver Salts in a Hospital Hot Water Network

**Luna Girolamini [1], Ada Dormi [2], Tiziana Pellati [3], Paolo Somaroli [3], Davide Montanari [4], Andrea Costa [4], Francesca Savelli [5], Andrea Martelli [5], Antonella Grottola [6], Giulia Fregni Serpini [6] and Sandra Cristino [1,*]**

1 Department of Biological, Geological, and Environmental Sciences, BiGeA, University of Bologna, 40126 Bologna, Italy; luna.girolamini2@unibo.it
2 Department of Medical and Surgical Science, DIMEC, University of Bologna, 40126 Bologna, Italy; ada.dormi@unibo.it
3 GVM Care & Research, Lugo di Ravenna, 48022 Ravenna, Italy; tpellati@gvm-engineering.it (T.P.); psomaroli@gvm-engineering.it (P.S.)
4 Eta-Beta S.r.l., 47122 Forlì, Forlì Cesena, Italy; manutenzione-mch@gvmnet.it (D.M.); costa.andrea@etabetasrl.com (A.C.)
5 Water Team S.r.l., 47522 Cesena, Forlì Cesena, Italy; francesca@waterteam.it (F.S.); andream@waterteam.it (A.M.)
6 Regional Reference Laboratory for Clinical Diagnosis of Legionellosis, Unit of Microbiology and Virology, Modena University Hospital, 41125 Modena, Italy; grottola.antonella@aou.mo.it (A.G.); fregniserpini.giulia@aou.mo.it (G.F.S.)
* Correspondence: sandra.cristino@unibo.it

Received: 27 September 2019; Accepted: 25 October 2019; Published: 29 October 2019

**Abstract:** *Legionella* surveillance is an important issue in public health, linked to the severity of disease and the difficulty associated with eradicating this bacterium from the water environment. Different treatments are suggested to reduce *Legionella* risk, however long-term studies of their efficiency are lacking. This study focused on the activity of a new formulation of hydrogen peroxide and silver salts, WTP828, in the hospital hot water network (HWN) to contain *Legionella* contamination during two years of treatment. The effectiveness of WTP828 was tested measuring physical-chemical and microbiological parameters such as *Legionella*, *Pseudomonas aeruginosa* (*P. aeruginosa*), and a heterotopic plate count (HPC) at 36 °C. *Legionella* isolates were identified by serotyping and genotyping. WTP 828 induced a reduction in *Legionella*–positive sites (60% to 36%) and contamination levels (2.12 to 1.7 $\log_{10}$ CFU/L), with isolates belonging to *L. pneumophila* SG1 (ST1 and ST104), *L. anisa* and *L. rubrilucens* widely distributed in HWN. No relevant contamination was found for other parameters tested. The long-term effect of WTP828 on *Legionella* containment suggest the easy and safe application of this disinfectant, that combined with knowledge of building characteristics, an adequate environmental monitoring and risk assessment plan, become the key elements in preventing *Legionella* contamination and exposure.

**Keywords:** WTP 828; *Legionella*; risk assessment plan; water quality; microbial analysis; chemical analysis

## 1. Introduction

Hot and cold water systems (e.g., tap water installations, distribution systems, and cooling towers) are important sources of nosocomial and community-acquired infections caused by opportunistic waterborne pathogens. Among them, *Legionella* spp. are water-based organisms that cause lung infections when inhaled in an aerosol form [1].

Several national standards have been established to ensure a high water quality using disinfection techniques that control and prevent the colonization of water systems by *Legionella* [2]. A wide variety of disinfection techniques, including chemical disinfection, ultraviolet (UV) light, and high temperature, have been employed worldwide to reduce the risk of legionellosis [3,4].

In Italy, Legionnaires' disease (LD) is a class II statutorily notifiable disease [5]; since 1983, it has also been subject to a reporting system designed to collect detailed information about contamination cases, which is held in a national register at the Istituto Superiore di Sanità (ISS), Italy. However, according to ISS annual reports, the number of LD cases is under-diagnosed and under-reported, leading to a significant underestimation of the real incidence of LD. In 2017, the incidence rate was 33.2 cases per million persons [6].

Following publication of the new Italian Guidelines for the Control and Prevention of legionellosis in May 2015 [7], the importance of a surveillance program encompassing all facilities at risk of LD (hospitals, healthcare facilities, dental units, hotels, tourist facilities, and spas) has been acknowledged, and this program has been implemented. The guidelines support the development of a risk assessment plan based on an evaluation of "risk" and also emphasize the need for an adequate environmental surveillance plan that includes an appropriate number of sites that are potential sources of *Legionella*.

A recent multicenter study performed by Montagna et al. [8] has demonstrated, as the main methods to perform *Legionella* prevention and control for the water network, were shock treatment and chlorination.

The shock treatment consists of a thermal disinfection of hot-water distribution systems performed at a temperature between 70–80 °C starting from the hot water storage heater. The temperature must be maintained in all outlets, faucets, and shower heads at least 30 min at 60–65 °C, for three consecutive days [7,9,10].

Several studies showed as the main disadvantage of shock treatment is its transitorily effect on bacterial community structure, e.g., biofilm, that was not removed preserving pathogenic *Legionella* niche [11–13].

Chlorine is the most common chemical disinfectant used in water (including drinking water), acts as an oxidizing agent, and reacts with several cellular constituents including the cell membrane of microbes. To perform *Legionella* control, plumbing water systems can be treated using chlorine as a shock hyperchlorination (residual chlorine concentration at distal outlets of 20–50 mg/L) or as continuous treatment using a concentration of 1–2 mg/L [10]. Although different studies have shown good performance using these methods to assess *Legionella* contamination, a reduction of effectiveness over a long-term period was consistently demonstrated [10,14–16]. However, increasing evidence suggests that humans are exposed to residual byproducts of water chlorination such as disinfection byproducts (DBPs) through drinking-water, oral, dermal, and inhalational contact. During the chlorination, especially by hypochlorous acid and hypobromous acid, the reaction with naturally occurring organic matter present in raw water supplies, create many water DBPs, including the four primary trihalomethanes: chloroform ($CHCl_3$), bromodichloromethane ($CHCl_2Br$), dibromochloromethane ($CHClBr_2$), and bromoform ($CHBr_3$), that can have adverse effects on human health [17–20].

Disinfection methods other than chlorination have been suggested for *Legionella* control in water, such as ozone treatment, copper and silver ($Ag^+$) ionization, monochloramine, point-of-use filters, and UV light. These measures have been tested over the last 30 years and are effective at controlling the growth of *Legionella*, all of them presented advantages and disadvantages that must be carefully considered [10,16].

Different studies have focused in the last years on the role of oxidizing agents, notably hydrogen peroxide ($H_2O_2$), as disinfection treatments. The use of $H_2O_2$ as a biocide is widespread, and it is increasingly used as a general surface disinfectant in the medical, food, and industrial fields, as well as for water treatment [21,22]. $H_2O_2$ is completely soluble in water and is stabilized in commercial formulation for disinfection treatment. It is compatible with different pipeline materials, and does not react with the organic constituents in the water to form dangerous residues with respect to chlorine, sodium hypochlorite (NaOCl), and monochloramine treatment. $H_2O_2$ decomposes rapidly in different environmental conditions due to microbial catalase and peroxidase, and other than abiotic action, the decomposition is promoted by heavy metal, oxidative, and reductive reactions. It shows a broad antimicrobial spectrum and has been shown to be active against bacteria, yeast, fungi, viruses, spores, proto-, and metazoans [23–25].

A disadvantage of using $H_2O_2$ is that its potency is influenced by several factors: pH, temperature, or the presence of substances that hamper its reactivity [26]. Since $H_2O_2$ is a renowned disinfectant, legislation [27] allows its use for the disinfection of water and in food; additionally, this compound is generally considered to have low eco-toxicity, as well as no odor or color [23,28].

To enhance its activity, $H_2O_2$ is sometimes used in combination with other oxidants such as ozone, $Ag^+$, or UV radiation [24]. Silver, a biologically non-essential metal, has been investigated and used as a biocide for many years [29], and multiple strategies have been proposed for its use to treat drinking water [30–32]. Indeed, the World Health Organization (WHO) allows its use in drinking water. It is thought that concentrations up to 50 µg/L (ppb) in drinking water pose no risk to health [33].

The literature contains several accounts of the properties, germicidal effectiveness, and potential uses for stabilized $H_2O_2$ in healthcare facilities [34–37]. In 2015, Martin et al. [24] have demonstrated that Huwa-San peroxide (HSP), a new generation peroxide stabilized with ionic silver and suitable for continuous disinfection of potable water, preferentially interacts with the bacterial cell surface in a mechanism likely mediated by silver. Furthermore, treatment of hospital hot water systems with various formulations of $H_2O_2/Ag^+$ compounds prevents contamination by *Legionella* and other microorganisms because of its bactericidal properties [38–40].

The $H_2O_2/Ag^+$ formulation is stable at high temperatures, and its disinfection power increases significantly as water temperature increases. In a hot water system, a temperature range of 40–50 °C and a residual disinfectant concentration of 20–25 mg/L, seems to be able to induce a *Legionella* control [41,42]. Casini et al. suggested therefore, how a continuous feed rate of approximately 25 mg/L, was able to control the planktonic population, and silver can be deposited on the piping system, promoting a bacteriostatic effect [42].

Different commercial formulations based on $H_2O_2/Ag^+$ are available to control *Legionella* contamination, but many studies lack data about the hospital settings and long-term applications.

Our study evaluated the effectiveness of a new disinfectant, Water Team Process 828 (WTP 828), based on $H_2O_2$ and $Ag^+$ salts in the hot water distribution networks at Maria Cecilia Hospital (MCH), Cotignola (RA), Italy, controlling *Legionella* contamination.

The hospital is comprised of three buildings connected to each other but were built and submitted to renovation works at different times. The plumbing system comprises a single cold water supply and three different hot water return lines. These characteristics permitted us to study the activity of WTP 828 as three separate hot water networks (HWNs), modulating the dosage with respect to the level of *Legionella* contamination found. *Legionella* level in response to disinfection treatment was also studied by taking into account the following water network characteristics: building area, annual water consumption, hospital activities involving the use of water, that can influence the *Legionella* contamination and the disinfectant exposure to distal outlets [16]. The isolates were typed using an agglutination and genotyping approach to assess the distribution of strains in the buildings. The effect of WTP 828 was also tested on *Pseudomonas aeruginosa* (*P. aeruginosa*), one of the main components of biofilm [43], and HPC at 36 °C, commonly used as an indicator of water quality, and to monitor the effectiveness of disinfection treatment [33,44].

The physical and chemical parameters were also measured during implementation of WTP 828 treatment in all buildings in order to maintain the water quality characteristics [45,46] and preserve the plumping system materials.

The purpose of the study is to perform an extended investigation on the effect of $H_2O_2/Ag^+$ treatment in a complex hospital water network system. The goal is to control *Legionella* infections throughout a risk assessment model based on the use of a low-cost disinfectant, easy to dose, and less aggressive on the material pipelines, with quick and safe monitoring of residual concentrations at distal outlets. This model associated to ordinary and extraordinary maintenance procedures (e.g., flushing, temperature control, and cleaning activities) could be extended to other hospitals, companies, and leisure facilities, where water represents a risk for public health.

## 2. Results

### 2.1. Legionella Contamination

The following results were obtained during the two phases of the study according to Italian Guidelines for prevention and control of legionellosis [7] that take into account the concentration of bacteria in relation to four levels of risk (<100, 101–100, 1001–10000, >10000 UFC/L) and the number of positive on the total number of samples collected. The measures to apply, in order to contain the risk, are different if the percentage of positive samples are <20% or >20%.

*WTP1 Phase (October 2013 to March 2015)*

The data were obtained from the analysis of 53 hot water samples spread throughout the three buildings. We observed different *Legionella* contamination trends in the MCH buildings (Table 1): 16/25 positive samples (64.0%) in Building 1, 13/23 positive samples (56.5%) in Building 2, and 3/5 positive samples (60.0%) in Building 3. The WTP1 phase was also compared with *Legionella* contamination data collected during the previous disinfection treatment involving the $ClO_2$ mixture. Although we observed a change in the percentage of *Legionella*-positive samples in MCH (Building 1-2-3) from 95.0% to 60.0%, no statistical differences were observed in terms of *Legionella* contamination levels following the introduction of WTP 828. The analysis of *Legionella* contamination inside each building revealed a significant change in the mean *Legionella* levels only in Building 2 ($p = 0.045$) (Table 1).

*WTP2 Phase (September 2014 to October 2015)*

In the second phase of the study, we observed a reduction in terms of the percentage of *Legionella*-positive samples in MCH with respect to the WTP1 phase (from 60% to 35.8%) (Table 1). The same trend was also observed for the *Legionella* contamination levels ($p = 0.0001$). The results inside each building show a marked reduction in the percentage of *Legionella*-contaminated sites in Building 2 (from 56.5% to 7.0%) and Building 3 (from 60.0% to 34.0%); by contrast, the percentage of positive samples was only slightly reduced in Building 1 (from 64.0% to 58.1%).

The *Legionella* contamination levels displayed a significant difference between the WTP1 and WTP2 phases for Building 2 ($p = 0.046$) and Building 3 ($p = 0.048$) (Table 1). No statistical difference between phases was observed for Building 1, in which *Legionella* contamination levels of 1000 CFU/L were detected and, in accordance with Italian Guidelines [7], two shock treatments, increasing up to 50–60 mg/L of WTP 828, were performed (from February to March and from July to August 2015), resulting in a concentration of 25–30 mg/L at distal outlets.

The data collected from the WTP2 phase were also compared with *Legionella* contamination data obtained during disinfection with the $ClO_2$ mixture. The comparison revealed significant differences for Buildings 2 ($p = 0.0001$) and 3 ($p = 0.045$) and no significant differences for Building 1 (Table 1).

During the study, the water reserves, softener, and tap water output sites were *Legionella*-free (below the detection limit of the culture technique used, i.e., 50 CFU/L).

To study the risk of *Legionella* disease that could be derived from the approach used during WTP 828 treatment (WTP1 and WTP2 phase) with the outcomes after $ClO_2$ mixture treatment, we used two measures of risk provided by epidemiological studies: OR in a retrospective approach, and RR in a prospective study. Comparison of the $ClO_2$ mixture with WTP1 (OR, 0.3) indicated that WTP 828 was not particularly effective in any of the MCH buildings ($p = 0.048$); however, comparison of the $ClO_2$

mixture with WTP2 (OR, 15.44) revealed a significant improvement in *Legionella* control with respect to the latter ($p = 0.0001$). A prospective study indicated that the level of contamination during the WTP1 phase was higher than during the WTP2 phase (RR, 0.36, $p = 0.002$), showing a decrease in *Legionella* risk (Table 2).

**Table 1.** *Legionella* concentration in three buildings of Maria Cecilia Hospital (MCH) for each study phase.

| | Treatment/Study Phase | Number of Samples | Number of *Legionella* Positive Samples (%) | Mean *Legionella* Levels ($log_{10}$ CFU/L) ± SD | Comparison between Phases | *p* Value |
|---|---|---|---|---|---|---|
| **MCH (Buildings 1–3)** | $ClO_2$ mixture | 120 | 114 (95.0) | 2.54 ± 0.74 | WTP1 *vs.* $ClO_2$ | 1 |
| | WTP1 | 53 | 32 (60.0) | 2.43 ± 0.95 | WTP1 *vs.* WTP2 | 0.0001 * |
| | WTP2 | 296 | 106 (35.8) | 1.67 ± 0.66 | WTP2 *vs.* $ClO_2$ | 0.0001 * |
| **Building 1** | $ClO_2$ mixture | 47 | 46 (98.0) | 2.47 ± 0.67 | WTP1 *vs.* $ClO_2$ | 0.623 |
| | WTP1 | 25 | 16 (64.0) | 2.80 ± 0.87 | WTP1 *vs.* WTP2 | 0.060 |
| | WTP2 | 141 | 82 (58.1) | 2.21 ± 0.55 | WTP2 *vs.* $ClO_2$ | 0.835 |
| **Building 2** | $ClO_2$ mixture | 58 | 53 (91.3) | 2.39 ± 0.63 | WTP1 *vs.* $ClO_2$ | 0.045 * |
| | WTP1 | 23 | 13 (56.5) | 1.81 ± 0.82 | WTP1 *vs.* WTP2 | 0.046 * |
| | WTP2 | 108 | 8 (7.0) | 1.26 ± 0.40 | WTP2 *vs.* $ClO_2$ | 0.0001 * |
| **Building 3** | $ClO_2$ mixture | 15 | 15 (100.0) | 3.12 ± 1.04 | WTP1 *vs.* $ClO_2$ | 1 |
| | WTP1 | 5 | 3 (60.0) | 2.97 ± 0.71 | WTP1 *vs.* WTP2 | 0.048 * |
| | WTP2 | 47 | 16 (34.0) | 1.47 ± 0.60 | WTP2 *vs.* $ClO_2$ | 0.01 * |

* Values are statistically significant at $p < 0.05$.

**Table 2.** Odds ratio and relative risk during the study phases.

| Study phases | Odds Ratio (OR) | 95% Confidence Intervals | *p* value |
|---|---|---|---|
| $ClO_2$ mixture versus WTP1 | 0.30 | 0.09–1.02 | 0.048 * |
| $ClO_2$ mixture versus WTP2 | 15.44 | 5.14–46.33 | 0.0001 * |
| **Study phases** | **Relative Risk (RR)** | **95% Confidence Intervals** | ***p* value** |
| WTP1 versus WTP2 | 0.36 | 0.18–0.71 | 0.002 * |

* Values are statistically significant at $p < 0.05$.

### 2.2. *Legionella Typing*

The isolates from the WTP1 and WTP2 phases were serotyped and genotyped using standard techniques. The agglutination test permitted us to identify *L. pneumophila* serogroup 1 (SG1) and two *Legionella* species in 138/349 positive samples (39.0%). The SBT method assigned ST1 and ST104 to *L. pneumophila* SG1 isolates in 74/138 (53.6%) of the samples, while *mip* gene sequencing identified, *L. anisa* and *L. rubrilucens*, in 35/138 (25.3%); the remaining 29/138 (21.0%) samples contained a mixture of the previously described strains.

The results revealed that each HWN building was colonized by a different mixture of *Legionella* spp. Accordingly, Building 1 isolates were the most diverse with *L. pneumophila* SG1 (ST1 and ST104) and *L. species* (*L. anisa* and *L. rubrilucens*). All Building 2 isolates belonged to *L. pneumophila* SG1 (ST1 and ST104), and Building 3 samples demonstrated the presence of *L. pneumophila* SG1 (ST1), with some samples containing a single *L. species* strain (*L. anisa* or *L. rubrilucens*).

The serotyping and genotyping data about the bacterial concentration ranges ($log_{10}$ CFU/L) are presented in Table 3.

**Table 3.** Serotyping and genotyping of *Legionella* isolates in MCH buildings (WTP1 and WTP2 phases).

| Building | Positive Samples | Serotyping | Genotyping | *Legionella* Isolates WTP1 Versus WTP2 | Isolates/Positive Samples | | Range of *Legionella* Concentration $log_{10}$ CFU/L (Min–Max) * |
|---|---|---|---|---|---|---|---|
| Building 1 | 98 | *L. pneumophila* SG1 | ST1 and ST104 | WTP1 | 36/98 | | <1.70–5.80 |
| | | *L. species* | *L. rubrilucens* | WTP2 | 11/98 | | <1.70–4.60 |
| | | *L. species* | *L. anisa* | WTP2 | 20/98 | | <1.70–3.77 |
| | | *L. species* | *L. rubrilucens* + *L. anisa* | WTP2 | 2/98 | | 2.83–3 |
| | | *L. pneumophila* SG1 + *L. species* | ST1 and ST104 + *L. rubrilucens* | WTP2 | 29/98 | 9/29 | 2–5.69 |
| | | | ST1 and ST104 + *L. anisa* | WTP2 | | 20/29 | <1.70–3.53 |
| Building 2 | 21 | *L. pneumophila* SG1 | ST1 and ST104 | WTP1 and WTP2 | 21/21 | | <1.70–4.50 |
| Building 3 | 19 | *L. pneumophila* SG1 | ST1 | WTP1 | 17/19 | | <1.70–4.18 |
| | | *L. species* | *L. rubrilucens* | WTP2 | 2/19 | 1/19 | 2.10 |
| | | | *L. anisa* | WTP2 | | 1/19 | 1.70 |

* (<1.70 Log UFC/L, detection limit).

During the study period, no significant association was found between *Legionella* colonization in the buildings and specific serogroups or strains. However, in Building 1, as after the two shocks treatment (February–March 2015 and July–August 2015), we observed a decrease in *L. pneumophila* SG1 levels, and detection of the other species, mainly *L. anisa* and *L. rubrilucens*. More experiments are still in progress.

### 2.3. *Pseudomonas aeruginosa and HPC Typing*

During the previous treatment by $ClO_2$ mixture, the risk assessment plan for *Legionella* surveillance was performed without control of HPC at 36 °C and *P. aeruginosa*. After the introduction of WTP 828 treatment, during the study period (WTP1 and WTP2 phase), 349 hot water and 65 cold-water samples distributed among distal outlets, water reserves, softener, and tap water outputs of MCH were also analyzed for the presence of *P. aeruginosa* and HPC at 36 °C.

*P. aeruginosa* was not detected (as prescribed in D. Lgs 31/2001) [46] in either cold or hot water samples.

The HPC at 36 °C results for each building expressed as the mean concentration ± SD ($log_{10}$ CFU/mL) were as follows: 0.82 ± 0.25 for Building 1 (0.48–1.20 $log_{10}$ CFU/mL), 0.77 ± 0.65 for Building 2 (0.30–0.90 $log_{10}$ CFU/mL), and 0.94 ± 0.35 (0.48–1.11 $log_{10}$ CFU/mL) for Building 3.

At all sites, the contamination range was lower than the D. Lgs 31/2001 [46] limit of 20 CFU/mL (1.3 $log_{10}$ CFU/mL).

### 2.4. *Physical and Chemical Parameters of Water*

The physical and chemical parameters linked to the quality of water after disinfection with WTP 828 were measured only during the WTP2 phase when relevant changes were made to the risk assessment plan. A total of 296 hot water and 65 cold water samples were analyzed. Physical and chemical data related to previous disinfection treatments and to the WTP1 phase are not reported because the *Legionella* surveillance during these phases took into account only bacteriological parameters.

The hardness, turbidity, and conductivity of the water (all of which are associated with the release of iron and total phosphorus) in the cold and hot water systems were not affected by WTP 828 treatment; these data are in agreement with recommendations established by Italian legislation [46]. In particular, the mean concentration of $Ag^+$ remained lower than the detection limit (3 μg/L) and in line with WHO Guidelines for drinking water [33]. These results are shown in Table 4.

**Table 4.** WTP2 phase: Physical and chemical parameters of water in MCH aqueduct and distal outlets.

| Parameters | Standardized Methods | Principle of the Method | U.M. | Sampling Points (Mean Value ± SD) | | | | | | |
|---|---|---|---|---|---|---|---|---|---|---|
| | | | | Aqueduct (n = 13) | Building 1 | | Building 2 | | Building 3 | |
| | | | | | Hot Water Return Line (1a) * (n = 13) | Distal Outlets (n = 128) | Hot Water Return Line (1b) * (n = 13) | Distal Outlets (n = 95) | Hot Water Return Line (1c) * (n = 3) | Distal Outlets (n = 34) |
| **pH** | APAT IRSA CNR 2060 Man 29 2003 | Potentiometric method | | 7.82 ± 0.23 | 7.88 ± 0.24 | 7.95 ± 0.07 | 7.87 ± 0.27 | 7.92 ± 0.04 | 7.95 ± 0.19 | 7.94 ± 0.09 |
| **Hardness** | APAT IRSA CNR 2040 Man 29 2003 | Complex metric titration | ° f | 12.10 ± 4.28 | 12.13 ± 3.14 | 11.07 ± 3.46 | 12.63 ± 3.50 | 9.75 ± 4.44 | 12.15 ± 2.44 | 11.92 ± 3.05 |
| **Conductivity** | APAT IRSA CNR 2030 Man 29 2003 | Conductimetric method | µS/cm | 407.71 ± 35.20 | 416.53 ± 41.80 | 374.91 ± 39.97 | 420.00 ± 40.70 | 447 ± 13.78 | 423.44 ± 29.75 | 424.80 ± 32.09 |
| **Turbidity** | APAT IRSA CNR 2110 Man 29 2003 | Spectrophotometric method | NTU | 0.40 ± 0.09 | 0.39 ± 0.12 | 0.63 ± 0.22 | 0.52 ± 0.25 | 1.69 ± 1.41 | 0.88 ± 0.97 | 0.94 ± 0.63 |
| **Total iron** | APAT IRSA CNR 3160A Man 29 2003 | Flame atomic absorption spectroscopy (FAAS) | mg/L | <0.04 | 0.04 | 0.05 ± 0.03 | <0.04 | <0.03 | 0.03 | 0.09 ± 0.10 |
| **Total phosphorus** | APAT IRSA CNR 4060 Man 29 2003 | Spectrophotometric method | mg/L $P_2O_5$ | <0.2 | 3.19 ± 1.31 | 3.05 ± 0.49 | 3.40 ± 1.40 | 3.52 ± 0.76 | 1.68 ± 0.85 | 2.05 ± 0.59 |
| **Silver** | EPA Method 272.2 | Electrothermal atomization atomic absorption spectrometry (ETA-AAS) | µg/L | <3 | <3 | <3 | <3 | <3 | <3 | <3 |
| **Temperature** | EPA Method 170.1 | Thermistor probe | °C | 15.20 ± 3.40 | 49.03 ± 2.42 | 49.75 ± 2.30 | 50.20 ± 0.61 | 50.38 ± 1.80 | 53.87 ± 4.13 | 51.07 ± 2.25 |
| **Peroxide** | Peroxide Test MQuant ™ | | mg/L | not detected | 7.42 ± 2.71 | 15.0 ± 7.07 | 8.46 ± 3.15 | 14.72 ± 4.75 | 5.83 ± 3.84 | 10 ± 4.63 |

* 1a = hot water return line of Building 1, 1b = hot water return line of building 2, 1c = hot water return line of building 3.

### 2.5. LD Surveillance

During the study (WTP1 and WTP2 phases), 32 patients underwent urine antigen testing and other diagnostic tests because of suspected pulmonary signs of pneumonia. The negative results obtained confirmed the absence of cases of nosocomial legionellosis.

## 3. Discussion

In this study, the effectiveness of WTP 828 was evaluated in a MCH water system because of its unique layout (i.e., constructed as three separate buildings). The water distribution system is characterized by a single tap water output, and each building is equipped with its own hot water return line and water disinfection treatment system. Before the introduction of WTP 828, MCH implemented a disinfection approach of continuous treatment with a $ClO_2$ mixture (dosage of 0.5 mg/L). This type of treatment led to corrosion of some parts of the plant and a visible decrement of the efficiency of *Legionella* colonization containment, as demonstrated by the high number of *Legionella*-positive samples in the three buildings (114/120, i.e., 95.0%) and the presence of *P. aeruginosa* in some water outlets, found during not routinely control (data not shown). In October 2013, the MCH Health Director decided to introduce WTP 828 into Building 2, as well as at available sampling points in Buildings 1 and 3.

The results obtained were discussed, for each MCH building, in relation to the period of WTP 828 introduction, as follows:

- Building 1: WTP1 from October 2013 to December 2014 and WTP2 from January 2015 to October 2015;
- Building 2: WTP1 from October 2013 to August 2014 and WTP2 from September 2014 to October 2015. This building was not subjected to any changes in disinfectant concentration or renovation works;
- Building 3: WTP1 October 2013 to March 2015 and WTP2 from April 2015 to October 2015.

Introduction of WTP 828 during the WTP1 phase (led to an overall reduction of the percentage of *Legionella*-positive samples (from 95% to 60.0%) when compared with the $ClO_2$ mixture. Preliminary results showed that WTP 828 treatment led to a marked reduction of contamination in Building 2 ($p = 0.046$); these results can be explained by the observation that Building 2 was the first building to undergo WTP 828 treatment. Additionally, this building has never been refurbished or otherwise altered since it was built. By contrast, Buildings 1 and 3 underwent an upgrade of the water distribution system and the construction of new accommodation sites, thus accounting for the small number of samples collected due to the absence of outlets.

These results are in line with Italian Guidelines and other authors [7,10], regarding the needs to have a broad knowledge of the buildings characteristics, the water distribution system, the pipelines material, and the disinfectant interaction with them, before to choose the disinfection method to use. As described by other authors, many actions undertaken during renovation works can induce a mobilization of biofilm and alter the flushing of disinfectant at distal outlets due to the lower water consumption and the closing of some outlets. The particulate and the increase of water turbidity, produced by structural works, can induce a decomposition of oxidants such as $H_2O_2$ [10,47,48]. Moreover, distal outlets in some parts of the hospital are seldom used: In particular areas of the hospital (surgeries room, intensive care units, etc.) the sterile water is preferred to the tap water, therefore the consumption of tap water is lower too.

The conclusion of accommodation works and the completion of the final structures within Buildings 1 and 3 allowed us to implement the risk assessment plan for *Legionella* control and to increase the number of sampling sites and the frequency of sampling, according to several studies indicating that routine cultures of the hospital water supply for *Legionella* may provide an important strategy for the prevention of legionellosis outbreaks [49].

To assess the effectiveness of WTP 828, we compared data obtained during the WTP1 phase with those obtained during the WTP2 phase. We observed a reduction both in the percentage of positive samples and the mean *Legionella* levels in all buildings during WTP2. In detail, a significant reduction in the amount of *Legionella* contamination was observed in Buildings 2 and 3 ($p$ = 0.001 and 0.037, respectively). *Legionella* control was then maintained for the entire duration of the study.

The observed differences in *Legionella* colonization between the buildings can be ascribed to the different uses and water consumption in these buildings. Risk factors that should not be overlooked are, in fact, the scale of the extension, connection of existing pipes within the newly constructed branched networks, presence of dead branches, pipe characteristics (e.g., materials, age), treatment of the water system (e.g., water softening and disinfection), intended utility, and maintenance procedures [20]. In light of these considerations, we also investigated our results in relation to data concerning annual water consumption in each building size, number of water outlets, pipe materials, and the timing of renovation works.

Building 1 has six levels and covers an area of 18,539.93 $m^2$. It mainly comprises offices, surgery rooms, operating rooms, and diagnostic rooms, some of which only require the use of sterile water; therefore, overall water consumption is limited. In this building, the third floor hosts a technical room for air treatment without water outlets; therefore, some closed pipes are present. The water consumption (1913 $m^3$/year) indicated a much lower use than in Building 2 (3017 $m^3$/year), suggesting lower water flushing from the outlets. It is evident that low use and stagnation of water may affect the activity and delivery of disinfectant, reducing its effect on the microorganisms [50,51]. The renovation works were completed in 2015. The pipelines that made up the water network comprised mainly multilayer PVC, which increases biofilm formation [52–57]. Our data revealed that, despite a reduction in the percentage of *Legionella*-positive sites and mean *Legionella* levels, WTP 828 was not completely effective in this building, demonstrating continuous fluctuations in the amount of *Legionella* spp. colonization. Corrective measures have since been implemented; these include two chemical shock treatments as above described and the implementation of maintenance hospital procedures such as increasing the flushing time once a week and during weekends, anti-scale procedures at each distal outlet (every fifteen days), and strictly cold and hot water temperature control weekly [12]. The long-term effects of our interventions resulted in the maintenance of *Legionella* contamination levels below the range of alert prescribed by the Italian Guidelines (101–1000 CFU/L); this will limit the risk of exposure and preserve the health of patients and workers.

In Building 2, the presence of multiple outlets (336) and some facilities with high water consumption (e.g., cafes, restaurants, and markets) suggested that water flushing facilitated the circulation of the disinfectant in the plumbing system, reducing the number of bacterium-positive samples and the *Legionella* concentration, in accordance with a study by Douterelo et al. [50]. The water distribution system consists of mainly galvanized iron, which, as suggested in the literature, on the contrary to plastic material, such as polyethylene (PE) and polyvinylchloride (PVC) [54,57], together with prolonged use of the WTP 828 disinfectant, may help to inhibit *Legionella* colonization and enable the maintenance of this inhibition over long periods.

Building 3 is the smallest structure of MCH, covering an area of 1271.06 $m^2$. The total annual water consumption in this building is 589 $m^3$ per 129 outlets. The services (two food preparation areas) and in-patient rooms allow the daily circulation of disinfectant within the plumbing system, thereby contributing to the effectiveness of WTP 828 in controlling *Legionella* contamination levels in this building.

The impact of the disinfectant used (WTP 828 or $ClO_2$ mixture) and the type of approach applied in MCH to reduce the risk of acquiring *Legionella* disease was therefore studied by calculating OR and RR epidemiological measures. Significant results were obtained comparing WTP 828 versus $ClO_2$ by calculating the OR measure, which showed that the introduction of WTP 828 after replacement of the $ClO_2$ mixture was a good strategy to decrease risk in MCH, increasing control of the *Legionella* contamination level. By comparing the two different phases of our study, WTP1 versus WTP2, in a prospective approach using the RR measure, showed that the approach implemented during the WTP2 phase characterized by a new monitoring plan, the increase in the number of samples and adoption of a new protocol of flushing, cleaning and disinfectant monitoring, can help to decrease the risk of acquiring the disease.

The risk of Legionellosis is linked to different factors such as personal characteristics and immunodeficiency status, but such personal risk factors can also enhance the risk of acquiring the disease when environmental control is not correctly performed or is underestimated [58,59].

The serotyping and genotyping data revealed different colonization patterns in MCH buildings, but we did not find a significant association between the presence of some *Legionella* strains in MCH buildings.

Different authors have suggested that changes in the disinfection treatment regime (e.g., the type of disinfectant) or the dose (e.g., shock treatment) might influence the type of *Legionella* strains that become prevalent [42,47,60–62]. In agreement with these observations, the increased WTP 828 dosage used during the shock treatment performed two times in Building 1 resulted in a reduction in *L. pneumophila* SG1 and increases in *L. species: L. anisa* and *L. rubrilucens* (data not shown).

The absence of *P. aeruginosa* from water samples during the study period seems to indicate a good effect of WTP 828 on the containment of these bacteria with respect to previous evidence provided by the MCH Health Director. During the previous treatments with $ClO_2$, the release of pipelines materials, the presence of accommodation works, and a lack of cleaning procedures favored the growth of *P. aeruginosa* (data not shown). The introduction of routine cultures of this bacterium in hot and cold-water samples suggested in the MCH study, also helped to control the efficiency of the cleaning procedures, other than evaluate the biofilm presence, where *Legionella* and other microorganisms become more resistant to antibiotics and disinfectants [48].

In detail, the protocol undertaken by the cleaning staff every fifteen days, i.e., cleaning and flushing procedures (e.g., disinfecting the taps and showers and flushing cold and hot water outlets) played an important role in preventing biofilm formation [63,64], which can support *Legionella* growth. Semiannual meetings with the stakeholders and hospital staff to inform them of the bacterial infection risk and the procedures undertaken to reduce such risks were also useful.

HPC is an indirect indicator of water quality and is often used to assess the efficacy of water treatment and to measure the amount of heterotrophic bacteria colonization in distribution systems. Despite some studies showing the absence of a correlation between levels of HPC bacteria and human infection, suggesting that HPC levels are not highly predictive of *Legionella* colonization, the control of this parameter could help to understand whether the water system contains potentially infectious organisms [33,44]. Our results indicated that WTP 828 performed well with respect to HPC containment during the entire study period, maintaining levels below directive limits [46].

A weakness of this study is that we were unable to demonstrate that WTP 828 treatment did not affect the physical and chemical parameters of the water in all study periods. Data regarding previous disinfection treatment with the $ClO_2$ mixture, along with data from the WTP1 phase, were missing; therefore, we could not compare changes in water quality that occurred during all study periods, underscoring the important role played by environmental monitoring of physical and chemical parameters when demonstrating the efficacy of a disinfectant. By contrast, the control of these parameters during the WTP2 phase allowed us to monitor the effect of the disinfectant on water pipes, take measures to prevent damage to the water network, and maintain the quality of "drinking water" to prevent risks to human health.

According to Borella et al. [10], the choice of WTP 828 has been carried out also on a careful evaluation of a cost-effective analysis, considering the system of disinfectant production (pump or others), the maintenance costs (disinfectant provision, service, etc.) and the potential dangerous effect on water pipelines, other than the possibility to safety measure disinfectant residues to outlets by colorimetric strips by staff.

## 4. Materials and Methods

### 4.1. MCH Structure and Water Outlet Characteristics

This study was conducted at Maria Cecilia Hospital (MCH), an Italian hospital founded in 1973 located in Cotignola (RA, Emilia Romagna).

The structure of the hospital is complex as it comprises three separate buildings (Buildings 1–3) built during different years and covering a total area of 27,989.64 m$^2$ (Supplementary Figure S1). Buildings 1 and 3 were constructed in 2001 and subjected to renovation or enlargement works until 2015. Building 2 is the main MCH building and did not undergo any changes in its structure during the whole study period.

These characteristics permit the study of WTP 828 activity as three separate HWNs and allow modulation of the dose with respect to the level of *Legionella* contamination, water demand, intended use, and renovation works (Supplementary Table S1).

At the end of the renovation works, the final structure of MCH had 122 in-patient rooms, each with one or two beds and an en-suite bathroom. There were 212 beds in total, mainly located in Building 2, and 769 water outlets (e.g., taps and showers) located in in-patient rooms, communal areas, diagnostic and operating rooms, offices and services, as follows:

- Building 1 covers an area of 18,539.93 m$^2$ and has six floors with communal areas for the guests (e.g., bar, restrooms), operating rooms, outpatient services (diagnostic and consulting rooms), intensive care units, and 27 in-patient rooms located on the second floor. In this building, 21 sampling points and one hot water return line point were identified. Two of the 21 sampling points in in-patient rooms were monitored monthly, on a rotational basis by room number (Supplementary Table S1).
- Building 2 covers an area of 8178.68 m$^2$, with six floors with 70 in-patient rooms distributed on floors one to four. Twenty-two sampling points were monitored (21 plus one hot water return line point) in this building, ten of which were in in-patient rooms, which were monitored monthly and rotated by room number (Supplementary Table S1).
- Building 3 covers an area of 1271.06 m$^2$ and was recently expanded, with a complete renovation in February 2015. The building has six floors, with 25 in-patient rooms located on the third and fourth floors. Due to their size and comfort, these rooms are designated as "suites" and are reserved for long-term guests. There were 13 sampling points (and one hot water return line point) in this building, six of which were in in-patient rooms, which were monitored monthly and rotated by room number (Supplementary Table S1).

### 4.2. Hospital Water Network (HWN)

The hospital plumbing system is very complex, partially antiquated, and (depending on age built) predominantly made up of galvanized iron and polyvinylchloride (PVC) multi-layers. The HWN was coated with an anti-scale treatment to create a protective film on the galvanized iron and PVC surface, as suggested by WHO guidelines in 2011 [32]. It consists of a product based on natural mineral salts such as orthophosphates, polyphosphates, and alkaline silicates dosed at 0.1 mg/L. The MCH structural characteristics, material pipelines, and water consumption for each building were kindly provided by Health Direction and described in Supplementary Table S1.

All buildings are supplied with the same municipal water aqueduct, which brings water from the Ridracoli dam located 53 km from Cotignola. The water is first collected in two 30 m$^3$ water reservoirs outside the buildings. After filtration through a 150 µm pore size filter, water is fed into two pipelines: one to the cooling towers and refrigerant circuit (a closed loop hydraulic system) and the other to the water treatment station (an open loop hydraulic system). A plan of the water distribution network is shown in Figure 1.

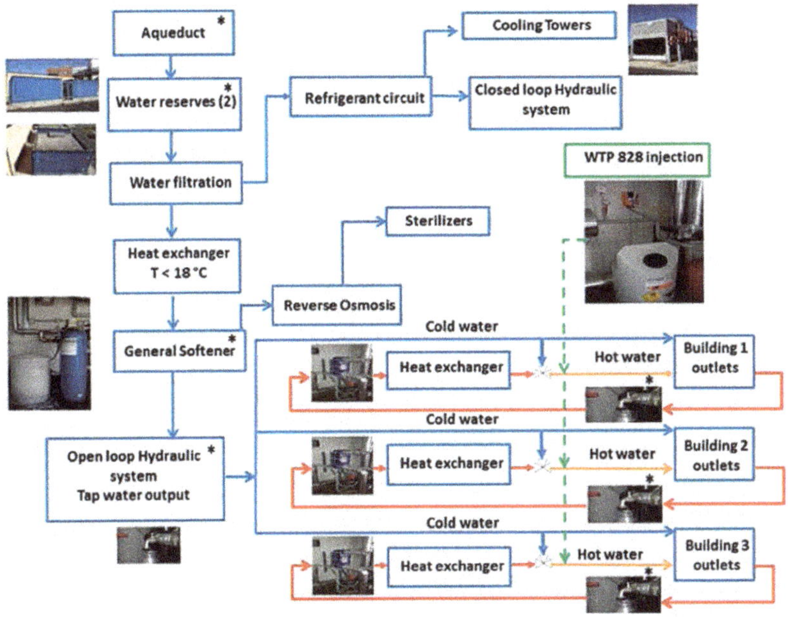


**Figure 1.** The scheme of the MCH water network with main sampling points in technical rooms (*).

A heat exchanger maintains the temperature of the cold water in the treatment station at <18 °C; the hardness of cold water is treated with a general softener to reduce its value between 12–15 °f (water moderately hard), which is in line with Italian and European Council directives [45,46]. Some of this water supplies the sterilizers after reverse osmosis treatment, and another portion is used as cold water by the hospital. The cold water is distributed to the substations within each building through a single tap water output. Three different heat exchangers (one at each substation) produce hot water. The cold and hot water circuits are independent of one another, and each building has its own hot water return line.

### 4.3. WTP 828

Water Team Process 828 (WTP 828) developed by an Italian Company involved in disinfectant production (Water Team S.r.l., Forlì (FC), Italy) is a multi-component oxidizing biocide formulated using a stabilized combination of $H_2O_2$ (34%, wt/wt) and $Ag^+$ salts (0.003%, wt/wt) in demineralized water, resulting in a highly effective disinfection solution. The formulation is covered by Italian regulation on intellectual property rights and actually is under investigation to acquire a patent. It is licensed by European and Italian legislation [27,65] for its application in drinking water. The synergistic action of $H_2O_2$ and $Ag^+$ salts renders the biocide more powerful than $H_2O_2$ alone [66,67]. $Ag^+$ was used to increase the activity of the peroxide, and $Ag^+$ forms an insoluble salt at distal points and is able to attach to pipes and exert bacteriostatic effects on biofilms [41,68].

The WTP 828 is injected into mixed water (hot/cold) after hot water output downstream from the heat exchangers and dosed proportionally to the volume of water supply.

WTP 828 was introduced into MCH for the first time in October 2013 after replacement of a previous disinfection system based on a continuous treatment performed with chlorine dioxide ($ClO_2$ mixture) at a dosage of 0.5 mg/L. This treatment, which was used from September 2009 to September 2013, had compromised the water pipelines and corroded some parts of the plant, thereby reducing efficacy with respect to *Legionella* colonization and supporting the presence of *P. aeruginosa* in some water outlets.

The WTP 828 concentrations during the study were modulated according to the microbiological results for each building. In particular, the initial dose of 30 mg/L resulting in a final concentration of 5–10 mg/L at distal outlets remained the same in Buildings 2 and 3 throughout the whole study period. By contrast, two shock treatments were required in Building 1 (from February to March and July to August 2015); at these times, the injected dose of WTP 828 increased up to 50–60 mg/L, which resulted in 25–30 mg/L of $H_2O_2$ at the distal outlets.

### 4.4. Study Design

This study was conducted in two experimental phases designated WTP1 and WTP2 in relation to the timing of the introduction of WTP 828, the renovation or enlargement works conducted in the buildings and the acquisition of a new risk assessment plan. The data collected during the WTP1 and WTP2 phases were then compared to evaluate differences in the efficacy of the WTP 828 treatment in the HWNs of the three buildings.

These data were then compared with the data obtained during disinfection with the $ClO_2$ mixture (i.e., $ClO_2$ mixture versus WTP1 phase and $ClO_2$ mixture versus WTP2 phase) to assess the effects of WTP 828 on *Legionella* contamination.

The details of the study period for each building are described below:

- Building 1: WTP1 from October 2013 to December 2014 and WTP2 from January 2015 to October 2015;
- Building 2: WTP1 from October 2013 to August 2014 and WTP2 from September 2014 to October 2015. This building was not subjected to any changes in disinfectant concentration or renovation works;
- Building 3: WTP1 October 2013 to March 2015 and WTP2 from April 2015 to October 2015.

During the WTP1 phase, disinfection with WTP 828 started in Building 2 in October 2013 and in some locations within Buildings 1 and 3, which were under construction or undergoing expansion in this period.

Sampling of hot water systems was performed according to the risk assessment plan, which was approved by the MCH Health Director and the Local Authority. There were 29 sampling points spread throughout the three buildings among consulting and diagnostic rooms, wards, common areas, and in-patient rooms, which were monitored every four months on a rotational basis. During this phase, a total of 53 samples were subjected to microbiological analysis for detection of *Legionella*, *P. aeruginosa* and heterotrophic plate count (HPC) bacteria at 36 °C, and data were collected.

During the WTP2 phase, renovation works of Buildings 1 and 3 were completed (January 2015 and April 2015, respectively). WTP 828 treatment was extended to all parts of these buildings and, based on preliminary results regarding WTP 828 efficacy (WTP1 phase), a new risk assessment and monitoring plan were adopted.

In accordance with Italian Guidelines [7], sampling points were chosen at the following three locations: in the vicinity of, mid-way to, and away from the technical room. The location of the sampling points took into account the size of the building, the number of in-patient rooms, the health services provided, the risk of patient, and worker exposure to bacteria and epidemiological data.

Every month, samples were collected from the technical room: one from the aqueduct, two from the cold water reserves, one downstream of the general softener treatment, one from a tap water output, and three from the hot water return lines (1a for Building 1, 1b for Building 2, and 1c for Building 3), and from another 55 sampling points in offices, consulting and diagnostic rooms, wards, common areas, and in-patient rooms (63 points in total). Despite the large number of in-patient rooms, the alternating sampling method enabled sampling of almost all in-patient rooms in the three buildings.

The increased time of monitoring (from once every four months to monthly), extension of the disinfection treatment, and development of a final MCH structure permitted the study of the modulation of microbiological and physical-chemical parameters in a total of 296 hot water and 65 cold water samples.

## 4.5. Sample Collection and Microbiological Analysis

Hot water and cold water (2 L) were collected in post-flushing modality (running water for 1 min) in sterile polytetrafluoroethylene (PTFE) bottles containing a sodium thiosulfate solution (10%, v/v). Microbiological analyses were performed in accordance with ISO11731:2017 [69] to detect and enumerate *Legionella*. During *Legionella* surveillance, according to Italian Guidelines [7], the level of risk took into account the concentration of bacteria and percentage of positive samples.

Samples were concentrated using 0.22 µm polycarbonate pre-sterilized filter membranes (Sartorius Stedim Biotech, Göttingen, Germany).

The concentrated samples (filtered, F) were then heated (for 30 min at 50 °C) to inhibit interfering microbiota (heated, H). Then, 0.1 mL of the untreated sample (UN) and 0.1 mL of each F and H sample were spread in duplicate onto GVPC agar plates (*Legionella* GVPC selective medium, Thermo Fisher Scientific, Oxoid Ltd., Basingstoke, UK), and incubated at 35.5 °C in a humid (2.5% $CO_2$) environment.

The plates were examined after four, eight, and 14 days, and colonies with a typical *Legionella* morphology (presumptive) were enumerated and confirmed by sub-culture on BCYE agar with and without cysteine. The isolates that grew on BCYE but failed to grow on the cysteine-free medium were verified serologically by an agglutination test (*Legionella* latex test kit; Thermo Fisher Scientific, Oxoid Ltd.). The data are expressed as the mean concentration ± standard deviation (SD) of the $\log_{10}$ colony forming units (CFU) per liter of water ($\log_{10}$ CFU/L) including all samples analyzed (positive + negative). The detection limit of the culture technique was 50 CFU/L. The samples with a value of <50 CFU/L were considered negative according to ISO 11731:2017 [69].

Other microorganisms can affect the growth of cultivable *Legionella*, and the samples were simultaneously analyzed for the presence of *P. aeruginosa*, a known competitor of *Legionella* that inhibits its growth on medium [70]. The analyses were performed according to UNI EN ISO 16266:2006 [71] using a selective Pseudomonas agar (Biolife, Milan, Italy). The detection limit of the culture technique was 1 CFU/100 mL.

The heterotrophic plate count (HPC) at 36 °C was used as an indicator of the actual level of bacterial contamination at the sampling points. The HPC is a useful indicator of increased microbial growth, increased biofilm activity, extended retention times, water stagnation, or breakdown of the integrity of the system [33,72]. The analyses were performed using a standard plate method based on tryptic glucose yeast agar (Biolife) in accordance with UNI EN ISO 6222:2001 [73]. The data are expressed as the mean concentration ± SD of the $\log_{10}$ CFU per milliliter of water ($\log_{10}$ CFU/mL) including all samples (positive + negative).

The detection limit of the culture technique was 1 CFU/mL.

## 4.6. Legionella Typing

Colonies identified by the agglutination test as belonging to the genus *Legionella* were subsequently analyzed by DNA sequencing. In particular, all strains identified as *L. pneumophila* were analyzed by sequence-based typing (SBT) to determine the sequence type (ST); strains identified as *Legionella* species were analyzed by *mip* sequencing. Genomic DNA was extracted from cultures using the InstaGene Purification Matrix (Bio-Rad, Hercules, CA, USA). SBT was performed according to an ELDSNet protocol (http://bioinforatics.phe.org.uk/legionella/legionella_sbt/php/sbt_homepage.php). The protocol was based on the sequencing of seven genes (flaA, pilE, asd, mip, mompS, proA, and neuA) and on the assignment of a ST allelic profile by the ELDSNet database (http://www.hpabioinformatics.org.uk/cgibin/legionella/sbt/seq_assemble_legionella1.cgi).

The strains that were serotyped by agglutination as *L. species* were then genotyped by *mip* gene amplification via the polymerase chain reaction (PCR) using degenerate primers, as described in 1998 by Ratcliff et al. [74] and modified by M13 tailing to avoid noise in the DNA sequence [75]. Gene amplification was carried out in a 50 µL reaction volume containing DreamTaq Green PCR Master Mix 2x (Thermo Fisher Scientific, OxoidLtd., Basingstoke, UK) and 40 pmol of each primer; 100 ng of DNA extracted from the presumptive colonies of *Legionella* was added as template. The same amounts of DNA from *Legionella pneumophila* (*L. pneumophila*) type strain EUL00137 provided by the European Working Group for *Legionella* Infections [76] and fetal bovine serum were used as positive and negative controls, respectively.

Following purification, DNA was sequenced using BigDye Chemistry and analyzed on an ABI PRISM 3100 Genetic Analyzer (Applied Biosystems, Foster City, CA, USA). Specifically, *mip* amplicons (661–715 bp) were sequenced using M13 forward and reverse primers (M13 FW, 5′-TGTAAAACGACGGCCAGT-3′; M13 RW, 3′-CAGGAAACAGCTATGACC-5′) to obtain complete coverage of the sequenced region of interest. Raw sequencing data were assembled using CLC Main Workbench 7.6.4 software (https://www.qiagenbioinformatics.com/). The sequences were compared with sequences deposited in the *Legionella mip* gene sequence database using a similarity analysis tool (http://bioinformatics.phe.org.uk/cgi-bin/legionella/mip/mip_id.cgi). The identification at the species level was conducted based on 98% similarity to a sequence in the database [77].

### 4.7. Physical and Chemical Parameters of Water

The physical and chemical parameters of water were analyzed only during the WTP2 phase, before this phase the hospital did not have any data on water quality as prescribed by WHO [49].

Cold water samples (1 L) were collected from each of the following locations: the aqueduct, water reserves, softener, and tap water output. Hot water samples (1 L) were collected from each of the three hot water return lines and distal outlets. The pH, hardness (°f), conductivity (μS/cm), turbidity (nephelometric turbidity units), total iron content (mg/L), total phosphorus content (mg/L of $P_2O_5$), and $Ag^+$ content (μg/L) were monitored monthly during the session sampling.

The analysis of total iron and phosphorus content (orthophosphate, condensed phosphate, and organic phosphate) allowed us to monitor the maintenance of anti-scale and corrosion treatment.

Temperature (°C) and residual WTP 828 levels [the peroxide component (mg/L)] were measured and recorded at distal outlets weekly in each building. WTP 828 (peroxide component) was measured using an MQuant™ Peroxide Test (Merck KGaA, Darmstadt, Germany) according to the manufacturer's instructions.

Other parameters were measured using different techniques according to standardized APAT CNR IRSA methods [78].

In our study the disinfection treatment was performed by a disinfectant based on $H_2O_2/Ag^+$, therefore the dosage of DBPs release in water is not necessary. The chemical water compounds measured are listed in Table 4. The results are expressed as the mean value ± SD.

### 4.8. Data Analyses

Bacteriological data were converted into log10 x values to normalize the distribution for the correlation analysis. The normality of continuous variables was assessed using the Shapiro-Wilk test, and data are presented as the mean ± SD. Continuous variables were evaluated using one-way ANOVA and a post-hoc test (Bonferroni), and categorical variables were compared using the $\chi 2$ and Mann Whitney test. One-way ANOVA and the post-hoc test (Bonferroni) were conducted to assess differences between disinfectant treatments and between buildings.

To test the changes in *Legionella* risk between treatments, we used odds ratios (ORs) in a retrospective analysis and relative risk (RR) in a prospective analysis. In detail, OR was calculated for WTP1 versus the $ClO_2$ mixture and WTP2 versus the $ClO_2$ mixture, and RR was calculated for the prospective treatments (WTP2 versus WTP1). Statistical analyses were performed using STATA version 10.0 (Stata Corp., College Station, TX). A *p*-value < 0.05 was accepted as significant.

### 4.9. Hospital LD Surveillance

MCH performed active legionellosis surveillance beginning in 2013. Data were collected throughout the entire study period (WTP1 and WTP2 phases). The symptoms of legionellosis are consistent with an acute infection of the lower airways, with clinical and/or radiological signs of focal pneumonia. A preliminary diagnosis was routinely confirmed by a urine antigen test (*Legionella* Urine Antigen EIA, Biotest, Milan, Italy) and a serological immunofluorescence test (*L. pneumophila* IFA, Meridian Diagnostic Europe, London, UK).

## 5. Conclusions

It is often difficult to guarantee the absence of *Legionella* from water distribution systems, even if a disinfection system is in place. Our data revealed that differences in three buildings belonging to the same structure were linked to building size, water consumption rates, the number of outlets, and their intended use. WTP 828 performed well in terms of reducing *Legionella* contamination, but only a change in the study approach (adequate risk assessment plan, increase in monitoring samples sites, and alteration of the WTP 828 dosage in relation to the *Legionella* levels) facilitated the discovery of differences in *Legionella* colonization and an understanding of disinfectant activity dynamics.

Further investigations are needed to elucidate how the dose of disinfectant affects the presence of specific strains in each building and to generate a risk map highlighting the phylogenetic correlations between strains. The assessment of changes in colonization dynamics will be useful for controlling the concentration and type of disinfectant that can be used in a water system (i.e., shock or continuous treatment, bacterial resistance development) in relation to accommodation works and technical operations in the water network that could support *Legionella* proliferation.

The low cost of WTP 828, the dosage by a pump, the easy maintenance procedures and simple and safe check of disinfectant residue at distal outlets, suggest that the approach used in this study could be a valid alternative to traditional disinfection methods.

**Supplementary Materials:** The following are available online at http://www.mdpi.com/2076-0817/8/4/209/s1. Figure S1. Site map of MCH, Cotignola (RA), Italy. A representative view of MCH and a map of its three buildings: MCH picture (A) and MCH planimetry (B); Table S1. MCH structure and water outlet characteristics.

**Author Contributions:** S.C. and L.G. conceived, designed the experiments, wrote the paper, and performed the experiments; A.G. and G.F.S. performed the sequence analysis; A.D. performed the data analysis; D.M., A.C., F.S., and A.M. provided disinfectants and performed the water sample collection; T.P. and P.S. provided technical support to study the MCH hot water network.

**Funding:** This work was supported by a research grant (RFO 2017) from University of Bologna, Italy.

**Acknowledgments:** The authors would like to thank the MCH staff for helping with the study, the technical staff at Water Team S.r.l. for assistance and provision of the WTP 828 disinfectant, Eta Beta S.r.l. for technical assistance, and Graziella Ciani for help with the microbiological analyses.

**Conflicts of Interest:** S.C., L.G., A.G., G.F.S., and A.D., declare no competing interest; T.P. and P.S. are employed by GVM Care & Research; D.M. and A.C. by Eta-Beta S.r.l. and F.S.; and A.M. by Water Team S.r.l. Water Team S.r.l. provided assistance and provision of the WTP 828 disinfectant and Eta Beta S.r.l. provided technical assistance for disinfection treatment.

**Ethics Approval and Consent to Participate:** Not applicable.

**Consent for Publication:** Not applicable.

**Availability of Data Materials:** The datasets generated and analyzed during the study are not publicly available due to ongoing further analysis on the data and protected by hospital privacy policies. The corresponding authors can response to any further questions about them.

## References

1. Hoffman, P.; Friedman, H.; Bendinelli, M. *Legionella Pneumophila: Pathogenesis and Immunity*; Springer: Berlin/Heidelberg, Germany, 2008; Volume XVIII, p. 208. [CrossRef]
2. Cervero-Aragó, S.; Rodríguez-Martínez, S.; Puertas-Bennasar, A.; Araujo, R.M. Effect of Common Drinking Water Disinfectants, Chlorine and Heat, on Free Legionella and Amoebae-Associated Legionella. *PLoS ONE* **2015**, *10*, e0134726. [CrossRef]
3. Blanc, D.; Carrara, P.; Zanetti, G.; Francioli, P.; Blanc, D. Water disinfection with ozone, copper and silver ions, and temperature increase to control Legionella: Seven years of experience in a university teaching hospital. *J. Hosp. Infect.* **2005**, *60*, 69–72. [CrossRef]

4. Canals, O.; Serrano-Suárez, A.; Salvadó, H.; Méndez, J.; Cervero-Aragó, S.; Ruiz de Porras, V.; Dellundè, J.; Araujo, R. Effect of chlorine and temperature on free-living protozoa in operational man-made water systems (cooling towers and hot sanitary water systems) in Catalonia. *Environ. Sci. Pollut. Res. Int.* **2015**, *22*, 6610–6618. [CrossRef]
5. Ministerial Decree 15 December 1990. Information System of Infectious and Diffusive Diseases (G.U. n.6 of 8 January 1991. Available online: http://www.salute.gov.it/imgs/C17normativa1357allegato.pdf (accessed on 27 June 2018).
6. Rota, M.C.; Caporali, M.G.; Bella, A.; Scaturro, M.; Giannitelli, S.; Ricci, M.L. Rapporto Annuale Sulla Legionellosi in Italia Nel 2017. Notiziario Ist. Super Sanità n.9, Volume 31. September 2018. Available online: http://www.legionellaonline.it/notiziario%20legionellosi%20_9_2018.pdf (accessed on 20 September 2019).
7. Italian Health Ministry. Guidelines for Prevention and Control of Legionellosis. Approvate in Conferenza Stato-Regioni Seduta Del 7 Maggio 2015. Available online: http://www.salute.gov.it/imgs/C_17_pubblicazioni_2362_allegato.pdf (accessed on 26 June 2018).
8. Montagna, M.T.; De Giglio, O.; Napoli, C.; Diella, G.; Rutigliano, S.; Agodi, A.; Auxilia, F.; Baldovin, T.; Bisetto, F.; Arnoldo, L.; et al. Control and prevention measures for legionellosis in hospitals: A cross-sectional survey in Italy. *Environ. Res.* **2018**, *166*, 55–60. [CrossRef] [PubMed]
9. Kim, B.; Anderson, J.; Mueller, S.; Gaines, W.; Kendall, A. Literature review—Efficacy of various disinfectants against Legionella in water systems. *Water Res.* **2002**, *36*, 4433–4444. [CrossRef]
10. Borella, P.; Bargellini, A.; Marchegiano, P.; Vecchi, E.; Marchesi, I. Hospital-acquired Legionella infections: An update on the procedures for controlling environmental contamination. *Ann. Ig.* **2016**, *28*, 98–108. [PubMed]
11. Farhat, M.; Moletta-Denat, M.; Frère, J.; Onillon, S.; Trouilhé, M.C.; Robine, E. Effects of disinfection on Legionella spp., eukarya, and biofilms in a hot water system. *Appl. Environ. Microbiol.* **2012**, *78*, 6850–6858. [CrossRef]
12. Cristino, S.; Legnani, P.P.; Leoni, E. Plan for the control of Legionella infections in long-term care facilities: Role of environmental monitoring. *Int. J. Hyg. Environ. Health* **2012**, *215*, 279–285. [CrossRef]
13. Bédard, E.; Boppe, I.; Kouamé, S.; Martin, P.; Pinsonneault, L.; Valiquette, L.; Racine, J.; Prévost, M. Combination of Heat Shock and Enhanced Thermal Regime to Control the Growth of a Persistent Legionella pneumophila Strain. *Pathogens* **2016**, *5*, 35. [CrossRef]
14. White, G.C. *The Handbook of Chlorination and Alternative Disinfectants*, 4th ed.; John Wiley & Sons: New York, NY, USA, 1999.
15. Florentin, A.; Hautemanière, A.; Hartemann, P. Health effects of disinfection by-products in chlorinated swimming pools. *Int. J. Hyg. Environ. Health* **2011**, *214*, 461–469. [CrossRef]
16. Lin, Y.E.; Stout, J.E.; Yu, V.L. Controlling Legionella in Hospital Drinking Water: An Evidence-Based Review of Disinfection Methods. *Infect. Control Hosp. Epidemiol.* **2011**, *32*, 166–173. [CrossRef] [PubMed]
17. WHO. Trihalomethanes in Drinking-Water. 2005. Available online: http://www.who.int/water_sanitation_health/dwq/chemicals/THM200605.pdf (accessed on 22 June 2018).
18. Richardson, S.D.; Plewa, M.J.; Wagner, E.D.; Schoeny, R.; DeMarini, D.M. Occurrence, genotoxicity, and carcinogenicity of regulated and emerging disinfection by-products in drinking water: A review and roadmap for research. *Mutat. Res.* **2007**, *636*, 178–242. [CrossRef] [PubMed]
19. Smith, E.M.; Plewa, M.J.; Lindell, C.L.; Richardson, S.D.; Mitch, W.A. Comparison of Byproduct Formation in Waters Treated with Chlorine and Iodine: Relevance to Point-of-Use Treatment. *Environ. Sci. Technol.* **2010**, *44*, 8446–8452. [CrossRef] [PubMed]
20. World Health Organization. Water Safety in Distribution Systems. WHO Library Cataloguing-in-Publication Data. 2014. Available online: http://www.who.int/water_sanitation_health/publications/WSH-distribution_system-20141114.pdf (accessed on 25 June 2018).
21. Schoenen, D. Role of disinfection in suppressing the spread of pathogens with drinking water: Possibilities and limitations. *Water Res.* **2002**, *36*, 3874–3888. [CrossRef]
22. Linley, E.; Denyer, S.P.; McDonnell, G.; Simons, C.; Maillard, J.-Y. Use of hydrogen peroxide as a biocide: New consideration of its mechanisms of biocidal action. *J. Antimicrob. Chemother.* **2012**, *67*, 1589–1596. [CrossRef]
23. Baldry, M. The bactericidal, fungicidal and sporicidal properties of hydrogen peroxide and peracetic acid. *J. Appl. Bacteriol.* **1983**, *54*, 417–423. [CrossRef]
24. Martin, N.L.; Bass, P.; Liss, S.N. Antibacterial Properties and Mechanism of Activity of a Novel Silver-Stabilized Hydrogen Peroxide. *PLoS ONE* **2015**, *10*, e0131345. [CrossRef]

25. Ricci, M.; Dell'Eva, I.; Scaturro, M.; Baruchelli, P.; De Ponte, G.; Losardo, M.; Ottaviani, M.; Guizzardi, F. Six-Month Experience of Silver-Hydrogen Peroxide Treatment for Legionella Control in Two Nursing Home Water Systems. In *Legionella. State of the Art 30 Years after Its Recognition*; Cianciotto, N., Kwaik, Y., Edelstein, P., Fields, B., Geary, D., Harrison, T., Joseph, C., Ratcliff, R., Stout, J., Swanson, M., Eds.; ASM Press: Washington, DC, USA, 2006; pp. 505–508. [CrossRef]
26. McDonnell, G.; Russell, A.D. Antiseptics and Disinfectants: Activity, Action, and Resistance. *Clin. Microbiol. Rev.* **2001**, *14*, 227. [CrossRef]
27. Regulation (EU) 528/2012 of the European Parliament and of the Council of 22 May 2012 Concerning the Making Available on the Market and Use of Biocide Products. Available online: http://eur-lex.europa.eu/legal-content/EN/TXT/PDF/?uri=CELEX:32012R0528&from=IT (accessed on 28 June 2018).
28. Juven, B.J.; Pierson, M.D. Antibacterial Effects of Hydrogen Peroxide and Methods for Its Detection and Quantitation. *J. Food Prot.* **1996**, *59*, 1233–1241. [CrossRef]
29. Slawson, R.M.; Van Dyke, M.I.; Lee, H.; Trevors, J.T. Germanium and silver resistance, accumulation, and toxicity in microorganisms. *Plasmid* **1992**, *27*, 72–79. [CrossRef]
30. Loo, S.-L.; Fane, A.G.; Lim, T.-T.; Krantz, W.B.; Liang, Y.-N.; Liu, X.; Hu, X. Superabsorbent Cryogels Decorated with Silver Nanoparticles as a Novel Water Technology for Point-of-Use Disinfection. *Environ. Sci. Technol.* **2013**, *47*, 9363–9371. [CrossRef] [PubMed]
31. Ehdaie, B.; Krause, C.; Smith, J.A. Porous Ceramic Tablet Embedded with Silver Nanopatches for Low-Cost Point-of-Use Water Purification. *Environ. Sci. Technol.* **2014**, *48*, 13901–13908. [CrossRef] [PubMed]
32. Tugulea, A.-M.; Berube, D.; Giddings, M.; Lemieux, F.; Hnatiw, J.; Priem, J.; Avramescu, M.-L. Nano-silver in drinking water and drinking water sources: Stability and influences on disinfection by-product formation. *Environ. Sci. Pollut. Res.* **2014**, *21*, 11823–11831. [CrossRef] [PubMed]
33. World Health Organization. Guidelines for Drinking-Water Quality. 2011. Available online: http://apps.who.int/iris/bitstream/10665/44584/1/9789241548151_eng.pdf (accessed on 28 June 2018).
34. Pedahzur, R.; Lev, O.; Fattal, B.; Shuval, H.I. The interaction of silver ions and hydrogen peroxide in the inactivation of E. coli: A preliminary evaluation of a new long acting residual drinking water disinfectant. *Water Sci. Technol.* **1995**, *31*, 123–129. [CrossRef]
35. Pedahzur, R. Silver and hydrogen peroxide as potential drinking water disinfectants: Their bactericidal effects and possible modes of action. *Water Sci. Technol.* **1997**, *35*, 87–93. [CrossRef]
36. Nabizadeh, R.; Samadi, N.; Sadeghpour, Z.; Beikzadeh, M. Feasibility study of using complex of hydrogen peroxide and silver for disinfecting swimming pool water and its environment. *Iran. J. Environ. Health* **2008**, *5*, 235–242.
37. Absalan, A.; Ehrampoush, M.; Davoudi, M.; Vakili, T.; Ebrahimi, A. Antibacterial effects of hydrogen peroxide and silver composition on selected pathogenic enterobacteriaceae. *Int. J. Environ. Health Eng.* **2012**, *1*, 23. [CrossRef]
38. Thomas, V.; Bouchez, T.; Nicolas, V.; Robert, S.; Loret, J.; Levi, Y. Amoebae in domestic water systems: Resistance to disinfection treatments and implication in Legionella persistence. *J. Appl. Microbiol.* **2004**, *97*, 950–963. [CrossRef]
39. Berry, D.; Xi, C.; Raskin, L. Microbial ecology of drinking water distribution systems. *Curr. Opin. Biotechnol.* **2006**, *17*, 297–302. [CrossRef]
40. Ditommaso, S.; Giacomuzzi, M.; Ricciardi, E.; Zotti, C.M. Efficacy of a Low Dose of Hydrogen Peroxide (Peroxy $Ag^+$) for Continuous Treatment of Dental Unit Water Lines: Challenge Test with Legionella pneumophila Serogroup 1 in a Simulated Dental Unit Waterline. *Int. J. Environ. Res. Public Health* **2016**, *13*, 745. [CrossRef]
41. Shuval, H.; Shenman, R.; Yarom, R. An Innovative Method for the Control of Legionella Infections in the Hospital Hot Water Systems with a Stabilized Hydrogen Peroxide-Silver Formulation. *Int. J. Infect. Control* **2009**, *5*. [CrossRef]
42. Casini, B.; Aquino, F.; Totaro, M.; Miccoli, M.; Galli, I.; Manfredini, L.; Giustarini, C.; Costa, A.L.; Tuvo, B.; Valentini, P.; et al. Application of Hydrogen Peroxide as an Innovative Method of Treatment for Legionella Control in a Hospital Water Network. *Pathogens* **2017**, *6*, 15. [CrossRef] [PubMed]
43. Rasamiravaka, T.; Labtani, Q.; Duez, P.; El Jaziri, M. The Formation of Biofilms by Pseudomonas aeruginosa: A Review of the Natural and Synthetic Compounds Interfering with Control Mechanisms. *BioMed Res. Int.* **2015**, *2015*, 759348. [CrossRef] [PubMed]

44. Duda, S.; Baron, J.L.; Wagener, M.M.; Vidic, R.D.; Stout, J.E. Lack of correlation between Legionella colonization and microbial population quantification using heterotrophic plate count and adenosine triphosphate bioluminescence measurement. *Environ. Monit. Assess.* **2015**, *187*, 393. [CrossRef]
45. Commission Regulation (EC) 1451/2007. Available online: http://eur-lex.europa.eu/legal-content/IT/TXT/PDF/?uri=CELEX:32007R1451&from=it (accessed on 28 June 2018).
46. Decreto Legislativo 2 febbraio 2001, n. 31, Attuazione Della Direttiva 98/83/CE Relativa Alla Qualità Delle Acque Destinate al Consumo Umano (G.U. n. 52 Del 3 Marzo 2001—s.o.n. 41). Available online: http://www.gazzettaufficiale.it/eli/id/2001/03/03/001G0074/sg (accessed on 22 June 2018).
47. Marchesi, I.; Ferranti, G.; Mansi, A.; Marcelloni, A.M.; Proietto, A.R.; Saini, N.; Borella, P.; Bargellini, A. Control of Legionella Contamination and Risk of Corrosion in Hospital Water Networks following Various Disinfection Procedures. *Appl. Environ. Microbiol.* **2016**, *82*, 2959–2965. [CrossRef] [PubMed]
48. Mahapatra, A.; Padhi, N.; Mahapatra, D.; Bhatt, M.; Sahoo, D.; Jena, S.; Dash, D.; Chayani, N. Study of Biofilm in Bacteria from Water Pipelines. *J. Clin. Diagn. Res.* **2015**, *9*, DC09–DC11. [CrossRef]
49. Sabria, M.; Yu, V.L. Hospital-acquired legionellosis: Solutions for a preventable infection. *Lancet Infect. Dis.* **2002**, *2*, 368–373. [CrossRef]
50. Douterelo, I.; Sharpe, R.; Boxall, J. Influence of hydraulic regimes on bacterial community structure and composition in an experimental drinking water distribution system. *Water Res.* **2013**, *47*, 503–516. [CrossRef]
51. Liu, S.; Gunawan, C.; Barraud, N.; Rice, S.A.; Harry, E.J.; Amal, R. Understanding, Monitoring, and Controlling Biofilm Growth in Drinking Water Distribution Systems. *Environ. Sci. Technol.* **2016**, *50*, 8954–8976. [CrossRef]
52. Rogers, J.; Dowsett, A.B.; Dennis, P.J.; Lee, J.V.; Keevil, C.W. Influence of Plumbing Materials on Biofilm Formation and Growth of Legionella pneumophila in Potable Water Systems. *Appl. Environ. Microbiol.* **1994**, *60*, 1842–1851.
53. Lehtola, M.J.; Laxander, M.; Miettinen, I.T.; Hirvonen, A.; Vartiainen, T.; Martikainen, P.J. The effects of changing water flow velocity on the formation of biofilms and water quality in pilot distribution system consisting of copper or polyethylene pipes. *Water Res.* **2006**, *40*, 2151–2160. [CrossRef] [PubMed]
54. Yu, J.; Kim, D.; Lee, T. Microbial diversity in biofilms on water distribution pipes of different materials. *Water Sci. Technol.* **2010**, *61*, 163–171. [CrossRef]
55. Rożej, A.; Cydzik-Kwiatkowska, A.; Kowalska, B.; Kowalski, D. Structure and microbial diversity of biofilms on different pipe materials of a model drinking water distribution systems. *World J. Microbiol. Biotechnol.* **2015**, *31*, 37–47. [CrossRef] [PubMed]
56. Schwake, D.O.; Alum, A.; Abbaszadegan, M. Impact of Environmental Factors on Legionella Populations in Drinking Water. *Pathogens* **2015**, *4*, 269–282. [CrossRef] [PubMed]
57. Wingender, J.; Flemming, H.-C. Contamination potential of drinking water distribution network biofilms. *Water Sci. Technol.* **2004**, *49*, 277–286. [CrossRef]
58. Fields, B.S.; Benson, R.F.; Besser, R.E. Legionella and Legionnaires' Disease: 25 Years of Investigation. *Clin. Microbiol. Rev.* **2002**, *15*, 506–526. [CrossRef]
59. Bartram, J.; Chartlier, Y.; Lee, J.V.; Pond, K.; Surman-Lee, S. *Legionella and the Prevention of Legionellosis*; World Health Organization: Geneva, Switzerland, 2007; Available online: http://www.who.int/water_sanitation_health/emerging/legionella.pdf (accessed on 21 June 2018).
60. Brenner, D.J.; Steigerwalt, A.G.; Gorman, G.W.; Wilkinson, H.W.; Bibb, W.F.; Hackel, M.; Tyndall, R.L.; Campbell, J.; Feeley, J.C.; Thacker, W.L.; et al. Ten new species of *Legionella*. *Int. J. Syst. Evol. Microbiol.* **1985**, *35*, 50–59. [CrossRef]
61. Casati, S.; Conza, L.; Bruin, J.; Gaia, V. Compost facilities as a reservoir of Legionella pneumophila and other Legionella species. *Clin. Microbiol. Infect.* **2010**, *16*, 945–947. [CrossRef]
62. Duda, S.; Kandiah, S.; Stout, J.E.; Baron, J.L.; Yassin, M.; Fabrizio, M.; Ferrelli, J.; Hariri, R.; Wagener, M.M.; Goepfert, J.; et al. Evaluation of a New Monochloramine Generation System for Controlling Legionella in Building Hot Water Systems. *Infect. Control Hosp. Epidemiol.* **2014**, *35*, 1356–1363. [CrossRef]
63. Husband, P.; Boxall, J. Asset deterioration and discolouration in water distribution systems. *Water Res.* **2011**, *45*, 113–124. [CrossRef]
64. El-Chakhtoura, J.; Saikaly, P.E.; Van Loosdrecht, M.C.M.; Vrouwenvelder, J.S. Impact of Distribution and Network Flushing on the Drinking Water Microbiome. *Front. Microbiol.* **2018**, *9*, 2205. [CrossRef] [PubMed]

65. Council of the European Union. European Council Directive 98/83/EC of 3 November 1998 on the Quality of Water Intended for Human Consumption. *Off. J. Eur. Community* **1998**, *330*, 32–54.
66. Imlay, J.A. The molecular mechanisms and physiological consequences of oxidative stress: Lessons from a model bacterium. *Nat. Rev. Genet.* **2013**, *11*, 443–454. [CrossRef] [PubMed]
67. Vatansever, F.; De Melo, W.C.; Avci, P.; Vecchio, D.; Sadasivam, M.; Gupta, A.; Chandran, R.; Karimi, M.; Parizotto, N.A.; Yin, R.; et al. Antimicrobial strategies centered around reactive oxygen species—Bactericidal antibiotics, photodynamic therapy, and beyond. *FEMS Microbiol. Rev.* **2013**, *37*, 955–989. [CrossRef]
68. Pedahzur, R.; Katzenelson, D.; Barnea, N.; Lev, O.; Shuval, H.; Fattal, B.; Ulitzur, S. The efficacy of long-lasting residual drinking water disinfectants based on hydrogen peroxide and silver. *Water Sci. Technol.* **2000**, *42*, 293–298. [CrossRef]
69. ISO 11731:2017 Water Quality—Enumeration of Legionella. Available online: https://www.iso.org/standard/61782.html (accessed on 22 June 2018).
70. Kimura, S.; Tateda, K.; Ishii, Y.; Horikawa, M.; Miyairi, S.; Gotoh, N.; Ishiguro, M.; Yamaguchi, K. Pseudomonas aeruginosa Las quorum sensing autoinducer suppresses growth and biofilm production in Legionella species. *Microbiology* **2009**, *155*, 1934–1939. [CrossRef]
71. UNI EN ISO 16266:2006 Water Quality—Detection and Enumeration of Pseudomonas Aeruginosa—Method by Membrane Filtration. 2006. Available online: https://www.iso.org/standard/39272.html (accessed on 24 June 2018).
72. National Research Council. *Drinking Water Distribution Systems: Assessing and Reducing Risks*; The National Academies Press: Washington, DC, USA, 2006. [CrossRef]
73. UNI EN ISO 6222:2001 Water Quality—Enumeration of Culturable Micro-Organisms—Colony Count by Inoculation in a Nutrient Agar Culture Medium. 2001. Available online: https://www.iso.org/standard/28960.html (accessed on 24 June 2018).
74. Ratcliff, R.M.; Lanser, J.A.; Manning, P.A.; Heuzenroeder, M.W. Sequence-Based Classification Scheme for the Genus Legionella Targeting the mip Gene. *J. Clin. Microbiol.* **1998**, *36*, 1560–1567.
75. Mentasti, M.; Fry, N.K.; Afshar, B.; Palepou-Foxley, C.; Naik, F.C.; Harrison, T.G.; Fry, N. Application of Legionella pneumophila-specific quantitative real-time PCR combined with direct amplification and sequence-based typing in the diagnosis and epidemiological investigation of Legionnaires' disease. *Eur. J. Clin. Microbiol. Infect. Dis.* **2012**, *31*, 2017–2028. [CrossRef]
76. Fry, N.K.; Alexiou-Daniel, S.; Bangsborg, J.M.; Bernander, S.; Castellani Pastoris, M.; Etienne, J.; Forsblom, B.; Gaia, V.; Helbig, J.H.; Lindsay, D.; et al. A multicenter evaluation of genotypic methods for the epidemiological typing of Legionella pneumophila serogroup 1: Results of a pan-European study. *Clin. Microbiol. Infec.* **1999**, *5*, 462–477. [CrossRef]
77. Fry, N.; Afshar, B.; Bellamy, W.; Underwood, A.P.; Ratcliff, R.; Harrison, T.G. Identification of Legionella spp. by 19 European reference laboratories: Results of the European Working Group for Legionella Infections External Quality Assessment Scheme using DNA sequencing of the macrophage infectivity potentiator gene and dedicated online tools. *Clin. Microbiol. Infect.* **2007**, *13*, 1119–1124.
78. APAT. Agenzia per la Protezione Dell'ambiente e per i Servizi Tecnici. Manuali e Linee Guida 29/2003: Metodi Analitici per le Acque. 2003. Available online: http://www.isprambiente.gov.it/it/pubblicazioni/manuali-e-linee-guida/metodi-analitici-per-le-acque (accessed on 25 June 2018).

*pathogens*

MDPI

*Article*

# Prevention and Control of *Legionella* and *Pseudomonas* spp. Colonization in Dental Units

**Benedetta Tuvo [1], Michele Totaro [1], Maria Luisa Cristina [2], Anna Maria Spagnolo [2], David Di Cave [3], Sara Profeti [1], Angelo Baggiani [1], Gaetano Privitera [1] and Beatrice Casini [1,*]**

1 Department of Translational Research, N.T.M.S., University of Pisa, 56123 Pisa, Italy; tuvobenedetta@hotmail.it (B.T.); micheleto@hotmail.it (M.T.); profeti.sara@gmail.com (S.P.); angelo.baggiani@med.unipi.it (A.B.); gaetano.privitera@med.unipi.it (G.P.)

2 Department of Health Sciences, University of Genova, Via Pastore 1, 16132 Genova, Italy; cristinamle@unige.it (M.L.C.); am.spagnolo@unige.it (A.M.S.)

3 Department of Clinical Sciences and Translational Medicine, University of Rome "Tor Vergata", 00133 Rome, Italy; dicave@uniroma2.it

* Correspondence: beatrice.casini@med.unipi.it

Received: 24 February 2020; Accepted: 17 April 2020; Published: 21 April 2020

**Abstract: Introduction:** Dental Unit Waterlines (DUWLs) have shown to be a source of *Legionella* infection. We report the experience of different dental healthcare settings where a risk management plan was implemented. **Materials and methods:** In a Hospital Odontostomatology Clinic (HOC) and three Private Dental Clinics (PDCs) housing 13 and six dental units (DUs), respectively, an assessment checklist was applied to evaluate staff compliance with guideline recommendations. DUWLs microbial parameters were investigated before and after the application of corrective actions. **Results:** In the HOC a poor adherence to good practices was demonstrated, whereas protocols were carefully applied in PDCs. *L. pneumophila* sg 2–15 was isolated in 31% (4/13) and 33% (2/6) of DUs in HOC and PDCs, respectively, mainly from handpieces (32%, 6/19) with counts >$10^2$ colony-forming units per milliliter (CFU/L), often associated with *P. aeruginosa* (68%, 13/19). The shock disinfection with 3% v/v hydrogen peroxide (HP) showed a limited effect, with a recolonization period of about 4 weeks. *Legionella* was eradicated only after 6% v/v HP shock disinfection and filters-installation, whilst *P. aeruginosa* after the third shock disinfection with a solution of 4% v/v HP and biodegradable surfactants. **Conclusions:** Our data demonstrate the presence and persistence of microbial contamination within the DUWLs, which required strict adherence to control measures and the choice of effective disinfectants.

**Keywords:** Dental unit waterlines; *Legionella* spp.; risk management; disinfection

## 1. Introduction

Several studies have demonstrated that water output from Dental Unit Waterlines (DUWLs) is often contaminated with high densities of microorganisms [1,2], ranging from $10^2$ to $10^6$ colony-forming units per milliliter (CFU/mL) [3–10]. Bacterial biofilm may be present on the inner surfaces of DUWLs due to contaminations coming from the proximal or distal portion of the circuit. In particular, the presence of small narrow-bore hydrophobic polymeric plastic tubing that facilitate the microorganisms' adhesion (2 mm diameter), electrical components that can heat the water (20–25 °C) and the discontinuous and low water flow are all factors that contribute to microbial growth and biofilm formation. The biofilm remains fixed to the tubing wall, but microbes keep spreading from the biofilm into the water as it flows through. For this reason, high microbial levels have been found in output water from handpieces and air/water syringes [11,12]. DUWLs are equipped with a dual water supply system that permits the system to be supplied with only municipal water or sterile water or with both types. Water supply is usually provided by public utilities and its quality must comply with the parameters required by law [13].

The suck-back of biological fluids from the oral cavities of patients (back-contamination) was also reported as important cause of DUWLs contamination [14,15].

*Pseudomonas* spp. are the prevalent bacteria found in DUWLs but a high incidence of *Legionella* was also reported, widely varying from 0% to 68% [16] (including *Legionella pneumophila* serogroup 1) [17]. DUWLs may also be important replication sites for free-living amoebae and protozoa that enable the maintenance of pathogenic intracellular bacteria, increasing their resistance to disinfectants [18]. There is evidence that amoebae huddle around microbial biofilm, and that their concentration is up to 300 times higher in DUWLs' output water than in tap water from the same source [15–17]. The microbial adhesion and biofilm on DUWLs surfaces remain very difficult to eradicate.

The Centers for Disease Control and Prevention (CDC) recommends that the maximum level of non-coliform bacteria emitted from dental handpieces and air/water syringes should be equal or less than 500 CFU/mL [19]. The Italian legislation on drinking water does not establish a limit for heterotrophic bacteria count; it requires this value not to undergo variations. The water in the operating theatre of healthcare facilities must comply with the target value established by the Italian Workers Compensation Authority guideline, 100 CFU/mL at 22 °C and 10 CFU/mL at 37 °C, respectively. In 2007, the French "*Ministère de la Santé et des Solidarités*" proposed the same target for drinking water. For this reason, we took it as a reference in our study [20,21].

Because of this contamination, dental units are recognized as a potential source of infection for human health, especially threatening dental staff and patients, who are regularly exposed to water and water-aerosol emitted by dental unit handpieces. A study reported the case of a patient who died after being infected with *L. pneumophila* serogroup 1 during a dental practice in Italy [22]. By different typing methods, it was demonstrated that the source of the *Legionella* infection was the DUWL. *L. pneumophila* sg 1 subgroup Benidorm ST593 was isolated in each sample collected from the cold tap water, tap of the DUWL, high-speedturbine and patient's bronchialaspirate. A study of Schönning et al. reported the case of a man diagnosed with leukaemia, who underwent a dental check-up and a high-dose chemotherapy and developed Legionnaires' disease in the following days. The analysis of the environmental samples, from the cup filler resulted to be positive for the same *L. pneumophila* sg 1 specimen detected in the patient's sputum collected through bronchoscopy [23].

The evidence linking exposure to *Pseudomonas aeruginosa* contaminated DUWL during dental treatment and subsequent infection is limited [24] and is based on the results from a single observational study reported in two cancer patients by Martin M.V. et al. [25].

Free-living amoebae were frequently isolated from DUWLs and *Vermamoeba* species were isolated from the throats of humans as long ago as 1967 [26] but it is unknown whether they are associated with a risk in the dental setting, through contaminated aerosols or droplets [27].

In literature there is a single documented fatal case of Legionnaires' disease regarding an American dentist. The infection was attributed to exposure to DUWL aerosols [28]. *L. pneumophilia* and *L. longbeachae* were detected in the dentist's lung tissue and in the DUWLs; however, the dentist's domestic water supplies also had very low levels of *Legionella* spp. Here, the evidence was not conclusive.

A meta-analysis conducted by Petti S et al. demonstrated that there was limited evidence of occupational risk for *Legionella* infection to dental healthcare workers and Legionnaires' disease outbreaks in dental healthcare settings were never reported despite billions of treatments provided each year [29].

To date, there are no known cases of Pontiac fever in patients, resulting from visits to, or treatment in, a dental clinic. This would suggest that the risk to patients posed by *Legionella* spp. from DUWLs is low; however, the risk is not absent [30].

The aim of this study was to report the experience of different dental healthcare settings where a risk management plan was implemented and an integrated filtration-disinfection strategy was applied to reduce the DUWLs microbial contamination.

## 2. Results

### 2.1. Risk Management

Using the checklist allowed us to verify that a water safety plan was in place in the hospital and a maintenance and control program was constantly applied to the water of the buildings and to the aeration systems. This should ensure a good quality of municipal water that can feed the DUWL when the switching is applied if the sterile water in the bottle runs out.

Despite the training activity for the correct adoption of the DU management procedure (using only sterile reverse osmosis water, flushing between patients, self-contained water bottles disinfection, etc.), a low adherence to good practices was found in the hospital.

On the contrary, a good compliance with manufacturer's instructions for DU management and use of biocides was observed in the private dental clinics, where the staff was not informed/formed neither on water risk assessment and management nor on good practices guidance.

### 2.2. Tap Water Results

The microbiological quality of tap water varied between hospitals and smaller premises. *Legionella* spp. was not detected in three Private Dental Clinic (PDC) tap water samples (0/48), whilst in Hospital Odontostomatology Clinics (HOC), housing 13 dental units, *Legionella* spp. was repeatedly isolated in all tap water samples (30/104, 28.8%). The strains were identified as *L. pneumophila* sg 1 and sg 2–15, with a geometric mean of 4.91 ± 0.69 Log CFU/L and 4.38 ± 0.72 Log CFU/L, respectively. *Pseudomonas aeruginosa* and coliform bacteria were not isolated despite the fact that the total microbial counts at 22 °C and 37 °C were higher than the values recommended by Italian regulations (geometric mean of 2.79 ± 0.40 Log CFU/ml and 1.88 ± 0.41 Log CFU/ml, respectively). Free-living amoebae were recovered in 23% (3/13) of hospital DUs, but never in DUs housed in the PDCs. Among all cells microscopically positive to the culture examination, all PCR positive isolates belonged to *Vermoamoeba vermiformis* (identity of 99%). Despite the continuous application of the water safety plan, the microbiological quality of the municipal water remained low in the hospital and an inadequate concentration of residual chlorine-dioxide was detected (0.07 ± 0.14 mg/L). The mean temperature in cold water was 22.4 ± 1.6 °C, and the values were demonstrated to be related to *Legionella* concentration ($R^2 = 0.51$).

### 2.3. Dental Unit Results before Shock Disinfection

During the first sampling, water samples collected from hospital DUWLs showed a high prevalence of *Legionella*, which was detected in 31% (4/13) of dental units with a geometric mean 3.99 ± 0.61 Log CFU/L, often at different sites on the device (Table 1). *Legionella* was isolated from inlets (3/4), spittoons (2/4) and from handpieces (4/4). Positive isolates were identified as *Legionella pneumophila* serogroup 2–15.

In PDCs, *Legionella pneumophila* serogroup 2–15 was isolated in 33% (2/6) of dental units during the first sampling, with a geometric mean of 4.15 ± 0.13Log CFU/L. *Legionella* was isolated from spittoons (2/2) and from handpieces (2/2).

All *Legionella* data collected before dental unit waterlines' disinfection are shown in Figure 1.

No coliform growth was detected in water samples. *P. aeruginosa* was isolated in high prevalence from handpieces and spittoons in 68% (13/19) of dental units. *P. aeruginosa* was detected in 33% (2/6) of collected samples, during the first sampling, from the spittoons of private clinics' dental units and in 85% (11/13) of dental units housed in the hospital clinic. *P. aeruginosa* was always associated with the presence of *Legionella*. In almost all water samples, the total microbial counts at 22 °C and 37 °C was ≥100 and 10 CFU/mL, respectively.

**Table 1.** Results of the microbiological analysis performed on water samples collected from dental units housed in a Private Dental Clinic and in the Hospital Dental Clinic before shock disinfection.

| | | TMC 22 °C (Log CFU/mL) | TMC 37 °C (Log CFU/mL) | *P. aeruginosa* (P/A) | *E. coli* (P/A) | *Legionella* spp. (Log CFU/L) | *B. vesicularis* (P/A) | Free Living Protozoa (P/A) |
|---|---|---|---|---|---|---|---|---|
| PDC | Inlet | 2.23 ± 0.30 | 2.06 ± 0.05 | A | A | 0 | P | A |
| | Spittoon | 2.43 ± 0.40 | 2.16 ± 0.07 | P | A | 4.06 ± 0.09 | A | A |
| | Handpiece | 2.52 ± 0.42 | 2.49 ± 0.19 | A | A | 4.24 ± 0.12 | A | A |
| | % of positive DUs | 100% (6/6) | 100% (6/6) | 33% (2/6) | 0% (0/6) | 33% (2/6) | 33% (2/6) | 0% (0/6) |
| HDC | Inlet | 3.14 ± 0.12 | 3.35 ± 0.23 | A | A | 3.54 ± 0.21 | A | P |
| | Spittoon | 3.25 ± 0.19 | 3.53 ± 0.32 | P | A | 4.30 ± 0.17 | A | A |
| | Handpiece | 3.42 ± 0.20 | 3.66 ± 0.27 | P | A | 4.59 ± 0.07 | A | A |
| | % of positive DUs | 85% (11/13) | 85% (11/13) | 85% (11/13) | 0% (0/13) | 31% (4/13) | 0% (0/13) | 46% (6/13) |

Note = TMC: Total Microbial Count; P/A: Presence or Absence; PDC: Private Dental Clinic; HDC: Hospital Dental Clinic.

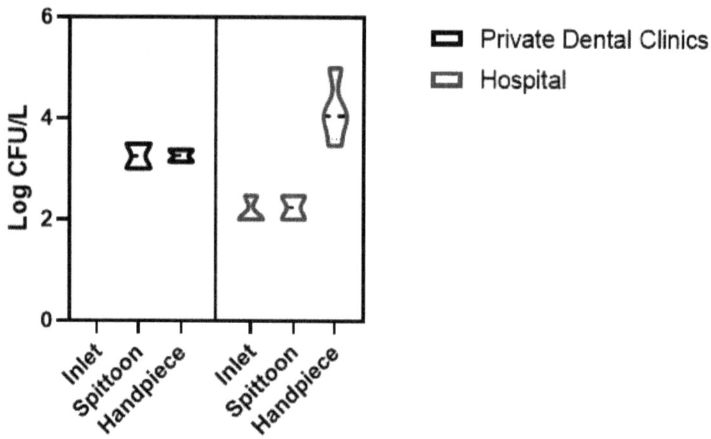


**Figure 1.** Violin plot of *Legionella pneumophila* sg 2–15 (Log colony-forming units per milliliter (CFU/L)) detected from each sampling site of the dental units before disinfection.

*Brevundimonas vesicularis* was identified in two/six of the dental units housed in private clinics. Colonies were selected from Cetrimide Agar medium. All the percentages of positive dental units to waterborne pathogens are shown in Table 1.

Free-living protozoa (FLA) were detected in 46% (6/13) of HOC dental units. Among all FLA microscopically positive to the culture examination, one isolate showed band with approximate sizes of *Valkampfia*. All PCR positive isolates showed bands with approximate sizes of 800 bp (expected for *Vermoamoeba vermiformis*). The analyses unambiguously identified all samples as *Vermoamoeba vermiformis*, the sequences showing the highest identity (99%) with those accessible in GenBank. No water sample analyzed in this study was characterized for the presence of other FLA species, including *Naegleria* spp. Concerning *V. vermiformis* sequences, a phylogenetic analysis was also performed.

### 2.4. Dental Unit Results after Shock Disinfection and Water Filtration

In only one Private Dental Clinic was shock disinfection with hydrogen peroxide (HP) 3% v/v performed for pathogens (*P. aeruginosa* and *Legionella* spp.) in the DUWL, and filters were simultaneously installed at the inlet of each dental unit. Disinfection and filter installation showed a good efficacy on *Legionella* spp., undetected after the treatment, as well as on *P. aeruginosa*. No other treatment was needed.

After 30 days from shock disinfection treatments with HP 3% v/v applied in 12 dental units in the HOC, total microbial counts at 22 °C and 37 °C resulted in the values being higher than the values recommended by Italian regulation (<100 CFU/mL and <10 CFU/mL, respectively) in almost all samples (67% and 92% respectively). *Legionella* spp. was detected in only one dental unit (8%), whereas *P. aeruginosa* was isolated in 83% of dental units (10/12), from handpieces and spittoons. The shock disinfection with HP 3% v/v showed a limited effect, with a recolonization period of about 4 weeks.

*Legionella* was eradicated after a shock disinfection with HP 6% v/v, applied after the installation of 31 days membrane filters at the inlet of each dental unit. The point-of-use water filtration showed good efficacy in containing the entrance of *Legionella*, while HP disinfection had a good performance in controlling the growth of pathogens. After 30 days from disinfection with HP 6% v/v, *P. aeruginosa* was still found in 100% of DUWLs' samples (10/10), showing the higher persistence of this bacteria even after shock disinfection treatment. *P. aeruginosa* was not detected only after the third shock disinfection applied with a solution containing HP 4% v/v and surfactants.

All the results of the microbiological analysis performed on water samples collected from dental units housed in a Private Dental Clinic and in the Hospital Dental Clinic before shock disinfection are shown in Table 2.

**Table 2.** Results of the microbiological analysis performed on water samples collected from dental units housed in a Private Dental Clinic and in the Hospital Dental Clinic after shock disinfection.

| | | TMC 22 °C (Log CFU/mL) | TMC 37 °C (Log CFU/mL) | *P. aeruginosa* (P/A) | *E. coli* (P/A) | *Legionella* spp. (Log CFU/L) | *B. vesicularis* (P/A) | Free Living PROTOZOA (P/A) |
|---|---|---|---|---|---|---|---|---|
| **PDC (HP 3%)** | Inlet | 1.60 ± 0.05 | 1.66 ± 0.36 | A | A | A | A | A |
| | Spittoon | 1.72 ± 0.17 | 1.87 ± 0.02 | A | A | A | A | A |
| | Handpiece | 2.03 ± 0.05 | 1.93 ± 0.04 | A | A | A | A | A |
| | % of positive DUs | 100% (2/2) | 100% (2/2) | 0% (0/2) | 0% (0/2) | 0% (0/2) | 0% (0/2) | 0% (0/2) |
| **HDC (HP 3%)** | Inlet | 2.53 ± 0.41 | 2.25 ± 0.31 | A | A | 1.18 ± 0.64 | A | A |
| | Spittoon | 2.62 ± 0.40 | 2.62 ± 0.25 | P | A | 1.79 ± 0.32 | A | A |
| | Handpiece | 2.74 ± 0.30 | 3.41 ± 0.58 | P | A | 2.47 ± 0.41 | A | A |
| | % of positive DUs | 67% (8/12) | 92% (11/12) | 83% (10/12) | 0% (0/12) | 8% (1/12) | 0% (0/12) | 0% (0/12) |
| **HDC (HP 6%)** | Inlet | 1.50 ± 0.43 | 2.20 ± 0.05 | A | A | 0 | A | A |
| | Spittoon | 1.69 ± 0.21 | 2.56 ± 0.23 | P | A | 0 | A | A |
| | Handpiece | 1.96 ± 0.03 | 2.84 ± 0.03 | P | A | 0 | A | A |
| | % of positive DUs | 90% (9/10) | 100% (10/10) | 100% (10/10) | 0% (0/10) | 0% (0/10) | 0% (0/10) | 0% (0/10) |
| **HDC (HP 4% and surfactants)** | Inlet | 0 | 0 | A | A | 0 | A | A |
| | Spittoon | 0 | 0 | A | A | 0 | A | A |
| | Handpiece | 0 | 0 | A | A | 0 | A | A |
| | % of positive DUs | 0% (0/10) | 0% (0/10) | 0% (0/10) | 0% (0/10) | 0% (0/10) | 0% (0/10) | 0% (0/10) |

Note = TMC: Total Microbial Count; P/A: Presence or Absence; HP: hydrogen peroxide; PDC: Private Dental Clinic; HDC: Hospital Dental Clinic.

## 3. Discussion

The dental unit consists of a complex water pipeline network connected to multiple pieces of equipment. Such an environment implicates a high risk of microbial contamination and transmission, especially through the contamination of water and bioaerosol by dental instruments, which are placed very close to patients and medical staff during dental treatments [31].

Even though the evidence associating DUWLs with risks for patients and staff is contradictory [32–35], exposing patients or dental healthcare personnel to water of uncertain microbiological quality is inconsistent with general accepted infection control principles [19]. Since patients and dental staff are regularly exposed to water and aerosols generated from the dental units, the microbial quality of the water in the DUWLs is extremely important. It is not acceptable that a DUWL should not meet the drinking water standards (<100 CFU/mL at 22 °C and <10 CFU/mL at 36 °C) [20,21].

To reduce the risk coming from contaminated DUWLs, it is necessary that dental healthcare workers routinely apply suggested infection prevention strategies [13]. The Italian guidelines suggest many approaches to reduce the microbial contamination and/or biofilm formation, including both non-chemical (using the anti-stagnation device, flushing, supplying the circuit with sterile solutions, antimicrobial filter installation, etc.) and chemical methods, which provide for the use of disinfectants continuously or in periodic shock treatment, following the manufacturer's instructions [36]. Not an available single method or device will eradicate the bio-contamination of DUWLs or exclude the risk of cross-infection. To reduce contamination risks, a combination of methods is desirable [36]. Where delivered water quality is in doubt, dental practice should consider adopting continuous dosing systems or shock disinfection, if permitted by the manufacturer's instructions [19]. Output water from DUs continuously treated with disinfection products is more compliant with the recommended standards (heterotrophic bacteria load) and it is reported to be remarkable in preventing the contamination by *Legionella* and *P. aeruginosa* [8,36,37]. The efficacy of the adopted measures depended on the strict adherence to the planned protocols [36].

The level of *Legionella* contamination in DUs is not established in the Italian Legislation. Nevertheless, all control actions to reduce the risk of contamination, biofilm formation in DUWLs and a risk assessment based on patients and clinical practices are required [13].

Besides the technical-practical measures and disinfection protocols, an integrated approach for microbial risk management in a dental health care setting should also include regular microbiological monitoring. Environmental surveillance for *Legionella* is useful not only to assess the efficacy of preventive measures but also as a guide for the choice of corrective strategies, under the principles of the internal control plan [38].

Our study confirms the literature evidences with the finding of *Legionella* or *P. aeruginosa* contamination in a large part of our samples before disinfection: 32% (6/19) and 68% (13/19) respectively. *P. aeruginosa* inside DUWLs may be related to the water quality or to the retro-contamination at the outlet of dental units. The competitive advantage of *P. aeruginosa* in the colonization of water lines is because of its ability to inhibit the growth of other bacteria by producing bacteriocins [39,40].

Frequent switchover from reverse osmotic to drinking water resulting from the intensive dental care activity in the hospital clinic, may have caused the higher *Legionella* spp. and *P. aeruginosa* isolation observed in the hospital DUWLs compared to private dental clinics. In large hospital buildings, drinking water quality may affect the DUWLs microbial contamination (more complex plumbing systems, large water storage tanks, multiple dead legs, cold water over 20 °C, etc.).

As reported by Lizzaro et al., the possibility of switching between two different water flows (municipal water and sterile water) reduces the risk of circuit contamination, but a mixed water supply is not recorded in water safety plans [31].

In the Hospital Dental Clinic (HDC), a Water Safety Plan (WSP), a maintenance and a control program were constantly applied to the building's water and aeration systems but there was a low adherence to good practices in DU management, despite the training activity of the dental healthcare staff on the correct adoption of the hospital procedure (exclusive use of sterile reverse osmosis water, flushing between patients, self-contained water bottles disinfection, etc.). A low adherence to the best practices guidance may also contribute to biofilm proliferation. On the contrary, a good compliance with manufacturer's instructions for DU management and the use of biocides was observed in PDCs, although the staff was not informed on either on water risk assessment and management or on good practices guidance. The medical staff's and manufacturers' poor knowledge about water quality and the role of biofilm formation was suggested as one of the main problems related to dental unit contamination [31].

In only one Private Dental Clinic was shock disinfection with hydrogen peroxide 3% v/v and filters simultaneously installed at the inlet of each dental unit proven effective in eliminating pathogens, although the bacterial loads remained too high. These results were in line with the study conducted by Ditommaso et al., that showed an increase to unacceptable levels of bacterial loads in the DUWL of a dental chair treated with HP 3%, getting a significant reduction of counts only after 9 months of treatment [41].

On the other hand, in the hospital dental clinic, the first shock disinfection with HP 3% v/v applied in 12 dental units reduced positive samples for *Legionella* to 8% (1/12 DUs), and after the second treatment with HP 6% v/v and filters installation *Legionella* was removed. Conversely, after HP 6% v/v disinfection, *P. aeruginosa* proved to be more resistant and was detected again after two cycles of treatment, giving no evidence of substantially decreasing (100%, 10/10). Only after a third cycle of HP 4% v/v and surfactants were the samples proved to be negative.

DU disconnected from the water supply and fed only with sterile water, as with the DU in the operating room of the HDC, are less contaminated and safer to use in invasive dental practices.

In conclusion, we assert that filter installation and shock disinfection with a solution of 4% v/v hydrogen peroxide and surfactants appear to be a promising alternative for decreasing *Legionella* colonization in DUWLs, although further field studies in other healthcare and community settings are required to confirm its effectiveness and its long-term efficacy in reducing biofilm.

## 4. Materials and Methods

### 4.1. Setting

A 24-month investigation was carried out in a Hospital Odontostomatology Clinics (HOC) and in three Private Dental Clinics (PDC) housing 13 and 6 dental units, respectively.

A validated checklist was used to identify the risk factors associated with contamination of DUWLs. The checklist included all the necessary items to create a schematic diagram showing the layout of the building's water systems and DUWLs, as well as the information related to the compliance with good practices in controlling *Legionella* colonization.

The DUs examined here varied in model and year of installation: in the HOC, 8 Puma ELI, Castellani S.p.A., were installed in 2010 and 5 Skema 4, Castellani S.p.A. were installed in 2005; in the PDCs 2 C4+, Sirona Dental Systems GmbH, were installed in 2014, 2 LINEA 90, World Health Organization (WHO) OMS, were installed in 2007; and, finally, 2 Puma EVO 5, Castellani S.p.A., were installed in 2010.

All devices were supplied through an independent system that received water from a 1-liter polypropylene bottle containing sterile water, manually filled. During the dental care activity, the municipal water fed the DUWLs by activating a switch when the bottle was running out of water. Contrastingly, the DU in the operating room was completely disconnected from the municipal water supply. In the hospital, the drinking water was softened and treated with a secondary chlorine dioxide disinfection (0.2–0.3 mg/L), while in private dental clinics the water was only softened.

In almost all DUs (Castellani and Sirona Dental Systems GmbH), at the end of the day, an automatic disinfection cycle was activated. According to the manufacturer's recommendations, a disinfectant product that generated peracetic acid (0.2%) was applied. Only in one PDC, housing 2

OMS dental units, was disinfection performed manually on a daily basis through the application of 2.5% quaternary ammonium compounds. In the hospital, a procedure was written and shared with the dental clinic staff that was trained for the correct adoption of the DU management procedure. According to the procedure, all DUWLs were flushed for at least two minutes at the beginning and end of the day and after any significant period of inactivity. In addition, flushing was performed for at least 20–30 s between patients. No private clinic had a procedure where the DUs' maintenance protocol was reported.

Following the results obtained from the microbiological investigations, a control program was started with more restrictive measures on the management of the quality of the supplied water, the periodic maintenance of the dental units (for example, the lubrication of the quick coupling seals on the handpiece, self-contained water bottles disinfection, etc.), the compliance with the flushing frequency, and the introduction of a shock disinfection procedure for the DUWL, where necessary.

A shock disinfection treatment was applied under the recommendations of the different dental unit manufacturers, when pathogens were isolated from the DUWL (the presence of *P. aeruginosa*, *Legionella* spp. or other pathogens). In particular, treatments were applied in all DUs except for the one of the operating room, which was disconnected from the water supply and fed only with sterile water.

### *4.2. Samplings and Microbiological Tests*

Before the disinfection treatment, every three months, for each DU, 1.5 L of water was collected from the inlet, spittoon and handpiece. To assess the effectiveness of the treatment, sampling was carried out after 30 days. Moreover, 1.5 L of tap water was collected from a sink located near the DU. All water samples were collected without flushing before the start of professional activity.

The total microbial count at 22 and 37 °C, *Legionella* spp., *P. aeruginosa* and coliform bacteria were determined in all water samples, in accordance to the Italian and French guidelines [20,21]. One more liter of water was collected from the same devices for the detection of free-living protozoa (FLA) according to international standards [42].

Water temperatures were tested in all samples; residual chlorine concentrations were determined only in water sampled from taps and dental units fed from the municipal water network.

*Legionella* spp. was detected as described by ISO11731:2017 [43]. One-liter samples were filtrated through a 0.2-μm membrane (Millipore, Billerica, MA, USA). The membrane was immersed in 10 mL of water and sonicated for 5 min to allow the bacterial cells to separate from the membrane. The resulting suspension was divided into three aliquots, the first one was brought to 50 °C to select *Legionella* spp. over the other bacterial species non heat resistant, the second one was added with 9 volumes of HCl-KCl acid solution, and the last one was plated with no treatment. A total of 0.1 mL of each aliquot was plated on Buffered Charcoal Yeast Extract (BCYE) Agar and Glycine Vancomycin Polymyxin Cycloheximide (GVPC) agar plates (Oxoid, UK), and incubated at 37 °C for 7–10 days in jars under an atmosphere containing 2.5% $CO_2$. The suspect *Legionella* colonies were tested for species and serogroup by polyvalent agglutination latex test (Legionella latex test—Oxoid, UK).

The total microbial count was performed by inclusion in Plate Count Agar (Oxoid, UK) according to ISO 6222:1999 [44]. Coliform bacteria and *Pseudomonas aeruginosa* detection was performed by filtrating 100 mL of water through 0.45-μm membranes (Nalgene, Rochester, NY, USA). The membranes were laid on Endo Agar Les plates (Liofilchem, Italy) and Cetrimide Agar (Oxoid, UK) according to ISO 9308-1:2014 [45] for coliforms growth and ISO 16266:2006 [46] for *Pseudomonas aeruginosa* growth. The species' confirmation of suspect colonies was obtained by Mini API galleries (bioMeriéux, Marcy-l'Étoile, France).

One-liter samples were filtrated through a 0.2-μm pore size of membrane (Millipore, Billerica MA, USA) for free-living protozoa search. The membrane filters were minced in 10 mL of sterile phosphate-buffered saline pH 7.2 (PBS), homogenized by vortex for 5 min and centrifuged at 1200 *g* for 15 min. A total of 200 μL of pellet was inoculated on Non-Nutrient Agar (NNA) with a lawn of heat-inactivated *Escherichia coli* in Page's Amoeba Saline solution (PAS) and incubated at 37 °C. The presence of FLA was investigated by examining the NNA culture plates by inverted microscope Dmi1 (Leica Wetzlar, Germany) using 20× and 40× objectives.

From all positive samples, the growing amoebae were harvested from culture plates, placed in Eppendorf tubes and washed two times with PBS, pH 7.4, before molecular procedure. DNA extraction was performed by using the QIAamp DNA Micro Kit (Qiagen, Milan, Italy). To identify FLA species, the 18S rDNA amplification with primers P-FLA-F and P-FLA-R was performed, according to the protocol published by Cristina ML et al. [18].

### 4.3. Shock Disinfection Treatment

Shock disinfection treatments were applied in DUs housed in PDC and HOC, when contaminated by pathogens. Subsequently, the treatment was carried out every three months until an acceptable result was obtained. The effectiveness of disinfection was assessed after one month and in the subsequent sampling, scheduled every three months.

In PDC, dental units the disinfection was performed with a galenic formulation of 3% v/v hydrogen peroxide applied in dental units' system for 1 h and followed by water flushing.

In HOC, a shock disinfection was performed by using a galenic formulation of 3% and 6% v/v hydrogen peroxide and a second treatment was applied with a solution containing 4% v/v hydrogen peroxide and surfactants (Green Line Hydrogen Peroxide Cleaner 4%, Corcraft Product, Comstock, NY, USA).

All disinfection treatments were carried out in a similar way, by using a biofilm removing system and a dye tracing to detect the disinfectant passage. The method of use was the following:

- Remove all hands and angle pieces to prevent blockages;
- Attach the provided connection adapter to the disinfectant bottle;
- Connect the device to the water inlet of the dental unit;
- Switch on the device and open the inlet valve to pump the disinfectant into the dental unit;
- As soon as purple liquid exits from the water-consuming units, switch off the device, close the valve and remove the device;
- Depending on the level of contamination of the dental unit, the action time may range between half an hour up to max 1 h;
- Upon completing the action time, turn on all water-consuming units until clean water runs out.

Membrane filters, 0.2 µm in size (Pall Corporation, New York, NY, USA), were set at the inlet of each dental unit to ensure a good microbial quality of the entrance water. The filters' installation was done simultaneously to the first disinfection of the dental unit waterline in the PDC where they were installed after the treatment with 6% v/v hydrogen peroxide in HOC.

**Author Contributions:** B.C., A.B., G.P. conceived and designed the experiments; B.T., B.C. and M.T. performed the water sampling; B.T., M.T. and B.C. performed laboratory tests; B.C., B.T., D.D.C., M.L.C., A.M.S., S.P. performed data elaboration and wrote the paper. All authors have read and agree to the published version of the manuscript.

**Funding:** This research received no external funding.

**Conflicts of Interest:** The authors declare no conflict of interest.

## References

1. Fujita, M.; Mashima, I.; Nakazawa, F. Monitoring the decontamination efficacy of the novel Poseidon-S disinfectant system in dental unit water lines. *J. Microbiol. Immunol. Infect.* **2017**, *50*, 270–276. [CrossRef] [PubMed]
2. Kohn, W.G.; Harte, J.A.; Malvitz, D.M.; Collins, A.S.; Cleveland, J.L.; Eklund, K.J. Centers for Disease Control and Prevention. Guidelines for infection control in dental health care settings-2003. *J. Am. Dent. Assoc.* **2004**, *135*, 33–47. [CrossRef] [PubMed]

3. Blake, G.C. The incidence and control of bacterial infection in dental spray reservoirs. *Br. Dent. J.* **1963**, *115*, 413416.
4. Kelstrup, J.; Funder Nielsen, T.D.; Theilade, J. Microbial aggregate contamination of water lines in dental equipment and its control. *Acta Pathol. Microbiol. Scand. B* **1977**, *85*, 177183. [CrossRef] [PubMed]
5. Szymańska, J.; Sitkowska, J. Bacterial contamination of dental unit waterlines. *Environ. Monit. Assess.* **2013**, *185*, 3603–3611. [CrossRef]
6. Szymanska, J.; Sitkowska, J. Opportunistic bacteria in dental unit waterlines: Assessment and characteristics. *Future Microbiol.* **2013**, *8*, 681–689. [CrossRef]
7. Sacchetti, R.; De Luca, G.; Zanetti, F. Influence of material and tube size on DUWLs contamination in a pilot plant. *New Microbiol.* **2007**, *30*, 29–34.
8. Leoni, E.; Dallolio, L.; Stagni, F.; Sanna, T.; D'Alessandro, G.; Piana, G. Impact of a risk management plan on Legionella contamination of dental unit water. *Int. J. Environ. Res. Public Health* **2015**, *12*, 2344–2358. [CrossRef]
9. Monteiro, R.M.; Marques, D.M.; Domingues, P.C.A. Evaluation of a Protocol for Reducing the Microbial Contamination of Dental Unit Water. *Acta Odontol. Latinoam.* **2018**, *31*, 138–143.
10. Xue-Yue, J.I.; Fei, C.; Zhang, Y. Three Key Factors Influencing the Bacterial Contamination of Dental Unit Waterlines: A 6-year Survey From 2012 to 2017. *Int. Dent. J.* **2019**, *69*, 192–199.
11. Coleman, D.C.; O'Donnell, M.J.; Shore, A.C.; Russell, R.J. Biofilm problems in dental unit water system and its practical control. *J. Appl. Microbiol.* **2009**, *106*, 1424–1437. [CrossRef] [PubMed]
12. Coleman, D.C.; O'Donnell, M.J.; Shore, A.C.; Swan, J.; Russell, R.J. The role of manufacturers in reducing biofilms in dental chair waterlines. *J. Dent.* **2007**, *35*, 701–711. [CrossRef]
13. Italian Health Ministry. Guidelines for Prevention and Control of Legionellosis. In Proceedings of the Approvate in Conferenza Stato-Regione, Roma, Italy, 7 May 2015. Available online: http://www.salute.gov.it/imgs/C_17_pubblicazioni_2362_allegato.pdf (accessed on 5 November 2018).
14. Cristina, M.L.; Spagnolo, A.M.; Sartini, M.; Dallera, M.; Ottria, G.; Lombardi, R.; Perdelli, F. Evaluation of the risk of infection through exposure to aerosols and spatters in dentistry. *Am. J. Infect. Control* **2008**, *36*, 304–307. [CrossRef] [PubMed]
15. Costa, D.; Mercier, A.; Gravouil, K.; Lesobre, J.; Delafont, V.; Bousseau, A. Pyrosequencing analysis of bacterial diversity in dental unit waterlines. *Water Res.* **2015**, *81*, 223–231. [CrossRef] [PubMed]
16. Pankhurst, C.L.; Coulter, W.A. Do contaminated dental unit waterlines pose a risk of infection? *J. Dent.* **2007**, *35*, 712–720. [CrossRef]
17. Arvand, M.; Hack, A. Microbial contamination of dental unit waterlines in dental practices in Hesse, Germany: A cross-sectional study. *Eur. J. Microbiol. Immunol.* **2013**, *3*, 49–52. [CrossRef] [PubMed]
18. Spagnolo, A.M.; Sartini, M.; Cave, D.D.; Casini, B.; Tuvo, B.; Cristina, M.L. Evaluation of Microbiological and Free-Living Protozoa Contamination in Dental Unit Waterlines. *Int. J. Environ. Res. Public Health* **2019**, *16*, 2648. [CrossRef]
19. Centers for Disease Control and Prevention. Guidelines for Infection Control in Dental Health-Care Setting. 2003. Available online: http://www.cdc.gov/oralhealth/InfectionControl/guidelines/index.htm (accessed on 20 October 2019).
20. Italian Workers Compensation Authority, INAIL. Linee Guida Sugli Standard Di Sicurezza E Di Igiene Nel Reparto Operatorio. Available online: https://www.inail.it/cs/internet/docs/linee-guida-igiene-reparto-operatorio.pdf?section=attivita (accessed on 20 December 2019).
21. Ministère de la Santé e des Solidarités. L'eau Dans les Etablissements de Santé. 2007. Available online: http://nosobase.chu-lyon.fr/Reglementation/2005/guide_eau_etabs.pdf (accessed on 30 March 2020).
22. Ricci, M.L.; Fontana, S.; Pinci, F.; Fiumana, E.; Pedna, M.F.; Farolfi, P. Pneumonia associated with a dental unit waterline. *Lancet* **2012**, *379*, 684. [CrossRef]
23. Schönning, C.; Jernberg, C.; Klingenberg, D. Legionellosis acquired through a dental unit—A case study. *J. Hosp. Infect.* **2017**, *96*, 89–92. [CrossRef]
24. Laheij, A.M.; Kistler, J.O.; Belibasakis, G.N.; Välimaa, H.; de Soet, J.J. European Oral Microbiology Workshop (EOMW) 2011. Healthcare-associated viral and bacterial infections in dentistry. *J. Oral Microbiol.* **2012**, *4*, 17659. [CrossRef]
25. Martin, M.V. The significance of the bacterial contamination of dental unit water systems. *Br. Dent. J.* **1987**, *163*, 152–154. [CrossRef]

26. Wang, S.S.; Feldman, H.A. Isolation of Hartmannella species from human throats. *N. Engl. J. Med.* **1967**, *277*, 1174–1179. [CrossRef]
27. Barbot, V.; Robert, A.; Rodier, M.H.; Imbert, C. Update on infectious risks associated with dental unit waterlines. *FEMS Immunol. Med. Microbiol.* **2012**, *65*, 196–204. [CrossRef]
28. Atlas, R.M.; Williams, J.F.; Huntington, M.K. Legionella contamination of dental-unit waters. *Appl. Environ. Microbiol.* **1995**, *61*, 1208–1213. [CrossRef]
29. Petti, S.; Vitali, M. Occupational Risk for Legionella Infection among Dental Healthcare Workers: Meta-Analysis in Occupational Epidemiology. *BMJ Open* **2017**, *7*, e015374. [CrossRef] [PubMed]
30. Cameron, G.E.; Gruninger, S.E.; Lipman, R.D. Rates and Predictors of Exposure to *Legionella pneumophila* in the United States among Dental Practitioners: 2002 Through 2012. *J. Am. Dent. Assoc.* **2017**, *148*, 164–171.
31. Lizzadro, J.; Mazzotta, M.; Girolamini, L.; Dormi, A.; Pellati, T.; Cristino, S. Comparison between Two Types of Dental Unit Waterlines: How Evaluation of Microbiological Contamination Can Support Risk Containment. *Int. J. Environ. Res. Public Health* **2019**, *16*, 328. [CrossRef] [PubMed]
32. Dutil, S.; Veillette, M.; Mériaux, A.; Lazure, L.; Barbeau, J.; Duchaine, C. Aerosolization of Mycobacteria and legionellae during dental treatment: Low exposure despite dental unit contamination. *Environ. Microbiol.* **2007**, *9*, 2836–2843. [CrossRef] [PubMed]
33. Mills, S.E. The dental unit waterline controversy: Defusing the myths, defining the solutions. *Am. Dent. Assoc.* **2000**, *131*, 1427–1441. [CrossRef] [PubMed]
34. Porteous, N. Dental unit waterline contamination—A review. *Text. Dent. J.* **2010**, *127*, 677–685.
35. Rowland, B.M. Bacterial contamination of dental unit waterlines: What is your dentist spraying into your mouth? *Clin. Microbiol.* **2003**, *25*, 73–77. [CrossRef]
36. Decontamination in Primary Care Dental Practices (HTM 01-05). Available online: https://www.gov.uk/government/publications/decontamination-in-primary-care-dental-practices (accessed on 20 March 2020).
37. Dallolio, L.; Scuderi, A.; Rini, M.S. Effect of different disinfection protocols on microbial and biofilm contamination of dental unit waterlines in community dental practices. *Int. J. Environ. Res. Public Health* **2014**, *11*, 2064–2076. [CrossRef] [PubMed]
38. Pasquarella, C.; Veronesi, L.; Napoli, C.; Castiglia, P. Microbial environmental contamination in Italian dental clinics: A multicenter study yielding recommendations for standardized sampling methods and threshold values. *Sci. Total Environ.* **2012**, *420*, 289–299. [CrossRef] [PubMed]
39. Wilson, R.; Dowling, R. Lung infections. Pseudomonas aeruginosa and other related species. *Thorax* **1998**, *53*, 213–219. [PubMed]
40. Kimura, S.; Tateda, K.; Ishii, Y.; Horikawa, M.; Miyairi, S.; Gotoh, N.; Ishiguro, M.; Yamaguchi, K. *Pseudomonas aeruginosa* Las quorum sensing autoinducer suppresses growth and biofilm production in Legionella species. *Microbiology* **2009**, *155*, 1934–1939. [CrossRef]
41. Ditommaso, S.; Giacomuzzi, M.; Ricciardi, E. The role of chemical products at low doses in preventing the proliferation of bacteria in dental unit waterlines: The ICX® experience. *J. Water Health* **2018**, *16*, 150–158. [CrossRef] [PubMed]
42. *American Public Health Association Standard Methods for the Examination of Water and Wastewater*, 21th ed.; American Public Health Association Inc.: New York, NY, USA, 2005.
43. *Water Quality—Enumeration of Legionella*; ISO 11731:2017; ISO: Geneva, Switzerland, 2017.
44. *Water Quality—Enumeration of Culturable Micro-Organisms—Colony Count by Inoculation in a Nutrient Agar Culture Medium*; ISO 6222:1999; ISO: Geneva, Switzerland, 1999.
45. *Water Quality—Enumeration of Escherichia Coli and coliform Bacteria—Part 1: Membrane Filtration Method for Waters with Low Bacterial Background Flora*; ISO 9308-1:2014; ISO: Geneva, Switzerland, 2014.
46. *Water Quality—Detection and Enumeration of Pseudomonas aeruginosa—Method by Membrane Filtration*; ISO 16266:2006; ISO: Geneva, Switzerland, 2006.

*pathogens*

MDPI

*Review*

# Critical Review: Propensity of Premise Plumbing Pipe Materials to Enhance or Diminish Growth of *Legionella* and Other Opportunistic Pathogens

**Abraham C. Cullom [1], Rebekah L. Martin [1,2], Yang Song [1], Krista Williams [3], Amanda Williams [4], Amy Pruden [1] and Marc A. Edwards [1,*]**

1 Civil and Environmental Engineering, Virginia Tech, 1145 Perry St., 418 Durham Hall, Blacksburg, VA 24061, USA; acullom@vt.edu (A.C.C.); martinrl@vmi.edu (R.L.M.); ys117@vt.edu (Y.S.); apruden@vt.edu (A.P.)
2 Civil and Environmental Engineering, Virginia Military Institute, Lexington, VA 24450, USA
3 TechLab, 2001 Kraft Drive, Blacksburg, VA 24060, USA; kwilli@vt.edu
4 c/o Marc Edwards, Civil and Environmental Engineering, Virginia Tech, 1145 Perry St., 418 Durham Hall, Blacksburg, VA 24061, USA; k78bass@gmail.com
* Correspondence: edwardsm@vt.edu

Received: 9 October 2020; Accepted: 13 November 2020; Published: 17 November 2020

**Abstract:** Growth of *Legionella pneumophila* and other opportunistic pathogens (OPs) in drinking water premise plumbing poses an increasing public health concern. Premise plumbing is constructed of a variety of materials, creating complex environments that vary chemically, microbiologically, spatially, and temporally in a manner likely to influence survival and growth of OPs. Here we systematically review the literature to critically examine the varied effects of common metallic (copper, iron) and plastic (PVC, cross-linked polyethylene (PEX)) pipe materials on factors influencing OP growth in drinking water, including nutrient availability, disinfectant levels, and the composition of the broader microbiome. Plastic pipes can leach organic carbon, but demonstrate a lower disinfectant demand and fewer water chemistry interactions. Iron pipes may provide OPs with nutrients directly or indirectly, exhibiting a high disinfectant demand and potential to form scales with high surface areas suitable for biofilm colonization. While copper pipes are known for their antimicrobial properties, evidence of their efficacy for OP control is inconsistent. Under some circumstances, copper's interactions with premise plumbing water chemistry and resident microbes can encourage growth of OPs. Plumbing design, configuration, and operation can be manipulated to control such interactions and health outcomes. Influences of pipe materials on OP physiology should also be considered, including the possibility of influencing virulence and antibiotic resistance. In conclusion, all known pipe materials have a potential to either stimulate or inhibit OP growth, depending on the circumstances. This review delineates some of these circumstances and informs future research and guidance towards effective deployment of pipe materials for control of OPs.

**Keywords:** non-tuberculous mycobacteria; *Pseudomonas; Acinetobacter*; amoebae; copper; iron; PEX; PVC; drinking water; disinfection

## 1. Introduction

Legionnaires' Disease is the "leading cause of reportable waterborne illness" in the United States [1,2], with 52,000–70,000 cases per year [1,3,4], 8000–18,000 hospitalizations [5], an overall mortality rate of 15% [4], and high healthcare and legal costs [2,6–8]. Bacteria belonging to the genus *Legionella* are the causative agent of Legionnaires' disease and Pontiac Fever, which infect the human respiratory system via inhalation or aspiration. *Legionella* is classified as "opportunistic" because it preferentially infects those with underlying illnesses or weakened immune systems [4,8,9]. To date more than 60 *Legionella* species have been identified [10], with *Legionella pneumophila* being the species most commonly attributed to human disease [11]. *Legionella* can be found even in "the most aggressively treated drinking water" [12]. Studies have confirmed that potable water is a key source of infection [1,4,13–17], for both hospital- and community-acquired cases [18–20]. Other opportunistic pathogens (OPs) such as nontuberculous mycobacteria (NTM), *Pseudomonas aeruginosa*, and *Acanthamoebae*, can similarly be transmitted via tap water and tend to infect individuals belonging to certain risk groups [8].

To infect humans, *Legionella* and other OPs must be present in tap water at the point of use. While *Legionella* can occasionally survive drinking water treatment and be transported through the main water distribution system, the primary environment for *Legionella* proliferation to numbers needed to infect humans generally occurs in building or "premise" plumbing [21,22]. Premise plumbing includes the service pipe that connects buildings to the water main, in addition to the full array of components comprising cold and hot portions of a building's potable water system [8]. Premise plumbing is characterized by high surface area to volume ratios, longer stagnation times, low disinfectant residual, areas with excess sediment and scale, chemically and biologically reactive plumbing materials, and water with relatively warm temperatures. Such conditions can create ideal micro- and macro-environmental niches for growth of various OPs [1,8,23].

Premise plumbing is a key conduit for human exposure via showering, handwashing, and other applications that create airborne aerosols [24]. *Legionella* has been detected in faucets, showerheads, decorative fountains, grocery store mist systems, ice machines, and cooling towers [13,14,16,25]. Larger buildings with more complex plumbing systems are more likely to create physicochemical conditions suited for *Legionella* proliferation, but it is also often detectable in water mains and residences with simple conventional hot and cold water plumbing systems [17,26,27]. A Centre for Disease Control (CDC) summary of Legionnaires' Disease potable water outbreak investigations from 2000–2014, concluded that 85% of the cases had "deficiencies" in water system maintenance within buildings as a contributing factor [28] and that water chemistry flowing into buildings is one, but not the only, predictor of *Legionella* incidence [29,30].

The mechanisms by which premise plumbing influences *L. pneumophila* and other OPs, as well as the broader premise plumbing microbiome, are varied and complex (Figure 1). The influent water chemistry has been found to influence *Legionella,* and also strongly shape the plumbing microbiome, especially through the delivery of growth-promoting nutrients, growth-inhibiting disinfectants, and influent microorganisms [31–34]. The ecological interactions among microorganisms in biofilms of building plumbing systems can also help overcome barriers to growth from low nutrient levels and disinfectants [24,35,36]. Conversely, other interactions, such as competition, exclusion, predation, or inactivation of symbiotic organisms, may inhibit the growth of OPs [37]. The selective pressures in premise plumbing might also alter the physiologies of resident microbes in a manner that influences infectivity [38]. All these phenomena are further complicated by the fact that premise plumbing configurations, hydraulics, temperature, and water use patterns including velocity, flow or stagnation events, all differ significantly from building to building. In particular, there is strong variability due to occupancy, building size, water heater design, water saving devices, storage and other factors [39,40]. Thus, while there are many overarching similarities, every premise plumbing system is at least as variable as the occupants' unique water use patterns and habits.

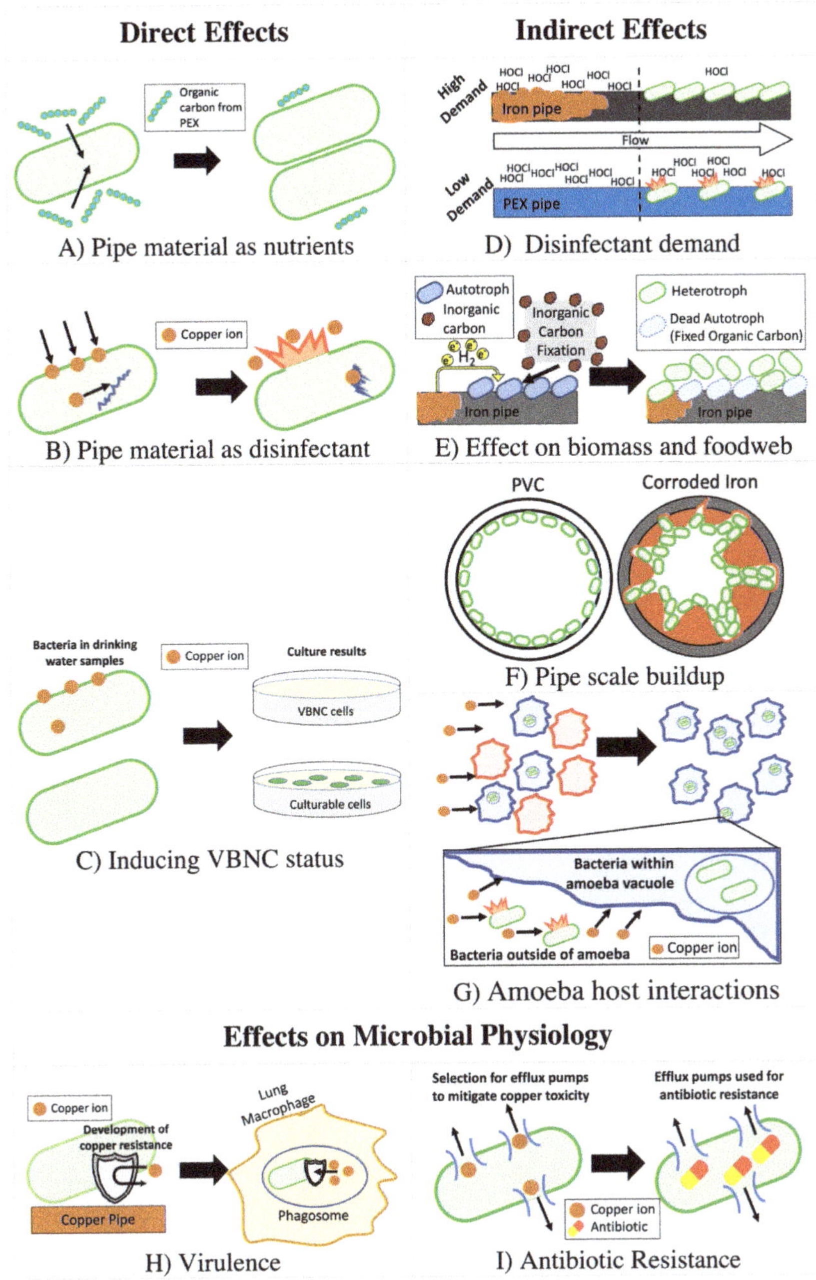


**Figure 1.** Overview of exemplar mechanisms by which pipe materials can affect OPs in premise plumbing. Depending on the circumstances, the pipe material itself can have direct effects on OPs growth by: (**A**) providing organic or inorganic nutrients that enhance growth, (**B**) acting as a growth-inhibiting antimicrobial, or (**C**) inducing viable-but-non-culturable (VBNC) status, from which microbes might recover in terms of infectivity and growth rates subsequent to exposure. Pipes can also indirectly affect OPs by: (**D**) consuming secondary disinfectants, allowing for microbial growth downstream, (**E**) evolving

hydrogen gas or enhance nitrification, fueling autotrophic growth, or (**F**) developing thick pipe scales, which provide additional surface area for microbial growth, or (**G**) selecting for certain types of amoebae that are preferred hosts for bacterial OPs and protect them from negative effects of copper and disinfectants. Finally, pipes may unfavorably alter the physiology of microbes by increasing (**H**) OP virulence by selecting for resistance to phago-somal copper overload, or (**I**) resistance to antibiotics.

The type of pipe material can also strongly influence the relationship between premise plumbing materials and OPs through both direct effects (interaction with chemical species released from pipe) and indirect effects (secondary consequences of released material from pipes) by altering the level of nutrients, disinfectants, and microbial biomass (Table 1, Figure 1). Selection of pipe material can therefore strongly affect chemistry, biological stability [41], and microbiome composition [42] of the drinking water.

**Table 1.** Positive (+), Negative (-, –), and Neutral (0) Pipe Material Effects on OPs Control as Mediated by Various Water Chemistry Attributes.

| Water Chemistry Attribute Influenced by Pipe Materials | Relevance to OPs | Effect of Pipe Materials on OPs Control as Mediated by the Indicated Water Chemistry Attribute | | | | |
|---|---|---|---|---|---|---|
| | | Copper | PVC | PEX | SS | Iron [1] |
| Chlorine | Disinfectant | - [43] | 0 [43–50] | - [43,51,52] | 0/- [43–45,48,53] | – [43–48] |
| Chloramine | Disinfectant | - [43,54] | 0 [43,50] | 0 [43,52] | 0 [43] | – [43,55,56] |
| Assimilable Organic Carbon | Carbon source | 0 | - [42,56,57] | – [42,56,58,59] | 0 | 0 |
| Hydrogen Gas (aq) | Food web | 0 | 0 | 0 | 0 | - [60,61] |
| Release of Metals | Release of metals | +/- [59,62–64] | 0 | 0 | 0 [65] | – [66] |

**Abbreviations:** OPs, opportunistic pathogens; PVC, Polyvinyl chloride; PEX, cross-linked polyethylene; SS, stainless steel; aq, aqueous. [1] Includes unlined iron and old galvanized iron pipes.

Motivations for this review include:

- Growing direct or indirect potable water reuse, which can sometimes alter levels of nutrients and $Cu^{+2}$ in the source water [67].
- Increased natural organic matter (NOM) in some source waters as an indirect consequence of improving sulfur and nitrogen air pollution controls under rules and regulations such as the U.S. Clean Air Act or Directive 2008/50/EU [68–70].
- Emphasis on and investment in green building design for water and energy efficiency and associated unintended consequences for in-building hydraulics (e.g., more stagnation, higher surface area to volume ratios of water to plumbing surfaces, required hot water recirculation systems) that alter water chemistry and delivery of nutrients or disinfectants [39,54,71,72].
- Greater use of plastic pipes (e.g., PEX, PVC, polyethylene), which vary in leaching potential by type of plastic and due to the presence of proprietary stabilizers and processes [73].
- Increasing awareness of viable-but-non-culturable (VBNC) bacteria, which are difficult to measure directly. Molecular and fluorescence-based techniques suggest that they can be prevalent under certain circumstances [8,74] and recent evidence indicates they can still cause disease [75,76].
- Heightened concern about an array of bacterial OPs besides *Legionella*, including *Pseudomonas aeruginosa*, *Acinetobacter baumannii*, and NTM, as well as amoebae (e.g. *Acanthamoeba*, *Vermamoeba*), which can themselves be pathogenic or can serve as host organisms for bacterial OP proliferation [8].

Here we critically examine existing knowledge with respect to the direct (Section 2) and indirect (Section 3) effects of common metallic (copper, iron, zinc, aluminum, magnesium) and plastic (PVC, PEX) building pipe materials on the growth of *Legionella* and other OPs, in addition to identifying the complex effects of plumbing system configuration (Section 4) and the characteristics of the drinking water microbiome (Section 5). This review is particularly timely, at a moment when societal expectations for public health protection are elevated and expanding aspirations for improved water/energy conservation will be a major drive of water system design and pipe material selection [39]. In executing this review, we aimed to holistically assess the effects of pipe materials, primarily focusing on *Legionella* while including other OPs, seeking to shed light on why various pipe materials appear to sometimes enhance and other times diminish OP proliferation under real-world premise plumbing conditions.

## 2. Direct Effects of Plumbing Material on Pathogen Growth

### 2.1. Copper Has Both Antimicrobial and Micronutrient Properties

Copper is sometimes present at trace levels in the source water or in distributed water mains, but the main sources in premise plumbing are copper pipes and brass fittings that are installed beginning at the service line connecting the building to the water main (Figure 2). Due to long-lasting life span, durability, and relatively few concerns about metal release when compared to those of antiquated lead and galvanized iron alternatives, copper and its alloys are common in premise plumbing systems [77]. Copper is a registered antimicrobial of the US Environmental Protection Agency (EPA) [78] and listed as a biocidal product in the European Union, but some countries require special approval for use of copper in drinking water for OP control [79]. It is also an essential nutrient for all living organisms, including humans and OPs [59,80]. Here we review the mechanisms by which copper plumbing may influence control of various OPs (Table 2).

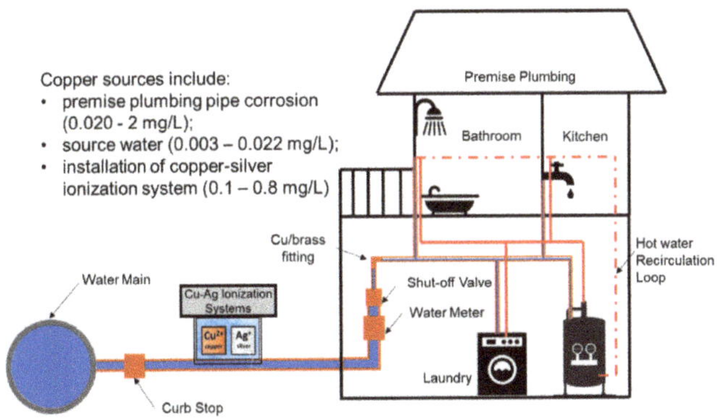


**Figure 2.** Copper sources in premise plumbing [81–84]. Note that Cu-Ag Ionization systems can be used in either point of entry or hot water distribution networks.

**Table 2.** Copper can be growth-promoting or -inhibiting to opportunistic pathogens.

| Opportunistic Pathogen | Associated Diseases | Exposure Route(s) | Inactivation via Copper | | Growth via Copper | |
|---|---|---|---|---|---|---|
| | | | *Antimicrobial Efficacy* * | *Evidence for Cu-Induced VBNC* | *Micronutrient Activity* | *Amoeba-Mediated Growth* |
| *Amoebae* | Encephalitis, Eye infections, Primary amebic meningoencephalitis [85–87] | Dermal, Inhalation, [85–87] | Moderate to Somewhat inhibited [59,88] | Unknown and unlikely | | NA |
| *Acinetobacter baumannii* | Bacteremia, Meningitis, Pneumonia, Urinary tract infections [89] | Dermal, Inhalation [89] | Moderate to Somewhat inhibited [58,90–92] | Unknown | | Yes [93,94] |
| *Staphylococcus aureus* | Bacteremia, Endocarditis, Osteomyelitis, Pneumonia, Sepsis, Skin infections [95] | Dermal, Inhalation [96,97] | Moderate [90,98,99] | Unknown | | Yes [100,101] |
| *Stenotrophomonas maltophilia* | Bacteremia, Endocarditis, Eye infections, Meningitis, Pneumonia, Sepsis, Skin infections, Urinary tract infections [102,103] | Dermal, Inhalation [102,103] | Moderate [91,92] | Limited [104] | | Yes [105] |
| *Nontuberculous Mycobacteria (NTM): Mycobacterium avium complex; Mycobacterium abscessus complex; Mycobacterium kansasii and other species* | Bacteremia, Pneumonia, Skin infections [106] | Dermal, Ingestion, Inhalation [107] | Moderate [108–111] | Limited [109] | Possible that organisms are copper deficient and additional copper could increase growth [59,80] | Yes [36,112] |
| *Aeromonas hydrophila* | Gastroenteritis, Meningitis, Peritonitis, Pneumonia, Skin infections [113] | Ingestion, Inhalation [113] | Unknown [114] | Unknown | | Yes [115,116] |
| *Legionella pneumophila* | Legionnaires' disease, Pontiac fever [117] | Inhalation [118] | Somewhat inhibited to High [62,83,119,120] | Moderate [121,122] | | Yes [123,124] |
| *Pseudomonas aeruginosa* | Bacteremia, Endocarditis, Eye infections, Gastroenteritis, Osteomyelitis, Pneumonia, Sepsis, Skin infections, Urinary tract infections [125] | Dermal, Ingestion, Inhalation [125,126] | Somewhat inhibited to High [90–92,98,99,127–132] | Strong [127,132,133] | | Yes [36,134] |

* Categorizations of efficacy based upon studies that showed planktonic phase growth inhibition at: <0.1 mg/L (High), 0.1–0.8 mg/L (Moderate), and >0.8 mg/L (somewhat inhibited) copper concentrations in water or media.

### 2.2. Copper Pipe as an Antimicrobial Material in Premise Plumbing

The antimicrobial properties of copper were first described more than 3000 years ago in the Hindu Vedas and are occasionally observed at least temporarily in modern plumbing systems [1,120,135–137]. The role of supplemental dosing of copper as disinfectants in building plumbing can be important, because *Legionella* and other premise-plumbing-associated OPs are more resistant to chlorine than traditional fecal-associated bacteria that are used for traditional water quality monitoring [8,24,138]. While there is no clear consensus on the primary mechanisms by which copper inactivates bacteria, two hypotheses have been put forward: (1) positively charged $Cu^{+2}$ ions interfere with negatively charged cell membranes, creating holes; and (2) $Cu^{+2}$ disrupts the replication and production of DNA, RNA, and proteins, potentially through metabolic cycling between $Cu^{1+}$ and $Cu^{2+}$ oxidation states, which generates radical oxidative species such as hydroxide radicals [139]. In potable water, copper passively released from plumbing materials can be present in the germicidal range for *Legionella* of 0.1–0.8 mg/L [62,119,120,140], even in some parts of plastic pipe systems connected with brass fittings [141,142]. Passive release or purposeful dosing that results in copper concentrations of 0.05–0.8 mg/L are thought to limit *Legionella* growth [62,83,119,120,143].

A number of studies have confirmed the efficacy of copper, either passively leached from premise plumbing materials [59,140,144] or actively added using copper-silver ionization (CSI) systems [62,83,145], as a *Legionella* antimicrobial. Biofilms grown at room temperature for 30 days in pre-sterilized reactors with copper, PVC, and stainless steel coupons were found to have lower total bacterial counts on copper than PVC surfaces [146]. Other batch reactor studies indicate similar results, demonstrating lower *L. pneumophila* numbers on copper plumbing than plastic plumbing [59,140,144,147]. Analogous responses to copper surfaces by other Ops, such as *Klebsiella* spp. [148], NTM [111,149], *P. aeruginosa* [128], and *Aeromonas hydrophila* [114], have been reported. Two different field studies found that copper concentrations were significantly lower in samples positive for *L. pneumophila* than samples negative for *L. pneumophila* [150,151]. Borella et al. [23,152] identified a threshold total copper level of 0.5 mg/L in one sample of water, above which samples were approximately two to seven times less likely to be positive for *L. pneumophila*.

Studies of CSI applications also demonstrate that copper can have direct antimicrobial effects. Lin et al. [83,109] showed that 0.5 and 48 h of exposure to 0.4/0.04 mg/L copper/silver achieved 99% inactivation of *L. pneumophila* and *Mycobacterium avium*, respectively, in bench-scale testing. Stout et al. [119] performed long-term monitoring of CSI systems in 16 hospitals and demonstrated their efficacy for *Legionella* control, as the numbers of hospitals with >30% *Legionella* positive samples dropped from 7/16 to 0/16, and no Legionnaire's disease cases were reported in 15 out of 16 hospitals after the implementation of CSI. Addition of copper ions to solution from pipes or via CSI, at the bench and building-scale, has also been shown to inhibit the growth or reduce the frequency of OPs such as *Staphylococcus* spp.[98,99], *Stenotrophomonas maltophilia* [91,92,104], *Acinetobacter baumannii* [58,91,92], NTM [108,109], and *P. aeruginosa* [91,92,98,99,127,130].

#### 2.2.1. Noteworthy Limitations to Copper's Antimicrobial Efficacy

Despite the encouraging examples presented in the previous section, the overall success of copper as a disinfectant for *Legionella* is mixed [110]. Several studies have found that the antimicrobial effects of copper were limited, or that copper even encouraged growth of *Legionella* in some instances [63,83,122,153]. In one study, *Legionella* was consistently detected in a hospital hot water plumbing system with average pH = 7.7, even when copper was present at concentrations of 1.1 ± 0.2 mg/L [153]. Other studies have shown similar trends. For instance, Giao et al. [121] found no significant difference between biofilm formed on plastic (PEX and PVC) coupons and biofilms formed on copper coupons when the biofilms contained a heterogeneous community or when the biofilms were purely *L. pneumophila*. *P. aeruginosa* has been found to persist in hospital copper plumbing [129] and the implementation of a CSI system in one hospital did not appear to fully eliminate patient *P. aeruginosa* infections associated with exposures from faucets [130].

Prominently, in one field study conducted in Germany with low or no chlorine residual, hot water systems containing copper pipes were colonized with *Legionella* much more often (>30x) than those with galvanized steel or plastic pipes, despite the fact that the temperature of the hot water in these systems was similar. Also, samples (n = 44) from hot water recirculation lines with >0.5 mg/L of copper displayed 2,4000 ± 15,000 (mean ± standard deviation) CFU *Legionella*/L, while samples (n = 153) with ≤0.5 mg/L of copper had 10 ± 100 CFU *Legionella*/L [63].

There are many possible explanations for the apparent contradictions in overall impacts of copper (Table 2). It is important to first recognize that the antimicrobial properties of copper can be almost completely controlled by water chemistry (Figure 3). Notably, the concentration of $Cu^{+2}$ and its associated inorganic ions tend to decrease in concentration in aged pipes, at higher pH, or in the presence of common corrosion inhibitors, such as orthophosphate. Unfortunately, studies frequently do not collect or report such relevant data [63,129,130,153], limiting the ability to trace differences in copper's antimicrobial efficacy to water quality parameters. There is also the likelihood of strain-to-strain differences in copper resistance, and the selection for copper resistant organisms in systems with copper pipes [154,155].

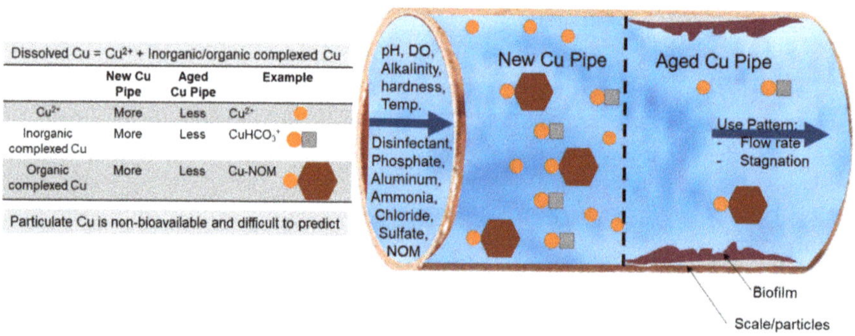


**Figure 3.** Copper pipe corrosion and speciation is controlled by influent water chemistry and pipe age. Water chemistry parameters, such as pH, dissolved oxygen (DO), disinfectants, inorganic complexing agents (e.g., alkalinity, phosphate, and ammonia), organic complexing agents (e.g., natural organic matter (NOM)), hardness, trivalent metal ions (e.g., aluminum, iron), sulfate, and chloride can influence copper pipe dissolution, speciation, and the precipitation process. Copper is categorized as either free copper ions and inorganic complexed copper (considered relatively bioavailable), or organically complexed or particulate copper (considered relatively non-bioavailable). The level of copper species in the premise plumbing systems are also affected by the pipe aging (new vs. old pipes) and the water use pattern, including flow rate, stagnation and temperature.

### 2.2.2. Water Chemistry Effects on Copper Bioavailability

The chemistry of the influent bulk water can reduce toxicity of copper by: (1) reducing overall solubility and the equilibrium level of $Cu^{+2}$ in the presence of copper rusts [156,157]; (2) forming copper complexes [158–160], (3) having elevated divalent ($Ca^{2+}$, $Mg^{2+}$) or trivalent ($Fe^{3+}$, $Al^{3+}$) cations, which compete with copper for uptake sites of organisms [161–163]. Therefore, water chemistry details are useful to explain the discrepancy of copper effects, but such information is often lacking in some studies [63,121,129,130,153].

Prior culture-based research demonstrated that precipitation of copper at pH 9 reduced toxicity of copper towards nascent *L. pneumophila* colonies by 16-fold relative to pH 7, where copper is more soluble [83]. Other compounds known to reduce levels of $Cu^{+2}$ by complexation and precipitation are logically expected to interfere with copper antimicrobial properties and include NOM and either ortho- or poly-phosphates [156–160]. Specifically, NOM and polyphosphate sequestrants can vary in concentration and complexation ability from water to water, can bind $Cu^{+2}$ and dramatically reduce its bioavailability. Orthophosphate added as a corrosion inhibitor can reduce metal pipe corrosion rates and lower free metal ion concentrations in drinking water. For example, our research has shown that the addition of 3 mg/L of phosphate and 5 mg/L NOM at pH = 7 reduced copper's antimicrobial effects towards *L. pneumophila* by four and seven times, respectively [164].

Copper's antimicrobial properties are expected to increase at lower pH, lower hardness, lower $Al^{+3}$ and $Fe^{+3}$, lower phosphate or polyphosphate, lower NOM, and colder temperatures due to known interactions with $Cu^{+2}$ ion. Studies of copper toxicity to algae and higher aquatic organisms have shown that $Mg^{2+}$, $Ca^{2+}$, $Al^{+3}$, and $Fe^{+3}$ compete with copper for binding sites, reducing the toxicity of copper [161–163]. For instance, Ebrahimpour et al. [161] reported that the 96-h median lethal concentration (LC50) values for *Capoeta fusca* increased roughly linearly (1.1 to 7.5 mg/L copper) over a hardness range of 40-380 mg/L as $CaCO_3$. Trivalent metal ions, such as $Al^{3+}$ and $Fe^{3+}$, can also form a layer of metal hydroxide gel around cells that can sorb copper and reduce its availability [165]. Free copper also tends to decrease at higher temperature and as pipe scales age [54,166].

#### 2.2.3. Copper as a Nutrient in Premise Plumbing

Copper (Cu) is an essential micronutrient used in protein synthesis, respiration, various oxidation/reduction reactions and other functions in prokaryotes [80,167]. Accordingly, it is reasonable to suspect that copper piping might sometimes act as a source of this essential nutrient in premise plumbing, thereby increasing microbial growth relative to other materials. Buse et al. [122] showed that effluent from CDC biofilm reactors equipped with coupons of different pipe materials at pH > 8 and $PO_4$ > 0.2 mg/L, had up to 20× more *L. pneumophila* gene copies when copper coupons were used relative to PVC coupons. Mullis et al. [111] indicated that copper surfaces supported two to four times more *Mycobacterium abscessus* than PVC. Mathys et al. [63] reported that hot water systems containing copper pipes were colonized significantly more often than those with galvanized steel or plastic pipes.

### 2.3. *Direct Release of Organic Carbon by Plastics*

Potable water is oligotrophic, because organic carbon is relatively scarce and often limiting to the growth of drinking water microorganisms [24,168,169]. Plastic premise plumbing pipes, which are made with polymeric organic compounds, including stabilizers, flexibilizers and plasticizers, can leach organic carbon to water [56,57,170] whereas metallic pipes do not. These organic carbon compounds can fuel the growth of *Legionella* [45,59] and presumably other OPs. In some cases, the organics leached to water are not the polymers themselves, but rather are additives (i.e., flexibilizers, plasticizers, stabilizers) to improve aspects of pipe performance [42,170,171].

New PEX pipes commonly leach 100-1800 μg/L of total organic carbon (TOC) as determined by temperature, stagnation, surface area to volume ratio, pipe brand and age [56,170,172]. These levels of carbon, are far above the commonly cited threshold of 100 μg/L suggested to spur microbial growth in potable water main distribution systems [173]. However, the proportion of this released organic carbon that is assimilable is not clear. Many studies have demonstrated that some PEX pipes increase biofilm growth [59,140,147] and OP growth [59,140] relative to copper and iron. Unfortunately, it is unclear how general these effects are because the formulation of PEX used (e.g., PEX-b) varies from one manufacturer to another [170,172] and is typically proprietary and thus not cited in the available literature [59,140,147]. An experiment in the Netherlands using small-scale recirculating water heater systems (eight gallon tanks) connected to copper or PEX pipes (19.4 ft) attributed over three times higher *Legionella* bulk water levels in PEX pipe systems as compared to copper pipe systems although the authors did not determine if the difference was due to copper antimicrobial effects or leached organic carbon growth-promotion [140].

PVC pipes can leach 60–50,000 μg/L of TOC under typical water use conditions [50,56,174], of which roughly 50% was estimated to be assimilable [42]. Other studies indicate that PVC can promote biofilm growth [175,176] and proliferation of OPs compared to copper, lined cement, iron, and stainless steel [111,177–179]. When copper, glass, PEX, and PVC were used as materials in a biofilm apparatus simulating premise plumbing, PVC and PEX materials maintained the highest *Legionella* growth potential in remineralized reverse osmosis water [178]. Other studies have drawn similar conclusions for other OPs compared to copper [111,128,148,149].

### 2.4. Iron Release from Pipes

Iron pipes may provide important niches and nutrients for OP growth. Antiquated cast iron, galvanized iron, and steel pipes in service lines and home plumbing can leach iron to water in a range of 0.2–18 mg/L dependent on factors including water chemistry, stagnation, surface area to volume ratio, and historical corrosion control [180,181]. Iron can also accumulate in loose deposit or biofilms and some studies have suggested that such locations are hotspots for growth of *Legionella* and other pathogens [40,182]. Studies examining *M. avium* have found that galvanized steel supported more growth than copper, PVC, and stainless steel [111,149].

Iron is an important nutrient for microorganisms involved in oxygen transfer, protein synthesis, and other essential metabolism [183] and some studies have shown that the presence of iron contributes to OP growth. Bench-scale studies have demonstrated that iron concentrations of up to 1 mg/L could enhance *L. pneumophila* growth in tap water while high concentrations (10, 100 mg/L) of iron produced toxic effects on *L. pneumophila* [184]. During the Legionnaires' Disease outbreak in Flint, MI, our research found that the median iron concentration was 0.11 mg/L in cold water samples during the outbreak, but the outbreak's end coincided with a water switch, dropping median iron in cold water samples down to less than 0.01 mg/L [26]. Other field studies have observed similar positive correlations between *L. pneumophila* levels and iron concentrations [15,185]. In a simulated household drinking water system with no chlorine, van der Lugt et al. [186] observed that colonization of stainless steel faucets by *Legionella* was enhanced in the presence of 0.09 mg/L cast iron rust. It is important to note that in any study employing chlorine, iron pipe corrosion will remove the chlorine, confounding simplistic attribution of the higher *Legionella* to either iron or chlorine [26,187,188]. One study specifically examined if iron addition increased *L. pneumophila* growth without any chlorine present, and showed that it did so in one water with naturally low iron, but had no effect in another water with relatively high ambient iron [187].

### 2.5. Zinc, Aluminum, Magnesium Plumbing Materials

Pipes and plumbing devices can be composed of other metals that might affect the growth of OPs, but their impacts are largely unexplored. Zinc is present in source waters in concentrations ranging from <0.011 to 0.04 mg/L [189,190] and is normally below 0.1 mg/L in finished water [191]. Zinc concentrations at the tap are largely driven by its addition in corrosion inhibitors, or release from brass fixtures and galvanized pipes [190–192], and concentrations can reach 5 mg/L or higher [193,194]. Analogous to copper, zinc is an essential nutrient for microbial growth [195–200]. Zinc addition has been shown to increase *L. pneumophila* and *P. aeruginosa* growth in culture media [201], and high soluble zinc has been correlated with NTM [202].

Zinc can be toxic to microorganisms [196,203–206], but is believed to have limited biocidal activity compared to other metals [207], especially as it is below the US EPA Secondary Drinking Water Regulation limit of 5 mg/L [207] and Chinese Standard for Drinking Water Quality of 1 mg/L [208]. Inhibitory concentrations of zinc for Ops such as *Pseudomonas* spp., *P. aeruginosa*, and *Aspergillus niger* range from 13 to 650 mg/L in nutrient broth [204–206]. While this is a relatively high concentration range, Zhang et al. [180] demonstrated that galvanized iron pipes can release zinc to these levels in the presence of nitrifying bacteria. Furthermore, the biocidal activity of zinc or any other trace metal in premise plumbing will be controlled by the same chemistry factors including pH, hardness and NOM mentioned previously for copper.

Aluminum or magnesium rods are also commonly present as sacrificial anodes in water heaters (Figure 4), elevating $Al^{+3}$ or $Mg^{+2}$ levels in the water. $Mg^{+2}$ is known to be an essential nutrient for *Legionella* [201], whereas no such criteria have been established for $Al^{+3}$. More research is needed to determine whether these additional trace metals encourage or discourage OP growth in plumbing systems.

## 3. Indirect Effects of Pipe Material on Pathogen Growth

### 3.1. Pipe Material Effect on Disinfectant Availability

Pipe material is a key factor affecting disinfectant decay in potable water systems. Maintaining relatively high levels of disinfectant residual is important to OP control because OPs are 20–600x more disinfectant resistant than the common indicator microorganisms such as *E. coli* [24] and are further protected in biofilms or host organisms [209–214]. Plastic pipe materials are generally non-reactive with chlorine and chloramine in terms of maintaining disinfectant residual levels, even though chlorine does sometimes slowly react with and degrade certain types of PEX and polyethylene pipe [44–49,51,215]. On the other hand, iron pipes have an extremely high disinfectant demand, as free chlorine cannot co-exist in equilibrium with ferrous or zero valent iron [44,46–48]. While chloramine is relatively non-reactive, iron oxide scale and associated nitrifying biofilms can cause relatively rapid monochloramine decay [216,217]. The reactivity of copper pipes and copper oxides is typically between plastics and iron and chemically catalyzes both chlorine and chloramine degradation [43,54,156,218–220]. Higher pH and the existence of phosphate can help maintain disinfectant residual levels in both iron and copper pipes [26,54].

### 3.2. Effect of Metallic Plumbing Materials on Nutrient Availability via Autotrophic Carbon Fixation

Although metallic plumbing does not leach assimilable organic carbon directly to water, certain metals can indirectly help OPs overcome carbon limitations by facilitating the growth of autotrophic microorganisms. Specifically, metallic pipes can encourage growth of hydrogen-oxidizing, ammonia-oxidizing, and ferrous-oxidizing autotrophic bacteria that fix inorganic carbon into new biomass [66,221].

#### 3.2.1. Hydrogen Oxidizing Bacteria

The corrosion of iron pipes and the galvanic corrosion of aluminum or magnesium sacrificial anodes protecting steel water heaters can evolve hydrogen gas, which is a strong electron donor for autotrophs [60,61,110,221]. Ishizaki et al. [222] indicated that hydrogen-oxidizing bacteria, *Alcaligenes eutrophus*, could fix 2300 μg C/mmol $H_2$ in biomass in closed circuit cultivation system at gas pressure slightly higher than atmosphere, which could practically translate into production of up to 80 μg/L organic carbon biomass per day in an 80-gallon water heater equipped with a magnesium anode [223]. A study by Dai et al. [224] of an experimental water heater plumbing rig at 39, 42, and 51 °C confirmed elevated levels of functional genes associated with hydrogen metabolism, demonstrating that hydrogen-oxidizing bacteria were able to proliferate in water heaters.

#### 3.2.2. Autotrophic Ammonia and Iron Oxidizing Bacteria

Iron and copper can catalyze the conversion of chloramine disinfectant to free ammonia, which can then serve as a substrate for autotrophic ammonia oxidizing bacteria. Ammonia-oxidizing bacteria can fix substantial amounts of organic carbon into the system, specifically 21 to 240 μg C/mg $NH_3$-N based on experimental growth yield values of pure or mixed cultures [225]. Ferrous iron, released as a natural by-product of iron corrosion, can also fix an average of 26 μg C/mg $Fe^{2+}$ under circumneutral condition measured in bioreactors [226].

#### 3.2.3. Copper Deposition Corrosion Accelerating $H_2$ Evolution

Although copper cannot corrode with evolution of $H_2$ gas, cupric ions in water can plate onto the less noble metals (zinc, aluminum, iron and magnesium) via deposition corrosion. This copper coating can dramatically accelerate corrosion of less noble metals and indirectly stimulate evolution of hydrogen ($H_2$) gas (Figure 4) [66,222,227,228]. A study using a combination of bench- and pilot-scale hot water system experiments demonstrated these effects [222].

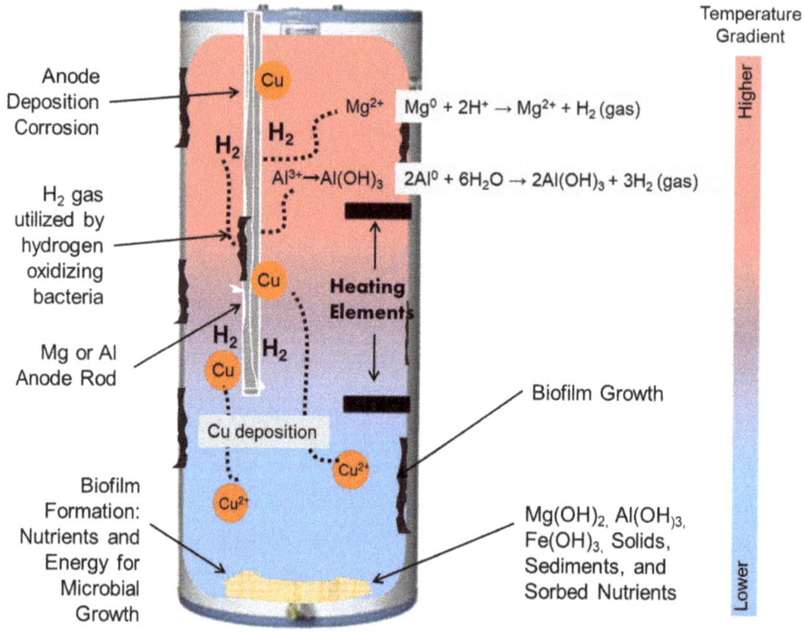


**Figure 4.** Water heater material interactions create multiple niches suitable for bacterial and opportunistic pathogen (OP) growth. Deposition of copper onto less noble metals (e.g., a water heater anode) can result in dramatically accelerated corrosion and release dissolved $H_2$ gas, which is an electron donor for autotrophs. If the anode rod consists of magnesium, then the pH will become elevated as well. Figure adapted from Brazeau et al. [229].

### 3.3. Pipe Scaling Effects

Scaling caused by pipe corrosion or higher pH can increase pipe surface roughness, which is known to enhance biofilm colonization and overall growth, creating an ideal environment for OP establishment and proliferation [112]. One study showed that copper coupons in a biofilm reactor formed extensive scales and promoted seven-fold more biofilm biomass than PVC pipes after three months of incubation [230]. Aged metal pipes may form very thick scales characterized by corrosion tubercles and extensive networks of pores [60,231–233], providing an area for not only additional biofilm growth, but also distinct microenvironments [233,234] with pH is as low as 2.0 or as high as 10 [235].

## 4. Influence of Plumbing System Design, Configuration and Operation

All of the direct and indirect interactions described in previous sections are further influenced by the specific premise plumbing design, configuration, and operation. Flow rate, water stagnation, temperature profile, secondary disinfectant concentration, and nutrient availability can all interact to create hot spots for OPs growth in buildings.

### 4.1. Water Stagnation

Water age is defined as the time it takes water to move from one point to another in the system, which may influence OP growth through a variety of mechanisms. This includes the time from when it is freshly produced at the treatment plant and travels to the service line, as well as the time from when it first enters the building's plumbing to the point of use [71]. High water age in buildings is increased by: (1) existence of dead ends/legs and stagnation in plumbing systems [182,236]; (2) use of low flow devices or presence of large storage tanks such as those used for solar water heating or onsite rainwater collection [39]; and (3) using low volumes of water in a building or at a particular outlet [192]. Stagnation and infrequent water use may concentrate and enhance release of organic matter in water in plastic pipes and metals in metallic pipes [181,237–240]. Zhang et al. [241] found a four-fold increase in bulk water TOC in unplasticized PVC pipes between 24 h and 72 h of stagnation. Fixtures in a green building with the fewest water use events (most stagnation) also had greater organic carbon, bacteria counts, and heavy metal (Zn, Fe, Pb) concentrations [192,242].

Stagnation and high water age also increases the likelihood and rate of disinfectant decay. High consumption of chlorine and chloramine during stagnant periods of 24–72 h have been observed for synthetic pipes (0.4 and 0.6 mg/L of chlorine loss, respectively), and stagnant periods of 2–8 h in metallic pipes (3 and 4 mg/L chlorine loss, 1.5 and 3.5 mg/L chloramine loss, respectively) [54,241]. In a green building study, six-hour stagnation almost fully eliminated monochloramine (>99%) within pipes [71].

Such water quality changes have been related to increased levels of OPs in premise plumbing systems [39,243–245]. In a field sampling study of main water distribution system, 120 water samples were taken throughout a drinking water distribution system. Only four samples were positive for cultivable *L. pneumophila* and all four samples were taken from dead end points at the end of streets with no chlorine residual remaining [246]. Another field study identified their most frequently *Legionella* positive sites as being located at the end of the distribution system and having the highest turbidity, iron, TOC, and water age, as well as the lowest flow [247]. The association between OPs and stagnation has created interest in strategies to reduce building water stagnation effects such as removing dead-legs, flushing, maintaining the hot water system, and shock disinfection [248–251]. The effectiveness of these strategies should be evaluated within the context of the specific pipe materials that are present.

### 4.2. Hot Water Recirculation Lines

Some plumbing codes require or suggest the use of recirculating hot water lines for water/energy conservation, convenience and comfort [1,252–254]. In these systems, water is circulated continuously between the water heater and the point of use, preventing cooling of the distal lines and allowing for nearly instant delivery of hot water at the point of use [255]. There are many important differences between hot water recirculating systems and conventional systems, which are stagnant during periods of disuse that can affect OP growth. The constantly flowing water can deliver more nutrients to biofilm and hypothetically increase OP growth [230]. On the other hand, continuous flow can deliver more disinfectants and more hot water, which are critical control measures for OPs [256,257]. The net effect depends on which of these factors is dominant.

Continuously recirculating water could also increase release of metals, increase deposition corrosion of anodes by constantly recirculating water through copper pipe, and result in greater accumulation of sediments and $H_2$ gas. One study showed that recirculating systems with copper piping had 3–13 times more aluminum and copper, 4–6 times more hydrogen in effluent water, and 9% more aluminum anode weight loss, compared with standard (non-recirculating) systems [222]. Recirculating systems can also accumulate 3–20 times more sediments [222] arising from corrosion of metallic pipe material and the anode rods [157,232–234,258]. These sediments, which also collect at the bottom of hot water tanks, may serve as an important growth niche within warm regions of hot water tanks where influent cold water depresses temperatures, and there are also relatively low levels of disinfectant and high levels of nutrients for *Legionella*, heterotrophs, and host organisms [17,259].

### 4.3. Pipe Aging

New plastic and copper pipes behave differently than older pipes. Specifically, corrosion and release of metals is strongly influenced by pipe age, with corrosion rates and metal release tending to decrease as thicker and more passivating pipe scales form. Aging can dramatically reduce levels of metal leaching from copper and other pipes [157,260,261]. The rate of aging, and whether it decreases release of pipe constituents at all, is highly affected by water chemistry and water use patterns [157]. Likewise, leaching of organics from plastic pipe may attenuate 50% to >99% after aging for a period of a few weeks with hot water exposure [51,170], but in other cases has been sustained for months [262] or even over a year [263]. Pipe aging is an important factor to consider when comparing PEX to copper's capacity for *Legionella* growth. One study showed that the *Legionella* numbers in bulk water of both PEX and copper pipes in a simulated warm water system were the same after two years [140]. We speculate that one possible cause for this convergence is that, as plastic pipes age, organic carbon migration to water decreases, whereas levels of antimicrobial copper released from copper pipe also tends to decrease. Hence, in some situations, it is expected that in very old copper and plastic pipe systems there would be little difference between these pipe materials.

### 4.4. Possible Mixed Material Interactions

Building plumbing is typically comprised of multiple pipe materials, either in the original design or after partial retrofits or renovations. It is anticipated that there are sometimes synergistic and other times antagonistic interactions between pipe materials that would influence growth of OPs. Copper deposition accelerating the evolution of $H_2$ from aluminum, zinc, magnesium and iron corrosion, as discussed in Section 3.2.3, is an important exemplar. Copper is also known to catalyze degradation of plastic pipes [264–268], and the presence of copper pipe upstream of plastic pipe might enhance organic carbon release [268], surface roughness for biofilm growth [264], and perhaps even disinfectant consumption due to copper in the scale. Iron pipes upstream of copper may produce mixed Fe-Cu oxides, which can be extraordinary catalysts for free chlorine decay [269]. Similarly, copper released upstream of iron pipes could increase iron release [270]. Any galvanic coupling between two metals in plumbing materials (copper/brass-lead [271,272], copper/iron [270,273,274] iron/zinc [275,276], copper/aluminum [277,278], copper/zinc [271,279], copper/magnesium [280]) has the potential to enhance corrosion and cause changes to water quality parameters relevant to corrosion and OP growth [235,281], dissolved oxygen (DO) [273], metal concentrations [271,272], and disinfectant residual concentration. These reactions also create microenvironments of very high or very low pH [235,238]. Given that in the 2017 American Housing Survey 10% of households that reported any home improvement projects also reported adding or replacing an interior water pipe [282], understanding the effects of mixing pipe materials during renovation appears to be a valuable research area as antiquated premise plumbing is increasingly replaced.

## 5. Mediating Role of Microbiome and other Microbiological Considerations

### 5.1. The Role of Pipe Material in Shaping the Premise Plumbing Microbiome and Resident Amoeba Host Organisms

Interactions between OPs and the microbial communities surrounding them are key to OP proliferation and are likely influenced by pipe materials. OPs can be parasitic to free-living amoebae that first prey upon them in drinking water biofilms, before they reproduce inside and eventually kill the host organism [24]. In fact, there is some doubt that *Legionella* actually reproduces significantly in drinking water outside of an amoeba host [283]. Amoebae can also protect OPs from disinfectants and provide access to nutrients. For example, *Legionella* exclusively use amino acids, which are abundant in amoeba vacuoles, as a carbon source [210–214,284,285]. Thus, although poorly studied, any factor altering growth of key host amoebae (including *Acanthamoeba*, *Vermamoeba*, and *Naegleria*) is expected to indirectly affect growth of OPs, including *L. pneumophila*, *P. aeruginosa*, and NTM [122,210–214,225,257,286,287]. In one experiment, copper coupons were found to host more *Acanthamoeba polyphaga* than PVC coupons [288], possibly because copper hosts less diverse eukaryotic communities [64,289] and limits competition for *A. polyphaga*. As a result, *L. pneumophila* grew and shed to the bulk waters in higher numbers on these copper coupons than on PVC coupons if co-inoculated with *A. polyphaga* [122].

Interbacterial interactions may also influence the growth of OPs. Broadly speaking, OPs benefit from the biofilm community through access to nutrients and protection from disinfectants [24,35,36,290]. Some studies have identified correlations between specific taxa and OPs in premise plumbing [291], cooling towers [292] and drinking water distribution systems [293]. However, the significance of these correlations to premise plumbing material selection is not well understood, as most studies examining differences in bacterial communities focus on very broad measures of community structure [48,59,64,216,289,294–296]. Certain waterborne bacteria are known to produce toxins that inhibit *L. pneumophila* growth [216,297] or exude other compounds that have secondary bacteriostatic effects on *Legionella* [298]. Intra-bacterial inhibition also may be mediated through amoebae by reducing host uptake [299,300] or killing the host population [134,301,302]. More research is needed to elucidate how the broad ecological differences resulting from pipe material influence these interactions. Integration of metagenomic or meta-transcriptomic analyses targeting the production of bacteriocins or other toxins with known effects on OPs could elucidate the ecological effects of taxonomic shifts resulting from pipe material. Interrupting OP-amoeba endosymbiosis through the enrichment of preferential non-OP amoeba prey [299,300] has been suggested as a probiotic means of controlling OPs [303], and pipe material could be explored as a means of enrichment of these taxa.

### 5.2. Variation in Copper Tolerance Among Species and Strains

Strain-to-strain differences in intrinsic tolerance of copper, acclimation to copper concentrations with time through induction of the appropriate genes, or acquisition of copper resistance via mutation or horizontal gene transfer in premise plumbing might explain some of the discrepancies in variable outcomes of copper on OPs (Table 2). *Legionella* [155] and other OPs [58] may acclimate to high copper levels through the expression of copper detoxification or efflux systems. Bedard et al. [155] reported four-fold differences in the copper tolerance of environmentally-isolated *L. pneumophila* strains, noting that more resistant strains showed increased copper ATPase *copA* expression, speculating that their increased tolerance may also be a result of higher biofilm production. Strikingly, Williams et al. [58] showed that, during exposure to 95 mg/L of copper over 6 h in liquid culture, culturable *A. baumannii* levels (CFU/mL) could increase by 2-logs or decrease by 2-logs, depending on the strain. The authors identified putative copper detoxification and efflux systems within the genome of the most resistant isolate and identified specific genes that were upregulated in response to copper exposure. However, a majority of the less tolerant strains tested also possessed these genes, leading the authors to suggest that further definition of the proteins involved in copper resistance is required. One recent study showed two environmentally-isolated *Legionella* strains reduced by less than one log in culturability, even after two weeks of exposure to 5 mg/L copper, which the authors attributed to adaptation to the high levels of copper (average 0.48 mg/L ) in the hot water system from which these isolates were collected [154]. A profile of *Fusarium* isolates revealed that tap water isolates were more copper-tolerant than soil isolates [303]. *P. aeruginosa* isolates isolated from a hospital with copper plumbing exhibited only slightly limited growth in the presence of 0.15 mg/L copper [129]. All of these strains were found to harbor GI-7, a mobile genetic element that confers copper resistance and that has also been identified in a *P. aeruginosa* strain associated with hospital outbreaks [304]. Limited data suggest that *A. baumannii* and mycobacteria are more difficult to inactivate with copper than other OPs, while *P. aeruginosa* is more readily inactivated [91,92,98,108,109]. *L. pneumophila* has been found both at the more resistant [98] and less resistant [91,108,109] ends of this spectrum. The wide variability among OPs and even strains of OPs in their intrinsic tolerance of copper, ability to acquire genetic resistance, and ability to acclimate to elevated levels of copper makes it difficult to precisely predict the efficacy of copper and other antimicrobials for OP control.

### 5.3. Confounding Effects of VBNC Bacteria

The discovery of VBNC bacteria has complicated prior understanding for all OP control strategies, including copper. Virtually all prior work relied on culture methods to determine copper's efficacy for killing OPs [62,63,83,91,92,98,108,109,120,137,153], but some microbes rendered not culturable might remain viable and still infect host amoebae or humans [74,76,305–307]. The existence of VBNC pathogens in premise plumbing has been demonstrated by comparing culture-based numbers with those enumerated via fluorescence (e.g., live/dead) and molecular-based (e.g., quantitative polymerase chain reaction) monitoring methods [308].

Bench-scale studies examining copper's antimicrobial efficacy have found discrepancies between culture-based and molecular-based numbers of *L. pneumophila* [121,122] that are also suggestive of a copper-induced VBNC state. Similar discrepancies have been noted for *P. aeruginosa, Stenotrophomonas maltophilia,* and *M. avium* [104,109,127,132,133]. Evidence of copper-induced VBNC activity is particularly strong in the case of *P. aeruginosa,* where one study applied multiple non-culture-based measures of viability [127,132]. Furthermore, VBNC *P. aeruginosa* have been shown to partially recover infectivity after removal of copper from solution [132,133]. To understand how VBNC bacteria contribute to OP infections, additional studies are needed to delineate the premise plumbing conditions more precisely that induce VBNC status and to confirm the range of functionality maintained in this state. A primary challenge in achieving this is that there are currently no reliable methods for confidently enumerating VBNC bacteria.

### 5.4. Virulence

The premise plumbing environment exhibits several features that could possibly contribute to the virulence of resident OPs. Wargo [38] describes features of drinking water plumbing that could prime OPs to infect cystic fibrosis patients, although the interactions described in this review could also pose risk to otherwise immunocompromised individuals. Such features that are relevant to pipe material include [38]:

- Elevated copper levels, selecting for resistance to copper overload within macrophage phagosomes, a component of the innate immune response [309].
- Elevated iron levels, influencing interactions between iron homeostasis and virulence.
- Exposure to lipids, which are generally not well removed by drinking water treatment, priming OPs for lipid-rich environments within hosts. Accumulation of phospholipid fatty acids has been shown to be greater in the biofilms of polyethylene pipes than copper pipes, though these lipids were putatively associated with bacteria [310].
- Low DO levels, selecting for OPs capable of survival in low DO regions of the biofilm in infected host tissue.
- Exposure to eukaryotic predation, selecting for resistance to the host's immune response (e.g., lung macrophages) or enhanced virulence.

Some studies suggest that the above types of interaction may increase the pathogenic potential of premise plumbing-associated OPs specifically. Copper resistance is important to mammalian host infection for *P. aeruginosa* [311] and *A. baumannii* [312,313], and other evidence suggests that exposure to copper in aquatic environments selects for greater copper resistance among certain OPs [129,303,304]. Copper and other divalent metals may also play a role in nutrient acquisition and pathogenesis even after infecting hosts [314].

The effects of iron exposure on OPs are not as apparent. *L. pneumophila* serogroup 1 grown in medium that was iron limited (0.017–0.056 mg/L) has been shown to lose its virulence [315], indicating that limiting adequate concentrations of iron could not only decrease the presence of

*Legionella* but also the likelihood of human infection. Iron also plays a role in modulating various behaviors, including modulating virulence factor production in *P. aeruginosa* and *A. baumannii* [316–321], but it is unclear what effects exposure to iron have on virulence in the premise plumbing environment. This subject is largely unexplored and more research is needed to determine the overall effects of the premise plumbing environment on OP virulence.

### 5.5. Antibiotic Resistance and Tolerance

Copper, among other heavy metals has been shown to exert selection pressure, leading to enhanced survival of antibiotic resistant bacteria. In fact, heavy-metal-associated co-selection and cross-selection has been proposed to be as much of a concern for environmental propagation of antibiotic resistance as antibiotics themselves [322]. Increases in antibiotic resistance genes at the community scale have been identified after long-term copper exposure in soil [323–326], sediment [327], and drinking water [327]. Bench-scale tests using bacterial isolates from biofilters [328] and wastewater [329] inoculated into growth media have shown that a selective or inductive effect of copper can take places within hours. However, these studies were performed with copper concentrations 5–77 times greater than the 1.3 mg/L US EPA copper action level and similarly in exceedance of the Chinese Standard for Drinking Water Quality of 1 mg/L [209] and WHO Guideline for Drinking-Water Quality of 2 mg/L [82]. Thus, these concentrations may not be representative of potable water systems. One study examining antibiotic resistant and sensitive strains of *Staphylococcus aureus* showed that the more antibiotic resistant strain survived longer in a copper container [90]. As discussed above, copper may also better support *Acanthamoeba* than other materials, while in one study *L. pneumophila* grown within *A. polyphaga* demonstrated increased tolerance to all antibiotics tested (rifampin, ciprofloxacin, and erythromycin) compared to those grown in culture media [330]. The role of copper plumbing and other pipe materials in these emerging areas of research is worthy of further investigation.

There is more limited evidence that the presence of iron may also induce or select for antibiotic resistance, as observed for *P. aeruginosa* using iron-amended growth media [330] and the gut microbiomes of mice supplied with iron-amended water [331]. The latter case, while using an iron concentration more than 25 times the EU drinking water standard of 0.2 mg/L [332] and 16 times both the US EPA National Secondary Drinking Water Standard and Chinese Standard for Drinking Water Quality of 0.3 mg/L, may be of particular concern, as it suggests that pipe corrosion products have the potential to select for antibiotic resistance inside the infected host organism.

## 6. Conclusions

Premise plumbing is a complex, temporally dynamic, and spatially diverse environment that is strongly influenced by pipe materials. Virtually all pipe materials have known benefits and/or detriments for OP growth. Plumbing materials are an important driver of the chemical and biological water quality parameters that influence the control of OPs and there are no silver (copper or plastic) bullets that will uniformly inhibit the growth of *Legionella* and other OPs under all circumstances.

Synthetic plastic pipe materials vary between type and manufacturer. They can act as a supply of organic carbon for the growth of microorganisms, but exert a lower chlorine demand and tend to form fewer scales that could provide more surface area for biofilm growth. Iron pipes supply nutrients for growth, exhibit a high disinfectant demand, produce hydrogen and other nutrients through corrosion, and tend to form thick scales with extremely high surface areas. While they may no longer be used in new construction, even short sections of pipe can affect an entire downstream premise plumbing distribution system. Stainless steel has few known effects on water quality, and correspondingly, OP control, perhaps because it is the least studied and is less commonly used as a result of its high cost. Copper pipes are known for their antimicrobial ability, but this is inconsistently realized in practice, and in some cases they seem to encourage OP growth relative to other pipes. Premise plumbing materials have a role to play in preventing OP infections and, at a minimum, should be examined more closely for their propensity to inhibit or stimulate OP proliferation during outbreak investigations. Research is needed to better define:

- Both the intra-species and inter-species variation of copper resistance amongst OPs, as well as environmental drivers of this variation.
- Effects of copper pipes on OPs in a more holistic sense, with identification of real-world conditions that are drivers for discrepancies in copper's antimicrobial capacity.
- Copper's possible micronutrient activity in OPs within premise plumbing contexts, including threshold concentrations required for various physiological functions, as well as physicochemical and ecological factors that influence those thresholds.
- The disease risk that VBNC OPs pose and conditions under which copper and other antimicrobials induce VBNC status in premise plumbing OPs
- The inhibitory action of trace metals on OP growth in premise plumbing, as well as growth requirements for other trace elements exhibited by OPs in premise plumbing.
- Potential mediating effects of the wider microbial community composition resulting from pipe material on OPs.
- Effect of mixed pipe materials on physicochemical parameters of bulk water and OP growth.
- The effects of plumbing materials on OP antibiotic resistance and virulence.
- The impact of stagnation, velocity, sediments, corrosion control, and consumer water use patterns on all of the above.

An improved understanding will provide actionable advice for multiple stakeholders. In addition to the obvious direct use of the results in the construction industry and by building water quality managers, water utilities can benefit from improved understanding of how the interplay of premise plumbing pipe materials with disinfectants, nutrients and corrosion control can be harnessed to reduce disease incidence.

**Author Contributions:** A.C.C., R.L.M., and Y.S. conceived the outline of review, performed literature review, and prepared the manuscript with equal contribution. K.W. and A.W. provided initial ideas on the key sections and feedbacks on review paper. M.A.E. and A.P. supervised the project and assisted in the manuscript preparation. All authors have read and agreed to the published version of the manuscript.

**Funding:** This work was supported by NSF RAPID Grant No. 1556258, NSF SusChEM GOALI No. 170673, NSF CBET 1336650, the Alfred P. Sloan Foundation Microbiology of the Built Environment program, an NSF Graduate Research Fellowship to Abraham Cullom (Grant No. 1840995), and funding from the Copper Development Association Inc. (CDA), McLean, Virginia, USA. Part of the article processing charge was subsidized by Virginia Tech Libraries.

**Conflicts of Interest:** The authors of this research have research support from the Copper Development Association Inc. There is no other conflict of interest to disclose.

## References

1. National Academies of Sciences, Engineering, and Medicine. *Management of Legionella in Water Systems*; The National Academies Press: Washington, DC, USA, 2019; ISBN 978-0-309-49947-7.
2. Ashbolt, N.J. Microbial Contamination of Drinking Water and Human Health from Community Water Systems. *Curr. Environ. Health Rep.* **2015**, *2*, 95. [CrossRef] [PubMed]
3. United States Occupational Safety & Health Administration Legionnaires' Disease: Facts and Frequently Asked Questions. Available online: https://www.osha.gov/dts/osta/otm/legionnaires/faq.html (accessed on 1 July 2020).
4. United States Occupational Safety & Health Administration Legionnaires' Disease. Available online: https://www.osha.gov/dts/osta/otm/otm_iii/otm_iii_7.html (accessed on 1 July 2020).
5. CDC Legionella (Legionnaires' Disease and Pontiac Fever) Surveillance and Reporting. Available online: http://www.cdc.gov/legionella/surv-reporting.html (accessed on 1 July 2020).

6. Singh, R.; Hamilton, K.A.; Rasheduzzaman, M.; Yang, Z.; Kar, S.; Fasnacht, A.; Masters, S.V.; Gurian, P.L. Managing Water Quality in Premise Plumbing: Subject Matter Experts' Perspectives and a Systematic Review of Guidance Documents. *Water* **2020**, *12*, 347. [CrossRef]
7. McCoy, W.F. *Preventing Legionellosis*; IWA Publishing: London, UK, 2005; ISBN 1843390949.
8. Falkinham, J.; Pruden, A.; Edwards, M. Opportunistic Premise Plumbing Pathogens: Increasingly Important Pathogens in Drinking Water. *Pathogens* **2015**, *4*, 373. [CrossRef]
9. Centers for Disease Control Legionella (Legionnaires' Disease and Pontiac Fever). Available online: https://www.cdc.gov/legionella/index.html (accessed on 1 July 2020).
10. Gomez-Valero, L.; Rusniok, C.; Rolando, M.; Neou, M.; Dervins-Ravault, D.; Demirtas, J.; Rouy, Z.; Moore, R.J.; Chen, H.; Petty, N.K.; et al. Comparative analyses of Legionella species identifies genetic features of strains causing Legionnaires' disease. *Genome Biol.* **2014**, *15*, 505. [CrossRef]
11. Association of Water Technologies Legionella 2019: A Position Statement and Guidance Document. Available online: https://www.awt.org/pub/035C2942-03BE-3BFF-08C3-4C686FB7395C (accessed on 15 November 2020).
12. Allen, M.J.; Edberg, S.C.; Clancy, J.L.; Hrudey, S.E. Drinking water microbial myths. *Crit. Rev. Microbiol.* **2015**, *41*, 366–373. [CrossRef]
13. Barrabeig, I.; Rovira, A.; Garcia, M.; Oliva, J.M.; Vilamala, A.; Ferrer, M.D.; SabriÀ, M.; DomÍNguez, A. Outbreak of Legionnaires' disease associated with a supermarket mist machine. *Epidemiol. Infect.* **2010**, *138*, 1823–1828. [CrossRef] [PubMed]
14. Mahoney, F.J.; Hoge, C.W.; Farley, T.A.; Barbaree, J.M.; Breiman, R.F.; Benson, R.F.; McFarland, L.M. Communitywide Outbreak of Legionnaires' Disease Associated with a Grocery Store Mist Machine. *J. Infect. Dis.* **1992**, *165*, 736–739. [CrossRef] [PubMed]
15. Stout, J.E.; Yu, V.L.; Yee, Y.C.; Vaccarello, S.; Diven, W.; Lee, T.C. Legionella pneumophila in residential water supplies: Environmental surveillance with clinical assessment for Legionnaires' disease. *Epidemiol. Infect.* **1992**, *109*, 49–57.
16. Stout, J.E.; Victor, L.Y.; Muraca, P. Isolation of Legionella pneumophila from the cold water of hospital ice machines: Implications for origin and transmission of the organism. *Infect. Control Hosp. Epidemiol.* **1985**, *6*, 141–146. [CrossRef]
17. Rhoads, W.J.; Bradley, T.N.; Mantha, A.; Buttling, L.; Keane, T.; Pruden, A.; Edwards, M.A. Residential water heater cleaning and occurrence of Legionella in Flint, MI. *Water Res.* **2020**, 115439. [CrossRef]
18. Gobin, I.; Newton, P.R.; Hartland, E.L.; Newton, H.J. Infections caused by nonpneumophila species of Legionella. *Rev. Med. Microbiol.* **2009**, *20*, 1–11. [CrossRef]
19. Joseph, C.A. Legionnaires' disease in Europe 2000–2002. *Epidemiol. Infect.* **2004**, *132*, 417–424. [CrossRef] [PubMed]
20. Stout, J.E.; Yu, V.L.; Muraca, P.; Joly, J.; Troup, N.; Tompkins, L.S. Potable water as a cause of sporadic cases of community-acquired legionnaires' disease. *N. Engl. J. Med.* **1992**, *326*, 151–155. [CrossRef]
21. Fliermans, C.B.; Cherry, W.B.; Orrison, L.H.; Smith, S.J.; Tison, D.L.; Pope, D.H. Ecological distribution of Legionella pneumophila. *Appl. Environ. Microbiol.* **1981**, *41*, 9–16. [CrossRef] [PubMed]
22. Lee, G.F.; Jones-Lee, A. Public Health Significance of Waterborne Pathogens in Domestic Water Supplies and Reclaimed Water. In *Report to State of California Environmental Protection Agency Comparative Risk Project*; California Environmental Protection Agency: Berkeley, CA, USA, 1993.
23. Borella, P.; Montagna, M.T.; Romano-Spica, V.; Stampi, S.; Stancanelli, G.; Triassi, M.; Neglia, R.; Marchesi, I.; Fantuzzi, G.; Tatò, D.; et al. Legionella Infection Risk from Domestic Hot Water. *Emerg. Infect. Dis.* **2004**, *10*, 457–464. [CrossRef] [PubMed]
24. Falkinham III, J.O. Common Features of Opportunistic Premise Plumbing Pathogens. *Int. J. Environ. Res. Public Health* **2015**, *12*, 4533–4545. [CrossRef] [PubMed]
25. Parr, A.; Whitney, E.A.; Berkelman, R.L. Legionellosis on the Rise: A Review of Guidelines for Prevention in the United States. *J. public Health Manag. Pract.* **2015**, *21*, E17–E26. [CrossRef]
26. Rhoads, W.J.; Garner, E.; Ji, P.; Zhu, N.; Parks, J.; Schwake, D.O.; Pruden, A.; Edwards, M.A. Distribution System Operational Deficiencies Coincide with Reported Legionnaires' Disease Clusters in Flint, Michigan. *Environ. Sci. Technol.* **2017**, *51*, 11986–11995. [CrossRef]
27. Mercante, J.W.; Winchell, J.M. Current and Emerging Legionella Diagnostics for Laboratory and Outbreak Investigations. *Clin. Microbiol. Rev.* **2015**, *28*, 95–133. [CrossRef]

28. Garrison, L.E. Vital Signs: Deficiencies in Environmental Control Identified in Outbreaks of Legionnaires' Disease—North America, 2000–2014. *MMWR. Morb. Mortal. Wkly. Rep.* **2016**, *65*, 576–584. [CrossRef]
29. Pierre, D.; Baron, J.L.; Ma, X.; Sidari, F.P.; Wagener, M.M.; Stout, J.E. Water Quality as a Predictor of Legionella Positivity of Building Water Systems. *Pathogens* **2019**, *8*, 295. [CrossRef]
30. Perrin, Y.; Bouchon, D.; Delafont, V.; Moulin, L.; Héchard, Y. Microbiome of drinking water: A full-scale spatio-temporal study to monitor water quality in the Paris distribution system. *Water Res.* **2019**, *149*, 375–385. [CrossRef] [PubMed]
31. Liu, G.; Zhang, Y.; Knibbe, W.-J.; Feng, C.; Liu, W.; Medema, G.; van der Meer, W. Potential impacts of changing supply-water quality on drinking water distribution: A review. *Water Res.* **2017**, *116*, 135–148. [CrossRef]
32. Pinto, A.J.; Xi, C.; Raskin, L. Bacterial community structure in the drinking water microbiome is governed by filtration processes. *Environ. Sci. Technol.* **2012**, *46*, 8851–8859. [CrossRef] [PubMed]
33. Wang, H.; Masters, S.; Falkinham, J.O.; Edwards, M.A.; Pruden, A. Distribution System Water Quality Affects Responses of Opportunistic Pathogen Gene Markers in Household Water Heaters. *Environ. Sci. Technol.* **2015**, *49*, 8416–8424. [CrossRef] [PubMed]
34. Baron, J.L.; Vikram, A.; Duda, S.; Stout, J.E.; Bibby, K. Shift in the Microbial Ecology of a Hospital Hot Water System following the Introduction of an On-Site Monochloramine Disinfection System. *PLoS ONE* **2014**, *9*. [CrossRef] [PubMed]
35. Schwering, M.; Song, J.; Louie, M.; Turner, R.J.; Ceri, H. Multi-species biofilms defined from drinking water microorganisms provide increased protection against chlorine disinfection. *Biofouling* **2013**, *29*, 917–928. [CrossRef] [PubMed]
36. Greub, G.; Raoult, D. Microorganisms Resistant to Free-Living Amoebae. *Clin. Microbiol. Rev.* **2004**, *17*, 413. [CrossRef]
37. Wang, H.; Edwards, M.A.; Falkinham, J.O.; Pruden, A. Probiotic Approach to Pathogen Control in Premise Plumbing Systems? A Review. *Environ. Sci. Technol.* **2013**, *47*, 10117–10128. [CrossRef]
38. Wargo, M.J. Is the Potable Water System an Advantageous Preinfection Niche for Bacteria Colonizing the Cystic Fibrosis Lung? *MBio* **2019**, *10*, e00883-19. [CrossRef]
39. Rhoads, J.W.; Pearce, A.; Pruden, A.; Edwards, M.A. Anticipating the Effects of Green Buildings on Water Quality and Infrastructure. *J. Am. Water Works Assoc.* **2015**, *107*, 50. [CrossRef]
40. Liu, G.; Tao, Y.; Zhang, Y.; Lut, M.; Knibbe, W.-J.; van der Wielen, P.; Liu, W.; Medema, G.; van der Meer, W. Hotspots for selected metal elements and microbes accumulation and the corresponding water quality deterioration potential in an unchlorinated drinking water distribution system. *Water Res.* **2017**, *124*, 435–445. [CrossRef] [PubMed]
41. Proctor, C.R.; Hammes, F. Drinking water microbiology - from measurement to management. *Curr. Opin. Biotechnol.* **2015**, *33*, 87–94. [CrossRef] [PubMed]
42. Neu, L.; Hammes, F. Feeding the Building Plumbing Microbiome: The Importance of Synthetic Polymeric Materials for Biofilm Formation and Management. *Water* **2020**, *12*, 1774. [CrossRef]
43. Zhang, Y.; Edwards, M. Accelerated chloramine decay and microbial growth by nitrification in premise plumbing. *J. Am. Water Works Assoc.* **2009**, *101*, 51–62. [CrossRef]
44. Al-Jasser, A.O. Chlorine decay in drinking-water transmission and distribution systems: Pipe service age effect. *Water Res.* **2007**, *41*, 387–396. [CrossRef]
45. Durand, M.L. *Disinfectants and Plumbing Materials: Effects on the Sensory and Chemical Characteristics of Drinking Water*; Virginia Tech: Blacksburg, VA, USA, 2005.
46. Haas, C.N.; Gupta, M.; Chitluru, R.; Burlingame, G. Chlorine Demand in disinfecting Water Mains. *J. Am. Water Works Assoc.* **2002**, *94*, 97–102. [CrossRef]
47. Clark, R.; Yang, Y.J.; Impellitteri, C.; Haught, R.; Schupp, D.; Panguluri, S.; Krishnan, R. Chlorine Fate and Transport in Distribution Systems: Experimental and modeling studies. *J. Am. Water Works Assoc.* **2010**, *102*, 144–156. [CrossRef]
48. Zhang, C.; Li, C.; Zheng, X.; Zhao, J.; He, G.; Zhang, T. Effect of pipe materials on chlorine decay, trihalomethanes formation, and bacterial communities in pilot-scale water distribution systems. *Int. J. Environ. Sci. Technol.* **2017**, *14*, 85–94. [CrossRef]

49. Lehtola, M.J.; Miettinen, I.T.; Lampola, T.; Hirvonen, A.; Vartiainen, T.; Martikainen, P.J. Pipeline materials modify the effectiveness of disinfectants in drinking water distribution systems. *Water Res.* **2005**, *39*, 1962–1971. [CrossRef]
50. Heim, T.H.; Dietrich, A.M. Sensory aspects and water quality impacts of chlorinated and chloraminated drinking water in contact with HDPE and cPVC pipe. *Water Res.* **2007**, *41*, 757–764. [CrossRef] [PubMed]
51. Mao, G.; Wang, Y.; Hammes, F. Short-term organic carbon migration from polymeric materials in contact with chlorinated drinking water. *Sci. Total Environ.* **2018**, *613*, 1220–1227. [CrossRef] [PubMed]
52. Durand, M.L.; Dietrich, A.M. Contributions of silane cross-linked PEX pipe to chemical/solvent odours in drinking water. *Water Sci. Technol.* **2007**, *55*, 153–160. [CrossRef] [PubMed]
53. Kim, H.; Koo, J.; Kim, S. A general framework of chlorine decay modeling at a pilot-scale water distribution system. *J. Water Supply Res. Technol. AQUA* **2015**, *64*, 543–557. [CrossRef]
54. Nguyen, C.; Elfland, C.; Edwards, M. Impact of advanced water conservation features and new copper pipe on rapid chloramine decay and microbial regrowth. *Water Res.* **2012**, *46*, 611–621. [CrossRef]
55. Westbrook, A.; Digiano, F.A. Rate of chloramine decay at pipe surfaces. *J. Am. Water Works Assoc.* **2009**, *101*, 59–70. [CrossRef]
56. Bucheli-Witschel, M.; Kötzsch, S.; Darr, S.; Widler, R.; Egli, T. A new method to assess the influence of migration from polymeric materials on the biostability of drinking water. *Water Res.* **2012**, *46*, 4246–4260. [CrossRef]
57. Skjevrak, I.; Due, A.; Gjerstad, K.O.; Herikstad, H. Volatile organic components migrating from plastic pipes (HDPE, PEX and PVC) into drinking water. *Water Res.* **2003**, *37*, 1912–1920. [CrossRef]
58. Williams, C.L.; Neu, H.M.; Gilbreath, J.J.; Michel, S.L.J.; Zurawski, D.V.; Merrell, D.S. Copper Resistance of the Emerging Pathogen Acinetobacter baumannii. *Appl. Environ. Microbiol.* **2016**, *82*, 6174–6188. [CrossRef]
59. Proctor, C.R.; Dai, D.; Edwards, M.A.; Pruden, A. Interactive effects of temperature, organic carbon, and pipe material on microbiota composition and Legionella pneumophila in hot water plumbing systems. *Microbiome* **2017**, *5*, 130. [CrossRef]
60. Rushing, J.C.; McNeill, L.S.; Edwards, M. Some effects of aqueous silica on the corrosion of iron. *Water Res.* **2003**, *37*, 1080–1090. [CrossRef]
61. Niu, L.; Cheng, Y.F. Corrosion behavior of X-70 pipe steel in near-neutral pH solution. *Appl. Surf. Sci.* **2007**, *253*, 8626–8631. [CrossRef]
62. Lin, Y.-S.E.; Vidic, R.D.; Stout, J.E.; Yu, V.L. Individual and combined effects of copper and silver ions on inactivation of Legionella pneumophila. *Water Res.* **1996**, *30*, 1905–1913. [CrossRef]
63. Mathys, W.; Stanke, J.; Harmuth, M.; Junge-Mathys, E. Occurrence of Legionella in hot water systems of single-family residences in suburbs of two German cities with special reference to solar and district heating. *Int. J. Hyg. Environ. Health* **2008**, *211*, 179–185. [CrossRef] [PubMed]
64. Lu, J.; Buse, H.Y.; Gomez-Alvarez, V.; Struewing, I.; Santo Domingo, J.; Ashbolt, N.J. Impact of drinking water conditions and copper materials on downstream biofilm microbial communities and Legionella pneumophila colonization. *J. Appl. Microbiol.* **2014**, *117*, 905–918. [CrossRef] [PubMed]
65. Merritt, K.; Brown, S.A. Release of hexavalent chromium from corrosion of stainless steel and cobalt—Chromium alloys. *J. Biomed. Mater. Res.* **1995**, *29*, 627–633. [CrossRef] [PubMed]
66. Morton, S.C.; Zhang, Y.; Edwards, M.A. Implications of nutrient release from iron metal for microbial regrowth in water distribution systems. *Water Res.* **2005**, *39*, 2883–2892. [CrossRef]
67. Garner, E.; Zhu, N.; Strom, L.; Edwards, M.; Pruden, A. A human exposome framework for guiding risk management and holistic assessment of recycled water quality. *Environ. Sci. Water Res. Technol.* **2016**, *2*, 580–598. [CrossRef]
68. Anderson, L.E.; Krkošek, W.H.; Stoddart, A.K.; Trueman, B.F.; Gagnon, G.A. Lake Recovery Through Reduced Sulfate Deposition: A New Paradigm for Drinking Water Treatment. *Environ. Sci. Technol.* **2017**, *51*, 1414–1422. [CrossRef]
69. Monteith, D.T.; Stoddard, J.L.; Evans, C.D.; De Wit, H.A.; Forsius, M.; Høgåsen, T.; Wilander, A.; Skjelkvåle, B.L.; Jeffries, D.S.; Vuorenmaa, J.; et al. Dissolved organic carbon trends resulting from changes in atmospheric deposition chemistry. *Nature* **2007**, *450*, 537–540. [CrossRef]
70. European Environment Agency. Air Quality Standards. Available online: https://www.eea.europa.eu/themes/air/air-quality-concentrations/air-quality-standards (accessed on 28 October 2020).

71. Rhoads, W.J.; Pruden, A.; Edwards, M.A. Survey of green building water systems reveals elevated water age and water quality concerns. *Environ. Sci. Water Res. Technol.* **2015**, *2*, 164–173. [CrossRef]
72. United Nations. Transforming Our World: The 2030 Agenda for Sustainable Development. Available online: https://sustainabledevelopment.un.org/post2015/transformingourworld/publication (accessed on 15 November 2020).
73. Lee, J.; Kleczyk, E.; Bosch, D.J.; Dietrich, A.M.; Lohani, V.K.; Loganathan, G.V. Homeowners' decision-making in a premise plumbing failure-prone area. *J. Am. Water Works Assoc.* **2013**, *105*, E236–E241. [CrossRef]
74. Ramamurthy, T.; Ghosh, A.; Pazhani, G.P.; Shinoda, S. Current Perspectives on Viable but Non-Culturable (VBNC) Pathogenic Bacteria. *Front. Public Health* **2014**, *2*, 103. [CrossRef] [PubMed]
75. Alleron, L.; Khemiri, A.; Koubar, M.; Lacombe, C.; Coquet, L.; Cosette, P.; Jouenne, T.; Frere, J. VBNC Legionella pneumophila cells are still able to produce virulence proteins. *Water Res.* **2013**, *47*, 6606–6617. [CrossRef] [PubMed]
76. Dietersdorfer, E.; Kirschner, A.; Schrammel, B.; Ohradanova-Repic, A.; Stockinger, H.; Sommer, R.; Walochnik, J.; Cervero-Aragó, S. Starved viable but non-culturable (VBNC) Legionella strains can infect and replicate in amoebae and human macrophages. *Water Res.* **2018**, *141*, 428–438. [CrossRef] [PubMed]
77. Copper Development Association Inc. Copper Facts. Available online: https://www.copper.org/education/c-facts/facts-print.html (accessed on 18 August 2020).
78. Copper Development Association Inc. Antimicrobial Copper Surfaces. Available online: https://www.copper.org/applications/antimicrobial/ (accessed on 18 August 2020).
79. European Chemicals Agency. Information on Biocides. Available online: https://www.echa.europa.eu/web/guest/information-on-chemicals/biocidal-active-substances?p_p_id=dissactivesubstances_WAR_dissactivesubstancesportlet&p_p_lifecycle=0&p_p_state=normal&p_p_mode=view&p_p_col_id=column-1&p_p_col_pos=2&p_p_col_count=3&_dissacti (accessed on 10 November 2020).
80. Samanovic, M.I.; Ding, C.; Thiele, D.J.; Darwin, K.H. Copper in microbial pathogenesis: Meddling with the metal. *Cell Host Microbe* **2012**, *11*, 106–115. [CrossRef] [PubMed]
81. Boulay, N.; Edwards, M. Copper in the Urban Water Cycle. *Crit. Rev. Environ. Sci. Technol.* **2000**, *30*, 297–326. [CrossRef]
82. World Health Organization. *Copper in Drinking-Water*; World Health Organization: Geneva, Switzerland, 2016.
83. Lin, Y.E.; Vidic, R.D.; Stout, J.E.; Victor, L.Y. Negative effect of high pH on biocidal efficacy of copper and silver ions in controlling Legionella pneumophila. *Appl. Environ. Microbiol.* **2002**, *68*, 2711–2715. [CrossRef]
84. Lin, Y.S.; Stout, J.E.; Yu, V.L.; Vidic, R.D. Disinfection of water distribution systems for Legionella. *Semin. Respir. Infect.* **1998**, *13*, 147–159.
85. Hajialilo, E.; Niyyati, M.; Solaymani, M.; Rezaeian, M. Pathogenic Free-Living Amoebae Isolated From Contact Lenses of Keratitis Patients. *Iran. J. Parasitol.* **2015**, *10*, 541–546.
86. Centers for Disease Control and Prevention, National Center for Emerging and Zoonotic Infectious Diseases (NCEZID), Division of Foodborne, Waterborne, and E.D. (DFWED) Parasites—Naegleria fowleri—Primary Amebic Meningoencephalitis (PAM)—Amebic Encephalitis. Available online: https://www.cdc.gov/parasites/naegleria/general.html (accessed on 21 July 2020).
87. Centers for Disease Control and Prevention, National Center for Emerging and Zoonotic Infectious Diseases (NCEZID), Division of Foodborne, Waterborne, and E.D. (DFWED) Parasites—Acanthamoeba—Granulomatous Amebic Encephalitis (GAE). Keratitis. Available online: https://www.cdc.gov/parasites/acanthamoeba/gen_info/acanthamoeba.html (accessed on 10 August 2020).
88. Thomas, V.; Bouchez, T.; Nicolas, V.; Robert, S.; Loret, J.F.; Lévi, Y. Amoebae in domestic water systems: Resistance to disinfection treatments and implication in Legionella persistence. *J. Appl. Microbiol.* **2004**, *97*, 950–963. [CrossRef] [PubMed]
89. Peleg, A.Y.; Seifert, H.; Paterson, D.L. Acinetobacter baumannii: Emergence of a successful pathogen. *Clin. Microbiol. Rev.* **2008**, *21*, 538–582. [CrossRef] [PubMed]
90. Cervantes, H.I.; Álvarez, J.A.; Muñoz, J.M.; Arreguín, V.; Mosqueda, J.L.; Macías, A.E. Antimicrobial activity of copper against organisms in aqueous solution: A case for copper-based water pipelines in hospitals? *Am. J. Infect. Control* **2013**, *41*, e115–e118. [CrossRef] [PubMed]
91. Huang, H.-I.; Shih, H.-Y.; Lee, C.-M.; Yang, T.C.; Lay, J.-J.; Lin, Y.E. In vitro efficacy of copper and silver ions in eradicating Pseudomonas aeruginosa, Stenotrophomonas maltophilia and Acinetobacter baumannii: Implications for on-site disinfection for hospital infection control. *Water Res.* **2008**, *42*, 73–80. [CrossRef]

92. Shih, H.-Y.Y.; Lin, Y.E. Efficacy of Copper-Silver Ionization in Controlling Biofilm- and Plankton-Associated Waterborne Pathogens. *Appl. Environ. Microbiol.* **2010**, *76*, 2032–2035. [CrossRef]
93. Cateau, E.; Delafont, V.; Hechard, Y.; Rodier, M.H. Free-living amoebae: What part do they play in healthcare-associated infections? *J. Hosp. Infect.* **2014**, *87*, 131–140. [CrossRef]
94. Cateau, E.; Verdon, J.; Fernandez, B.; Hechard, Y.; Rodier, M.-H. Acanthamoeba sp. promotes the survival and growth of Acinetobacter baumanii. *FEMS Microbiol. Lett.* **2011**, *319*, 19–25. [CrossRef]
95. Centers for Disease Control and Prevention, National Center for Emerging and Zoonotic Infectious Diseases (NCEZID), D. of H.Q.P. (DHQP) Staphylococcus aureus in Healthcare Settings|HAI|CDC. Available online: https://www.cdc.gov/hai/organisms/staph.html (accessed on 10 August 2020).
96. Kozajda, A.; Jeżak, K.; Kapsa, A. Airborne Staphylococcus aureus in different environments-a review. *Environ. Sci. Pollut. Res. Int.* **2019**, *26*, 34741–34753. [CrossRef]
97. Plipat, N.; Spicknall, I.H.; Koopman, J.S.; Eisenberg, J.N. The dynamics of methicillin-resistant Staphylococcus aureus exposure in a hospital model and the potential for environmental intervention. *BMC Infect. Dis.* **2013**, *13*, 595. [CrossRef]
98. Landeen, L.K.; Yahya, M.T.; Kutz, S.M.; Gerba, C.P. Microbiological Evaluation of Copper: Silver Disinfection Units for Use in Swimming Pools. *Water Sci. Technol.* **1989**, *21*, 267–270. [CrossRef]
99. Yahya, M.T.; Landeen, L.K.; Messina, M.C.; Kutz, S.M.; Schulze, R.; Gerba, C.P. Disinfection of bacteria in water systems by using electrolytically generated copper:silver and reduced levels of free chlorine. *Can. J. Microbiol.* **1990**, *36*, 109–116. [CrossRef] [PubMed]
100. Hopkin, M. MRSA "hiding in hospital sinks and vases". *Nature* **2006**. [CrossRef]
101. Huws, S.A.; Smith, A.W.; Enright, M.C.; Wood, P.J.; Brown, M.R.W. Amoebae promote persistence of epidemic strains of MRSA. *Environ. Microbiol.* **2006**, *8*, 1130–1133. [CrossRef] [PubMed]
102. Brooke, J.S. Stenotrophomonas maltophilia: An emerging global opportunistic pathogen. *Clin. Microbiol. Rev.* **2012**, *25*, 2–41. [CrossRef] [PubMed]
103. Denton, M.; Kerr, K.G. Microbiological and clinical aspects of infection associated with Stenotrophomonas maltophilia. *Clin. Microbiol. Rev.* **1998**, *11*, 57–80. [CrossRef]
104. Gomes, I.B.; Simões, L.C.; Simões, M. Influence of surface copper content on Stenotrophomonas maltophilia biofilm control using chlorine and mechanical stress. *Biofouling* **2020**, *36*, 1–13. [CrossRef]
105. Thomas, J.M.; Ashbolt, N.J. Do free-living amoebae in treated drinking water systems present an emerging health risk? *Environ. Sci. Technol.* **2011**, *45*, 860–869. [CrossRef]
106. Centers for Disease Control and Prevention, National Center for Emerging and Zoonotic Infectious Diseases (NCEZID), D. of H.Q.P. (DHQP) Nontuberculous Mycobacteria (NTM) Infections|HAI|CDC. Available online: https://www.cdc.gov/hai/organisms/nontuberculous-mycobacteria.html (accessed on 10 August 2020).
107. Johnson, M.M.; Odell, J.A. Nontuberculous mycobacterial pulmonary infections. *J. Thorac. Dis.* **2014**, *6*, 210–220. [CrossRef]
108. Kusnetsov, J.; Iivanainen, E.; Elomaa, N.; Zacheus, O.; Martikainen, P.J. Copper and silver ions more effective against Legionellae than against mycobacteria in a hospital warm water system. *Water Res.* **2001**, *35*, 4217–4225. [CrossRef]
109. Lin, Y.E.; Vidic, R.D.; Stout, J.E.; McCartney, C.A.; Yu, V.L. Inactivation of Mycobacterium avium by copper and silver ions. *Water Res.* **1998**, *32*, 1997–2000. [CrossRef]
110. Rhoads, W.J.; Pruden, A.; Edwards, M.A. Interactive Effects of Corrosion, Copper, and Chloramines on Legionella and Mycobacteria in Hot Water Plumbing. *Environ. Sci. Technol.* **2017**, *51*, 7065–7075. [CrossRef] [PubMed]
111. Mullis, S.N.; Falkinham, J.O. Adherence and biofilm formation of *Mycobacterium avium*, *Mycobacterium intracellulare* and *Mycobacterium abscessus* to household plumbing materials. *J. Appl. Microbiol.* **2013**, *115*, 908–914. [CrossRef] [PubMed]
112. Cirillo, J.D.; Falkow, S.; Tompkins, L.S.; Bermudez, L.E. Interaction of Mycobacterium avium with environmental amoebae enhances virulence. *Infect. Immun.* **1997**, *65*, 3759–3767. [CrossRef] [PubMed]
113. Igbinosa, I.H.; Igumbor, E.U.; Aghdasi, F.; Tom, M.; Okoh, A.I. Emerging Aeromonas species infections and their significance in public health. *ScientificWorldJournal.* **2012**, *2012*, 625023. [CrossRef]
114. Assant, M.A.; Roy, D.; Montpetit, D. Adhesion of Aeromonas hydrophila to Water Distribution System Pipes after Different Contact Times. *J. Food Prot.* **1998**, *61*, 1321–1329. [CrossRef]

115. Rahman, M.; Abd, H.; Romling, U.; Sandstrom, G.; Möllby, R. Aeromonas-Acanthamoeba interaction and early shift to a viable but nonculturable state of Aeromonas by Acanthamoeba. *J. Appl. Microbiol.* **2008**, *104*, 1449–1457. [CrossRef]
116. Delafont, V.; Perraud, E.; Brunet, K.; Maisonneuve, E.; Kaaki, S.; Rodier, M.H. Vermamoeba vermiformis in hospital network: A benefit for Aeromonas hydrophila. *Parasitol. Res.* **2019**, *118*, 3191–3194. [CrossRef]
117. National Center for Immunization and Respiratory Diseases, D. of B.D. Legionnaires Disease Signs and Symptoms|Legionella|CDC. Available online: https://www.cdc.gov/legionella/about/signs-symptoms.html (accessed on 10 August 2020).
118. National Center for Immunization and Respiratory Diseases, D. of B.D. Legionnaires Disease Cause and Spread|Legionella|CDC. Available online: https://www.cdc.gov/legionella/about/causes-transmission.html (accessed on 10 August 2020).
119. Stout, J.E.; Lin, Y.-S.E.; Goetz, A.M.; Muder, R.R. Controlling Legionella in hospital water systems: Experience with the superheat-and-flush method and copper-silver ionization. *Infect. Control Hosp. Epidemiol.* **1998**, *19*, 911–914. [CrossRef] [PubMed]
120. Miuetzner, S.; Schwille, R.C.; Farley, A.; Wald, E.R.; Ge, J.H.; States, S.J.; Libert, T.; Wadowsky, R.M. Efficacy of thermal treatment and copper-silver ionization for controlling Legionella pneumophila in high-volume hot water plumbing systems in hospitals. *Am. J. Infect. Control* **1997**, *25*, 452–457. [CrossRef]
121. Gião, M.S.; Wilks, S.A.; Keevil, C.W. Influence of copper surfaces on biofilm formation by Legionella pneumophila in potable water. *Biometals* **2015**, *28*, 329. [CrossRef] [PubMed]
122. Buse, H.Y.; Lu, J.; Struewing, I.T.; Ashbolt, N.J. Preferential colonization and release of Legionella pneumophila from mature drinking water biofilms grown on copper versus unplasticized polyvinylchloride coupons. *Int. J. Hyg. Environ. Health* **2014**, *217*, 219–225. [CrossRef] [PubMed]
123. Rowbotham, T.J. Preliminary report on the pathogenicity of Legionella pneumophila for freshwater and soil amoebae. *J. Clin. Pathol.* **1980**, *33*, 1179–1183. [CrossRef]
124. Grossi, M.; Dey, R.; Ashbolt, N. Searching for Activity Markers that Approximate (VBNC) Legionella pneumophila Infectivity in Amoeba after Ultraviolet (UV) Irradiation. *Water* **2018**, *10*, 1219. [CrossRef]
125. Bodey, G.P.; Bolivar, R.; Fainstein, V.; Jadeja, L. Infections Caused by Pseudomonas aeruginosa. *Rev. Infect. Dis.* **1983**, *5*, 279–313. [CrossRef] [PubMed]
126. Williams, B.J.; Dehnbostel, J.; Blackwell, T.S. Pseudomonas aeruginosa: Host defence in lung diseases. *Respirology* **2010**, *15*, 1037–1056. [CrossRef] [PubMed]
127. Bédard, E.; Charron, D.; Lalancette, C.; Déziel, E.; Prévost, M. Recovery of Pseudomonas aeruginosa culturability following copper- and chlorine-induced stress. *FEMS Microbiol. Lett.* **2014**, *356*, 226–234. [CrossRef]
128. Moritz, M.M.; Flemming, H.C.; Wingender, J. Integration of Pseudomonas aeruginosa and Legionella pneumophila in drinking water biofilms grown on domestic plumbing materials. *Int. J. Hyg. Environ. Health* **2010**, *213*, 190–197. [CrossRef]
129. Jeanvoine, A.; Meunier, A.; Puja, H.; Bertrand, X.; Valot, B.; Hocquet, D. Contamination of a hospital plumbing system by persister cells of a copper-tolerant high-risk clone of Pseudomonas aeruginosa. *Water Res.* **2019**, *157*, 579–586. [CrossRef]
130. Petignat, C.; Francioli, P.; Nahimana, I.; Wenger, A.; Bille, J.; Schaller, M.-D.; Revelly, J.-P.; Zanetti, G.; Blanc, D.S. Exogenous Sources of Pseudomonas aeruginosa in Intensive Care Unit Patients Implementation of Infection Control Measures and Follow-Up With Molecular Typing. *Infect. Control Hosp. Epidemiol.* **2006**, *27*, 953–957. [CrossRef]
131. Teitzel, G.M.; Parsek, M.R. Heavy Metal Resistance of Biofilm and Planktonic Pseudomonas aeruginosa. *Appl. Environ. Microbiol.* **2003**, *69*, 2313. [CrossRef]
132. Dwidjosiswojo, Z.; Richard, J.; Moritz, M.M.; Dopp, E.; Flemming, H.-C.; Wingender, J. Influence of copper ions on the viability and cytotoxicity of Pseudomonas aeruginosa under conditions relevant to drinking water environments. *Int. J. Hyg. Environ. Health* **2011**, *214*, 485–492. [CrossRef] [PubMed]
133. Dopp, E.; Richard, J.; Dwidjosiswojo, Z.; Simon, A.; Wingender, J. Influence of the copper-induced viable but non-culturable state on the toxicity of Pseudomonas aeruginosa towards human bronchial epithelial cells in vitro. *Int. J. Hyg. Environ. Health* **2017**, *220*, 1363–1369. [CrossRef] [PubMed]

134. Pukatzki, S.; Kessin, R.H.; Mekalanos, J.J. The human pathogen Pseudomonas aeruginosa utilizes conserved virulence pathways to infect the social amoeba Dictyostelium discoideum. *Proc. Natl. Acad. Sci. USA* **2002**, *99*, 3159. [CrossRef] [PubMed]
135. US EPA Office of Water. *Technologies for Legionella Control in Premise Plumbing Systems*; US Environmental Protection Agency: Washington, DC, USA, 2016.
136. Hans, M.; Erbe, A.; Mathews, S.; Chen, Y.; Solioz, M.; Mücklich, F. Role of Copper Oxides in Contact Killing of Bacteria. *Langmuir* **2013**, *29*, 16160–16166. [CrossRef] [PubMed]
137. Cachafeiro, S.P.; Naveira, I.M.; García, I.G. Is copper-silver ionisation safe and effective in controlling legionella? *J. Hosp. Infect.* **2007**, *67*, 209–216. [CrossRef]
138. Kim, B.R.; Anderson, J.E.; Mueller, S.A.; Gaines, W.A.; Kendall, A.M. Literature review—Efficacy of various disinfectants against Legionella in water systems. *Water Res.* **2002**, *36*, 4433–4444. [CrossRef]
139. Rakshit, A.; Khatua, K.; Shanbhag, V.; Comba, P.; Datta, A. Cu2+ selective chelators relieve copper-induced oxidative stress: In vivo. *Chem. Sci.* **2018**, *9*, 7916–7930. [CrossRef]
140. Van der Kooij, D.; Veenendaal, H.R.; Scheffer, W.J.H. Biofilm formation and multiplication of Legionella in a model warm water system with pipes of copper, stainless steel and cross-linked polyethylene. *Water Res.* **2005**, *39*, 2789–2798. [CrossRef]
141. Dodrill, D.M.; Edwards, M. Corrosion control on the basis of utility experience. *J. Am. Water Works Assoc.* **1995**, *87*, 74–85. [CrossRef]
142. Kimbrough, D.E. Brass corrosion as a source of lead and copper in traditional and all-plastic distribution systems. *J. Am. Water Works Assoc.* **2007**, *99*, 70–76. [CrossRef]
143. June, S.G.; Dziewulski, D.M. Copper and Silver Biocidal Mechanisms, Resistance Strategies, and Efficacy for Legionella Control. *Journal-American Water Work. Assoc.* **2018**, *110*, E13–E35. [CrossRef]
144. Assaidi, A.; Ellouali, M.; Latrache, H.; Mabrouki, M.; Hamadi, F.; Timinouni, M.; Zahir, H.; El Mdaghri, N.; Barguigua, A.; Mliji, E.M. Effect of temperature and plumbing materials on biofilm formation by Legionella pneumophila serogroup 1 and 2-15. *J. Adhes. Sci. Technol.* **2018**, *32*, 1471–1484. [CrossRef]
145. Ashbolt, N.J. Environmental (Saprozoic) Pathogens of Engineered Water Systems: Understanding Their Ecology for Risk Assessment and Management. *Pathog.* **2015**, *4*, 390–405. [CrossRef] [PubMed]
146. Morvay, A.A.; Decun, M.; Scurtu, M.; Sala, C.; Morar, A.; Sarandan, M. Biofilm formation on materials commonly used in household drinking water systems. *Water Sci. Technol. Water Supply* **2011**, *11*, 252–257. [CrossRef]
147. Dai, D.; Proctor, C.R.; Williams, K.; Edwards, M.A.; Pruden, A. Mediation of effects of biofiltration on bacterial regrowth, Legionella pneumophila, and the microbial community structure under hot water plumbing conditions. *Environ. Sci. Water Res. Technol.* **2018**, *4*, 183–194. [CrossRef]
148. Soothill, J.S. Carbapenemase-bearing Klebsiella spp. in sink drains: Investigation into the potential advantage of copper pipes. *J. Hosp. Infect.* **2016**, *93*, 152–154. [CrossRef]
149. Norton, C.D.; LeChevallier, M.W.; Falkinham, J.O. Survival of Mycobacterium avium in a model distribution system. *Water Res.* **2004**, *38*, 1457–1466. [CrossRef]
150. Leoni, E.; De Luca, G.; Legnani, P.P.; Sacchetti, R.; Stampi, S.; Zanetti, F. Legionella waterline colonization: Detection of Legionella species in domestic, hotel and hospital hot water systems. *J. Appl. Microbiol.* **2005**, *98*, 373–379. [CrossRef]
151. Marrie, T.; Green, P.; Burbridge, S.; Bezanson, G.; Neale, S.; Hoffman, P.S.; Haldane, D. Legionellaceae in the potable water of Nova Scotia hospitals and Halifax residences. *Epidemiol. Infect.* **1994**, *112*, 143–150. [CrossRef]
152. Borella, P.; Montagna, M.T.; Stampi, S.; Stancanelli, G.; Romano-Spica, V.; Triassi, M.; Marchesi, I.; Bargellini, A.; Tatò, D.; Napoli, C. Legionella contamination in hot water of Italian hotels. *Appl. Environ. Microbiol.* **2005**, *71*, 5805–5813. [CrossRef]
153. Mathys, W.; Hohmann, C.P.; Junge-Mathys, E. Efficacy of Copper-Silver Ionization in Controlling Legionella in a Hospital Hot Water Distribution System: A German Experience. In *Legionella*; Marre, R., Kwaik, Y.A., Bartlett, C., Cianciotto, N.P., Fields, B.S., Frosch, M., Hacker, J., Luck, P.C., Eds.; ASM Press: Washington, DC, USA, 2002; pp. 419–424. ISBN 1-55581-230-9.
154. Bédard, E.; Paranjape, K.; Lalancette, C.; Villion, M.; Quach, C.; Laferrière, C.; Faucher, S.P.; Prévost, M. Legionella pneumophila levels and sequence-type distribution in hospital hot water samples from faucets to connecting pipes. *Water Res.* **2019**, *156*, 277–286. [CrossRef] [PubMed]

155. Bédard, E.; Trigui, H.; Liang, J.; Doberva, M.; Paranjape, K.; Lalancette, C.; Faucher, S.P.; Prévost, M. Local adaptation of Legionella pneumophila within a hospital hot water system increases tolerance to copper. *bioRxiv* **2020**. [CrossRef]
156. Lytle, D.A.; Liggett, J. Impact of water quality on chlorine demand of corroding copper. *Water Res.* **2016**, *92*, 11–21. [CrossRef] [PubMed]
157. Edwards, M.; Powers, K.; Hidmi, L.; Schock, M.R. *The role of Pipe Ageing in Copper Corrosion by-Product Release*; IWA Publishing: London, UK, 2001; Volume 1, pp. 25–32.
158. Zevenhuizen, L.P.T.M.; Dolfing, J.; Eshuis, E.J.; Scholten-Koerselman, I.J. Inhibitory effects of copper on bacteria related to the free ion concentration. *Microb. Ecol.* **1979**, *5*, 139–146. [CrossRef]
159. Garvey, J.E.; Owen, H.A.; Winner, R.W. Toxicity of copper to the green alga, Chlamydomonas reinhardtii (Chlorophyceae), as affected by humic substances of terrestrial and freshwater origin. *Aquat. Toxicol.* **1991**, *19*, 89–96. [CrossRef]
160. Meador, J.P. The interaction of pH, dissolved organic carbon, and total copper in the determination of ionic copper and toxicity. *Aquat. Toxicol.* **1991**, *19*, 13–31. [CrossRef]
161. Ebrahimpour, M.; Alipour, H.; Rakhshah, S. Influence of water hardness on acute toxicity of copper and zinc on fish. *Toxicol. Ind. Health* **2010**, *26*, 361–365. [CrossRef]
162. Pourkhabbaz, A.; Kasmani, M.E.; Kiyani, V.; Hosynzadeh, M.H. Effects of water hardness and Cu and Zn on Lc50 in Gambusia holbrooki. *Chem. Speciat. Bioavailab.* **2011**, *23*, 224–228. [CrossRef]
163. Riethmuller, N.; Markich, S.; Parry, D.; van Dam, R. *The Effect of True Water Hardness and Alkalinity on the Toxicity of Cu and U to Two Tropical Australian Freshwater Organisms*; Supervising Scientist, Environment Australia: Darwin NT, Australia, 2000.
164. Yang, S.; Amy, P.; Edwards, M.; Rhoads, W. Natural Organic Matter, Orthophosphate, pH, and Growth Phase Can Limit Copper Antimicrobial Efficacy for Legionella in Drinking Water. *Environ. Sci. Technol.* **2020**. *Submitted*.
165. Stauber, J.L.; Florence, T.M. Mechanism of toxicity of ionic copper and copper complexes to algae. *Mar. Biol.* **1987**, *94*, 511–519. [CrossRef]
166. Ives, D.J.G.; Rawson, A.E. Copper Corrosion. *J. Electrochem. Soc.* **1962**, *109*, 447. [CrossRef]
167. Galai, S.; Touhami, Y.; Marzouki, M.N. Response surface methodology applied to laccases activities exhibited by Stenotrophomonas maltophilia AAP56 in different growth conditions. *BioResources* **2012**, *7*, 706–726.
168. Wingender, J.; Flemming, H.C. Biofilms in drinking water and their role as reservoir for pathogens. *Int. J. Hyg. Environ. Health* **2011**, *214*, 417–423. [CrossRef] [PubMed]
169. Van der Kooij, D.; van der Wielen, P.W.J.J.; Rosso, D.; Shaw, A.; Borchardt, D.; Ibisch, R.; Apgar, D.; Witherspoon, J.; Di Toro, D.M.; Paquin, P.R. *Microbial Growth in Drinking Water Supplies*; IWA Publishing: London, UK, 2013; ISBN 1780400403.
170. Connell, M.; Stenson, A.; Weinrich, L.; LeChevallier, M.; Boyd, S.L.; Ghosal, R.R.; Dey, R.; Whelton, A.J. PEX and PP Water Pipes: Assimilable Carbon, Chemicals, and Odors. *J. Am. Water Work Assoc.* **2016**, *108*, E192–E204. [CrossRef]
171. Tsuchida, D.; Kajihara, Y.; Shimidzu, N.; Hamamura, K.; Nagase, M. Hydrogen sulfide production by sulfate-reducing bacteria utilizing additives eluted from plastic resins. *Waste Manag. Res.* **2011**, *29*, 594–601. [CrossRef]
172. Kelley, K.M. *The Impact of Cross-linked Polyethylene (PEX) Pipe on Drinking Water Chemical and Odor Quality*; University of South Alabama: Mobile, AL, USA, 2014.
173. Corfitzen, C.B. *Afgivelse af Organisk Stof Fra Polymere Materialer—Mikrobiel Vækst [Release of Organic Matter from Polymeric Materials—Microbial Growth]*; The Danish Environmental Protection Agency: Odense, Denmark, 2002.
174. Kowalska, B.; Kowalski, D.; Rozej, A. Organic compounds migrating from plastic pipes into water. *J. Water Supply Res. Technol. AQUA* **2011**, *60*, 137–146. [CrossRef]
175. Pedersen, K. Biofilm development on stainless steel and pvc surfaces in drinking water. *Water Res.* **1990**, *24*, 239–243. [CrossRef]
176. Hallam, N.B.; West, J.R.; Forster, C.F.; Simms, J. The potential for biofilm growth in water distribution systems. *Water Res.* **2001**, *35*, 4063–4071. [CrossRef]
177. Thomson, R.M.; Carter, R.; Tolson, C.; Coulter, C.; Huygens, F.; Hargreaves, M. Factors associated with the isolation of Nontuberculous mycobacteria (NTM) from a large municipal water system in Brisbane, Australia. *BMC Microbiol.* **2013**, *13*, 89. [CrossRef]

178. Learbuch, K.L.G.; Lut, M.C.; Liu, G.; Smidt, H.; van der Wielen, P.W.J.J. Legionella growth potential of drinking water produced by a reverse osmosis pilot plant. *Water Res.* **2019**, *157*, 55–63. [CrossRef]
179. Buse, H.Y.; Morris, B.; Struewing, I.T.; Szabo, J.G. Chlorine and monochloramine disinfection of Legionella pneumophila colonizing copper and PVC drinking water biofilms. *Appl. Environ. Microbiol.* **2019**. [CrossRef] [PubMed]
180. Zhang, Y.; Griffin, A.; Edwards, M. Effect of nitrification on corrosion of galvanized iron, copper, and concrete. *J. Am. Water Works Assoc.* **2010**, *102*, 83–93. [CrossRef]
181. Sarin, P.; Snoeyink, V.L.; Bebee, J.; Jim, K.K.; Beckett, M.A.; Kriven, W.M.; Clement, J.A. Iron release from corroded iron pipes in drinking water distribution systems: Effect of dissolved oxygen. *Water Res.* **2004**, *38*, 1259–1269. [CrossRef]
182. WHO. *Legionella and the Prevention of Legionellosis*; World Health Organization: Geneva, Switzerland, 2007.
183. Faraldo-Gómez, J.D.; Sansom, M.S.P. Acquisition of siderophores in gram-negative bacteria. *Nat. Rev. Mol. Cell Biol.* **2003**, *4*, 105–116. [CrossRef]
184. States, S.J.; Conley, L.F.; Ceraso, M.; Stephenson, T.E.; Wolford, R.S.; Wadowsky, R.M.; McNamara, A.M.; Yee, R.B. Effects of metals on Legionella pneumophila growth in drinking water plumbing systems. *Appl. Environ. Microbiol.* **1985**, *50*, 1149–1154. [CrossRef] [PubMed]
185. Rakić, A.; Perić, J.; Foglar, L. Influence of temperature, chlorine residual and heavy metals on the presence of *Legionella pneumophila* in hot water distribution systems. *Ann. Agric. Environ. Med.* **2012**, *19*, 431–436. [PubMed]
186. van der Lugt, W.; Euser, S.M.; Bruin, J.P.; Den Boer, J.W.; Walker, J.T.; Crespi, S. Growth of Legionella anisa in a model drinking water system to evaluate different shower outlets and the impact of cast iron rust. *Int. J. Hyg. Environ. Health* **2017**, *220*, 1295–1308. [CrossRef] [PubMed]
187. Martin, R.; Strom, O.; Pruden, A.; Edwards, M. Copper Pipe, Stagnant Conditions, Corrosion Control, and Trace Disinfectant Residual Enhanced Reduction of Legionella pneumophila during Simulations of the Flint Water Crisis. *Pathogens* **2020**, in press. [CrossRef]
188. Wang, H.; Masters, S.; Hong, Y.; Stallings, J.; Falkinham, J.O.; Edwards, M.A.; Pruden, A. Effect of Disinfectant, Water Age, and Pipe Material on Occurrence and Persistence of Legionella, mycobacteria, Pseudomonas aeruginosa, and Two Amoebas. *Environ. Sci. Technol.* **2012**, *46*, 11566–11574. [CrossRef]
189. Oyem, H.H.; Oyem, I.M.; Usese, A.I. Iron, manganese, cadmium, chromium, zinc and arsenic groundwater contents of Agbor and Owa communities of Nigeria. *Springerplus* **2015**, *4*, 104. [CrossRef]
190. World Health Organization. *Zinc in Drinking Water: Background Document for Development of Guidelines for Drinking-Water Quality. Health Criteria and Other Supporting Information*; World Health Organization: Geneva, Switzerland, 2003; Volume 1.
191. World Health Organization Water Sanitation. *Guidelines for Drinking-Water Quality: Incorporating First Addendum. Vol. 1, Recommendations*; World Health Organization: Geneva, Switzerland, 2006.
192. Salehi, M.; Odimayomi, T.; Ra, K.; Ley, C.; Julien, R.; Nejadhashemi, A.P.; Hernandez-Suarez, J.S.; Mitchell, J.; Shah, A.D.; Whelton, A. An investigation of spatial and temporal drinking water quality variation in green residential plumbing. *Build. Environ.* **2020**, *169*, 106566. [CrossRef]
193. Sharrett, A.R.; Carter, A.P.; Orheimt, R.M.; Feinleib, M. Daily intake of lead, cadmium, copper, and zinc from drinking water: The seattle study of trace metal exposure. *Environ. Res.* **1982**, *28*, 456–475. [CrossRef]
194. Howard, C.D. Zinc Contamination in Drinking Water. *J. Am. Water Works Assoc.* **1923**, *10*, 411–414. [CrossRef]
195. Silver, S.; Lusk, J.E. Bacterial Magnesium, Manganese, and Zinc Transport. In *Ion Transport in Prokaryotes*; Rosen, B.P., Silver, S., Eds.; Academic Press: Cambridge, MA, USA, 1987; pp. 165–180. ISBN 978-0-12-596935-2.
196. Choudhury, R.; Srivastava, S. Zinc resistance mechanisms in bacteria. *Curr. Sci.* **2001**, *81*, 768–775.
197. Ma, L.; Terwilliger, A.; Maresso, A.W. Iron and zinc exploitation during bacterial pathogenesis. *Metallomics* **2015**, *7*, 1541–1554. [CrossRef]
198. Li, Y.; Sharma, M.R.; Koripella, R.K.; Yang, Y.; Kaushal, P.S.; Lin, Q.; Wade, J.T.; Gray, T.A.; Derbyshire, K.M.; Agrawal, R.K.; et al. Zinc depletion induces ribosome hibernation in mycobacteria. *Proc. Natl. Acad. Sci. USA* **2018**, *115*, 8191. [CrossRef]
199. Suryawati, B. Zinc homeostasis mechanism and its role in bacterial virulence capacity. *AIP Conf. Proc.* **2018**, *2021*, 070021. [CrossRef]
200. Nies, D.H. Resistance to cadmium, cobalt, zinc, and nickel in microbes. *Plasmid* **1992**, *27*, 17–28. [CrossRef]

201. Reeves, M.W.; Pine, L.; Hutner, S.H.; George, J.R.; Harrell, W.K. Metal requirements of Legionella pneumophila. *J. Clin. Microbiol.* **1981**, *13*, 688–695. [CrossRef]
202. Kirschner, R.A.; Parker, B.C.; Falkinham, J.O. Epidemiology of Infection by Nontuberculous Mycobacteria: Mycobacterium avium, Mycobacterium intracellulare, and Mycobacterium scrofulaceum in Acid, Brown-Water Swamps of the Southeastern United States and Their Association with Environmental Variables. *Am. Rev. Respir. Dis.* **1992**, *145*, 271–275. [CrossRef]
203. Norberg, A.B.; Molin, N. Toxicity of cadmium, cobalt, uranium and zinc to Zoogloea ramigera. *Water Res.* **1983**, *17*, 1333–1336. [CrossRef]
204. Nweke, C.; Chibuogwu, O.; Alisi, C. Response of planktonic bacteria of New Calabar River to zinc stress. *Afr. J. Biotechnol.* **2006**, *5*, 653–658.
205. Nweke, C.; Alisi, C.; Chibuogwu, O.; Nwanyanwu, C. Toxicity of Zinc to heterotrophic bacteria from a tropical river sediment. *Appl. Ecol. Environ. Res.* **2007**, *5*, 123–132. [CrossRef]
206. Babich, H.; Stotzky, G. Toxicity of zinc to fungi, bacteria, and coliphages: Influence of chloride ions. *Appl. Environ. Microbiol.* **1978**, *36*, 906. [CrossRef]
207. United States Environmental Protection Agency. *Secondary Drinking Water Standards: Guidance for Nuisance Chemicals*; United States Environmental Protection Agency: Washington, DC, USA, 2015.
208. Standardization Administration of China Guobiao 5749-2006. Available online: https://www.chinesestandard.net/PDF.aspx/GB5749-2006 (accessed on 10 November 2020).
209. Hwang, M.G.; Katayama, H.; Ohgaki, S. Effect of Intracellular Resuscitation of Legionella pneumophila in Acanthamoeba polyphage Cells on the Antimicrobial Properties of Silver and Copper. *Environ. Sci. Technol.* **2006**, *40*, 7434–7439. [CrossRef]
210. Barker, J.; Brown, M.R.; Collier, P.J.; Farrell, I.; Gilbert, P. Relationship between Legionella pneumophila and Acanthamoeba polyphaga: Physiological status and susceptibility to chemical inactivation. *Appl. Environ. Microbiol.* **1992**, *58*, 2420–2425. [CrossRef]
211. García, M.T.; Jones, S.; Pelaz, C.; Millar, R.D.; Abu Kwaik, Y. Acanthamoeba polyphaga resuscitates viable non-culturable Legionella pneumophila after disinfection. *Environ. Microbiol.* **2007**, *9*, 1267–1277. [CrossRef]
212. Adékambi, T.; Ben Salah, S.; Khlif, M.; Raoult, D.; Drancourt, M. Survival of Environmental Mycobacteria in Acanthamoeba polyphaga. *Appl. Environ. Microbiol.* **2006**, *72*, 5974. [CrossRef]
213. Cervero-Aragó, S.; Rodríguez-Martínez, S.; Puertas-Bennasar, A.; Araujo, R.M. Effect of Common Drinking Water Disinfectants, Chlorine and Heat, on Free Legionella and Amoebae-Associated Legionella. *PLoS ONE* **2015**, *10*, e0134726. [CrossRef]
214. Marciano-Cabral, F.; Jamerson, M.; Kaneshiro, E.S. Free-living amoebae, Legionella and Mycobacterium in tap water supplied by a municipal drinking water utility in the USA. *J. Water Health* **2009**, *8*, 71–82. [CrossRef]
215. Inkinen, J.; Jayaprakash, B.; Ahonen, M.; Pitkänen, T.; Mäkinen, R.; Pursiainen, A.; Santo Domingo, J.W.; Salonen, H.; Elk, M.; Keinänen-Toivola, M.M. Bacterial community changes in copper and PEX drinking water pipeline biofilms under extra disinfection and magnetic water treatment. *J. Appl. Microbiol.* **2018**, *124*, 611–624. [CrossRef] [PubMed]
216. Vikesland, P.J.; Valentine, R.L. Reaction Pathways Involved in the Reduction of Monochloramine by Ferrous Iron. *Environ. Sci. Technol.* **2000**, *34*, 83–90. [CrossRef]
217. Zhang, Z.; Stout, J.E.; Yu, V.L.; Vidic, R. Effect of pipe corrosion scales on chlorine dioxide consumption in drinking water distribution systems. *Water Res.* **2008**, *42*, 129–136. [CrossRef] [PubMed]
218. Nguyen, C.K.; Powers, K.A.; Raetz, M.A.; Parks, J.L.; Edwards, M.A. Rapid free chlorine decay in the presence of Cu(OH)2: Chemistry and practical implications. *Water Res.* **2011**, *45*, 5302–5312. [CrossRef] [PubMed]
219. Fu, J.; Qu, J.; Liu, R.; Zhao, X.; Qiang, Z. The influence of Cu(II) on the decay of monochloramine. *Chemosphere* **2009**, *74*, 181–186. [CrossRef] [PubMed]
220. Edwards, M.A.; Parks, J.; Griffin, A.; Raetz, M.A.; Martin, A.K.; Scardina, P.; Elfland, C. *Lead and Copper Corrosion Control in New Construction*; The Water Research Foundation: Denver, CO, USA, 2011; ISBN 978-1-60573-137-7.
221. Brazeau, R.H.; Edwards, M.A. Role of Hot Water System Design on Factors Influential to Pathogen Regrowth: Temperature, Chlorine Residual, Hydrogen Evolution, and Sediment. *Environ. Eng. Sci.* **2013**, *30*, 617–627. [CrossRef]
222. Ishizaki, A.; Tanaka, K.; Taga, N. Microbial production of poly-D-3-hydroxybutyrate from CO2x. *Appl. Microbiol. Biotechnol.* **2001**, *57*, 6–12. [CrossRef]

223. Edwards, M.; Pruden, A.; Falkinham III, J.O.; Brazeau, R.; Williams, K.; Wang, H.; Martin, A.; Rhoads, W. Relationship between Biodegradable Organic Matter and Pathogen Concentrations in Premise Plumbing. Available online: https://www.researchgate.net/publication/282650492_Relationship_Between_Biodegradable_Organic_Matter_and_Pathogen_Concentrations_in_Premise_Plumbing (accessed on 7 October 2020).
224. Dai, D.; Rhoads, W.J.; Edwards, M.A.; Pruden, A. Shotgun Metagenomics Reveals Taxonomic and Functional Shifts in Hot Water Microbiome Due to Temperature Setting and Stagnation. *Front. Microbiol.* **2018**, *9*, 2695. [CrossRef]
225. González-Cabaleiro, R.; Curtis, T.P.; Ofiţeru, I.D. Bioenergetics analysis of ammonia-oxidizing bacteria and the estimation of their maximum growth yield. *Water Res.* **2019**, *154*, 238–245. [CrossRef]
226. Neubauer, S.C.; Emerson, D.; Megonigal, J.P. Life at the energetic edge: Kinetics of circumneutral iron oxidation by lithotrophic iron-oxidizing bacteria isolated from the wetland-plant rhizosphere. *Appl. Environ. Microbiol.* **2002**, *68*, 3988–3995. [CrossRef]
227. Martin, A.K.; Edwards, M.A.; Pruden, A.J.; Falkinham, J.O. *Organic Carbon Generation Mechanisms in Main and Premise Distribution Systems*; Virginia Tech: Blacksburg, VA, USA, 2012.
228. Clark, B.; Clair, J.S.; Edwards, M. Copper Deposition Corrosion Elevates Lead Release to Potable Water. *J. Am. Water Works Assoc.* **2015**, *107*, E627–E637. [CrossRef]
229. Brazeau, R.H.; Edwards, M.A.; Falkinham, J.O.; Pearce, A.R.; Pruden, A.J. *Sustainability of Residential Hot Water Infrastructure: Public Health, Environmental Impacts, and Consumer Drivers*; Virginia Tech: Blacksburg, VA, USA, 2012.
230. Fox, P.; Abbaszadegan, M. Impact of Scale Formation on Biofilm Growth in Premise Plumbing. Available online: https://www.semanticscholar.org/paper/Impact-of-Scale-Formation-on-Biofilm-Growth-in-Fox-Abbaszadegan/b615ff0df2516a1bac2ac11854019ab0554e6fc2 (accessed on 31 August 2020).
231. Sarin, P.; Snoeyink, V.L.; Lytle, D.A.; Kriven, W.M. Iron Corrosion Scales: Model for Scale Growth, Iron Release, and Colored Water Formation. *J. Environ. Eng.* **2004**, *130*, 364–373. [CrossRef]
232. Yang, F.; Shi, B.; Gu, J.; Wang, D.; Yang, M. Morphological and physicochemical characteristics of iron corrosion scales formed under different water source histories in a drinking water distribution system. *Water Res.* **2012**, *46*, 5423–5433. [CrossRef] [PubMed]
233. Tuovinen, O.H.; Button, K.S.; Vuorinen, A.; Carlson, L.; Mair, D.M.; Yut, L.A. Bacterial, chemical, and mineralogical characteristics of tubercles in distribution pipelines. *J. Am. Water Works Assoc.* **1980**, *72*, 626–635. [CrossRef]
234. Lee, W.; Lewandowski, Z.; Nielsen, P.H.; Allan Hamilton, W. Role of sulfate-reducing bacteria in corrosion of mild steel: A review. *Biofouling* **1995**, *8*, 165–194. [CrossRef]
235. Nguyen, C.K.; Stone, K.R.; Dudi, A.; Edwards, M.A. Corrosive Microenvironments at Lead Solder Surfaces Arising from Galvanic Corrosion with Copper Pipe. *Environ. Sci. Technol.* **2010**, *44*, 7076–7081. [CrossRef]
236. Hasit, Y.J.; Anderson, J.L.; Anthony, J.; Parolari, T.; Rockaway, T.D.; Frenc, M.L. *Distribution Water Quality Issues Related to New Development or Low Usage*; Awwa Research Foudation: Denver, CO, USA, 2006.
237. McNeill, L.S.; Edwards, M. Iron pipe corrosion in distribution systems. *J. Am. Water Works Assoc.* **2001**, *93*, 88–100. [CrossRef]
238. Nawrocki, J.; Raczyk-Stanisławiak, U.; Świetlik, J.; Olejnik, A.; Sroka, M.J. Corrosion in a distribution system: Steady water and its composition. *Water Res.* **2010**, *44*, 1863–1872. [CrossRef]
239. Merkel, T.H.; Groß, H.J.; Werner, W.; Dahlke, T.; Reicherter, S.; Beuchle, G.; Eberle, S.H. Copper corrosion by-product release in long-term stagnation experiments. *Water Res.* **2002**, *36*, 1547–1555. [CrossRef]
240. Lytle, D.A.; Schock, M.R. Impact of stagnation time on metal dissolution from plumbing materials in drinking water. *J. Water Supply Res. Technol. AQUA* **2000**, *49*, 243–257. [CrossRef]
241. Zhang, L.; Liu, S. Investigation of organic compounds migration from polymeric pipes into drinking water under long retention times. *Procedia Eng.* **2014**, *70*, 1753–1761. [CrossRef]
242. Richard, R.; Hamilton, K.A.; Westerhoff, P.; Boyer, T.H. Tracking copper, chlorine, and occupancy in a new, multi-story, institutional green building. *Environ. Sci. Water Res. Technol.* **2020**, *6*, 1672–1680. [CrossRef]
243. Haig, S.J.; Kotlarz, N.; Lipuma, J.J.; Raskin, L. A high-throughput approach for identification of nontuberculous mycobacteria in drinking water reveals relationship between water age and Mycobacterium avium. *MBio* **2018**, *9*. [CrossRef] [PubMed]

244. Ley, C.; Proctor, C.; Singh, G.; Ra, K.; Noh, Y.; Odimayomi, T.; Salehi, M.; Julien, R.; Mitchell, J.; Nejadhashemi, A.P.; et al. Drinking water microbiology in a water-efficient building: Stagnation, seasonality, and physiochemical effects on opportunistic pathogen and total bacteria proliferation. *Environ. Sci. Water Res. Technol.* **2020**. [CrossRef]
245. Hozalski, R.M.; Lapara, T.M.; Zhao, X.; Kim, T.; Waak, M.B.; Burch, T.; Mccarty, M. Flushing of stagnant premise water systems after the COVID-19 shutdown can reduce infection risk by Legionella and Mycobacterium spp. *medRxiv* **2020**. [CrossRef]
246. Sánchez-Busó, L.; Olmos, M.P.; Camaró, M.L.; Adrián, F.; Calafat, J.M.; González-Candelas, F. Phylogenetic analysis of environmental Legionella pneumophila isolates from an endemic area (Alcoy, Spain). *Infect. Genet. Evol.* **2015**, *30*, 45–54. [CrossRef] [PubMed]
247. Pryor, M.; Springthorpe, S.; Riffard, S.; Brooks, T.; Huo, Y.; Davis, G.; Sattar, S.A. Investigation of opportunistic pathogens in municipal drinking water under different supply and treatment regimes. *Water Sci. Technol.* **2004**, *50*, 83–90. [CrossRef]
248. Guidance for Reopening Buildings After Prolonged Shutdown or Reduced Operation|CDC. Available online: https://www.cdc.gov/coronavirus/2019-ncov/php/building-water-system.html (accessed on 26 October 2020).
249. Proctor, C.; Rhoads, W.; Keane, T.; Salehi, M.; Hamilton, K.; Pieper, K.J.; Cwiertny, D.M.; Prévost, M.; Whelton, A. Considerations for Large Building Water Quality after Extended Stagnation. *AWWA Water Sci.* **2020**, *2*, e1186. [CrossRef]
250. United States Environmental Protection Agency. Information on Maintaining or Restoring Water Quality in Buildings with Low or No Use. Available online: https://www.epa.gov/coronavirus/information-maintaining-or-restoring-water-quality-buildings-low-or-no-use (accessed on 26 October 2020).
251. Rhoads, W.J.; Prévost, M.; Pieper, K.J.; Keane, T.; Whelton, A.J.; Rölli, F.; Proctor, C.R.; Grimard-Conea, M. *Responding to Water Stagnation in Buildings with Reduced or No Water Use*; American Water Works Association: Denver, CO, USA, 2020.
252. *2021 Uniform Plumbing Code*; Twenty-Nin.; International Association of Plumbing and Mechanical Officials: Ontario, CA, USA, 2020.
253. *2018 International Plumbing Code*; International Code Council, Inc.: Country Club Hills, IL, USA, 2017; ISBN 978-1-60983-745-7.
254. *2015 International Energy Conservation Code*; International Code Council, Inc: Country Club Hills, IL, USA, 2014; ISBN 978-1-60983-486-9.
255. Ally, M.R. Water and Energy Savings using Demand Hot Water Recirculating Systems in Residential Homes: A Case Study of Five Homes in Palo Alto, California. Available online: http://www.osti.gov/servlets/purl/885864-jNK2Il/ (accessed on 26 August 2020).
256. Rhoads, W.J.; Ji, P.; Pruden, A.; Edwards, M.A. Water heater temperature set point and water use patterns influence Legionella pneumophila and associated microorganisms at the tap. *Microbiome* **2015**, *3*, 1–13. [CrossRef]
257. Flannery, B.; Gelling, L.B.; Vugia, D.J.; Weintraub, J.M.; Salerno, J.J.; Conroy, M.J.; Stevens, V.A.; Rose, C.E.; Moore, M.R.; Fields, B.S.; et al. Reducing Legionella colonization of water systems with monochloramine. *Emerg. Infect. Dis.* **2006**, *12*, 588. [CrossRef]
258. Oh, S.J.; Cook, D.C.; Townsend, H.E. Characterization of Iron Oxides Commonly Formed as Corrosion Products on Steel. *Hyperfine Interact.* **1998**, *112*, 59–66. [CrossRef]
259. Stout, J.E.; Yu, V.L.; Best, M.G. Ecology of Legionella pneumophila within water distribution systems. *Appl. Environ. Microbiol.* **1985**, *49*, 221–228. [CrossRef] [PubMed]
260. Zhu, Z.; Wu, C.; Zhong, D.; Yuan, Y.; Shan, L.; Zhang, J. Effects of pipe materials on chlorine-resistant biofilm formation under long-term high chlorine level. *Appl. Biochem. Biotechnol.* **2014**, *173*, 1564–1578. [CrossRef] [PubMed]
261. Sancy, M.; Gourbeyre, Y.; Sutter, E.M.M.; Tribollet, B. Mechanism of corrosion of cast iron covered by aged corrosion products: Application of electrochemical impedance spectrometry. *Corros. Sci.* **2010**, *52*, 1222–1227. [CrossRef]
262. Salehi, M.; Abouali, M.; Wang, M.; Zhou, Z.; Nejadhashemi, A.P.; Mitchell, J.; Caskey, S.; Whelton, A.J. Case study: Fixture water use and drinking water quality in a new residential green building. *Chemosphere* **2018**. [CrossRef]

263. Lund, V.; Anderson-Glenna, M.; Skjevrak, I.; Steffensen, I.-L. Long-term study of migration of volatile organic compounds from cross-linked polyethylene (PEX) pipes and effects on drinking water quality. *J. Water Health* **2011**, *9*, 483–497. [CrossRef] [PubMed]
264. Tanemura, D.; Yamada, K.; Nishimura, H.; Igawa, K.; Higuchi, Y. Investigation of degradation mechanism by copper catalytic activity and mechanical property of polyethylene pipes for hot water supply. *Annu. Tech. Conf. ANTEC Conf. Proc.* **2014**, *3*, 2022–2026.
265. Wright, D. *Failure of Plastics and Rubber Products: Causes, Effects and Case Studies Involving Degradation*; Rapra Technology Limited: Shawbury, UK, 2001; ISBN 185957517X.
266. Plastic Pipe Institute. *Proper Integration of Copper Tubing and Components with PP-R Piping Materials for Plumbing Applications*; The Plastic Pipe Institute: Irving, TX, USA, 2018.
267. Bulletin, A.T. *Intermixing Copper Tube and Aquatherm*; Aquatherm GmbH: Attendom, Germany, 2012.
268. Huang, X.; Pieper, K.J.; Cooper, H.K.; Diaz-Amaya, S.; Zemlyanov, D.Y.; Whelton, A.J. Corrosion of upstream metal plumbing components impact downstream PEX pipe surface deposits and degradation. *Chemosphere* **2019**, *236*, 124329. [CrossRef]
269. Lewis, J.R. The Catalytic Decomposition of Sodium Hypochlorite Solutions.II. Iron Oxide as Promoter in the Copper Oxide Catalysis of Sodium Hypochlorite. *J. Phys. Chem.* **1928**, *32*, 1808–1819. [CrossRef]
270. Cruse, H. Dissolved-Copper Effect on Iron Pipe. *J. Am. Water Works Assoc.* **1971**, *63*, 79–81. [CrossRef]
271. Cartier, C.; Nour, S.; Richer, B.; Deshommes, E.; Prévost, M. Impact of water treatment on the contribution of faucets to dissolved and particulate lead release at the tap. *Water Res.* **2012**, *46*, 5205–5216. [CrossRef] [PubMed]
272. DeSantis, M.K.; Triantafyllidou, S.; Schock, M.R.; Lytle, D.A. Mineralogical Evidence of Galvanic Corrosion in Drinking Water Lead Pipe Joints. *Environ. Sci. Technol.* **2018**, *52*, 3365–3374. [CrossRef] [PubMed]
273. Zhang, X.G. Chapter 10: Galvanic Corrosion. In *Uhlig's Corrosion Handbook: Third Edition*; Revie, R.W., Ed.; The Electrochemical Society: Pennington, NJ, USA, 2011; pp. 123–143. ISBN 978-0-8031-0981-0.
274. Hack, H.P. *Galvanic Corrosion—Google Books*; American Society for Testing and Materials: Philadelphia, PA, USA, 1988.
275. Souto, R.M.; González-García, Y.; Bastos, A.C.; Simões, A.M. Investigating corrosion processes in the micrometric range: A SVET study of the galvanic corrosion of zinc coupled with iron. *Corros. Sci.* **2007**, *49*, 4568–4580. [CrossRef]
276. Marques, A.G.; Taryba, M.G.; Panão, A.S.; Lamaka, S.V.; Simões, A.M. Application of scanning electrode techniques for the evaluation of iron-zinc corrosion in nearly neutral chloride solutions. *Corros. Sci.* **2016**, *104*, 123–131. [CrossRef]
277. Kuntyi, O.I.; Zozulya, H.I.; Dobrovets'ka, O.Y.; Kornii, S.A.; Reshetnyak, O.V. Deposition of Copper, Silver, and Nickel on Aluminum by Galvanic Replacement. *Mater. Sci.* **2018**, *53*, 488–494. [CrossRef]
278. Jorcin, J.-B.; Blanc, C.; Pébère, N.; Tribollet, B.; Vivier, V. Galvanic Coupling Between Pure Copper and Pure Aluminum. *J. Electrochem. Soc.* **2008**, *155*, C46. [CrossRef]
279. Clark, B.N.; Masters, S.V.; Edwards, M.A. Lead Release to Drinking Water from Galvanized Steel Pipe Coatings. *Environ. Eng. Sci.* **2015**, *32*, 713–721. [CrossRef]
280. Zhou, M.; Liu, C.; Xu, S.; Gao, Y.; Jiang, S. Accelerated degradation rate of AZ31 magnesium alloy by copper additions. *Mater. Corros.* **2018**, *69*, 760–769. [CrossRef]
281. Tada, E.; Sugawara, K.; Kaneko, H. Distribution of pH during galvanic corrosion of a Zn/steel couple. *Electrochim. Acta* **2004**, *49*, 1019–1026. [CrossRef]
282. United States Bureau of the Census. American Housing Survey, 2017 National Data. Available online: https://www.census.gov/programs-surveys/ahs/data/2017/ahs-2017-public-use-file--puf-/ahs-2017-national-public-use-file--puf-.html (accessed on 10 October 2020).
283. Ewann, F.; Hoffman, P.S. Cysteine Metabolism in Legionella pneumophila: Characterization of an L-Cystine-Utilizing Mutant. *Appl. Environ. Microbiol.* **2006**, *72*, 3993–4000. [CrossRef]
284. Shaheen, M.; Scott, C.; Ashbolt, N.J. Long-term persistence of infectious Legionella with free-living amoebae in drinking water biofilms. *Int. J. Hyg. Environ. Health* **2019**, *222*, 678–686. [CrossRef] [PubMed]
285. Richards, A.M.; Von Dwingelo, J.E.; Price, C.T.; Abu Kwaik, Y. Cellular microbiology and molecular ecology of Legionella–amoeba interaction. *Virulence* **2013**, *4*, 307–314. [CrossRef] [PubMed]

286. Michel, R.; Burghardt, H.; Bergmann, H. Acanthamoeba, naturally intracellularly infected with Pseudomonas aeruginosa, after their isolation from a microbiologically contaminated drinking water system in a hospital. *Zent. Hyg. Umweltmed.* **1995**, *196*, 532–544.
287. Ji, P.; Rhoads, W.J.; Edwards, M.A.; Pruden, A. Impact of water heater temperature setting and water use frequency on the building plumbing microbiome. *Isme J.* **2017**, *11*, 1318–1330. [CrossRef] [PubMed]
288. Buse, H.Y.; Lu, J.; Lu, X.; Mou, X.; Ashbolt, N.J. Microbial diversities (16S and 18S rRNA gene pyrosequencing) and environmental pathogens within drinking water biofilms grown on the common premise plumbing materials unplasticized polyvinylchloride and copper. *FEMS Microbiol. Ecol.* **2014**, *88*, 280–295. [CrossRef] [PubMed]
289. Falkinham, J.O.; Hilborn, E.D.; Arduino, M.J.; Pruden, A.; Edwards, M.A. Epidemiology and Ecology of Opportunistic Premise Plumbing Pathogens: Legionella pneumophila, Mycobacterium avium, and Pseudomonas aeruginosa. *Environ. Health Perspect.* **2015**, *123*, 749–758. [CrossRef] [PubMed]
290. Ma, X.; Pierre, D.; Bibby, K.; Stout, J.E. Bacterial community structure correlates with Legionella pneumophila colonization of New York City high rise building premises plumbing systems. *Environ. Sci. Water Res. Technol.* **2020**, *6*, 1324–1335. [CrossRef]
291. Paranjape, K.; Bédard, É.; Whyte, L.G.; Ronholm, J.; Prévost, M.; Faucher, S.P. Presence of Legionella spp. in cooling towers: The role of microbial diversity, Pseudomonas, and continuous chlorine application. *Water Res.* **2020**, *169*, 115252. [CrossRef]
292. Garner, E.; McLain, J.; Bowers, J.; Engelthaler, D.M.; Edwards, M.A.; Pruden, A. Microbial Ecology and Water Chemistry Impact Regrowth of Opportunistic Pathogens in Full-Scale Reclaimed Water Distribution Systems. *Environ. Sci. Technol.* **2018**, *52*, 9056–9068. [CrossRef]
293. Ji, P.; Parks, J.; Edwards, M.A.; Pruden, A. Impact of Water Chemistry, Pipe Material and Stagnation on the Building Plumbing Microbiome. *PLoS ONE* **2015**, *10*, e0141087. [CrossRef]
294. Lin, W.; Yu, Z.; Chen, X.; Liu, R.; Zhang, H. Molecular characterization of natural biofilms from household taps with different materials: PVC, stainless steel, and cast iron in drinking water distribution system. *Appl. Microbiol. Biotechnol.* **2012**, *97*. [CrossRef] [PubMed]
295. Proctor, C.R.; Gächter, M.; Kötzsch, S.; Rölli, F.; Sigrist, R.; Walser, J.-C.; Hammes, F. Biofilms in shower hoses–choice of pipe material influences bacterial growth and communities. *Environ. Sci. Water Res. Technol.* **2016**, *2*, 670–682. [CrossRef]
296. Corre, M.-H.; Delafont, V.; Legrand, A.; Berjeaud, J.-M.; Verdon, J. Exploiting the Richness of Environmental Waterborne Bacterial Species to Find Natural Legionella pneumophila Competitors. *Front. Microbiol.* **2019**, *9*, 3360. [CrossRef] [PubMed]
297. Kimura, S.; Tateda, K.; Ishii, Y.; Horikawa, M.; Miyairi, S.; Gotoh, N.; Ishiguro, M.; Yamaguchi, K. Pseudomonas aeruginosa Las quorum sensing autoinducer suppresses growth and biofilm production in Legionella species. *Microbiology* **2009**, *155*, 1934–1939. [CrossRef] [PubMed]
298. Declerck, P.; Behets, J.; Delaedt, Y.; Margineanu, A.; Lammertyn, E.; Ollevier, F. Impact of Non-Legionella Bacteria on the Uptake and Intracellular Replication of Legionella pneumophila in Acanthamoeba castellanii and Naegleria lovaniensis. *Microb. Ecol.* **2005**, *50*, 536–549. [CrossRef]
299. Berry, D.; Horn, M.; Xi, C.; Raskin, L. Mycobacterium avium Infections of Acanthamoeba Strains: Host Strain Variability, Grazing-Acquired Infections, and Altered Dynamics of Inactivation with Monochloramine. *Appl. Environ. Microbiol.* **2010**, *76*, 6685–6688. [CrossRef]
300. Weitere, M.; Bergfeld, T.; Rice, S.A.; Matz, C.; Kjelleberg, S. Grazing resistance of Pseudomonas aeruginosa biofilms depends on type of protective mechanism, developmental stage and protozoan feeding mode. *Environ. Microbiol.* **2005**, *7*, 1593–1601. [CrossRef]
301. Matz, C.; Moreno, A.M.; Alhede, M.; Manefield, M.; Hauser, A.R.; Givskov, M.; Kjelleberg, S. Pseudomonas aeruginosa uses type III secretion system to kill biofilm-associated amoebae. *ISME J.* **2008**, *2*, 843–852. [CrossRef]
302. Steinberg, C.; Laurent, J.; Edel-Hermann, V.; Barbezant, M.; Sixt, N.; Dalle, F.; Aho, S.; Bonnin, A.; Hartemann, P.; Sautour, M. Adaptation of Fusarium oxysporum and Fusarium dimerum to the specific aquatic environment provided by the water systems of hospitals. *Water Res.* **2015**, *76*, 53–65. [CrossRef]
303. Petitjean, M.; Martak, D.; Silvant, A.; Bertrand, X.; Valot, B.; Hocquet, D. Genomic characterization of a local epidemic Pseudomonas aeruginosa reveals specific features of the widespread clone ST395. *Microb. Genom.* **2017**, *3*, e000129. [CrossRef]

304. Oliver, J.D. Recent findings on the viable but nonculturable state in pathogenic bacteria. *FEMS Microbiol. Rev.* **2010**, *34*, 415–425. [CrossRef] [PubMed]
305. Williams, K.; Pruden, A.; Falkinham, J.O.; Edwards, M. Relationship between Organic Carbon and Opportunistic Pathogens in Simulated Glass Water Heaters. *Pathogens* **2015**, *4*, 355–372. [CrossRef] [PubMed]
306. Dusserre, E.; Ginevra, C.; Hallier-Soulier, S.; Vandenesch, F.; Festoc, G.; Etienne, J.; Jarraud, S.; Molmeret, M. A PCR-based method for monitoring Legionella pneumophila in water samples detects viable but noncultivable legionellae that can recover their cultivability. *Appl. Environ. Microbiol.* **2008**, *74*, 4817–4824. [CrossRef] [PubMed]
307. Wang, H.; Bédard, E.; Prévost, M.; Camper, A.K.; Hill, V.R.; Pruden, A. Methodological approaches for monitoring opportunistic pathogens in premise plumbing: A review. *Water Res.* **2017**, *117*, 68–86. [CrossRef] [PubMed]
308. Rowland, J.L.; Niederweis, M. Resistance mechanisms of Mycobacterium tuberculosis against phagosomal copper overload. *Tuberculosis* **2012**, *92*, 202–210. [CrossRef]
309. Lehtola, M.J.; Miettinen, I.T.; Keinänen, M.M.; Kekki, T.K.; Laine, O.; Hirvonen, A.; Vartiainen, T.; Martikainen, P.J. Microbiology, chemistry and biofilm development in a pilot drinking water distribution system with copper and plastic pipes. *Water Res.* **2004**, *38*, 3769–3779. [CrossRef]
310. Schwan, W.R.; Warrener, P.; Keunz, E.; Kendall Stover, C.; Folger, K.R. Mutations in the cueA gene encoding a copper homeostasis P-type ATPase reduce the pathogenicity of Pseudomonas aeruginosa in mice. *Int. J. Med. Microbiol.* **2005**, *295*, 237–242. [CrossRef]
311. Abdollahi, S.; Rasooli, I.; Mousavi Gargari, S.L. The role of TonB-dependent copper receptor in virulence of Acinetobacter baumannii. *Infect. Genet. Evol.* **2018**, *60*, 181–190. [CrossRef]
312. Alquethamy, S.; Khorvash, M.; Pederick, V.; Whittall, J.; Paton, J.; Paulsen, I.; Hassan, K.; McDevitt, C.; Eijkelkamp, B. The Role of the CopA Copper Efflux System in Acinetobacter baumannii Virulence. *Int. J. Mol. Sci.* **2019**, *20*, 575. [CrossRef]
313. Huston, W.M.; Jennings, M.P.; McEwan, A.G. The multicopper oxidase of Pseudomonas aeruginosa is a ferroxidase with a central role in iron acquisition. *Mol. Microbiol.* **2002**, *45*, 1741–1750. [CrossRef]
314. James, B.W.; Mauchline, W.S.; Fitzgeorge, R.B.; Dennis, P.J.; Keevil, C.W. Influence of iron-limited continuous culture on physiology and virulence of Legionella pneumophila. *Infect. Immun.* **1995**, *63*, 4224–4230. [CrossRef] [PubMed]
315. Jimenez, P.N.; Koch, G.; Papaioannou, E.; Wahjudi, M.; Krzeslak, J.; Coenye, T.; Cool, R.H.; Quax, W.J. Role of PvdQ in Pseudomonas aeruginosa virulence under iron-limiting conditions. *Microbiology* **2010**, *156*, 49–59. [CrossRef] [PubMed]
316. Vasil, M.L.; Ochsner, U.A. The response of Pseudomonas aeruginosa to iron: Genetics, biochemistry and virulence. *Mol. Microbiol.* **1999**, *34*, 399–413. [CrossRef] [PubMed]
317. Sokol, P.A.; Woods, D.E. Relationship of iron and extracellular virulence factors to Pseudomonas aeruginosa lung infections. *J. Med. Microbiol.* **1984**, *18*, 125–133. [CrossRef] [PubMed]
318. Bjorn, M.J.; Iglewski, B.H.; Ives, S.K.; Sadoff, J.C.; Vasil, M.L. Effect of iron on yields of exotoxin A in cultures of Pseudomonas aeruginosa PA-103. *Infect. Immun.* **1978**, *19*, 785–791. [CrossRef]
319. Penwell, W.F.; Arivett, B.A.; Actis, L.A. The Acinetobacter baumannii entA Gene Located Outside the Acinetobactin Cluster Is Critical for Siderophore Production, Iron Acquisition and Virulence. *PLoS ONE* **2012**, *7*, e36493. [CrossRef]
320. Fiester, S.E.; Arivett, B.A.; Schmidt, R.E.; Beckett, A.C.; Ticak, T.; Carrier, M.V.; Ghosh, R.; Ohneck, E.J.; Metz, M.L.; Sellin Jeffries, M.K.; et al. Iron-Regulated Phospholipase C Activity Contributes to the Cytolytic Activity and Virulence of Acinetobacter baumannii. *PLoS ONE* **2016**, *11*, e0167068. [CrossRef]
321. Pruden, A. Balancing Water Sustainability and Public Health Goals in the Face of Growing Concerns about Antibiotic Resistance. *Environ. Sci. Technol.* **2014**, *48*, 5–14. [CrossRef]
322. Berg, J.; Tom-Petersen, A.; Nybroe, O. Copper amendment of agricultural soil selects for bacterial antibiotic resistance in the field. *Lett. Appl. Microbiol.* **2005**, *40*, 146–151. [CrossRef]
323. Hu, H.-W.; Wang, J.-T.; Li, J.; Li, J.-J.; Ma, Y.-B.; Chen, D.; He, J.-Z. Field-based evidence for copper contamination induced changes of antibiotic resistance in agricultural soils. *Environ. Microbiol.* **2016**, *18*, 3896–3909. [CrossRef]
324. Knapp, C.W.; McCluskey, S.M.; Singh, B.K.; Campbell, C.D.; Hudson, G.; Graham, D.W. Antibiotic Resistance Gene Abundances Correlate with Metal and Geochemical Conditions in Archived Scottish Soils. *PLoS ONE* **2011**, *6*, e27300. [CrossRef] [PubMed]

325. Berg, J.; Thorsen, M.K.; Holm, P.E.; Jensen, J.; Nybroe, O.; Brandt, K.K. Cu Exposure under Field Conditions Coselects for Antibiotic Resistance as Determined by a Novel Cultivation-Independent Bacterial Community Tolerance Assay. *Environ. Sci. Technol.* **2010**, *44*, 8724–8728. [CrossRef] [PubMed]
326. Graham, D.W.; Olivares-Rieumont, S.; Knapp, C.W.; Lima, L.; Werner, D.; Bowen, E. Antibiotic Resistance Gene Abundances Associated with Waste Discharges to the Almendares River near Havana, Cuba. *Environ. Sci. Technol.* **2011**, *45*, 418–424. [CrossRef] [PubMed]
327. Zhang, M.; Chen, L.; Ye, C.; Yu, X. Co-selection of antibiotic resistance via copper shock loading on bacteria from a drinking water bio-filter. *Environ. Pollut.* **2018**, *233*, 132–141. [CrossRef]
328. Becerra-Castro, C.; Machado, R.; Vaz-Moreira, I.; Manaia, C. Assessment of copper and zinc salts as selectors of antibiotic resistance in Gram-negative bacteria. *Sci. Total Environ.* **2015**, *530*. [CrossRef]
329. Barker, J.; Scaife, H.; Brown, M.R. Intraphagocytic growth induces an antibiotic-resistant phenotype of Legionella pneumophila. *Antimicrob. Agents Chemother.* **1995**, *39*, 2684. [CrossRef]
330. Oglesby-Sherrouse, A.G.; Djapgne, L.; Nguyen, A.T.; Vasil, A.I.; Vasil, M.L. The complex interplay of iron, biofilm formation, and mucoidy affecting antimicrobial resistance of Pseudomonas aeruginosa. *Pathog. Dis.* **2014**, *70*, 307–320. [CrossRef]
331. Guo, X.; Liu, S.; Wang, Z.; Zhang, X.; Li, M.; Wu, B. Metagenomic profiles and antibiotic resistance genes in gut microbiota of mice exposed to arsenic and iron. *Chemosphere* **2014**, *112*, 1–8. [CrossRef]
332. European Commisssion. Council Directive 98/83/EC of 3 November 1998 on the quality of water intended for human consumption. Available online: https://eur-lex.europa.eu/legal-content/EN/TXT/?uri=CELEX%3A31998L0083 (accessed on 10 November 2020).

**Publisher's Note:** MDPI stays neutral with regard to jurisdictional claims in published maps and institutional affiliations.

pathogens

MDPI

Article

# Interactive Effects of Copper Pipe, Stagnation, Corrosion Control, and Disinfectant Residual Influenced Reduction of *Legionella pneumophila* during Simulations of the Flint Water Crisis

**Rebekah L. Martin [1,2], Owen R. Strom [3], Amy Pruden [1] and Marc A. Edwards [1,*]**

1 Department of Civil and Environmental Engineering, Virginia Tech, 418 Durham Hall, Blacksburg, VA 24061, USA; hrebek1@vt.edu (R.L.M.); apruden@vt.edu (A.P.)
2 Department of Civil and Environmental Engineering, Virginia Military Institute, Lexington, VA 24450, USA
3 Elson S. Floyd College of Medicine, Washington State University, Spokane, WA 99202, USA; owen_strom@wsu.edu
* Correspondence: edwardsm@vt.edu

Received: 29 July 2020; Accepted: 31 August 2020; Published: 4 September 2020

**Abstract:** Flint, MI experienced two outbreaks of Legionnaires' Disease (LD) during the summers of 2014 and 2015, coinciding with use of Flint River as a drinking water source without corrosion control. Using simulated distribution systems (SDSs) followed by stagnant simulated premise (i.e., building) plumbing reactors (SPPRs) containing cross-linked polyethylene (PEX) or copper pipe, we reproduced trends in water chemistry and *Legionella* proliferation observed in the field when Flint River versus Detroit water were used before, during, and after the outbreak. Specifically, due to high chlorine demand in the SDSs, SPPRs with treated Flint River water were chlorine deficient and had elevated *L. pneumophila* numbers in the PEX condition. SPPRs with Detroit water, which had lower chlorine demand and higher residual chlorine, lost all culturable *L. pneumophila* within two months. *L. pneumophila* also diminished more rapidly with time in Flint River SPPRs with copper pipe, presumably due to the bacteriostatic properties of elevated copper concentrations caused by lack of corrosion control and stagnation. This study confirms hypothesized mechanisms by which the switch in water chemistry, pipe materials, and different flow patterns in Flint premise plumbing may have contributed to observed LD outbreak patterns.

**Keywords:** flint; *Legionella pneumophila*; copper; PEX; iron; chlorine; premise plumbing

## 1. Introduction

The Flint Water Crisis began when the City of Flint, Michigan switched from purchasing its long-term Detroit municipal water supply (sourced from Lake Huron) to the local Flint River in April 2014. Although the Flint River water was predictably much more corrosive than Detroit water, no federally-mandated corrosion control program was implemented, resulting in rampant corrosion of lead plumbing and iron water mains, low chlorine residuals, elevated bacteria, and high levels of lead [1–4]. During the summers of 2014 and 2015, Flint also experienced two outbreaks of Legionnaires' Disease (LD), with 91 cases and 12 deaths documented in Genesee County (the county which Flint is located), compared to the 6–13 cases per year and no deaths during 2009–2013 while on Detroit water [1,3,5,6]. The period of Flint River water use was characterized by high levels of LD incidence [6] and *L. pneumophila* gene marker levels [1] associated with large buildings. However, lower LD incidence associated with residential exposure was noted from August 2015 onwards and our sampling during this period revealed undetectable or very low levels of *L. pneumophila* in residential plumbing [1,3]. While temporal associations between the switch in water supply with reduced levels of chlorine, high levels of iron, elevated temperature for Flint River water, and the resulting outbreak of LD were predictable based on prior work [7–10], and duly noted for the Flint outbreak [1,3,5], precise patterns of the outbreak in relation to large health care facilities versus residential single family homes are still the subject of scientific and public interest [5,11].

We recently examined *L. pneumophila* growth in simulated glass water heaters with either cross-linked polyethylene (PEX) or copper pipe, Detroit tap water (sourced from Lake Huron) or treated Flint River water, and an initial influent pH 7.3 with continuous mixing representing hot water recirculation often present in large buildings [12]. After one year, very high levels of *L. pneumophila* (2.6–3.0 log CFU/mL) were observed in all treated Flint River water conditions with PEX pipe and with copper pipe when there was even low levels of phosphate corrosion control present (2.9 $\log_{10}$ CFU/mL). By contrast, all treated Flint River water conditions with copper pipe and no phosphate corrosion control had 1–2 log lower levels of *L. pneumophila*, likely due to biotoxicity of copper, as directly evidence by an inverse correlation ($R^2$ = 0.85–0.95) between *L. pneumophila* and measured soluble copper or $Cu^{+2}$ [12].

Premise plumbing, and its resident microbiome, is highly sensitive to variation in pipe material, pH, disinfection conditions, and stagnation. All of these factors can produce important synergistic or antagonistic effects [13]. For example, a 0.5-unit higher pH in the influent, would be expected to reduce bacteriostatic effects of copper pipe on resident microbes due to reduced concentrations of $Cu^{+2}$ and soluble copper [14,15]. Likewise, complete stagnation has sometimes been associated with greatly reduced growth of *Legionella* versus either completely-mixed (i.e., water recirculation) or more frequent flow [16–19] conditions, whereas the opposite effect is expected if the water has high levels of disinfectant or very high temperature [20]. The presence of free chlorine has also been associated with lower levels of *Legionella* in general [21–23] and with LD incidence, in particular, during the Flint Water Crisis [1,3,5]. Thus, it is of interest to evaluate the extent to which recent phenomena observed to be at play in Flint's premise plumbing [12] hold true under a broader range of relevant conditions. Specifically, conditions with a higher influent pH (7.8–8.5), stagnation (which is more common in residences than large building hot water systems), and the free chlorine levels representative of before, during, and after the Flint Water Crisis.

Here, we evaluated how the corrosive treated Flint River water and the less corrosive Detroit tap water interacted with unlined iron water mains, and then subsequently with the premise plumbing pipe materials into which that water flowed, to influence levels of disinfectant and the propensity for *Legionella* growth (Figure 1). The overarching hypothesis was that the lack of corrosion control of Flint River water would cause higher iron and lower chlorine after contact with unlined iron pipe mains (Figure 1), creating conditions less likely to disinfect *Legionella* when this water flowed into stagnant PEX plumbing. Copper pipe, which has the potential to either catalyze chlorine decay and thereby hinder disinfection [24–26], or release antimicrobial soluble copper ions and enhance disinfection, was compared to a control with PEX pipe for all conditions. The expectation was that *Legionella* would survive best in treated Flint River water with PEX versus copper due to little or no chlorine delivery, but that the converse would be true in Detroit tap water if copper pipe catalyzed chlorine decay and the high levels of corrosion control would virtually eliminate passive disinfection by copper (Figure 1). This study provides important insight into interactive effects of water chemistry and pipe material in affecting the trajectory of community-wide LD outbreak.

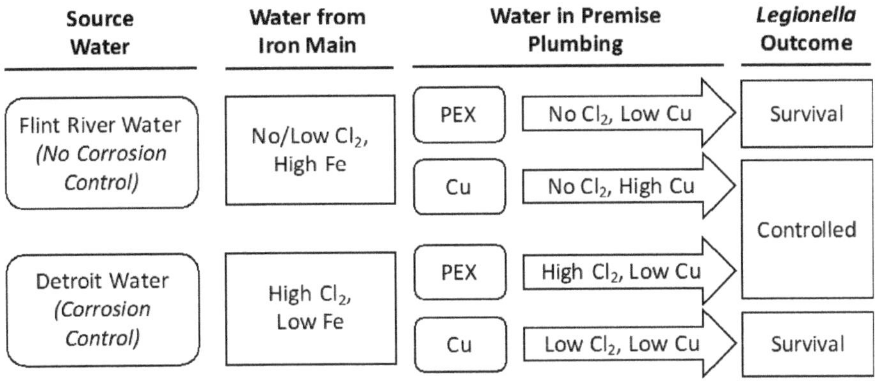


**Figure 1.** Experimental framework and specific hypotheses for this study. Corrosivity of the municipal water supply influences levels of chlorine and iron in the water delivered to the premise plumbing. Premise plumbing materials, cross-linked polyethylene (PEX) or copper (Cu), further alter the water chemistry and overall propensity for *Legionella* to be controlled or to survive. Corrosion of copper and iron will consume free chlorine, whereas plastic materials have little or no chlorine demand. Corrosive water will also release soluble copper ions from copper pipe, especially in stagnant premise plumbing conditions. Elevated levels of either copper or chlorine can control *Legionella*.

## 2. Materials and Methods

### 2.1. Source Water Treatment

Raw water was directly collected from the Flint River at GPS coordinates 43.018230, −83.693944. Lake Huron-sourced drinking water (Detroit tap water) was collected after > 5 min flushing from the tap of a residential Flint home. Raw Flint River water and Detroit tap water were both collected on 18 August 2016, 21 September 2016, 11 October 2016, 21 November 2016, and 27 January 2017 and express shipped to Blacksburg, Virginia in 30-L containers. Additional raw Flint River water samples were collected and shipped express on 8 February 2017 and 11 March 2017. All collected water was stored at 4 °C prior to preparation for experiments.

Water treatments applied to raw Flint River water during the crisis were simulated in the laboratory. These included 56 mg/L ferric chloride for coagulation, 10 min of stirring for flocculation, 159 mg/L lime as $Ca(OH)_2$ for softening, followed by another 15 min of flocculation. The water was subsequently settled for 4 h and filtered through a column of glass wool to simulate sand filtration. Water treated in this manner was designated as "treated Flint River" water (Figure 2). Working stocks (10–20 L) of treated Flint River water and Detroit tap water were stored at room temperature (23 °C) until the supply was exhausted.

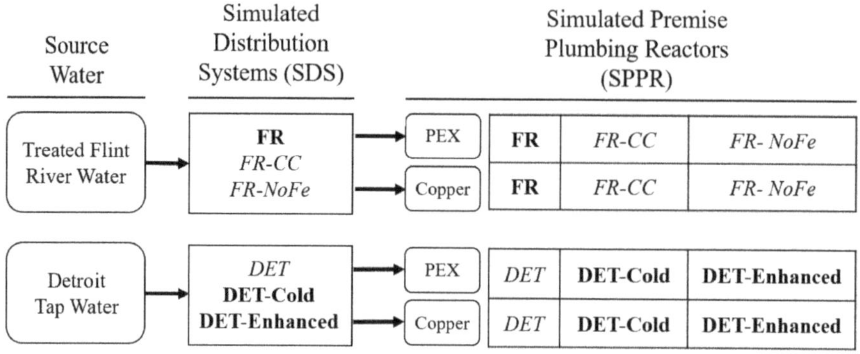


**Figure 2.** Experimental design from source water to simulated distribution systems (SDSs) to simulated premise plumbing reactors (SPPRs). Source water was treated and stored in 10–30 L batches until fed to SDSs. Each SDS condition was set up in a flask containing 330 mL of source water, an iron wire (except for *FR-no Fe*), a stir bar, and 3.0 or 3.5 mg/L chlorine. Bold conditions were designed to replicate scenarios found before (**DET-Cold**), during (**FR**), or after the Flint Water Crisis (**DET-Enhanced**). Conditions in *italics* were designed to simulate hypothetical scenarios if corrosion control had been implemented or if water had not flowed through unlined iron pipe. After the SDSs simulation was completed, the water was fed to corresponding SPPRs containing either PEX (n = 3) or copper (n = 3). The total number of SPPRs was 36.

## 2.2. Simulated Distribution Systems: Chlorination and Corrosion

Six SDS conditions served to reproduce distributed waters that either occurred (conditions designated in **bold** font) under conditions relevant to the Flint Water Crisis or its aftermath or hypothetical scenarios if corrosion control had been implemented or if water had not flowed through unlined iron pipe (conditions designated in *italics*) (Figure 2). In five of the six conditions, the practical influence of unlined iron distribution system pipe was simulated by addition of an iron wire to flasks mixing each water for 3 h. Treated Flint River water conditions included a condition with the omission of phosphate corrosion control (as was the case during the crisis) (**FR**), a hypothetical condition if 1 mg/L as $PO_4$-P orthophosphate corrosion control had been implemented (*FR-CC*), and a condition without any phosphate corrosion control or iron corrosion (i.e., no iron wire) (*FR-NoFe*) representing some sections of Flint served by newer concrete lined iron or PVC distribution system pipe during the crisis. Detroit tap water conditions examined the pre-crisis effect of Lake Huron-sourced water with lower distribution system temperature (**DET-Cold**) containing 2.5 mg/L orthophosphate $PO_4$-P, the post-crisis water with enhanced doses of chlorine and additional phosphate (3.5 mg/L chlorine and 4.0 mg/L orthophosphate) to assist with system recovery once Flint switched back to Detroit-sourced water (**DET-Enhanced**), and a hypothetical condition if normal Detroit distribution water with 2.5 mg/L orthophosphate had been as warm as treated Flint River water during summer months (*DET*).

## 2.3. General SDSs Water Preparation

Sodium hypochlorite (10% diluted Clorox™ bleach, the Clorox Company, Oakland, CA, USA) was added to 330 mL of each water condition until an initial stable target of 3 mg/L free chlorine residual was obtained (the only exception being a higher residual of 3.5 mg/L in **DET-Enhanced**), followed by the SDSs in 500 mL glass flasks containing magnetic stir bars and mixing 400 rpm for three hours. In all conditions, except *FR-NoFe*, the presence of iron pipe was simulated in the SDS with a 12 cm length of 99% 2 mm diameter iron wire (approximately, 7.6 $cm^2$ Fe surface per liter of water) and orthophosphate was added to achieve corrosion control targets of 1 (*FR-CC*), 2.5 (*DET*, **DET-Cold**), or 4.0 mg/L (**DET-Enhanced**).

### 2.4. Premise Plumbing

#### 2.4.1. Simulated Premise Plumbing Reactors (SPPRs)

Following the SDS step, waters were transferred to 100 mL borosilicate glass bottles (36 total) designed to simulate changes occurring in water as it ages in premise plumbing (SPPR, Figure 2). Each SPPR was equipped with either eight pieces of 20 mm × 10 mm cross-sectional PEX (n = 18) or solid copper (n = 18) pipe material. Pipe coupons had been aged in the bottles for six years in prior experiments, described elsewhere [8–10,27], which provided a benefit of well-aged premise plumbing pipe materials and mature biofilms at the start of the experiment.

#### 2.4.2. Initializing the SPPRs

All 36 SPPRs were conditioned prior to the experiment, by dosing a homogeneous aliquot of reactor effluents according to pipe material, followed by an acclimation phase of 50% water volume changes with treated Flint River water every three days for 101 days. This water change frequency and volume simulated a low use, high-stagnation condition considered to be conducive to *Legionella* growth in premise plumbing [8]. On Day 14, the SPPRs were inoculated with three environmental *L. pneumophila* isolates from Flint, MI buildings at a total concentration of 10,000 colony forming units per milliliter (CFU/mL). The inoculum was composed of a mixture of *L. pneumophila* serogroup 1 as well as two non-serogroup 1 isolates. Inoculum concentration was determined by optical density readings of *L. pneumophila* colonies scraped from agar plates, resuspended in Nanopure water, and measured at 600 nm using a 4500 HACH spectrophotometer (Hach Company, Loveland, CO, USA).

To avoid introduction of *Legionella* spp. that may have been present in the water shipments once the experiment was in progress, effluent SDS waters were monitored for survival of culturable *Legionella* prior to their addition to corresponding SPPRs. In no case was detectable culturable *Legionella* present after chlorination of the water and incubation in the SDSs.

#### 2.4.3. Water Changes with SDS Conditions

Following inoculation and a 101-day conditioning period with treated Flint River water, 50% water changes were performed every 3–4 days for 175 days. SPPRs were reproducibly inverted five times for each water change to resuspend any settled material, and 50% of the volume was decanted and replaced with water from one of the six SDS conditions (Figure 2). Each of the six SDS conditions were tested in triplicate copper or PEX SPPRs. Reactors were incubated under static conditions at 37 °C between water changes. Thus, the experimental design included 6 SDS conditions × 2 pipe materials × 3 replicates = 36 total SPPRs.

Culturable *L. pneumophila* were enumerated as colony forming units per deciliter (CFU/mL) on Buffered Charcoal Yeast Extract (BYCE) agar (Remel, Lenexa, KS, USA) supplemented with 3 g/L glycine, 0.4 g/L L-cysteine, 80,000 units/L of polymyxin B sulfate, 0.001 g/L vancomycin, and 0.08 g/L cycloheximide. Initially, water was directly taken from SPPRs and plated onto BYCE; however, once CFUs dropped below detection of direct plating of 1 mL, 50 mL of effluent SPPR water was filter concentrated using 0.22 μM pore size mixed-cellulose ester membranes (Millipore, Billerica, MA, USA) and resuspended in 5 mL of Nanopure water prior to plating (1 mL). Water from each reactor was plated in triplicate. Plates were incubated at 37 °C for 5 days, after which *L. pneumophila* colonies were counted and CFU/mL were calculated. Direct plates with 0.02 CFU/mL were considered below detection. When no *L. pneumophila* colonies were detected from 50 mL concentrates, counts were considered below detection, resulting in a detection limit of 0.001 CFU/mL.

### 2.5. Culture Confirmation

To gain insight into the types of *Legionella* that persisted through the experiment, colonies visually determined as *Legionella* and non-*Legionella* species were picked from plates after 5 days of incubation at 37 °C for polymerase chain reaction confirmation. Polymerase chain reaction was used to confirm *Legionella* spp. (i.e., genus), *L. pneumophila*, and serogroup 1 using established primers and protocols [28,29].

## 2.6. Water Quality Analyses

Influent SDSs, effluent SDSs (influent SPPRs), and effluent SPPR waters were analyzed on Days 0, 9, 20, 72, 87, 126, 131, and 153. Inorganics, including dissolved and particulate iron and copper, were measured by inductively coupled plasma–mass spectrometry (ICP-MS) following 2% acidification with nitric acid. Total organic carbon (TOC) was measured according to standard method 5310 C using a persulfate-ultraviolet detection by a Sievers Model 5300 C (General Electric Company, Boston, MA, USA). pH was measured using an Oakton 110 series meter (Cole Parmer, Count Vernon Hills, IL, USA).

Free chlorine was measured using a 4500 HACH spectrophotometer (Loveland, CO) according to 4500-Cl standard method. To examine the kinetics of chlorine in the various water conditions used in this study, chlorine decay tests were performed on source waters (treated Flint River water and Detroit tap water) in non-reactive glass containers, on SDS water conditions with iron wire according to the experimental design (Figure 2), and after the SDSs waters were added to the SPPRs.

## 2.7. Data Analysis

Statistical tests were performed using R Studio (Version 1.0.153). A Shapiro–Wilk normality test was performed and none of the data were normally distributed. Arithmetic means were calculated for displaying results due to the high proportion of non-detect values in the dataset. Wilcoxon rank sum and Kruskal–Wallis rank sum tests with post-hoc Tukey tests were performed to determine statistical correlations. Wilcoxon tests were used for *Legionella* culture data (log transformed), pipe material, iron, and copper data, whereas the Kruskal–Wallis test was used to determine significance of chlorine data between SPPRs. Significance was set at a $p$ value $\leq 0.05$.

# 3. Results and Discussion

## 3.1. Simulated Treatment and Distribution Reproduced Key Factors of Pre-, During-, and Post-Crisis Flint Water

### 3.1.1. Treated Source Waters Employed in this Experiment

To recreate water quality conditions in Flint, influent water conditions were simulated by treating raw Flint River water in the lab and collecting Lake Huron-sourced water from a well-flushed tap in Flint post crisis (Detroit tap water). The unaltered pH of treated Flint River water ranged between 7.84 and 8.57, while Detroit tap water ranged from 7.96 to 8.06, which recreated the stable pH observed when Flint was using Detroit water and the more variable pH when using Flint River water in 2014 [3,12].

The source water was added in 300-mL aliquots to six glass flasks (3 with Detroit tap water, 3 with treated Flint River water) with iron wire and mixed for 3 h to simulate six different conditions in distribution systems (SDSs). Just prior to being added to the SDSs, the source waters (treated Flint River water and Detroit tap water) were chlorinated, achieving an initial disinfectant residual of 3.10 mg/L $Cl_2$ (Table 1 section B). Additional chlorine was added to only the **DET-Enhanced** SDS condition to achieve a higher average initial residual of 3.80 ± 0.19 mg/L (Table 1 section B). The possible short-term role of cooler temperature during distribution while on Detroit tap water was tested in this work with the **DET-Cold** SDS condition, held at an average of 18.3 ± 1.4 °C compared to an average 21.8 ± 1.3 °C of the other five SDS conditions (**FR**, *FR-NoFe*, *FR-CC*, *DET*, **DET-Enhanced**) (Table 1 section B,C). This ~3 °C difference served to recreate the reported average summer water temperature of 19.9 ± 2.24 °C (pre-crisis, Detroit) and 22.6 ± 2.14 °C (during crisis, Flint River) (Table 1 section A) [3].

**Table 1.** Relevant Flint Water Crisis data compared to simulated treatment, distribution, and premise plumbing conditions.

| **A: Flint Water Crisis Field Data** * | | | | **B: Simulated Distribution System Influent** ^ | | | | | | **C: Simulated Distribution System (T, $Cl_2$, Fe) or Simulated Premise Plumbing Reactor (Cu) Effluent** + | | | | | |
|---|---|---|---|---|---|---|---|---|---|---|---|---|---|---|---|
| Condition | Pre-Crisis | Crisis | Post-Crisis | **DET -Cold** | **FR** | **DET -Enhanced** | *FR-CC* | *FR-NoFe* | *DET* | **DET -Cold** | **FR** | **DET -Enhanced** | *FR-CC* | *FR-NoFe* | *DET* |
| Temp (°C) | 20 ± 2.2 | 23 ± 2.1 | 18 ± 2.4 | 18 | 22 | 22 | 22 | 22 | 22 | 18 ± 1.4 | 22 ± 1.3 | 22 ± 1.3 | 22 ± 1.3 | 22 ± 1.3 | 22 ± 1.3 |
| Chlorine (mg/L $Cl_2$) | 0.50 ± 0.19 | 0.28 ± 0.24 | 0.38 ± 0.16 | 3.10 ± 0.20 | 3.10 ± 0.21 | 3.80 ± 0.19 | 3.10 ± 0.21 | 3.10 ± 0.21 | 3.10 ± 0.20 | 1.28 ± 0.23 | 0.26 ± 0.23 | 1.87 ± 0.70 | 0.17 ± 0.20 | 1.27 ± 0.49 | 1.07 ± 0.52 |
| Flushed Total Iron (μg/L) | UNK | 208 (21–340) | 42.4 (0–130) | 92.1 (80–110) | 15.4 (2.0–47) | 92.1 (80–110) | 15.4 (2.0–47) | 15.4 (2.0–47) | 92.1 (80–110) | 1300 (490–2500) | 2300 (590–5200) | 900 (360–2000) | 3100 (690–9500) | 15.4 (2.0–47) | 2400 (710–6000) |
| Total Copper (μg/L) | UNK | 129 (14–380) | 46 (3.0–140) | 12 (5.7–20) | 8.0 (4.1–15) | 8.0 (5.7–20) | 8.0 (4.1–15) | 8.0 (4.1–15) | 12 (5.7–20) | 962 (510–1390) | 2380 (1790–3120) | 510 (170–1100) | 1200 (520–2000) | 650 (530–850) | 880 (640–1400) |

* Section (**A**): Representative chemical mean, ± standard deviation, (5–95 percentile range where available) for peak Legionnaires' Disease (LD) months of June–September for the indicated stage of the Flint Water Crisis. Representative distribution system temperature data are reported in Rhoads et al. (2017), chlorine data are from monitoring station 6, iron data are from citizen science sampling of flushed water from the same 150 homes in August 2015 (crisis) and August 2017 (post-crisis), and copper water crisis data are from a subset of first draw samples from homes that records indicate had at least partial copper service lines (n = 79). UNK = unknown. ^ Section (**B**): Same parameters as in Section (A) were measured in the influent to the SDSs mean (5th–95th percentile) (n = 10 samples over a 10-month period) (Ambient laboratory set point reported for temperature data). Bold conditions SDSs were designed to simulate actual conditions found during the crisis. Conditions in italics were designed to simulate hypothetical scenarios. "DET" conditions received Detroit tap water influent. "FR" conditions received treated Flint River influent. + Section (**C**): Mean and standard deviation of temperature and chlorine and mean (5th–95th percentile) of iron effluent from SDSs (i.e., influent to the SPPRs) and mean (5th–95th percentile) of copper in the effluent of copper pipe SPPRs.

3.1.2. SDSs Chlorine

The effluent water collected following the 3-h SDSs reaction time (Figure 2) successfully replicated known trends in chlorine residuals observed in the Flint water distribution system before, during, and after the water crisis. To assess inherent chlorine demand prior to the SDSs step, treated Flint River and Detroit tap waters were aliquoted to non-reactive glass containers without iron. The chlorine residual in treated Flint River water dropped from ~3 to ~1 mg/L in 180 min, presumably due to relatively high levels of organic matter (5.2 ± 0.03 mg/L TOC), whereas there was little to no decay occurred in the Detroit tap water (1.2 ± 0.03 mg/L TOC) over the same time period (Figure 3A). The addition of iron wire to simulate unlined iron pipe corrosion during distribution further reduced chlorine residuals in conditions with both treated Flint River water and Detroit tap water as influents (Figure 3B). However, while some residual was consistently detected in the Detroit tap water effluents after simulated distribution (DET, **DET-cold, DET-Enhanced**; 0.5–1 mg/L $Cl_2$ after 180-min exposure), treated Flint River water conditions (**FR**, FR-CC, FR-no Fe) generally had no detectable residual (Figure 3B). Condition *FR-NoFe* is not shown in Figure 3B because no iron wire was added to the SDSs for that condition.

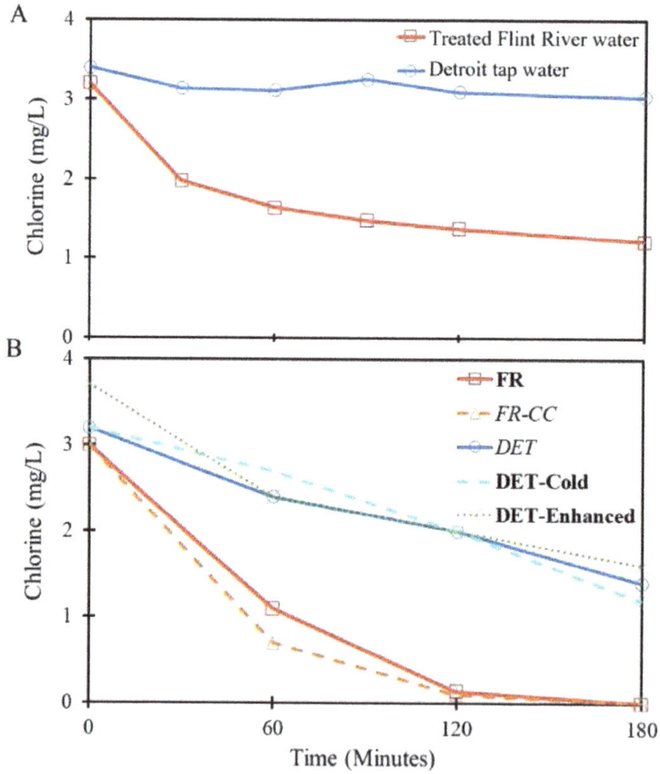


**Figure 3.** Representative $Cl_2$ decay in source water and simulated distribution systems (SDSs). (**A**) Control experiment of chlorine decay of treated Flint River and Detroit tap water in non-reactive glass reactors without iron. (**B**) Representative results in different SDSs conditions: **FR**, treated Flint river water aged with iron wire; *FR-CC*, treated Flint River water with added corrosion control and aged with iron wire; *DET*, Detroit tap water aged with iron wire; **DET-Cold**, Detroit tap water incubated at cooler temperature with iron wire; **DET-Enhanced**, Detroit tap water with additional corrosion control and initial elevated chlorine levels.

While there was variability due to seasonal changes in the source water and variable iron wire corrosion rates throughout the experiment, the mean chlorine concentration (n = 43) after incubation in the SDSs exhibited a general trend of (lowest to highest): *FR-CC* ≈ **FR** < *FR-NoFe* ≈ *DET* ≈ **DET-Cold** < textbfDET-Enhanced (Table 1). Based on a Kruskal–Wallis rank sum test, the mean chlorine concentrations across the SDS conditions were significantly different ($p$ value $< 2 \times 10^{-16}$), while a pairwise post-hoc Tukey test further confirmed specific differences between conditions indicated by a "<" sign in the above trend analysis (all $p$ values ≤ 0.009).

Overall, key expectations were also recreated with respect to known trends resulting from water chemistry and corresponding chlorine residual in SDSs effluent. Specifically, the SDSs successfully reproduced chlorine residuals comparable to those during the crisis of 0.28 ± 0.24 mg/L (at Flint city monitoring station 6) [30], compared to levels of 0.26 ± 0.23 mg/L in our treated Flint River water simulation (**FR** condition, Table 1 section C). SDS conditions also successfully simulated pre-crisis (**DET-Cold**) and post-crisis (**DET-Enhanced**) high chlorine, with actual values only 1 mg/L higher than measured during the pre- or post-crisis conditions (Table 1 section C). Both conditions with treated Flint River water and iron present (**FR** and *FR-CC*) occasionally had undetectable chlorine residuals under the conditions tested, whereas *FR-NoFe* and all conditions with Detroit tap water consistently had a measurable chlorine disinfectant residual following simulated distribution, as hypothesized (Figure 1). Iron has been shown to decay chlorine residual in typical drinking waters [31], but the chlorine decay observed in the SDS step was accelerated beyond what is typical due to the corrosivity of the treated Flint River water and lack of corrosion control.

#### 3.1.3. SDSs Iron and Corrosion Control

Known benefits of corrosion control (**FR** vs. *FR-CC*; **FR** vs. *DET*) in terms of hindered iron release and maintenance of higher chlorine residuals in the actual Flint distribution system (Table 1) were not achieved in these simplistic simulations. Based on a prior study [2], the addition of phosphate corrosion control to treated Flint River water reduced iron weight loss by 5.1 times compared to that observed in treated Flint River water without phosphate, while also reducing chlorine decay rates. Further, iron corrosion rates were 8.6 times lower in Detroit tap water with corrosion control versus treated Flint River water without corrosion control, a trend confirmed by our citizen science field sampling throughout Flint in August 2015 versus August 2017 (Figure 1) [3,5]. However, the corrosion control simulation applied to the SDSs in this study did not produce known significant differences in mean effluent iron (i.e., **FR**, *FR-CC*, and *DET*; Table 1 section C). The only condition with relatively low iron in this work was treated Flint River water without any iron present (*FR-NoFe*), in which mean iron was 15.4 ± 19.4 µg/L compared to the 60.5 ± 212 µg/L observed in August 2017 flushed water samples collected in Flint (Table 1).

We were aware that the simple approach applied here would not effectively replicate impacts of iron corrosion control, given that phosphate inhibition of iron corrosion and associated chlorine decay can sometimes require 6–12 months to produce expected benefits even under continuous-flow conditions in water mains, and even longer under more stagnant conditions [32,33]. In this seven-month simulation, the iron was only exposed to the water approximately 6 h each week, which translates into seven days total exposure of iron to the target water over the entire study. Thus, the analysis that follows considers that this particular aspect of the simulation is not representative of what occurred in the field.

### 3.2. *Simulated Premise Plumbing Reactors Reproduced Key Water Chemistry Trends of Pre-, During, and Post-Crisis Flint Water*

#### 3.2.1. SPPRs Chlorine

After the effluents from the SDSs were transferred to the SPPRs, the 50% water change produced an immediate dilution of chlorine. Beyond dilution, there is an immediate chlorine demand from the combination of aged pipe material, pipe surface area, biofilm, and water within each reactor [24,26]. Notably, all SDS conditions, except **DET-Enhanced**, experienced an immediate chlorine demand within the first 10 min, which exceeded the 50% loss expected from dilution. **FR** and *FR-CC* never received any initial chlorine residual flowing into the SPPRs (Figure 4), whereas *FR-NoFe* retained a low, but detectable, chlorine residual (>0.1 mg/L) for a period of 60–120 min in the PEX SPPRs and 1–10 min in the copper SPPRs (data not shown). Chlorine was reduced in the Detroit tap water SPPRs to below 0.1 mg/L within 30–60 min in both PEX and copper SPPRs, while chlorine was maintained above 0.1 mg/L for up to 120 min in **DET-Cold with** PEX (Figure 4) versus just 10–30 min in the corresponding copper SPPRs. Chlorine residuals in the **DET-Enhanced** conditions after 120 min were 0.92 and 0.38 mg/L in the PEX and copper SPPRs, respectively (Figure 3). In some instances, chlorine was still detectable in **DET-Enhanced** SPPRs after 24 h. As a general rule, when detectable chlorine residual was present in the influent to the SPPRs, levels were higher in the system with PEX after 10 min than in the equivalent system with copper, consistent with the overall hypothesis of this work and our prior research [25] (Figure 1).

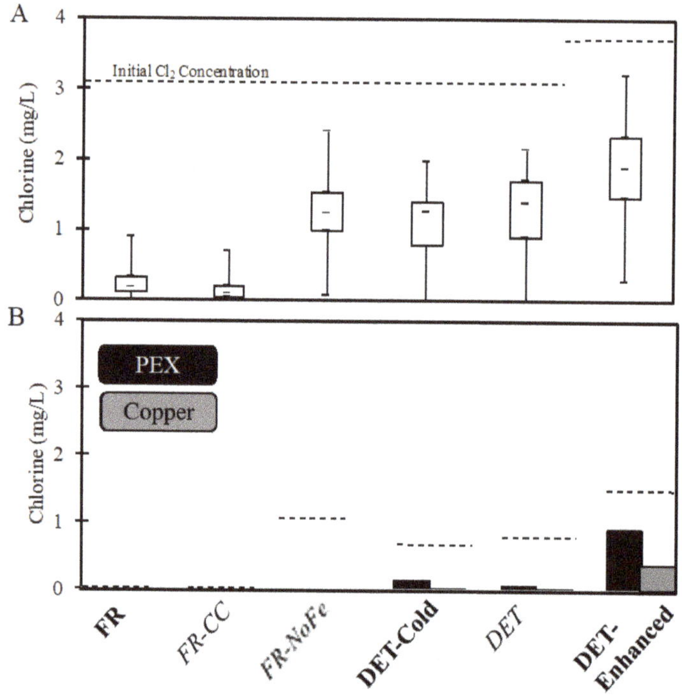


**Figure 4.** Chlorine residuals (**A**) after 3 h contact time in the simulated distribution systems (SDSs) and (**B**) 120 min after the effluent from the SDSs were fed to the simulated premise plumbing reactors (SPPRs) (50% fresh SDSs water with 50% remaining SDSs following incubation in the SPPRs the previous cycle). Dashed lines indicate the calculated initial chlorine level added for each water or reactor type. Bars represent the maximum and minimum, the upper and lower bounds of the box are the first and third quartiles, and the median is indicated by the internal dash. The detection limit was 0.02 mg/L.

Overall, these results illustrate quick and drastic decay of the chlorine disinfectant residual in premise plumbing systems (Figure 4) that added to decay in the distribution systems (Figure 3). The U.S. Environmental Protection Agency (EPA) recommends that a free chlorine disinfectant residual be detectable (often, >0.1 or >0.2 mg/L) in 95% of distribution system samples [34], which has previously been acknowledged not to be adequate for the reduction of *Legionella* in large buildings, single-family homes, or small buildings [35]. The results from the Detroit tap water SPPRs (*DET*, **DET-Cold**, **DET-Enhanced**) demonstrate that the residual was detectable (>0.1 mg/L $Cl_2$) after 120 min in the SPPRs only when the disinfectant residual entering much higher than 0.2 mg/L $Cl_2$ (Figure 4).

#### 3.2.2. SPPRs Copper

Mean total copper in influent water (Table 1B) to all SPPRs was consistently < 15 μg/L and mean effluent copper from PEX reactors was consistently < 100 μg/L, in accordance with the assumption that the only source of copper was traces from plumbing used to collect well-flushed raw water samples in the field. However, SPPRs containing copper pipe consistently produced effluent with total mean copper concentrations > 700 μg/L (Table 1 section C) and were statistically higher than the copper concentrations from the PEX SPPRs effluent ($p$ value = $2 \times 10^{-16}$). Further, effluents from SPPRs receiving treated Flint River water, simulating Flint water during the crisis (**FR**), contained higher total copper than each of the corresponding conditions representing Detroit water (**DET-Cold**, **DET-Enhanced**, $p$ values < $4 \times 10^{-5}$), consistent with the lack of copper corrosion control during this time period. Thus, the laboratory simulation successfully reproduced the trends in copper levels characteristic of pre-/during and post-crisis conditions in Flint, where mean, 5th, and 95th percentiles of first draw copper during the crisis were approximately three times higher than post-crisis (Table 1 section A,C).

### 3.3. *Legionella pneumophila Response to Simulation of Water Chemistry and Premise Plumbing Material*

After 101 days of SPPRs acclimation to the SDSs water and the cross-inoculation period, culturable *L. pneumophila* numbers were greatest in SPPRs receiving treated Flint River waters, particularly the PEX condition (Figure 5A). Among all SPPRs containing PEX material, *L. pneumophila* CFU/mL were significantly higher in treated Flint River water-sourced (Figure 5A) compared to Detroit-sourced (Days 25–175; $p < 0.05$; Figure 5C) water. This demonstrated the main hypothesized effect of treated Flint River water being more conducive than Detroit tap water to maintaining viable *L. pneumophila*, at least in the absence of copper pipe (Figure 1).

Throughout the study, *L. pneumophila* persisted at low numbers in the copper SPPRs fed with treated Flint River waters, but at levels significantly lower than in the PEX SPPRs ($p$ value = 0.03). In particular, the copper SPPRs receiving *FR-NoFe* influent water sustained little to no culturable *L. pneumophila* beyond 75 days (Figure 5B). During our field sampling at the height of the summer 2015 LD outbreak, the pH was 7.0 in Flint homes, in which case the higher acidity likely caused much higher levels of $Cu^{+2}$ in premise plumbing [3,12,14] than in this study at a pH of 7.8. Together, the results from the treated Flint River water copper SPPRs at a pH 7.8 suggest that, under conditions of corrosive influent water (including the *FR-CC* water as evidenced by chlorine decay tests; Figure 3B), the elevated copper concentrations can enhance reduction of *L. pneumophila*, consistent with the overarching hypotheses of this study (Figure 1).

The Detroit tap water conditions provide a simulation of what occurred before the city of Flint switched to the treated Flint River water (April 2014), and after they switched treated Flint River water back to the Detroit municipal water supply on 16 October 2015. Initial culturable counts of *L. pneumophila* declined under all Detroit tap water conditions within the first month (Figure 5C,D), which was consistent with our field data [1,3] and the corresponding drop in LD incidence after switching back to Detroit water [5]. The loss of culturable *L. pneumophila* was greatest in PEX SPPRs for all three Detroit tap water SDS conditions, with culturable *L. pneumophila* falling below detection by Day 25, with a single exception (Figure 5C). Interestingly, *L. pneumophila* fared better in Detroit SPPRs containing copper pipe relative to those containing PEX pipe material, consistent with a previous study in the same Detroit tap water PEX reactors with no chlorine and additional mixing [12]. Plate counts in Detroit SPPRs containing copper remained near 1 CFU/mL from Day 25 until the end of the experiment. This suggests the additional chlorine demand and reduced chlorine levels caused by the presence of copper can actually increase growth of *Legionella* as hypothesized (Table 1 and Figure 1). Notably, the colony counts in the **DET-Enhanced** condition dropped below detection after just 100 days (Figure 5D), illustrating that extra chlorine can overcome the demand exerted by copper and more effectively control *L. pneumophila*. The presence of orthophosphate in Detroit tap water also would have reduced toxicity of the *L. pneumophila* to copper as reported in earlier studies [12,24,36].

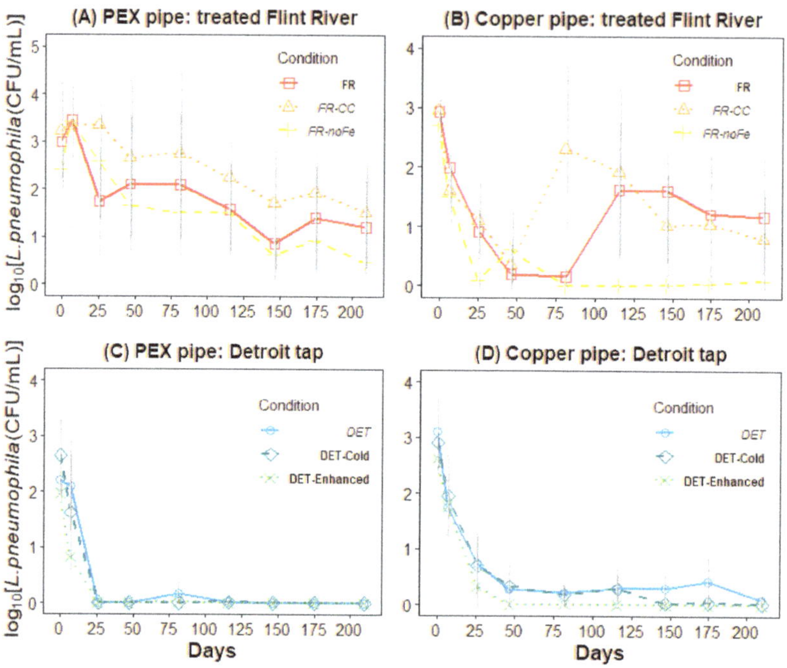


**Figure 5.** The effects of water source, distribution conditions, and pipe material on culturable *L. pneumophila* in simulated premise plumbing reactors (SPPRs). Effluent log transformed average *L. pneumophila* numbers (CFU/mL) and standard deviations from SPPRs receiving simulated distribution system (SDS) water corresponding to: (**A**) treated Flint River water with PEX pipe coupons; (**B**) treated Flint River water with copper pipe coupons; (**C**) Detroit tap water with PEX pipe coupons; and (**D**) Detroit tap water with copper pipe coupons. All SPPRs influent waters were spiked to an initial target concentration of 3 mg $Cl_2$/L and aged three hours under completely-mixed conditions at 23 °C in the presence of an iron wire (SDS step), except *FR-NoFe* conditions, which had no iron wire, and **DET-Cold** condition, which was incubated at 17 °C. *FR-CC* additional orthophosphate corrosion control agent added at 1 mg/L, **DET-Enhanced** additional $CL_2$ added at 3.5 mg/L and orthophosphate at 2.5 mg/L.

### 3.3.1. Isolate Analysis

To gain insight into whether a single strain or mixture of strains of *Legionella* persisted under the various conditions, 56 representative isolates collected from the SPPRs on Days 0, 47, 82, 175, and 210 were subject to genotypic screening by PCR. Interestingly, it was observed that *L. pneumophila* survived through the end of the experiment across all simulations, except **DET-Enhanced** (Table S1). Of the *L. pneumophila* strains recovered from SPPRs fed with treated Flint River water by Day 150, 56/56 were characterized as serogroup 1. By Day 210, **FR** and all other water conditions supporting *Legionella* were confirmed to contain a mixture of serogroup 1 and non-serogroup 1 *L. pneumophila*, based on PCR detection of the wzm gene. The ability to multiple serogroups of *L. pneumophila* to persist under the various conditions of this experiment suggests that the trends observed in response to the water conditions employed in this study were robust across multiple serogroups.

3.3.2. Chlorine Disinfectant

Chlorine is by far the most widely implemented and relied upon secondary disinfectant residual applied in the U.S. and around the world. Previous studies have indicated that concentrations >0.5 mg/L are sometimes sufficient for limiting detectable levels of *Legionella* in large building plumbing water systems [37,38]. As demonstrated above, once added to SPPRs, chlorine concentrations immediately decreased and were often no longer detectable after 60 min following a water change. Further, in all SPPRs, except those receiving **DET-Enhanced** water, chlorine concentrations decreased below 0.5 mg/L within the first 10 min. The persistence of culturable *Legionella* in all conditions, except the **DET-Enhanced**, essentially proves that adequate disinfection was not achieved for the other 10 conditions

3.3.3. Pipe Material: PEX and Copper

Comparing copper versus PEX pipe materials further illustrated the potential for premise plumbing conditions to mediate the effects of the distribution system water chemistry. Copper is of particular interest because of its known antimicrobial properties towards *L. pneumophila* [28,39–41]. Ironically, the lack of corrosion control, which triggered higher iron and lower chlorine residuals and exacerbated *Legionella* problems in some portions of the Flint distribution system during the crisis [3], also released high levels of copper that might have helped to control *Legionella* growth in some buildings and homes. In this study, regardless of influent water sources, copper SPPRs displayed rapid initial loss of culturable *L. pneumophila* subsequent to feeding the SDS waters. However, low colony counts persisted in most SPPR conditions containing copper at the higher pH, representative of the summer 2014 outbreak, for nearly six months. The initial reduction of *Legionella* CFUs in copper SPPRs may have occurred through limited antimicrobial properties of aged copper at the relatively high pH employed in this study, as observed by others [12,24,42,43].

As *Legionella* are facultative intracellular bacteria, they are capable of residing in biofilms and replicating in more than 20 species of amoebae [44,45]. In harsh environments, such as the surface of copper pipe material or variable disinfection levels of chlorine, existing biofilms may serve as a protective environment for *Legionella* to shelter from disinfectants [26,43,45]. However, this study suggests that such protective mechanisms can be overcome by higher chlorine residual and contact time. This is demonstrated by the **DET-Enhanced** copper condition, which included increased chlorine and corrosion control agent and contained no culturable *L. pneumophila* from Day 50 forward (with the exception of a single colony obtained from a 50 mL sample of water on Day 82) (Figure 5D).

Interactions between copper and influent chlorine residuals may provide an environment that allows for the persistence of *Legionella* in premise plumbing, especially if corrosion control maintains $Cu^{+2}$ below thresholds controlling *Legionella* [14,19,46]. A previous study determined that disinfectant potential of free chlorine can be affected by the age of copper pipe material [28]. In both treated Flint River water and Detroit tap water, the presence of aged copper reduced the capacity of the SPPRs to maintain measurable free chlorine residuals (Figure 5B), but the presence of orthophosphate corrosion control may have allowed *Legionella* to persist at higher levels in *DET* or **DET-Cold** water, whereas the corresponding condition with higher copper in **FR** helped reduce *Legionella*.

*3.4. Experimental Conditions: Hypothetical Effects of Stagnation, Chlorine, and Elevated pH*

Comparing results of the current study to those obtained to those from a companion study using the same SPPRs, but with continuous mixing, slightly lower pH, and no disinfectants (Figure 6) [12] can provide insight into overarching effects of experimental conditions selected for this study. Notably, *L. pneumophila* in the treated Flint River water with PEX was 2.5-log higher in the previous study. We hypothesize that this is attributable to continuous mixing versus stagnation, since the small pH change is not expected to be influential for PEX, and chlorine levels delivered to this reactor from the SDSs in this work were consistently undetectable. A much smaller increase of 0.4–1.0 log was observed for treated Flint River water copper conditions in the prior study versus this work, mostly likely due to the lower pH significantly enhancing bacteriostatic impacts of cupric ion. In any case, the higher *L. pneumophila* with more mixing is consistent with prior results in recirculating versus non-recirculating systems [17,19,47,48] in which there were warm temperatures and low disinfectant residuals. In Detroit tap water conditions, comparable conditions between the two studies consistently led to non-detectable *L. pneumophila*, except for copper pipe with both mixing and corrosion control, where relatively low levels of *L. pneumophila* persisted.

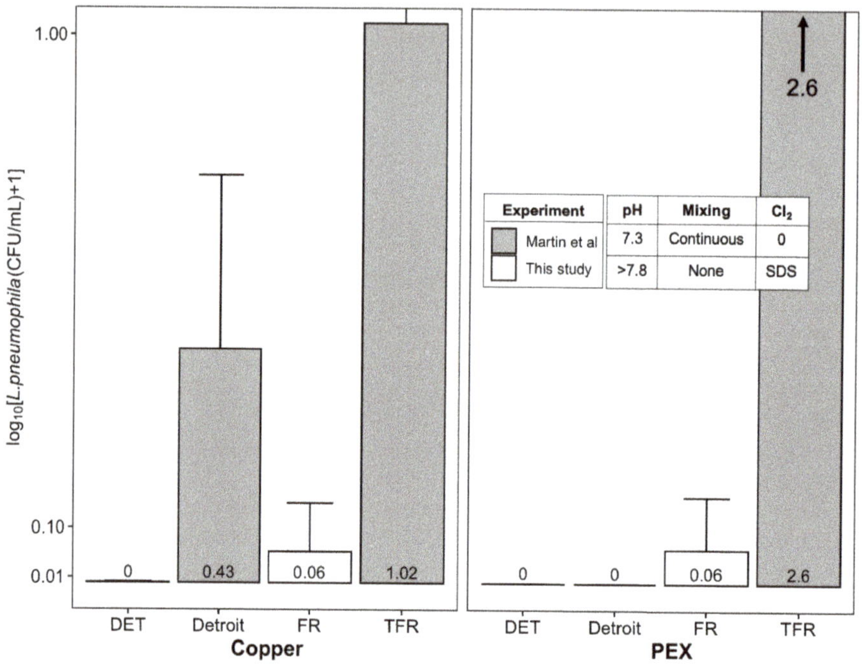


**Figure 6.** Comparison of culturable *L. pneumophila* (CFU/mL) from control reactors in this study (Day 210 samples only) compared to Martin et al. (sampled at one year). Each bar in the plot represents the average of triplicate reactors with the error bars showing the standard deviation. The Detroit condition contained the same influent as DET, with no iron and 2.5 mg/L of orthophosphate. The treated Flint River (TFR) condition contained the same influent as FR except without the iron aging step.

## 4. Conclusions

The findings of this study are consistent with the understanding that *L. pneumophila* is not uncommon in municipal water flowing into buildings [49], but a range of water chemistry and premise plumbing conditions and disinfectant residuals can prevent their proliferation. Persistent disinfectant, including chlorine, is known to be a critical factor in reducing *Legionella* risk [35,37,39], and a recent study confirmed predicted associations between low levels of chlorine in Flint's distribution system and observed incidence of LD [5]. Consistent with our prior in-field observations [1,3], we further demonstrate under controlled laboratory conditions the importance of considering interactive effects with flow and pipe materials, particularly with respect to relative water corrosivity and influence on residual chlorine levels, in keeping *Legionella* levels low. Indeed, many individual factors can act as "two-edged swords" in terms of their net effect of controlling versus enhancing *Legionella* growth, depending on the status of other factors. For example, copper pipe achieves its best antimicrobial efficacy without corrosion control, but absence of corrosion control also leads to elevated iron and depleted chlorine residual, which in turn enhance *Legionella* growth. Such interactive effects can help explain why prior studies reported relatively low levels of *Legionella* in single family homes, which tend to have greater stagnation and more copper in water from copper service lines and plumbing, compared to large multi-story buildings during the Flint Water Crisis.

**Supplementary Materials:** The following are available online at http://www.mdpi.com/2076-0817/9/9/730/s1, Table S1: PCR confirmation of unique morphologies that were identified morphologically as *Legionella*.

**Author Contributions:** Conceptualization, O.R.S., M.A.E., and A.P.; methodology, O.R.S., M.A.E., and A.P.; software, O.R.S. and R.L.M.; validation, O.R.S. and R.L.M.; formal analysis, R.L.M. and O.R.S.; investigation, O.R.S., R.L.M., M.A.E., and A.P.; resources, O.R.S.; data curation, R.L.M. and O.R.S.; writing—original draft preparation, R.L.M., O.R.S., M.A.E. and A.P.; writing—review and editing, O.R.S., R.L.M., M.A.E., and A.P.; visualization, O.R.S. and R.L.M.; supervision, M.A.E. and A.P.; project administration, M.A.E. and A.P.; and funding acquisition, M.A.E. and A.P. All authors have read and agreed to the published version of the manuscript.

**Funding:** This research was funded by National Science Foundation for CBET via Awards 1336650, 1556258, and 1706733. The Alfred P. Sloan Foundation Microbiology of the Built Environment Program, and the Science and Engineering of the Exposome (SEE) Center supported by the Institute for Critical Technology and Applied Science at Virginia Tech, also provided financial support on this project.

**Acknowledgments:** We would like to thank Stephen Wolbert and LeeAnne Walters for their cooperation in collecting and shipping Flint water, as well as Laurel Strom, D. Otto Schwake, Emily Garner, Jacob Seymour, and Jeffrey Parks for their technical and laboratory assistance, and the Virginia Tech Library Open Access Subvention Fund.

**Conflicts of Interest:** The authors declare no conflict of interest.

## References

1. Schwake, D.O.; Garner, E.; Strom, O.R.; Pruden, A.; Edwards, M.A. Legionella DNA Markers in Tap Water Coincident with a Spike in Legionnaires' Disease in Flint, MI. *Environ. Sci. Technol. Lett.* **2016**, *3*, 311–315. [CrossRef]
2. Pieper, K.J.; Tang, M.; Edwards, M.A. Flint Water Crisis Caused By Interrupted Corrosion Control: Investigating "Ground Zero" Home. *Environ. Sci. Technol.* **2017**, *51*, 2007–2014. [CrossRef] [PubMed]
3. Rhoads, W.J.; Garner, E.; Ji, P.; Zhu, N.; Parks, J.; Schwake, D.O.; Pruden, A.; Edwards, M.A. Distribution System Operational Deficiencies Coincide with Reported Legionnaires' Disease Clusters in Flint, MI. *Environ. Sci. Technol.* **2017**, *51*, 11986–11995. [CrossRef] [PubMed]
4. Masten, S.J.; Davies, S.H.; Mcelmurry, S.P. Flint Water Crisis: What Happened and Why? *J. Awwa* **2016**, *108*, 22–34. [CrossRef]
5. Zahran, S.; McElmurry, S.P.; Kilgore, P.E.; Mushinski, D.; Press, J.; Love, N.G.; Sadler, R.C.; Swanson, M.S. Assessment of the Legionnaires' disease outbreak in Flint, Michigan. *Proc. Natl. Acad. Sci. USA* **2018**. [CrossRef]
6. Smith, A.F.; Huss, A.; Dorevitch, S.; Heijnen, L.; Arntzen, V.H.; Davies, M.; Robert-Du Ry van Beest Holle, M.; Fujita, Y.; Verschoor, A.M.; Raterman , B.; et al. Sources of the Outbreak of Legionnaires' Disease in Genesee County, Michigan, in 2014 and 2015. *Environ. Health Perspect.* **2019**, *127*, 127001. [CrossRef]
7. Garrison, L.E. Vital Signs: Deficiencies in Environmental Control Identified in Outbreaks of Legionnaires' Disease—North America, 2000–2014. *Mmwr. Morb. Mortal. Wkly. Rep.* **2016**, *65*.
8. Proctor, C.R.; Dai, D.; Edwards, M.A.; Pruden, A. Interactive effects of temperature, organic carbon, and pipe material on microbiota composition and Legionella pneumophila in hot water plumbing systems. *Microbiome* **2017**, *5*, 130. [CrossRef]
9. Williams, K.; Pruden, A.; Falkinham, J.O.; Edwards, M. Relationship between Organic Carbon and Opportunistic Pathogens in Simulated Glass Water Heaters. *Pathogens* **2015**, *4*, 355–372. [CrossRef]
10. Wang, H.; Masters, S.; Falkinham, J.O.; Edwards, M.A. Distribution System Water Quality Affects Responses of Opportunistic Pathogen Gene Markers in Household Water Heaters. *Environ. Sci. Technol.* **2015**, *49*, 8416–8424. [CrossRef]
11. Michigan Department of Human and Health Services (MDHHS). MDHHS Response to Flint Area Community Health and Environment Partnership Proceedings of the National Academy of Sciences Article. 2018. Available online: https://content.govdelivery.com/attachments/MIDHHS/2018/02/05/file_attachments/953946/MDHHS%2BResponse%2Bto%2BFACHEP%2BProceedings%2Bof%2Bthe%2BNational%2BAcademy%2Bof%2BSciences%2BArticle%2BFINAL.pdf (accessed on 3 September 2020).

12. Martin, R.L.; Strom, O.R.; Song, Y.; Mena-Aguilar, D.; Rhoads, J.W.; Pruden, A.; Edwards, M. Copper Pipe, Lack of corrosion control, and uncontrolled pH influenced the trajectory of the Flint Legionnaires' Disease outbreak. *Environ. Sci. Technol.* (under review).
13. Ji, P.; Parks, J.; Edwards, M.A.; Pruden, A. Impact of Water Chemistry, Pipe Material and Stagnation on the Building Plumbing Microbiome. *PLoS ONE* **2015**, *10*, e0141087. [CrossRef] [PubMed]
14. Lin, Y.-s.E.; Vidic, R.D.; Stout, J.E.; Victor, L.Y. Negative effect of high pH on biocidal efficacy of copper and silver ions in controlling Legionella pneumophila. *Appl. Environ. Microbiol.* **2002**, *68*, 2711–2715. [CrossRef] [PubMed]
15. Zhang, Y.; Griffin, A.; Edwards, M. Nitrification in premise plumbing: Role of phosphate, pH and pipe corrosion. *Environ. Sci. Technol.* **2008**, *42*, 4280–4284. [CrossRef] [PubMed]
16. Liu, Z.; Lin, Y.E.; Stout, J.E.; Hwang, C.C.; Vidic, R.D.; Yu, V.L. Effect of flow regimes on the presence of Legionella within the biofilm of a model plumbing system. *J. Appl. Microbiol.* **2006**, *101*, 437–442. [CrossRef] [PubMed]
17. Rhoads, W.J.; Ji, P.; Pruden, A.; Edwards, M.A. Water heater temperature set point and water use patterns influence Legionella pneumophila and associated microorganisms at the tap. *Microbiome* **2015**, *3*, 1–13. [CrossRef]
18. Shen, Y.; Guillermo, L.M.; Nicolas, D.; Dao, J. Role of Biofilm Roughness and Hydrodynamic Conditions in Legionella pneumophila Adhesion to and Detachment from Simulated Drinking Water Biofilms. *Environ. Sci. Technol.* **2015**, *49*, 4274. [CrossRef]
19. Mathys, W.; Stanke, J.; Harmuth, M.; Junge-Mathys, E. Occurrence of Legionella in hot water systems of single-family residences in suburbs of two German cities with special reference to solar and district heating. *Int. J. Hyg. Environ. Health* **2008**, *211*, 179–185. [CrossRef]
20. Rhoads, W.; Pruden, A.; Edwards, M. Convective Mixing in Distal Pipes Exacerbates Legionella pneumophila Growth in Hot Water Plumbing. *Pathogens* **2016**, *5*, 29. [CrossRef]
21. Mouchtouri, V.A.; Goutziana, G.; Kremastinou, J.; Hadjichristodoulou, C. Legionella species colonization in cooling towers: Risk factors and assessment of control measures. *Am. J. Infect. Control* **2010**, *38*, 50–55. [CrossRef]
22. Kim, B.R.; Anderson, J.E.; Mueller, S.A.; Gaines, W.A.; Kendall, A.M. Literature review—efficacy of various disinfectants against Legionella in water systems. *Water Res.* **2002**, *36*, 4433–4444. [CrossRef]
23. Muraca, P.; Stout, J.E.; Yu, V.L. Comparative assessment of chlorine, heat, ozone, and UV light for killing Legionella pneumophila within a model plumbing system. *Appl. Environ. Microbiol.* **1987**, *53*, 447–453. [CrossRef] [PubMed]
24. Nguyen, C.; Elfland, C.; Edwards, M. Impact of advanced water conservation features and new copper pipe on rapid chloramine decay and microbial regrowth. *Water Res.* **2012**, *46*, 611–621. [CrossRef] [PubMed]
25. Nguyen, C.K.; Powers, K.A.; Raetz, M.A.; Parks, J.L.; Edwards, M.A. Rapid free chlorine decay in the presence of $Cu(OH)_2$: Chemistry and practical implications. *Water Res.* **2011**, *45*, 5302–5312. [CrossRef] [PubMed]
26. Zheng, M.; He, C.; He, Q. Fate of free chlorine in drinking water during distribution in premise plumbing. *Ecotoxicology* **2015**, *24*, 2151–2155. [CrossRef] [PubMed]
27. Dai, D.; Proctor, C.R.; Williams, K.; Edwards, M.A.; Pruden, A. Mediation of effects of biofiltration on bacterial regrowth, Legionella pneumophila, and the microbial community structure under hot water plumbing conditions. *Environ. Sci. Water Res. Technol.* **2018**, *4*, 183–194. [CrossRef]
28. Merault, N.; Rusniok, C.; Jarraud, S.; Gomez-Valero, L.; Cazalet, C.; Marin, M.; Brachet, E.; Aegerter, P.; Gaillard, J.; Etienne, J. Specific real-time PCR for simultaneous detection and identification of Legionella pneumophila serogroup 1 in water and clinical samples. *Appl. Environ. Microbiol.* **2011**, *77*, 1708–1717. [CrossRef]
29. Kozak, N.A.; Benson, R.F.; Brown, E.; Alexander, N.T.; Taylor, T.H.; Shelton, B.G.; Fields, B.S. Distribution of lag-1 alleles and sequence-based types among Legionella pneumophila serogroup 1 clinical and environmental isolates in the United States. *J. Clin. Microbiol.* **2009**, *47*, 2525–2535. [CrossRef]
30. Flint Water Treatment Plant. Monthly Operating Reports. Available online: https://www.michigan.gov/flintwater/0,6092,7-345-76292_76364-377816--,00.html (accessed on 28 January 2020).
31. Miller, C.M.; Bower, K.C. Iron Oxide Enhanced Chlorine Decay and Disinfection By-Product Formation. *J. Environ. Eng.* **2006**, *132*, 1609–1616. [CrossRef]

32. McNeill, L.S.; Edwards, M. Iron pipe corrosion in distribution systems. *J. Am. Water Work. Assoc.* **2001**, *93*, 88–100. [CrossRef]
33. McNeill, L.S.; Edwards, M. Phosphate inhibitor use at US utilities. *J. Am. Water Work. Assoc.* **2002**, *94*, 57–63. [CrossRef]
34. Wahman, D.G.; Pressman, J.G. Distribution System Residuals—Is "Detectable" Still Acceptable for Chloramines? *J. Awwa* **2015**, *107*, 53–63. [CrossRef]
35. National Academies of Sciences and Engineering. *Management of Legionella in Water Systems*; The National Academies Press: Washington, DC, USA, 2019. [CrossRef]
36. Lin, Y.E.; Vidic, R.D. Possible phosphate interference with copper–silver ionization for Legionella control. *J. Hosp. Infect.* **2006**, *62*, 119. [CrossRef] [PubMed]
37. Orsi, G.B.; Vitali, M.; Marinelli, L.; Ciorba, V.; Tufi, D.; Del Cimmuto, A.; Ursillo, P.; Fabiani, M.; De Santis, S.; Protano, C.; et al. Legionella control in the water system of antiquated hospital buildings by shock and continuous hyperchlorination: 5 years experience. *Bmc Infect. Dis.* **2014**, *14*, 394. [CrossRef] [PubMed]
38. Alary, M.; Joly, J.R. Factors contributing to the contamination of hospital water distribution systems by Legionellae. *J. Infect. Dis.* **1992**, *165*, 565–569. [CrossRef] [PubMed]
39. Rhoads, W.J.; Pruden, A.; Edwards, M.A. Interactive Effects of Corrosion, Copper, and Chloramines on Legionella and Mycobacteria in Hot Water Plumbing. *Environ. Sci. Technol.* **2017**. [CrossRef]
40. Triantafyllidou, S.; Lytle, D.; Muhlen, C.; Swertfeger, J. Copper-silver ionization at a US hospital: Interaction of treated drinking water with plumbing materials, aesthetics and other considerations. *Water Res.* **2016**, *102*, 1–10. [CrossRef]
41. Lin, Y.E.; Stout, J.E.; Vidic, R.D. Disinfection of water distribution systems for Legionella. *Semin. Respir. Infect.* **1998**, *13*, 147–159.
42. Van der Kooij, D.; Veenendaal, H.R.; Scheffer, W.J.H. Biofilm formation and multiplication of Legionella in a model warm water system with pipes of copper, stainless steel and cross-linked polyethylene. *Water Res.* **2005**, *39*, 2789–2798. [CrossRef]
43. Gião, M.S.; Wilks, S.A.; Keevil, C.W. Influence of copper surfaces on biofilm formation by Legionella pneumophila in potable water. *Biometals* **2015**, *28*, 329. [CrossRef]
44. Falkinham, J.O.; Pruden, A.; Edwards, M. Opportunistic Premise Plumbing Pathogens: Increasingly Important Pathogens in Drinking Water. *Pathogens* **2015**, *4*, 373–386. [CrossRef] [PubMed]
45. Lau, H.; Ashbolt, N. The role of biofilms and protozoa in Legionella pathogenesis: Implications for drinking water. *J. Appl. Microbiol.* **2009**, *107*, 368–378. [CrossRef] [PubMed]
46. Buse, H.Y.; Lu, J.; Struewing, I.T.; Ashbolt, N.J. Preferential colonization and release of Legionella pneumophila from mature drinking water biofilms grown on copper versus unplasticized polyvinylchloride coupons. *Int. J. Hyg. Environ. Health* **2014**, *217*, 219. [CrossRef] [PubMed]
47. Bédard, E.; Paranjape, K.; Lalancette, C.; Villion, M.; Quach, C.; Laferrière, C.; Faucher, S.P.; Prévost, M. Legionella pneumophila levels and sequence-type distribution in hospital hot water samples from faucets to connecting pipes. *Water Res.* **2019**, *156*, 277–286. [CrossRef] [PubMed]
48. Moore, M.R.; Pryor, M.; Fields, B.; Lucas, C.; Phelan, M.; Besser, R.E. Introduction of monochloramine into a municipal water system: Impact on colonization of buildings by Legionella spp. *Appl. Environ. Microbiol.* **2006**, *72*, 378–383. [CrossRef]
49. Donohue, M.J.; O'Connell, K.; Vesper, S.J.; Mistry, J.H.; King, D.; Kostich, M.; Pfaller, S. Widespread Molecular Detection of Legionella pneumophila Serogroup 1 in Cold Water Taps across the United States. *Environ. Sci. Technol.* **2014**, *48*, 3145–3152. [CrossRef]

*pathogens*

MDPI

*Article*

# Biogeography and Environmental Drivers of *Legionella pneumophila* Abundance and Genotype Composition across the West Bank: Relevance of a Genotype-Based Ecology for Understanding *Legionella* Occurrence

**Ashraf R. Zayed [1,2], Suha Butmeh [2], Marina Pecellin [1], Alaa Salah [2], Hanna Alalam [2], Michael Steinert [3], Manfred G. Höfle [1], Dina M. Bitar [2] and Ingrid Brettar [1,*]**

1 Department of Vaccinology and Applied Microbiology, Helmholtz Centre for Infection Research (HZI), Inhoffenstrasse 7, 38124 Braunschweig, Germany; ashraf.rashadlab@gmail.com (A.R.Z.); marinapecellin@gmail.com (M.P.); manfred.hoefle@helmholtz-hzi.de (M.G.H.)

2 Department of Microbiology and Immunology, Al-Quds University, Abu-Dis, University Street, Jerusalem 19356, Palestine; salbutmeh@staff.alquds.edu (S.B.); alaa.hisham@uks.de (A.S.); Hanna.Alalam@gu.se (H.A.); dbitar@staff.alquds.edu (D.M.B.)

3 Department of Life Sciences, Institute of Microbiology, Technical University of Braunschweig, Universitätsplatz 2, 38106 Braunschweig, Germany; m.steinert@tu-bs.de

* Correspondence: ibrettar@web.de

Received: 15 October 2020; Accepted: 27 November 2020; Published: 1 December 2020

**Abstract:** The West Bank can be considered as a high-risk area for *Legionella* prevalence in drinking water due to high ambient temperature, intermittent water supply, frequent pressure loss, and storage of drinking water in roof containers. To assess occurrence of *Legionella* species, especially *L. pneumophila*, in the drinking water of the West Bank, the drinking water distribution systems of eight hospitals were sampled over a period of 2.3 years covering the seasonal cycle and the major geographic regions. To gain insight into potential environmental drivers, a set of physico-chemical and microbiological parameters was recorded. Sampling included drinking water and biofilm analyzed by culture and PCR-based methods. Cultivation led to the isolation of 180 strains of *L. pneumophila* that were genotyped by Multi-Locus Variable Number of Tandem Repeat Analysis (MLVA). Surprisingly, the abundance of culturable *L. pneumophila* was low in drinking water of the sampling sites, with only three out of eight sites where *Legionella* was observed at all (range: 30–500 CFU/Liter). By contrast, biofilm and PCR-based analyses showed a higher prevalence. Statistical analyses with physico-chemical parameters revealed a decrease of *L. pneumophila* abundance for water and biofilm with increasing magnesium concentrations (>30 mg/L). MLVA-genotype analysis of the *L. pneumophila* isolates and their spatial distribution indicated three niches characterized by distinct physico-chemical parameters and inhabited by specific consortia of genotypes. This study provides novel insights into mechanisms shaping *L. pneumophila* populations and triggering their abundance leading to an understanding of their genotype-specific niches and ecology in support of improved prevention measures.

**Keywords:** MLVA-genotypes; ecotype; groundwater; environmental factors; magnesium; niche

## 1. Introduction

*Legionella* is a genus comprising about 60 species mostly of aquatic origin and with a large fraction of pathogenic species [1]. The most relevant species for human health and artificial freshwater systems is *L. pneumophila*. It is the most frequent causative agent of an atypical pneumonia, Legionnaires' disease (LD), and Pontiac fever, a self-limiting flu-like disease [2]. Anthropogenic fresh water systems are considered as the major source for *Legionella* infections [3,4]. Co-infections with aquatic bacteria of LD patients hint on a co-transfer of bacteria from freshwater to the patient presumably via protozoa and/or their bacteria containing vesicles [5,6]. Phin et al. [7] concluded in a review on the worldwide epidemiology on LD, that the lack of thorough knowledge on the ecology on *Legionella* is a major obstacle for management and prevention measures against *Legionella* infections.

The West Bank is a semi-arid region in the Middle East with hot and dry summers and cool winters with substantial water scarcity problems. Main precipitation falls in winter leading to an often only partial recharge of groundwater aquifers [8]. The source water for drinking water is mostly groundwater that is pumped into a storage reservoir and chlorinated, before delivered to the drinking water distribution system (DWDS). Due to frequent water shortage and supply interruption, water is stored in private containers, mostly on the roof, by the end users. All these factors may cause hygienic water problems in general and may lead to high *Legionella* abundance in the drinking water as consumed by the end user.

In general, the abundance of *Legionella* is considered to be enhanced by high water temperature and low chlorine concentrations [9–11]. However, recent studies have indicated that the factors triggering *Legionella* abundance might be more complex and that it is of high value to study the ecology of *L. pneumophila* on the genotype level [12,13]. The study by Rodriguez-Martinez et al. [13] performed in a similar climatic region (around Haifa, Israel) hints to a link between the genotypic composition of *L. pneumophila* and the abundance of *Legionella* species. In their study, a specific genotype (MLVA-Gt 4), closely related to *L. pneumophila* strain Paris, showed a correlation with very high *Legionella* plate counts and surprisingly low water temperature (mean value of 20.6 °C). The authors suggested that specific genotypes may act as triggers of high *Legionella* abundance, even at low temperature. In addition, genotype assessment is also of interest for tracking the source of *Legionella* infections, and the virulence itself is considered to be genotype-dependent [14,15].

The standard method for genotype assessment of *L. pneumophila* is sequence based typing (SBT) [1]. Sobral et al. and Visca et al. [12,16] showed that multi-locus variable number of repeat analysis (MLVA), a less laborious method, can be very well matched with the sequence types (ST) generated by SBT. In addition, the genotypes of ST1 are better resolved by MLVA. ST1 resolution had been shown to be of special relevance for drinking water [13], with respect to the environment as well as for virulence aspects [13,14]. Furthermore, ST1 is considered as of high relevance for artificial freshwater systems and human health on a global scale [17,18]. To achieve the needed high-resolution genotyping of *L. pneumophila* isolates, a MLVA-method with 13 loci was used by combining the loci of the MLVA-methods of Sobral et al. and Pourcel et al. [12,19]. Details on the results of MLVA-genotyping of the West Bank strains are given by Zayed et al. [20].

The overall aim of the study was to understand the relationship of the *L. pneumophila* genotypes and the environmental drivers determining their niches and abundances. This aim was pursued by seasonal assessment of *Legionella* abundance in water and biofilm in eight drinking water sampling sites, i.e., the DWDS of eight hospitals, covering the whole West Bank. The abundance of *L. pneumophila* was assessed by cultivation and PCR concomitantly with a record of relevant bacteriological and physico-chemical parameters of the drinking water during a period of 2.3 years. The 180 *L. pneumophila* isolates obtained from water and biofilm were identified by 16S rRNA partial sequencing and high resolution MLVA-genotyping in a previous study [16]. Correlation analysis and principal component analysis (PCA) were used to identify the niches of relevant *L. pneumophila* MLVA-genotypes and to identify environmental drivers of *L. pneumophila* abundance in water and biofilm. The study advocates for a genotype-based ecology of *L. pneumophila* and sheds light on so far not yet considered mechanisms of *L. pneumophila* prevalence at the level of individual genotypes.

## 2. Materials and Methods

### 2.1. Study Sites, Water, and Biofilm Sampling

Drinking water in the West Bank is derived from groundwater, mainly well water, and some provided by springs. Water was provided to most of the sampled sites by the Palestinian Water Authority, except for Ramallah (sampling site D) with Mekorot as provider. Except for site D, water treatment consisted of chlorination of the water in storage sites before provided to the end user. All hospitals had drinking water reservoirs for water storage.

Water samples and biofilm swabs were sampled six times during the period from October 2012 to December 2014 from eight hospitals across the West Bank (Supplementary Figure S1). The six samplings covered twice the main seasons, i.e., spring (March–May), summer (June–August), and autumn (October–December). It should be noted that site D could only be sampled once for spring, summer, and autumn, while all other seven sites were sampled twice for these seasons [21]. Thus, the overall sampling campaign of the eight hospitals comprised 45 samplings.

Sampling was achieved on tap water and biofilm of faucets and shower heads of the drinking water distribution system (DWDS) of eight hospitals (Hospital A-H, Supplementary Figure S1, Table 1) of five cities covering the main regions across the West Bank. Furthermore, samples were taken occasionally from Al-Quds University (AQU) main campus, Abu Dies, Jerusalem (31°45′18.07″ N, 35°15′37.614″ E). These samples from Al-Quds were not included in the overall comparison on *Legionella* ecology and only used for comparison.

**Table 1.** Average *Legionella spp.* and *L. pneumophila* abundance per sampling site in water and biofilm as determined by cultivation and PCR (mean (SD)).

| Sampling Site (North to South) | Coordinates | Water/Culture: *Legionella spp.* CFU/L | Water/PCR: % *L. pneum-ophila* Positive | Biofilm/Culture: % *Legionella spp.* Positive | Biofilm/PCR: % *L. pneumophila* Positive |
|---|---|---|---|---|---|
| Hospital A (Jenin) | 32°27′ N, 35°17′ E | 43.3 (106.1) | 66.7 (51.6) | 14.6 (20.0) | 73.3 (24.2) |
| Hospital B (Nablus) | 32°13′ N, 35°14′ E | 0 (0) | 33.3 (40.8) | 21.4 (10.8) | 63.3 (8.2) |
| Hospital C (Nablus) | 32°13′ N, 35°15′ E | 0 (0) | 16.7 (40.8) | 2.8 (4.5) | 40.0 (25.3) |
| Hospital D (Ramallah) | 31°53′ N, 35°12′ E | 0 (0) | 33.3 (28.9) | 14.5 (21.0) | 66.7 (11.5) |
| Hospital E (East Jerusalem) | 31°46′ N, 35°14′ E | 0 (0) | 0 (0) | 6.4 (7.2) | 60.0 (28.3) |
| Hospital F (Bethlehem) | 31°42′ N, 35°11′ E | 148.0 (229.7) | 100.0 (0.0) | 29.9 (25.8) | 90.0 (11.0) |
| Hospital G (Hebron) | 31°33′ N, 35° 04′ E | 8.3 (20.4) | 100.0 (0.0) | 23.8 (18.8) | 93.3 (16.3) |
| Hospital H (Hebron) | 31°31′ N, 35° 05′ E | 0 (0) | 16.7 (40.8) | 4.5 (6.4) | 60.0 (17.9) |
| **Mean ± *(SD)*** | | **25.0 *(51.9)*** | **45.8 *(38.6)*** | **14.8 *(9.8)*** | **68.3 *(17.3)*** |

Cold and hot water (if available) was collected from faucets of the hospitals drinking water distribution system (DWDS). For cold and hot water sampling specific preselected faucets in the vicinity to the reservoir were used. Hot water was only available at five of the eight hospitals and not at all sampling dates (hot water was not available during sampling in May 2013; total set of 24 hot water samples). There was no water sampling directly from the hospital's reservoir, but only from the DWDS. Biofilm sampling was achieved from a predefined set of biofilm swabs from faucets, showerheads, and hoses. The sampled drinking water from the hospitals was considered as representative for the cities of Jenin, Nablus, Ramallah, Jerusalem, Bethlehem, and Hebron, going from the north towards the south across the West Bank.

### 2.2. Physico-Chemical Analyses of Bulk Water

Cold and hot water samples were analyzed for temperature, pH, conductivity (using probes), and chlorine (Quantofix, Macherey-Nagel GmbH, Düren, Germany) directly upon collection. After being returned to the laboratory, water samples were analyzed for total iron, nitrate, nitrite, ammonia, copper, phosphate, zinc, carbonate hardness, and total hardness using Quantofix sticks. Magnesium and calcium concentrations were measured photometrically using Macherey–Nagel Nanocolor assays. Data on turbidity, bicarbonate, chloride, sulphate, total dissolved solids (TDS), and fluoride were provided by the Palestinian Water Authority.

### 2.3. Cultivation Dependent Analysis of Water and Biofilm

Per sampling date and site, one cold and one hot water sample was collected from the preselected faucets of the DWDS (vicinity to water reservoir) in sterile bottles after a flushing time of 2 min: One liter of each cold and hot water was collected for Heterotrophic Plate Counts (HPC), and one liter of each cold and hot water was collected for *Legionella* counts. To neutralize residual free chlorine, 0.5 mL of 0.1N sodium thiosulphate was added in the sterile bottles for *Legionella* count determination [22].

Concerning *Legionella* plate counts, a 100 mL water sample was filtered onto a membrane filter (membrane solutions, pore size 0.45 μm, diameter 47 mm, Whatman, England) using a sterile filtration unit (Nalgene, Germany). A vacuum of 200 mbar was applied. After filtration, 30 mL of acid buffer (3.9 mL of 0.2 mol/L HCl and 25 mL of sterile 0.2 mol/L KCl were mixed, pH 2.2 ± 0.2) was placed on top of the membrane filter and left for 5 min. The filter was rinsed with 20 mL Page's saline (1.20 g NaCl, 0.04 g $MgSO_4{\cdot}7H_2O$, 0.04 g $CaCl_2{\cdot}2H_2O$), and 1.42 g $Na_2HPO_4$ and 1.36 g $KH_2PO_4$ were dissolved in ten liters of distilled water and autoclaved. The membrane filter was removed from the filtration unit with sterile forceps and placed onto the relevant agar plate. Duplicates of BCYE and GVPC (M809, Himedia, Mumbai India) agar plates were used according to the manufacturer's instructions. The plates were incubated inverted at 37 °C for 10 days. Plates were checked for growth twice (after three and ten days). Final counts of the triplicates were done after ten days with descriptions of the colonies. Detection limit was five CFU/L.

Biofilm swabs were sampled from the anterior surfaces of faucets, showerheads, or shower hoses in all hospital wards. Per sampling and sampling site 20 biofilm swabs were taken, except for the first sampling when 44 swabs were sampled per site to check the variability per sampling site. Biofilm swabs were obtained using transport medium (Copan, Culture swab transport system, Brescia, Italy). Swabs for *Legionella* identification were processed immediately by culturing on GVPC agar (medium M809, Himedia, India) according to ISO 11731:2004 [23]. More details on cultivation dependent analyses are given by Zayed et al. [20].

From all water and biofilm samples with visible growth of *Legionella*-like colonies on agar plates, representative isolates were chosen and purified. Isolates were later characterized by *L. pneumophila*-specific PCR (Primer L1) [24], 16S rRNA gene sequence, serogroup and genotype assignment using MLVA, and a representative subset by sequence-based typing (SBT) (see below). Please note that *Legionella*-like colonies were mostly confirmed by PCR as *L. pneumophila*. As it was a rather rare event that non-*L. pneumophila* colonies were detected, the *Legionella* plate counts can be considered as reflecting the culturable fraction of *L. pneumophila*.

### 2.4. Cultivation-Independent Analysis of Water and Biofilm

Per sampling site and date, five liters of both cold and hot water were collected per sampling and site from the predefined faucet from the DWDS in the vicinity of the reservoir for DNA extraction. Water samples were filtered onto sandwich membrane filters composed of a nucleopore-filter (Nucleopore Track-Etch Membrane, 90 mm diameter, 0.2 μm pore size, Whatman, England) and a glass fiber-microfilter (GF/F) (GFF, 90 mm, Whatman, England) [25]. Filters were stored frozen at −80 °C for later DNA extraction. More details on cultivation-independent analyses are given by Zayed et al. [20].

For biofilms, 5 swabs were taken per sampling and site from the anterior surfaces of faucets, showerheads, or shower hoses using sterile cotton swabs (Cotton Tipped Applicator, Beijing, China). Swabs were stored frozen at −20 °C for later DNA extraction. For the extraction of DNA from the filter sandwiches and the swabs, a modified DNAeasy protocol (Qiagen No.69506, Hilden, Germany) was used [26].

Using the DNA of the extracted water and biofilm samples and strains, PCRs with different targets were carried out as described by [24]: (i) for the detection of any bacteria, universal 16S rRNA gene primers (Com1F, Com2R) were used, and (ii) a *Legionella* genus-specific PCR (primer set Lgsp17F, Lgsp28R) (iii) and a *L. pneumophila*-specific PCR (primer set Lp-16S_246-248F, Lp-16S_246-248R) were applied. On all samples, used for PCR-based analysis, cultivation-dependent analysis for *Legionella* was additionally performed to allow a direct comparison. Sequencing of the 16S rRNA gene of six representative isolates confirmed the identification of *L. pneumophila* (≥99.8% 16S rRNA gene similarities) [20].

### 2.5. Genotyping of *L. pneumophila* Isolates

For molecular genotyping of *L. pneumophila* at the strain level, MLVA-8(12) analysis was performed for 180 isolates. For all details see Zayed et al. [20]. Briefly, DNA extraction was done either directly from living biomass using DNAeasy (Qiagen No. 69504, Hilden, Germany) according to the manufacturer protocol or from biomass applied to FTA cards (Whatman, Sigma-Aldrich, Germany). For the final MLVA thirteen loci were used, i.e., the twelve loci of MLVA-12 (12) plus the one additional locus of MLVA-8 [19] not used in MLVA-12. A subset of MLVA-genotypes was characterized by sequence-based typing (SBT) [1]. The MLVA-8(12) genotypes of the West Bank were compared to the International MLVA database (http://microbesgenotyping.i2bc.paris-saclay.fr/as performed and described in more detail by Pecellín [27].

### 2.6. Statistical Analyses

The GraphPad Prism software v7.0 (Graph-Pad, California, USA), SPSS 20, and multivariate analyses using PRIMER software v7.0.7 were used to perform all statistical analyses. Data are presented as means ± standard deviation (SD). Non-normalized data were normalized. Then, repeated ANOVA tests with post hoc analysis using the Bonferroni test were conducted for determining site differentiation. Out of the 20 determined water parameters, eight were distinct between sampling sites, i.e., *Legionella* plate counts, water turbidity, chloride, sulphate, total dissolved solids (TDS), magnesium, calcium, and calcium/magnesium ratio.

Associations between MLVA-genotypes were calculated using the Similarity Profile Analysis (SIMPROF) [28] based on Spearman rank correlation. To determine the effect of physico-chemical parameters on *L. pneumophila* genotypes, Principal Component Analysis (PCA) was used for visualization of cluster identification. PCA included the 8 parameters distinct for the sampling sites. Only MLVA-genotypes represented by at least three strains were included in the cluster analyses and PCA.

## 3. Results

A period of 2.3 years was covered by six sampling campaigns targeting eight drinking water sites, i.e., DWDS of eight local hospitals, representing different geographic regions of the West Bank (Supplementary Figure S1). The six sampling campaigns targeted the main seasons in Palestine, i.e., spring (March to May), summer (June to August), and autumn (October to December). These seasons were sampled twice from 2012 to 2014. Data on sampling per site and sampling campaign are given in detail in Zayed [21].

The occurrence of *Legionella* species, with emphasis on *L. pneumophila*, in the drinking water of the West Bank was determined by cultivation and molecular approaches. To gain insight into potential environmental drivers, a set of physico-chemical and microbiological parameters were determined. Sampling included bulk water of the DWDS and biofilm analyzed by culture- and PCR-based methods using environmental DNA extracted directly from the sample material.

### 3.1. Physico-Chemical and Microbiological Characteristics of the Drinking Water

The drinking water supplied to the hospitals in the West Bank was mostly derived from groundwater either by wells or springs and stored in a reservoir of each hospital. The supplied drinking water was characterized by high hardness (210–350 mg/L $CaCO_3$-equivalents), high bicarbonate level (170–250 mg/L), high conductivity (400–1000 µS/cm), and a high content of total dissolved solids (TDS) (260–470 mg/L). It contained variable amounts of chloride (20–110 mg/L) and sulfate (10–40 mg/L). On-site analyses of the drinking water retrieved from the hospitals' DWDS showed that a high Mg concentration (21–40 mg/L) in addition to the high Ca concentration (75–100 mg/L) contributed to the high hardness. All water components showed a high regional variability) [21].

In hospital drinking water, pH ranged from 7.6 to 8.4, with an average pH per hospital between 7.6 and 8.0. Temperature of the cold water ranged from 18 to 26 °C and only rarely reached values higher than 27 °C (<9% of total sampling). On average (mean value for all samplings per hospital) cold water temperature per hospital ranged between 21 and 25 °C. Hot water temperature ranged from 30 to 70 °C and had a mean temperature per hospital between 39 °C and 52 °C. Chlorine ranged from 0.1 to 1 mg/L and was only rarely (<7% of total sampling) elevated above these values. On average per hospital chlorine ranged between 0.2 and 0.7 mg/L. Heterotrophic plate counts (HPC) were assessed at 22 and 37 °C. HPC at 37 °C varied from $3 \times 10^2$ to $4 \times 10^5$ CFU/L, with averages per hospital ranging from $1.5 \times 10^4$ to $1.6 \times 10^5$. At 22 °C, HPC were about one order of magnitude lower, with an average per hospital of $1.4 \times 10^3$ to $6.3 \times 10^4$.

### 3.2. Abundance of L. pneumophila in Water and Biofilm as Assessed by Cultivation and PCR-Based Methods

An overview on the *Legionella* abundances in the bulk water and biofilm is given in Table 1. For water samples, *Legionella* plate counts were mostly below detection level, with only three sampling sites out of eight where *Legionella* were detected in a range of 8 to 148 CFU/L (mean value per site), and only one site with more frequent observation of *Legionella* in summer and autumn (hospital F). In biofilm, culturable *Legionella* were detected at all sampling sites. On average, *Legionella* positive swabs per sampling site were 15% ranging from 3 to 30% (mean value per site) (Table 1).

PCR-based detection of *Legionella* spp. and *L. pneumophila* showed a higher fraction of *Legionella*-positive samples. In water samples, at seven out of eight sampling sites *L. pneumophila* was detected with an average detection rate per sampling site of 46% and a range from 0 to 100%. In biofilm samples, *Legionella* spp. was regularly detected at all sites at an average detection rate of 68% ranging from 40 to 93% per site (Table 1.).

For water samples, *L. pneumophila*-specific PCR was more sensitive than plate counts. The observations by culture and PCR were consistent in a way that whenever plate counts were observed, PCR gave positive results, whereas a large set of PCR-positive samples did not show any plate counts (Supplementary Figure S2A).

Culture based detection of *L. pneumophila* in biofilms was much more successful than in water samples. *Legionella* plate counts from water samples were only positive when about half or more of the biofilm swabs were positive for *L. pneumophila* cultivation (Supplementary Figure S2B).

In general, the detection of non-*L. pneumophila* colonies on the agar media used for *Legionella* spp. counts was very rare as assessed by species-specific PCR. The number of *Legionella* spp. counts can therefore be approximately addressed as *L. pneumophila* counts.

### 3.3. Comparison of Abundance of Culturable Legionella in Hot and Cold Water

Hot water was available at five of the eight sampling sites except for spring 2013. Hot water temperature ranged from 30 to 70 °C and had a mean value around 45 °C [21]. At the five sites, sampling was achieved in parallel for hot and cold water. There was no significant difference observed for the *Legionella* plate counts (Supplementary Figure S3). These counts were usually below detection limit in hot water as in cold water. Only for site F (Bethlehem), there was an increased level of *Legionella* counts in summer 2013 and 2014: in cold water 467 CFU/L in 2013 and 421 CFU/L in 2014; in hot water plate counts were similar (508 CFU/L) in 2013, while in 2014 no *Legionella* were detected. Thus, a low level of culturable *Legionella spp.* was observed also for the hot water in this sampling comparison.

### 3.4. Seasonal Dynamics of L. pneumophila in Biofilm and Water

*L. pneumophila* showed an increase in biofilm samples from spring to autumn across all sampling sites. Both culture- and PCR-based methods showed this tendency (Figure 1). For PCR-based detection, the percentage of positive swabs was in general higher than for culture-based detection. Culture-based methods and *L. pneumophila*-specific PCR-based methods showed a good correlation ($r^2$ = 0.83, Figure 1A; correlation with *Legionella* genus-specific PCR: $r^2$ = 0.78). The relative increase was higher for culture from spring to autumn (increase from about 10 to 20% of positive swabs, Figure 1C) compared to the PCR-based detection (increase from about 60 to 80% of positive swabs, Figure 1B). The comparison of culture-based and PCR-based detection of *L. pneumophila* indicated an increase of culturability of *L. pneumophila* from biofilm from spring to autumn (and the respective exposure to higher temperature) from about 15% in spring to about 27% in autumn as shown by the respective ratios (Figure 1D).

In contrast, *L. pneumophila* abundance in bulk water showed a maximum in summer as detected by cultivation and, as a tendency, by *L. pneumophila*-specific PCR (Figure 2B,C). Detection by PCR was more sensitive than plate counts; the correlation between PCR detection and plate counts (CFU/L) in a seasonal comparison was $r^2$ = 0.69 (Figure 2A).

On a seasonal basis, there was no correlation between biofilm and water samples neither by culture nor by PCR-based detection; this is consistent with the observation of different seasonal maxima for biofilm and water. In summary, the seasonal dynamics of the *Legionella* abundance was not strongly pronounced, especially not as detected by PCR.

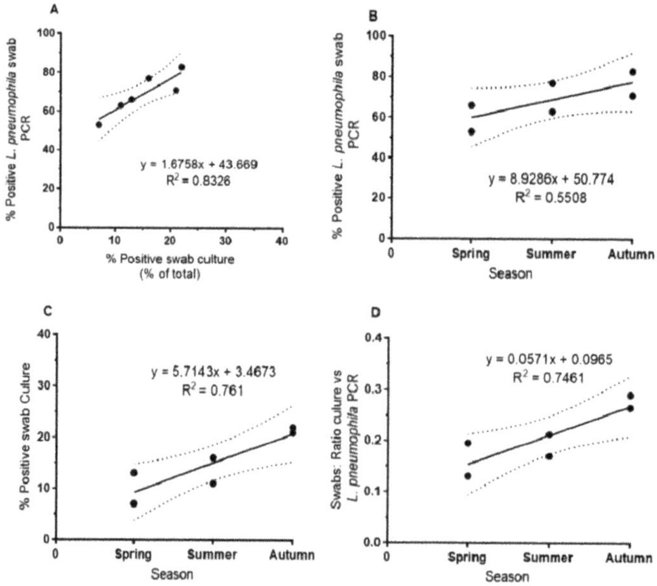


**Figure 1.** Seasonal variation of abundances of *L. pneumophila* in biofilm of eight sampling sites (mean values) of the West Bank sampled from 2012 to 2014. (**A**) Swabs positive by culture vs. swabs positive by *L. pneumophila*-specific (L1-primer) PCR ($p < 0.05$), (**B**) swabs positive by *L. pneumophila*-specific (L1-primer) PCR vs. seasons (NS: Not Significant), and (**C**) swabs positive by culture vs. seasons ($p < 0.05$). (**D**) Ratio of swabs positive by culture vs. swabs positive by *L. pneumophila*-specific PCR vs. seasons ($p < 0.05$) ($n = 45$, mean values $n = 6$).

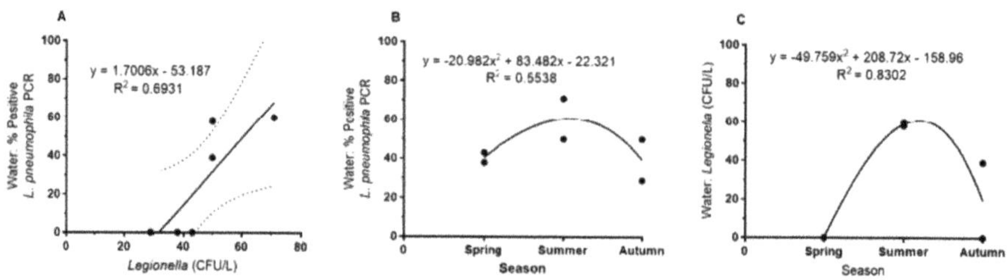


**Figure 2.** Seasonal variation of abundances of *L. pneumophila* in water of eight sampling sites (mean values) of the West Bank sampled from 2012 to 2014. (**A**) *Legionella* plate counts vs. water samples positive by *L. pneumophila*-specific (L1-primer) PCR ($p < 0.05$). (**B**) Water samples positive *by L. pneumophila*-specific (L1-primer) PCR vs. seasons (NS, not significant). (**C**) *Legionella* plate counts of water samples vs. seasons ($n = 45$, mean values $n = 6$) ($p < 0.05$).

### *3.5. Influence of Physico-Chemical and Bacteriological Parameters on Legionella Abundance in Water and Biofilm*

For the assessment of the relationship between *Legionella* abundance in water and biofilm with bacteriological and physico-chemical parameters in water, these parameters were pairwise compared and displayed in a correlation matrix (Supplementary Table S1). Eighteen quantitatively determined parameters were used to define the physico-chemical background of the sampling sites. Heterotrophic plate counts incubated at 25 °C and 37 °C were used as general bacteriological parameters. Culturable *Legionella* counts were used for water and biofilm. PCR-based detection of *L. pneumophila* and the *Legionella* genus were added for biofilm swabs. For the clonal level of *L. pneumophila*, the incidence of the seven most abundant MLVA-genotypes (Gt) and four clonal complexes (VACC) were included in the correlation matrix.

Correlation analyses of the abundance of culturable *L. pneumophila* ("*Legionella* count") in water and biofilm vs. the physico-chemical parameters showed a correlation with the magnesium concentration and the Ca/Mg ratio in the sampled tap water, but not with any other of the analyzed physico-chemical parameters (Supplementary Table S1). Surprisingly, there was no significant correlation with chlorine concentration or temperature. Abundance of culturable *Legionella* in water and biofilm were correlated; furthermore, both parameters were correlated with occurrence of specific *L. pneumophila* genotypes and clonal complexes (VACC). The magnesium concentration in water showed a relationship not only with the abundance of *Legionella* in water and biofilm, but also with some genotypes and clonal complexes. The findings of the correlation analysis are elaborated in more detail in the following paragraphs.

As shown in Figure 3, there was a tight correlation between the Ca/Mg ratio and the Mg concentration for the whole data set (Figure 3A) as well as for the mean of the eight sampling sites (Figure 3B). There was no correlation between Ca and Mg; Ca varied between 103 and 78 mg/L with no correlation with the Mg concentration.

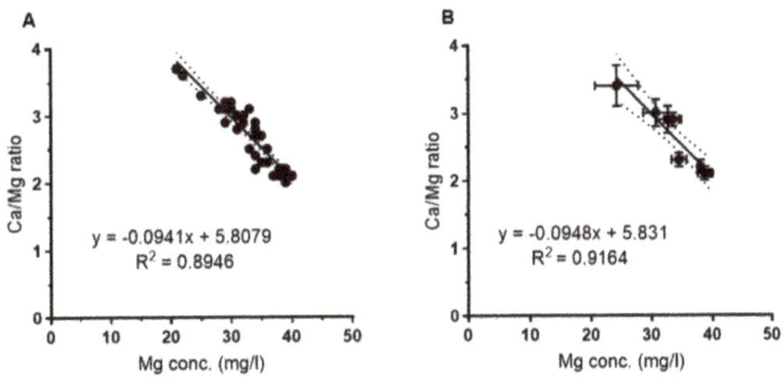


**Figure 3.** (**A**) Correlation between the Ca/Mg ratio and the Mg concentration for the total data set ($n = 45$) and (**B**) for the mean values of the eight sampling sites ($n = 45$, mean values $n = 8$) $p < 0.001$.

For the water samples (Figure 4), the plate counts of *L. pneumophila* showed a negative correlation with increasing Mg concentrations. This was observed for the overall set of samples (regression not shown) as well as for the mean for each sampling site (Figure 4A). For both data sets, the correlation could be best described by a power function with a correlation coefficient of $r^2 = 0.54$ and 0.78 (0.68 for the linear correlation, Figure 4A) respectively. Figure 4B shows the correlation with the Ca/Mg ratio that is best described by an exponential function ($r^2 = 0.63$). Abundance of culturable *L. pneumophila* increased with increasing Ca/Mg ratio as expected from the above described (Figure 3) relationship of Mg with the Ca/Mg ratio. *L. pneumophila*-specific PCR of water samples showed the same tendency, i.e., a negative correlation with increasing Mg concentrations, but was not significant (Figure 4C).

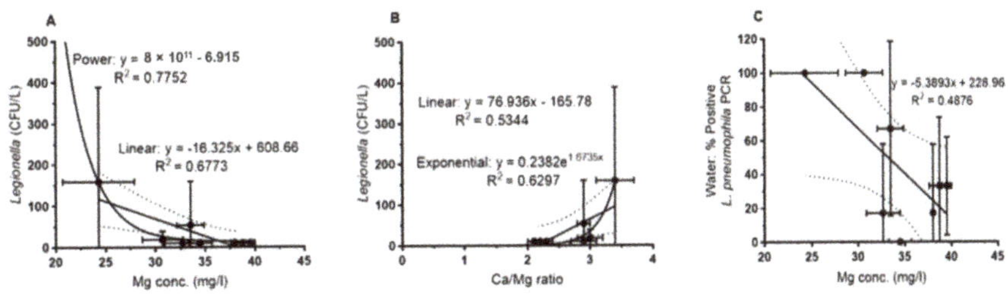


**Figure 4.** Correlation of *L. pneumophila* abundance with Mg and the Ca/Mg ratio of the water samples as detected by plate counts (**A**,**B**) ($p < 0.05$) and *L. pneumophila* specific PCR (**C**) ($n = 45$, mean values $n = 8$) (NS: not significant).

For the biofilm samples, a correlation between the magnesium concentration and the PCR-based detection of *Legionella* spp. and *L. pneumophila* was observed as shown for the mean of each sampling site in Figure 5. By contrast, culture-based analyses of swabs did not show a clear trend. As for the water samples, increasing Mg concentrations yielded a lower percentage of positive biofilm samples (swabs), both for *Legionella* genus-specific PCR (Figure 5A) and *L. pneumophila*-specific PCR (Figure 5B).

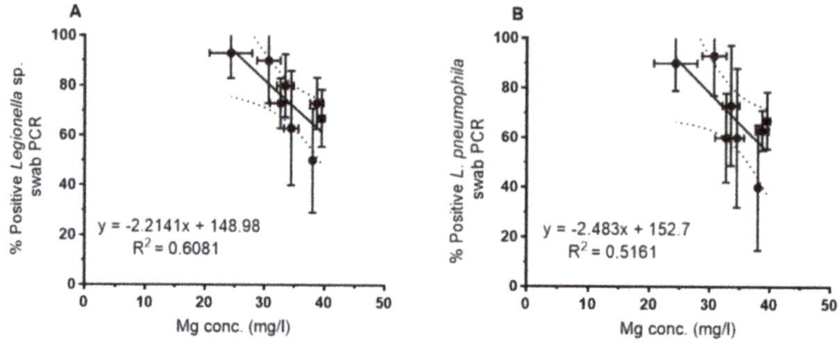


**Figure 5.** (**A**) PCR-based detection of the genus *Legionella* ($p < 0.05$) and (**B**) *L. pneumophila* in biofilm swabs vs. magnesium concentration in bulk water ($p < 0.05$) ($n = 45$, mean values $n = 8$).

For culture-based and PCR-based detection of *L. pneumophila* in water and biofilm, correlations with the Ca/Mg ratio yielded similar correlation coefficients as with Mg, but with an increase of *Legionella* abundance with an increasing Ca/Mg ratio as expected from the correlation between Mg and the Ca/Mg ratio (see Figure 4B and Supplementary Table S1).

### 3.6. *Prevalence and Biogeography of* L. pneumophila *Genotypes Across the West Bank*

During the six sampling campaigns from 2012 to 2014, 180 strains were obtained and successfully genotyped by MLVA-8(12) using 13 loci resulting in 27 different genotypes (Table 2). Twelve genotypes were represented by three strains up to a maximum of 74 strains per genotype. The remaining 15 genotypes were represented by two or one strains. Details on the results of the MLVA analysis and the clonal structure of the *L. pneumophila* population of the West Bank are given by Zayed et al. [20].

The 27 genotypes were affiliated with four VNTR clonal complexes (VACC 1, 2, 5, and 11) indicating the genetic relatedness among the respective genotypes. Seventeen of the MLVA-genotypes are affiliated with nine different sequence types (ST), meaning that some of the MLVA-genotypes pertained to the same ST. In the following analyses, MLVA genotyping will be used as a classification scheme and as a basis to study the ecology of *L. pneumophila* on a clonal level.

Though sampling was achieved in the same way for all hospital DWDS, the yield of isolates per site was rather variable and in accordance with the isolation success from water and biofilm (Table S2). Most isolates were retrieved from biofilm swabs (175 strains) and only a minor fraction from water samples (five strains in total, three from Bethlehem (sampling site F), one from Jenin (sampling site A) and Hebron (sampling site G)). The genotypes isolated from water were among the most abundant genotypes in general, and of high relevance for the respective sampling site (Table 2). For a comparison, 15 isolates retrieved during the same sampling period from biofilm samples of the Al Quds-University were added and indicated for the respective genotypes (Table 2), sampling sites (Supplementary Table S2), and the biogeographic distribution (Figure 6).

The isolates comprised serogroups (Sg) 1, 6, 8, 10, and 2–14 (Table 2). Sg 1 comprised most of the isolates (62%) and a total of seven MLVA-genotypes. Sg 6 was the second most important serogroup (30%), comprising the largest diversity with 11 MLVA-genotypes [20].

**Table 2.** MLVA-Genotype composition of isolated *L. pneumophila* strains and their affiliation with VNTR clonal complexes (VACC), sequence types (ST), and serogroup (Sg).

| MLVA-8(12) Genotype (Gt) | VACC | No. of Strains | Frequency (%) | Sequence Type (ST) | Serogroup (Sg) |
|---|---|---|---|---|---|
| Gt4(17)[AQ] | 1 | 74 | 41.1 | ST 1 | 1 |
| Gt6(18)[W(G)] | 1 | 30 | 16.6 | ST 1 * | 1 |
| Gt10(93)[W(F)] | 11 | 16 | 8.9 | ST 461 | 6 |
| Gt13(72) | 2 | 7 | 3.9 | ST 1326 | 6 |
| Gt9(92) | 11 | 8 | 4.4 | ST 461 | 6 |
| Gt10(141)[W(F)] | 11 | 6 | 3.3 | ST 461 * | 6 |
| Gt12(84)[AQ] | 1 | 5 | 2.8 | ST 1358 | 8 |
| Gt16(1)[W(A)] | 5 | 5 | 2.8 | ST 1438 | 6 |
| Gt40(47) | 5 | 3 | 1.7 | ST 292 * | 6 |
| Gt63(83) | 1 | 3 | 1.7 | NA | 1 |
| Gt64(72) | 2 | 3 | 1.7 | NA | 6 |
| Gt64(74) | 2 | 3 | 1.7 | NA | 6 |
| Gt13(143) | 2 | 2 | 1.1 | ST 1326 [e] | 10 |
| Gt8(7) | 5 | 2 | 1.1 | ST 1482 | 2-14 |
| Gt11(87)[AQ] | 1 | 1 | 0.6 | ST 1358 | 8 |
| Gt13(106)[AQ] | 2 | 1 | 0.6 | ST 1326 [e] | 6 |
| Gt16(3) | 5 | 1 | 0.6 | ST1438 [e] | 2-14 |
| Gt16(6) | 5 | 1 | 0.6 | ST1438 [e] | 2-14 |
| Gt24(68) | 2 | 1 | 0.6 | ST 93 * | 2-14 |
| Gt38(109) | 2 | 1 | 0.6 | NA | 1 |
| Gt4(16) | 1 | 1 | 0.6 | ST 1 * | 1 |
| Gt4(20)[AQ] | 1 | 1 | 0.6 | ST 1 * | 1 |
| Gt55(94) | 11 | 1 | 0.6 | NA | 6 |
| Gt6(15) | 1 | 1 | 0.6 | ST 1 * | 1 |
| Gt8(142) | 5 | 1 | 0.6 | ST1482 [e] | 2-14 |
| Gt8(23) | 5 | 1 | 0.6 | ST1482 [e] | 2-14 |
| Gt84(106)[AQ] | 2 | 1 | 0.6 | ST 187* | 6 |
| **Total No. 27** | **4** | **180** | **100** | **9 STs vs. 22 Gts** | **Sg 1: 7 Gt Sg 2-14: 20 Gt (including Sg 6: 11 Gt)** |

NA: not available; *, ST was assessed for strains of the same MLVA-8(12) genotype, and not directly for the West Bank strains; [e], ST was estimated from the MLVA-8 pattern; W, genotype retrieved from water, in brackets the site of isolation is indicated; AQ, contains strains retrieved from biofilm of the AQU.

The biogeographic distribution of the strains according to their MLVA-genotype and clonal VNTR complex (VACC) is indicated in Figure 6. On the genotype (Gt) level, Figure 6A shows a genotype pattern that varied on the regional level. In the north of the West Bank, Gt 4(17) was highly prevalent. In the south, the pattern showed high divergence from site to site. For example, site G is dominated by Gt 6(18) that was not retrieved from any other site; similarly, site F showed a high prevalence of Gt 10(141) that was endemic for this site. Furthermore, on the level of the clonal complexes the regional variability was well pronounced (Figure 6B). In general, there was a high prevalence of VACC1 in the West Bank, except for sites E, F, and H in the south. Site F showed a high prevalence of VACC11, site E for VACC 2, and site H for VACC 5.

The richness, i.e., the number of *L. pneumophila* MLVA-genotypes, varied from 2 to 7 per sampling site, with a mean of 4.5 genotypes for the eight sampling sites. The ratio of number of genotypes vs. number of strains retrieved per sampling site was added as an indicator of the "genotype diversity" ranging from 0.11 to 0.67 and a mean of 0.30 (Supplementary Table S2). There was no significant correlation ($r^2$ = 0.28) between the number of strains and the number of genotypes retrieved per sampling site. In contrast, correlation analysis revealed a negative correlation ($r^2$ = 0.53) between the average percentage of biofilm swabs positive for *Legionella* culture per sampling site and the "genotype diversity", i.e., sites with low *Legionella* incidence on biofilm swabs showed a high diversity compared to a low diversity in case of high *Legionella* incidence (Supplementary Figure S4).

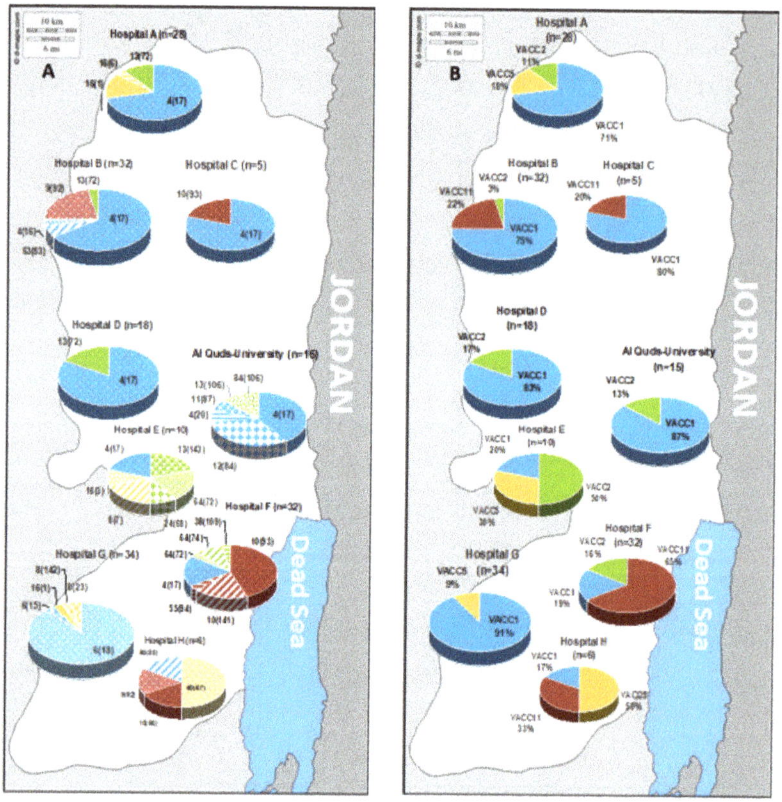


**Figure 6.** Biogeographic distribution of the *L. pneumophila* strains according to their MLVA-8(12) genotype (**A**) and their clonal complex (VACC) (**B**). In Figure 6A, the respective MLVA-8(12) genotype is indicated in the following way; genotype MLVA-8 plus genotype MLVA-12 in brackets, e.g., the genotype MLVA-8(12) 4(17) is indicated as "4(17)".

### 3.7. Environmental Factors Correlating with Genotype Abundance and Composition

The variable pattern of genotype prevalence and composition across the West Bank raised the question concerning the influencing environmental factors. There was a broad set of physico-chemical parameters recorded or obtained from the water authorities of the West Bank [21]. The eight quantitively measured parameters that showed differences between the sampling sites were used for detailed statistical analysis, i.e., PCA and cluster analysis. The selection criterion for the genotypes to be included in these analyses was the number of strains available per genotype, i.e., only genotypes represented by at least three isolates were included. This selection resulted in ten genotypes subjected to PCA and cluster analyses.

Three groups of MLVA-genotypes could be separated by PCA and cluster analysis based on the eight parameters, i.e., concentrations of chloride, sulfate, Ca, Mg, total dissolved solids (TDS), turbidity, Ca/Mg-ratio ("Ratio"), and *Legionella* plate counts (Figure 7). Cluster analyses revealed three groups for the ten genotypes (Figure 7A). By PCA, eight of the ten genotypes were assigned to three groups, with two genotypes (Gt 9(92) and Gt 63(83)) being close the PCA-group B1 comprising Gt 4(17) and Gt 13(72) (Figure 7B). These four genotypes (Gt 9(92), 63(83), 4(17), and 13(72)) were included in one group (cluster group B) by cluster analyses. For further considerations, genotypes Gt 9(92) and Gt 63(83) were combined with Gt 4(17) and Gt 13(72) in group B based on the cluster results, and because they can be considered to live in a similar environment with respect to the chosen parameters.

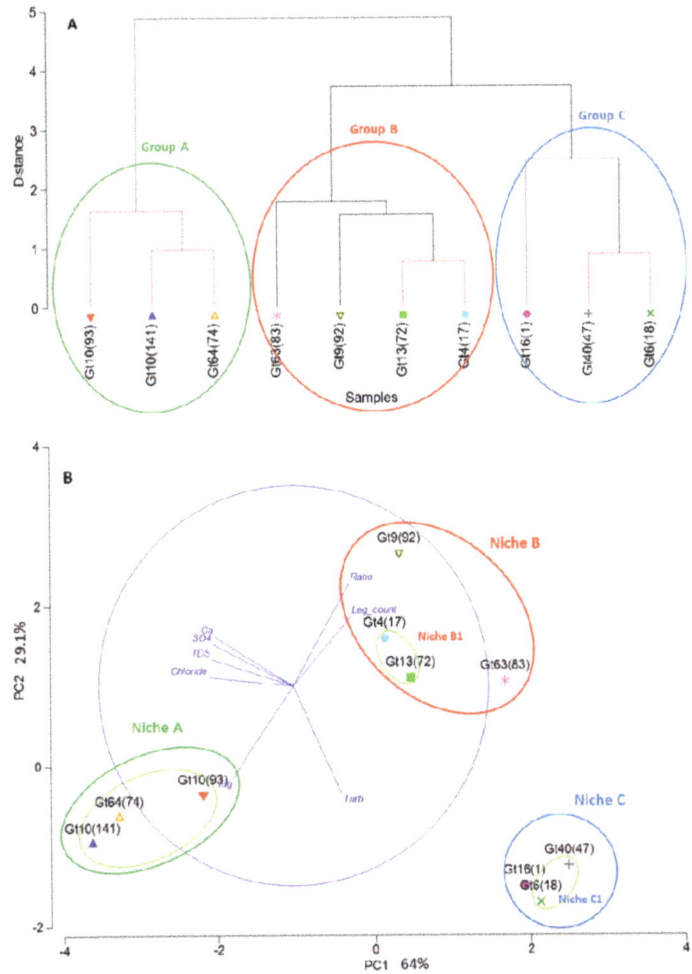


**Figure 7.** (**A**) Dendrogram showing group average hierarchical clustering of MLVA-8(12) genotypes. (**B**) Principal component analysis (PCA) of MLVA-8(12) genotypes showing the effect of biological and physicochemical parameters. Light green ellipses represent the log normal distributions of principal component values for genotype groups. Genotype groups were suggested as pertaining to the same niche. The resulting niches are A (green), niche B (red), niche C (blue). Niches A and C are consistent with the calculated grouping A and C; the calculated group B1 was enlarged to include two more genotypes in a suggested group B resulting in an enlarged niche B. Niches are generally considered as larger than the calculated genotype groups. Legend: Ratio, Ca:Mg ratio; Leg. count, *Legionella* plate counts in water samples; TDS, total dissolved solids; Turb, turbidity; Ca, Calcium; Mg, Magnesium; SO4, Sulfate.

The environment of these three groups could be characterized by the respective parameters in summary and assigns distinct niches to the three groups. The characterization of the three groups by the individual parameters is given in Table 3. While a comprehensive and stable distinction is shown for the sum of the parameters, many of the individual parameters also allow a distinction between two or three groups.

**Table 3.** Summary of environmental parameters describing the different niches of *L. pneumophila* MLVA-genotypes and their significant differences based on a correlation analysis.

| Niche Designation | VACC: Genotypes (Gt) | No. isolates | Statistics | *Legionella* (CFU/L) | Turb (mg/L) | Chloride (mg/L) | SO4 (mg/L) | TDS (mg/L) | Mg (mg/L) | Ca (mg/L) | Ca/Mg ratio |
|---|---|---|---|---|---|---|---|---|---|---|---|
| A | VACC11: Gt10(93), Gt10(141), VACC2: Gt64(74) | 25 | Mean | 263 | 1.3 | 27.6 | 13.8 | 284.3 | 23.8 | 80.5 | 3.4 |
| | | | SD | 235 | 0.2 | 6.7 | 3.8 | 5.9 | 4.2 | 2.9 | 0.4 |
| B | VACC1: Gt4(17), Gt63(83), VACC2: Gt13(72), VACC11: Gt9(92) | 95 | Mean | 61 | 1.3 | 52 | 14.9 | 314.3 | 34.2 | 81.8 | 2.3 |
| | | | SD | 122 | 0.4 | 21.1 | 5.6 | 91.8 | 9.2 | 21.2 | 0.7 |
| C | VACC1: Gt6(18) VACC5: Gt16(1), Gt40(47) | 38 | Mean | 41 | 0.9 | 62.3 | 32.7 | 405.7 | 31.2 | 93.8 | 3 |
| | | | SD | 79 | 0.1 | 4.7 | 6.2 | 37.8 | 2.1 | 3.8 | 0.2 |
| A vs. B | | | $p$ * | *** | NS | *** | ** | ** | *** | * | *** |
| A vs. C | | | $p$ * | *** | *** | *** | *** | *** | *** | *** | ** |
| B vs.C | | | $p$ * | NS | *** | ** | *** | *** | ** | *** | *** |

**Legend:** Independent $t$-test: NS, not significant; *, $p \leq 0.05$; **, $p \leq 0.01$; ***, $p \leq 0.001$.

The three groups of genotypes were considered to co-occur in their respective environment as described by the above-mentioned parameters. The respective habitats as described by these environmental parameters for the three groups of genotypes could be considered as their "niche". This means that to each niche three to four genotypes were assigned. More genotypes may be sharing these niches, but they were not included in the analysis due to the low number of strains per genotype.

## 4. Discussion

### 4.1. *Legionella* Abundance in Water and Biofilm

The number of culturable *L. pneumophila* was surprisingly low in the drinking water of the eight sampling sites in hospitals of the West Bank, with only three positive sites out of eight and a low average abundance (mean values for the three positive sites: 8–150 CFU/L). Hot water followed this trend and did not show higher concentrations than cold water. Using *L. pneumophila*-specific PCR, *L. pneumophila* was detected at all sampling sites except for site E. PCR-based detection was thus more sensitive with a broad range of site specific variability from 0 to 100% of water samples being positive on average per site. In biofilm, culturable *L. pneumophila* was detected at all sites, but with high sampling site specific variability. The incidence rate of culturable *L. pneumophila* in biofilm ranged on average per site from 3 to 30%, whereas *L. pneumophila*-specific PCR was positive on average for 40 to 93% of the biofilm samples per site. Due to the rare incidence of non-*L. pneumophila* colonies on agar media, *Legionella* counts on agar media will be addressed in the discussion as the culturable fraction of *L. pneumophila* in water and biofilm.

Culture-based detection was always lower compared to *L. pneumophila*-specific PCR detection in bulk water, but culture-based data were consistent with PCR-based data, i.e., PCR-based detection was always positive when cultivation was successful. This higher sensitivity of the PCR-based detection compared to cultivation was valid for water and biofilm, an observation often reported [29]. Bonetta et al. [30] suggested the viable-but-not-culturable state and the PCR detection of DNA of dead *Legionella* cells as reasons for the higher PCR detection.

Culture-based detection in biofilm was far more successful than in water. Culturable *L. pneumophila* was only detected in water when about 50% of the biofilm samples of the respective site and sampling were positive. This is consistent with observations in other studies showing that a major fraction of the microbial biomass in a DWDS is found in the biofilms attached to the pipe walls presumably due to improved nutrient conditions and shelter from stressing agents [31].

Compared to an annual study in a comparable climatic region, i.e., a water network at an university campus (Oranim) close to Haifa, Israel, much higher *Legionella* counts in drinking water were observed [13]. The level of culturable *Legionella* in water at Oranim campus was in the range of 10 to 5800 CFU/L, with more than 60% positive water samples. In the Oranim study, samples with culturable *L. pneumophila* in water usually showed culture positive biofilm samples, as it was the case in the West Bank study.

With respect to Mediterranean climate, there were several drinking water studies performed in buildings in Italy including hospitals and hot water systems [32,33]. Leoni et al. reported an incidence of 93.7% in hospital water in Bologna with average *Legionella* counts of 2400 CFU/L [32]. Borella et al. report for 60% of the water samples sampled in hotels across Italy a contamination of >1000 CFU/L, with 20% exceeding even 10,000 CFU/L [33]. Based on an extensive study of the drinking water supply system of a city in Northwestern France (Rennes), Sobral et al. (2011) reported an abundance of culturable *Legionella* of 50–200,000 CFU/L with a mean of 800 CFU/L for hot water systems [12].

Compared to the above-mentioned studies, the drinking water in the West Bank seemed to have a relatively low contamination with culturable *Legionella* despite a set of risk factors, such as frequent water supply interruption and storage in containers, raising the question for the drivers behind this unexpected observation.

### 4.2. Seasonal Dynamics of *L. pneumophila*

The seasonal dynamics of *L. pneumophila* abundance was not strongly pronounced and showed a different pattern for water and biofilm. Abundance in water had a maximum in summer, and thus followed the temperature regime over the year. Abundance in biofilm increased from spring to autumn, indicating a maximum of *L. pneumophila* prevalence in biofilm in autumn. These seasonal patterns of water and biofilm were better observed by culture-based analyses than by *L. pneumophila*-specific PCR. The observation that in biofilm the ratio of culturable *L. pneumophila* vs. PCR-based detection approximately doubled from spring to autumn, may have contributed to the better detection of this seasonal trend by culture-based analysis. The weak seasonal dynamics of *Legionella* abundance may have been due to the site-dependent variability of the temperature regime. In addition, there was a lack of correlation between temperature and *Legionella* abundance by culture and PCR (Table S1).

In contrast, maxima in spring and summer of culturable *Legionella* for both water and biofilm were observed in the water network of a campus close to Haifa [13]. Reasons for these divergent observations might be manifold. A main difference was the presence of water reservoirs in the West Bank hospitals. Depending on size and management, reservoirs may serve as a buffer for water temperature and thus may change seasonal effects on *L. pneumophila* in water and biofilm.

### 4.3. Influence of Physico-Chemical and Bacteriological Factors on Legionella Abundance

To determine factors that influence the prevalence of *L. pneumophila* at the different sampling sites a set of physico-chemical and bacteriological parameters was analysed with respect to their relationship with *Legionella* abundance in water and biofilm of the eight sampling sites. In addition, the abundance of relevant *L. pneumophila* MLVA-genotypes was included in this analysis.

In terms of physico-chemical parameters, only the magnesium concentration showed a significant negative correlation with *Legionella* abundance in water and biofilm (Figures 4 and 5). Magnesium concentration showed a very close negative correlation with the Ca/Mg ratio but was not correlated with the calcium concentration. Moreover, the calcium concentration did not correlate with *Legionella* abundance. Due to this Mg vs. Ca/Mg relationship, *Legionella* abundance showed a positive correlation with the Ca/Mg ratio. Therefore, we hypothesize that either Mg, the Ca/Mg ratio or a factor closely related to Mg had an influence on *Legionella* abundance in the drinking water of the West Bank. In terms of bacteriological factors, there was no correlation of *Legionella* abundance in water and biofilm with heterotrophic plate counts, but with the prevalence of specific *L. pneumophila* genotypes and their clonal complexes. The magnesium concentration also showed a negative correlation with the respective specific genotypes and clonal complexes of *L. pneumophila* (Supplementary Table S1).

The magnesium concentrations for all sampling sites were rather high ranging from 21 to 40 mg/L. High magnesium concentrations were observed in a hydrological study of spring and ground water in the West Bank with especially high values in the Eastern part of the West Bank [34]. To the best of our knowledge, no study on *Legionella* abundance in drinking water is available that deals with drinking water with magnesium concentrations of this high level which is due to the specific geological conditions.

In general, an impact of magnesium on *Legionella* abundance has not yet been demonstrated so far. There were large-scale investigations in residential drinking water distribution systems (DWDS) that included magnesium in the overall analysis of environmental drivers for *Legionella* abundance [35,36]. However, due to low magnesium concentrations in these DWDS (<3 mg/L), an effect of magnesium was not observed. On the other hand, some *L. pneumophila* studies in drinking water provide data on higher magnesium concentrations [32,33]. Borella et al. showed in the Italian hotel study that samples with no culturable *Legionella* had the highest Mg concentrations (mean 19 mg/L) [33]. Leoni et al. (2005) found in their study of buildings in Bologna based on the same public water supply, that absence of culturable *Legionella* significantly correlated with a higher Mg content of 16 mg/L vs. 11 mg/L for their presence [32]. The smaller the studied hot water systems were, the lower the *Legionella* counts presumably due to water softening devices in larger hot water systems. The observation of reduced *Legionella* abundance at sites with higher Mg concentrations may indicate that Mg may also have played a role at the Italian sites at low concentrations, in comparison to the West Bank study. However, the aspect of Mg impact on *Legionella* was not further studied, nor was there an analysis of *L. pneumophila* genotypes for these studies.

In summary, there are quite a few indications that Mg could play a role for suppression of *L. pneumophila* abundance. However, future more detailed studies including growth studies with Mg using a set of different genotypes of *L. pneumophila* are needed to elucidate this aspect in more detail.

### 4.4. Biogeography of L. pneumophila Genotype Prevalence

A large set of 27 MLVA-genotypes was retrieved from 180 *L. pneumophila* isolates with most isolates obtained from biofilm due to the low abundance of culturable *L. pneumophila* strains in the bulk water. Only five strains were obtained from water, assigned to four different genotypes (Table 2) that were frequently isolated from biofilm. As shown in detail by Zayed et al. [20], the *L. pneumophila* population had a high uniqueness, i.e., the major fraction of the MLVA-genotypes (20 out of 27 genotypes) has been described so far only for the West Bank. In addition, the distribution of genotypes among the four clonal clusters (VACC) indicates a high genetic diversity of the whole strain set.

The prevalence of the genotypes in the West Bank showed a site-specific regional diversity (Figure 6). Moreover, a large fraction of the genotypes unique for the West Bank occurred only in one site. Zayed et al. [20] suggested that the site-specific groundwater-based individual water sources may have contributed to the differences among the different sites.

As an index of genotype diversity per sampling site, the ratio of the number of genotypes vs. the number of strains retrieved from the respective site was used. A correlation analysis showed that sites where high numbers of strains were isolated had a low genotype diversity compared to sites with low isolate numbers (Supplementary Table S2, Supplementary Figure S4). Only from sites with low genotype diversity were isolates obtained from water. This may indicate that higher numbers of *L. pneumophila* in the bulk water reduce the diversity in biofilm, or, that high *L. pneumophila* counts in water are often due to a rather restricted number of genotypes or even a single genotype. The observations in the water network near Haifa support the hypothesis that a single/few specific genotype(s) may cause high abundance in drinking water and biofilm [13].

On the other hand, this aspect could also have contributed to the obtained diversity of genotypes in the West Bank: the low abundance of *L. pneumophila* in the drinking water may have supported or conserved a larger diversity in the biofilms.

### 4.5. Environmental Drivers of Genotype Consortia or Do L. pneumophila Genotypes Prefer Specific Niches?

MLVA-genotypes with more than three representatives were analyzed by average hierarchical cluster analysis and PCA with respect to their co-occurrence and positioning in the frame of bacteriological and physico-chemical parameters. Both statistical analyses showed comparable groupings of the genotypes in three groups characterized by a distinct set of environmental parameters describing the niche of each group (Figure 7, Table 3).

The highest abundance of culturable *Legionella* in water was associated with group A, comprising two VACC11-genotypes (Gt 10(93), Gt 10(141)) and one VACC2-genotype (Gt 64(74)). This group was also characterized by low chloride and low magnesium concentrations compared to the other two groups. Group B comprised four genotypes: two of VACC1, the highly abundant Gt 4(17) and Gt 63(83), one genotype of VACC2 (Gt 13(72)), and one of VACC11 (Gt 9(93)). This group was distinct from groups A and C with respect to most parameters but was with most parameters in between the other two groups. Groups B and C had higher average Mg concentrations. This may indicate that these genotypes are more tolerant towards Mg. Group C comprised the VACC1-genotype Gt 6(18) and the VACC5-genotypes Gt 16(1) and Gt 40(47) that had a high prevalence in Hebron in the South of the West Bank. The environmental parameters of this group were characterized by high sulphate, chloride, and calcium concentrations and a resulting high content of TDS.

The impact of the environment in DWDS on *L. pneumophila* was usually analyzed with respect to the species level or to the serogroup (Sg) level [32,33]. Borella et al. (2005) showed evidence that hard water selected against Sg 1, but in favor of Sg 2-14 [33]. These observations are not supported by our study, where sites with the highest hardness were inhabited by group B and C, that both comprised members of Sg 1 and Sg 6 (Tables 2 and 3); group A inhabiting the softest water comprised only members of Sg 6. However, all of these sites of the West Bank had relatively hard water, and the high Mg concentrations may have had an additional impact.

On the MLVA-genotype level, Rodriguez-Martinez et al. showed for the first time-distinct niche preferences for *L. pneumophila* genotypes most pronounced with respect to temperature in a campus water network (Haifa, Israel) [13]. Niche preferences were previously shown for many aquatic bacterial species with a few studies tackling the subspecies or clonal level [37,38]. For *L. pneumophila*, this is the first assessment of niching of its genotypes as characterized by a set of environmental parameters in drinking water systems of a larger region such as the West Bank. We think that the information on niche preferences in combination with a genotyping at an adequate resolution (such as MLVA with 8 to 13 loci) could be helpful in order to better understand and model abundance of *L. pneumophila* in DWDS. These genotypic groups occupying environmental niches could be considered as ecotypes, i.e., a set of strains of a bacterial species inhabiting the same niche [38].

Though a broad set of environmental factors were analyzed for describing the niches of *L. pneumophila* genotypes, more studies including more conceivable factors are needed to complement the niche understanding, e.g., the abundance of protozoa and their species composition and their interaction with drinking water bacteria and especially the different genotypes of *L. pneumophila* could be of high relevance [6]. As shown by Sharaby et al. [39], the interaction of *L. pneumophila* with protozoa may be genotype-dependent in addition to being temperature-dependent in a genotype-dependent way.

Studies on physiological and autecological traits of the relevant genotypes are considered crucial to allow better predictions and modeling [39]. Overall, we observed a correlation of specific genotypes with specific environmental niches which might be stemming from the different types of local groundwater used as source water for the respective drinking water.

### *4.6. Relevance of the Observed Genotypes for Human Health and Environmental Issues in the West Bank and Worldwide*

MLVA-genotyping of *L. pneumophila* can be considered as an economic genotyping method that has a good level of resolution for addressing clinical and environmental aspects [20]. MLVA-genotyping was shown to be comparable to sequenced-based typing (SBT) but has a higher resolution, which is especially of relevance for a better resolution of the highly important large STs such as ST1. Comparability and the increased resolution was shown for the method analyzing 8 loci (MLVA-8) as for the here used MLVA-8(12) using 13 loci [12,19,20,27] Using 13 loci increased the resolution compared to eight loci significantly; however, the used nomenclature allows a direct comparison of MLVA-8 with MLVA-8(12): the genotype indicated before brackets indicates the genotype according to MLVA-8, while the number in brackets refers to the genotype obtained from MLVA-12. Compared to SBT, MLVA subdivided strains from the West Bank of the larger STs into several MLVA-genotypes) [20].

Based on the International MLVA data base and larger studies [12,16,27], the worldwide distribution and relevance for health and environment of the strains from the West Bank were estimated [20]. Most of the genotypes (20 out of 27) were considered as unique for the West Bank. Having high relevance worldwide as clinical and environmental MLVA-genotypes, Gt 4(17) and Gt 64(74) can be regarded as having MLVA-profiles identical to *L. pneumophila* strain Paris and strain Philadelphia-1, respectively. In Israel, Gt 6(18) played an eminent role as clinical and environmental genotype in addition to GT 4(17). For more details on the occurrence of the remaining genotypes not exclusive to the West Bank see Zayed et al. [20].

More specifically for the West Bank, some of the highly abundant MLVA-genotypes can be of special health relevance. PCR-based direct analysis of the sequence types in respiratory specimens of pneumonia patients of the West Bank revealed that genotypes belonging to ST1 and ST461 were present in half of the *L. pneumophila* positive clinical specimens [40]. We assume that the highly abundant representatives of these sequence types of our study are the relevant pathogenic genotypes (Table 2), meaning that Gt 4(17) may have been relevant for the detected ST1 infections, and the highly abundant genotypes in the South of the West Bank, i.e., Gt 9(92), Gt 10(93), and Gt 10(141), may have been responsible for the ST461 infections [18].

The highly abundant MLVA-genotypes Gt 4(17) and Gt 6(18) from the West Bank were also of high relevance in drinking water distribution systems and clinical isolates in Israel [13,14,39]. In contrast to observations in the West Bank, Gt 4(17) and Gt 6(18) had a high abundance in water and biofilm of a campus drinking water network in Northern Israel. The presence of Gt 4 was associated with average *Legionella* counts in water of 2500 CFU/L at an average water temperature of 20.6 °C. The presence of Gt 6 was associated with average *Legionella* counts in water of 240 CFU/L at an average water temperature of 27.9 °C. In the West Bank, Gt 4(17) was only detected in biofilm with no detection of culturable *Legionella* in bulk water; Gt 6(18) was endemic at one site, isolated regularly from biofilm and only once from water.

Rodriguez-Martinez et al. [13] concluded that the presence of Gt 4 could be considered as an indicator of high *Legionella* presence in drinking water, and they suggested Gt 4 as indicator genotype. Based on the observations in the West Bank where Gt 4 was very frequently observed in biofilms in the Northern part without co-occurrence of high *Legionella* counts in water, we suggest that the presence in biofilm might be not an indicator for high *Legionella* counts in water [13]. Due to the worldwide occurrence of Gt 4 and the observations in Israel and the West Bank, we suggest that Gt 4 may be regarded as an indicator of high *Legionella* abundance when showing up in the water phase. Furthermore, the presence of Gt 4 in biofilm might be considered as warning that, if conditions change, a "*Legionella* bloom" may be at risk.

### 4.7. Relevance of the Findings of this Study for Drinking Water Management Strategies

*L. pneumophila* is a major water-based pathogen, and its high abundance in drinking water is therefore a significant health risk. On the other hand, there are substantial uncertainties in the assessment of this risk [41,42]. Our study revealed that *L. pneumophila* populations in drinking water are composed of a set of genotypes sharing similar niche characteristics. Physico-chemical factors seemed to determine these niches and may also have shaped the *L. pneumophila* community, i.e., consortia of genotypes. The response of specific MLVA-genotypes to the environmental factors seemed to determine the abundance of *Legionella* in drinking water. Understanding the relationship between *L. pneumophila* genotypes and their environmental drivers might be crucial for understanding *L. pneumophila* abundance and the design of management concepts. If there are generally important genotypes (such as Gt 4(17)) and environmental drivers (such as Mg) or if there is a high individuality between the drinking water sites, future studies have to elucidate for a broad set of climatic regions and in more detail. Furthermore, a broader set of conceivable niche-relevant parameters and environmental drivers should be included in future studies. The overall finding of our study is that a genotype-based ecology for understanding *L. pneumophila* abundance in artificial drinking water distribution systems can be considered as of high relevance for their management and to this end also to human health. Therefore, we emphasize the need of a genotype-based ecology for *L. pneumophila* enabling the definition of niches for specific genotypes, their co-occurrence, and interactions.

## 5. Conclusions

- Analysis of *L. pneumophila* ecology at the genotype level allows a better insight into the environmental drivers triggering their abundance. The study analyzed the environmental drivers determining the niches of abundant MLVA-genotypes. This may be in support of better prediction and management of *L. pneumophila* abundance. However, more studies at sites differing with respect to climate and water quality are needed to provide a more thorough insight into environmental drivers relevant for different *L. pneumophila* ecotypes.

- Magnesium was observed as an environmental factor correlating with low *L. pneumophila* abundance at high concentrations (> 30 mg/L). If and how Mg has a suppressing effect on *L. pneumophila* abundance needs further detailed studies, e.g., for other regions, site-specific inhibiting Mg concentrations may vary depending on the susceptibility of the present *L. pneumophila* genotypes.
- The diversity of *L. pneumophila* genotypes and *L. pneumophila* abundance showed an inverse correlation. It is hypothesized that "blooms" of *L. pneumophila* are caused by a few or even a single well-adapted genotype (e.g., Gt 4, *L. pneumophila* Paris). Thus, the diversity and/or presence of a specific genotype may indicate if high *Legionella* abundance is at risk, e.g., if relevant environmental drivers will change. More studies are needed to investigate the value of potential indicator genotypes for *L. pneumophila* blooms.
- In summary, the diversity of *L. pneumophila* in DWDS on the subspecies/clonal level is considered too high to not be included in studies on drinking water. An ecology on the genotype level is needed to get insight into *L. pneumophila* "behavior" in DWDS providing the basis for better modeling and prediction.

**Supplementary Materials:** The following are available online at http://www.mdpi.com/2076-0817/9/12/1012/s1. Table S1: Correlation Matrix of physico-chemical and bacteriological parameters in water and biofilm of the eight sampling sites; Table S2: Sampling location and number of isolates obtained from each site from six sampling campaigns in 2012–2014; Figure S1: Sampling sites of the eight hospitals (A-H) and Al-Quds University (AQU) in the West Bank, Palestine; Figure S2: *Legionella* plate counts vs. *L. pneumophila* specific PCR-detection in cold water samples and vs. culture-based positive biofilm swabs; Figure S3: Correlation of Legionella plate counts of cold versus hot water; Figure S4: MLVA 8(12)-genotype diversity (number of genotypes/number of strains retrieved per sampling site) vs. the average percentage of *Legionella* positive biofilm swabs (by cultivation) per sampling site.

**Author Contributions:** A.R.Z. performed sample acquisition and analyses, strain isolation, data analyses and contributed to drafting of the manuscript. D.M.B.; M.S.; I.B.; and M.G.H. designed the work, contributed to data evaluation and drafting of the manuscript. S.B.; M.P.; A.S. and H.A. contributed to sampling, and sample and strain analyses. All authors read and agreed to the published version of the manuscript.

**Funding:** This work was supported by DFG grants (HO 930/5-1; HO 930/5-2) and the Hildegard-Elisabeth-Foundation.

**Acknowledgments:** We thank the administration of the hospitals for allowing us to collect the samples. We thank the Palestinian Water Authority (PWA) for helpful support, information, and data. We are grateful to Dalia Abu Hilal, Verena Maiberg and Josefin Koch for technical assistance. Christian Lück and the National Consulting Laboratory for *Legionella* kindly provided the Sequence Types of the isolates.

**Conflicts of Interest:** The authors declare no conflict of interest.

## References

1. Lück, P.C.; Fry, N.K.; Helbig, J.H.; Jarraud, S.; Harrison, T.G. Typing Methods for Legionella. *Methods Mol. Biol.* **2012**, *954*, 119–148. [CrossRef]
2. Steinert, M.; Hentschel, U.; Hacker, J. Legionella pneumophila: An aquatic microbe goes astray. *FEMS Microbiol. Rev.* **2002**, *26*, 149–162. [CrossRef]
3. Mercante, J.W.; Winchell, J.M. Current and Emerging Legionella Diagnostics for Laboratory and Outbreak Investigations. *Clin. Microbiol. Rev.* **2015**, *28*, 95–133. [CrossRef]
4. Kirschner, A.K.; Rameder, A.; Schrammel, B.; Indra, A.; Farnleitner, A.H.; Sommer, R. Development of a new CARD-FISH protocol for quantification of Legionella pneumophila and its application in two hospital cooling towers. *J. Appl. Microbiol.* **2012**, *112*, 1244–1256. [CrossRef] [PubMed]
5. Mizrahi, H.; Peretz, A.; Lesnik, R.; Aizenberg-Gershtein, Y.; Rodríguez-Martínez, S.; Sharaby, Y.; Pastukh, N.; Brettar, I.; Höfle, M.G.; Halpern, M. Comparison of sputum microbiome of legionellosis-associated patients and other pneumonia patients: Indications for polybacterial infections. *Sci. Rep.* **2017**, *7*, 40114. [CrossRef] [PubMed]
6. Delafont, V.; Brouke, A.; Bouchon, D.; Moulin, L.; Héchard, Y. Microbiome of free-living amoebae isolated from drinking water. *Water Res.* **2013**, *47*, 6958–6965. [CrossRef] [PubMed]

7. Phin, N.; Parry-Ford, F.; Harrison, T.; Stagg, H.R.; Zhang, N.; Kumar, K.; Lortholary, O.; Zumla, A.; Abubakar, I. Epidemiology and clinical management of Legionnaires' disease. *Lancet Infect. Dis.* **2014**, *14*, 1011–1021. [CrossRef]
8. Palestinian Water Authority (PWA). Status report of water sources in the occupied state of Palestine. Annual Water Resources Status Report 2012. 1. 18. 2013. Available online: http://www.pwa.ps/userfiles/file/تصنيف/تقارير%201/WR%20STATUS%20Report-final%20draft%202014-04-01.pdf (accessed on 15 October 2016).
9. Bargellini, A.; Marchesi, I.; Righi, E.; Ferrari, A.; Cencetti, S.; Borella, P.; Rovesti, S. Parameters predictive of Legionella contamination in hot water systems: Association with trace elements and heterotrophic plate counts. *Water Res.* **2011**, *45*, 2315–2321. [CrossRef]
10. Rakić, A.; Perić, J.; Foglar, L. Influence of temperature, chlorine residual and heavy metals on the presence of Legionella pneumophila in hot water distribution systems. *Ann. Agric. Environ. Med.* **2012**, *19*, 431–436.
11. Schwake, D.O.; Alum, A.; Abbaszadegan, M. Impact of Environmental Factors on Legionella Populations in Drinking Water. *Pathogens* **2015**, *4*, 269–282. [CrossRef]
12. Sobral, D.; Le Cann, P.; Gerard, A.; Jarraud, S.; Lebeau, B.; Loisy-Hamon, F.; Vergnaud, G.; Pourcel, C. High-Throughput Typing Method To Identify a Non-Outbreak-Involved Legionella pneumophila Strain Colonizing the Entire Water Supply System in the Town of Rennes, France. *Appl. Environ. Microbiol.* **2011**, *77*, 6899–6907. [CrossRef] [PubMed]
13. Rodríguez-Martínez, S.; Sharaby, Y.; Pecellín, M.; Brettar, I.; Höfle, M.; Halpern, M. Spatial distribution of Legionella pneumophila MLVA-genotypes in a drinking water system. *Water Res.* **2015**, *77*, 119–132. [CrossRef] [PubMed]
14. Sharaby, Y.; Rodríguez-Martínez, S.; Pecellin, M.; Sela, R.; Peretz, A.; Höfle, M.G.; Halpern, M.; Brettar, I. Virulence Traits of Environmental and Clinical Legionella pneumophila Multilocus Variable-Number Tandem-Repeat Analysis (MLVA) Genotypes. *Appl. Environ. Microbiol.* **2018**, *84*. [CrossRef] [PubMed]
15. Khodr, A.; Kay, E.; Gomez-Valero, L.; Ginevra, C.; Doublet, P.; Buchrieser, C.; Jarraud, S. Molecular epidemiology, phylogeny and evolution of Legionella. *Infect. Genet. Evol.* **2016**, *43*, 108–122. [CrossRef] [PubMed]
16. Visca, P.; D'Arezzo, S.; Ramisse, F.; Gelfand, Y.; Benson, G.; Vergnaud, G.; Fry, N.K.; Pourcel, C. Investigation of the population structure of Legionella pneumophila by analysis of tandem repeat copy number and internal sequence variation. *Microbiology* **2011**, *157*, 2582–2594. [CrossRef] [PubMed]
17. David, S.; Rusniok, C.; Mentasti, M.; Gomez-Valero, L.; Harris, S.R.; Lechat, P.; Lees, J.; Ginevra, C.; Glaser, P.; Ma, L.; et al. Multiple major disease-associated clones of Legionella pneumophila have emerged recently and independently. *Genome Res.* **2016**, *26*, 1555–1564. [CrossRef]
18. Borchardt, J.; Helbig, J.H.; Lück, P.C. Occurrence and distribution of sequence types among Legionella pneumophila strains isolated from patients in Germany: Common features and differences to other regions of the world. *Eur. J. Clin. Microbiol. Infect. Dis.* **2007**, *27*, 29–36. [CrossRef]
19. Pourcel, C.; Visca, P.; Afshar, B.; D'Arezzo, S.; Vergnaud, G.; Fry, N.K. Identification of Variable-Number Tandem-Repeat (VNTR) Sequences in Legionella pneumophila and Development of an Optimized Multiple-Locus VNTR Analysis Typing Scheme. *J. Clin. Microbiol.* **2007**, *45*, 1190–1199. [CrossRef]
20. Zayed, A.R.; Pecellin, M.; Salah, A.; AlAlam, H.; Butmeh, S.; Steinert, M.; Lesnik, R.; Brettar, I.; Höfle, M.G.; Bitar, D.M. Characterization of *Legionella pneumophila* Populations by Multilocus Variable Number of Tandem Repeats (MLVA) Genotyping from Drinking Water and Biofilm in Hospitals from Different Regions of the West Bank. *Pathogens* **2020**, *9*, 862. [CrossRef]
21. Zayed, A.R. Ecology, virulence and genomics of Legionella pneumophila isolates from the West Bank, Palestine. Ph.D. Thesis, Technical University of Braunschweig, Braunschweig, Germany, October 2018. Available online: https://publikationsserver.tu-braunschweig.de/receive/dbbs_mods_00066061 (accessed on 13 October 2020).
22. Centers for Disease Control and Prevention CDC. Procedures for the Recovery of Legionella from the Environment. Centers for Disease Control and Prevention (CDC). 1–13. 2005. Available online: http://www.flenviro.com/CDC_legionella_method.pdf (accessed on 13 October 2020).

23. International Organization for Standardization. Detection and Enumeration of Legionella ISO 11731. 2004. Available online: https://www.evs.ee/products/iso-11731-2017 (accessed on 15 July 2016).
24. Kahlisch, L.; Henne, K.; Draheim, J.; Brettar, I.; Höfle, M.G. High-Resolution In Situ Genotyping of Legionella pneumophila Populations in Drinking Water by Multiple-Locus Variable-Number Tandem-Repeat Analysis Using Environmental DNA. *Appl. Environ. Microbiol.* **2010**, *76*, 6186–6195. [CrossRef]
25. Henne, K.; Kahlisch, L.; Höfle, M.G.; Brettar, I. Seasonal dynamics of bacterial community structure and composition in cold and hot drinking water derived from surface water reservoirs. *Water Res.* **2013**, *47*, 5614–5630. [CrossRef]
26. Henne, K.; Kahlisch, L.; Brettar, I.; Höfle, M.G. Analysis of Structure and Composition of Bacterial Core Communities in Mature Drinking Water Biofilms and Bulk Water of a Citywide Network in Germany. *Appl. Environ. Microbiol.* **2012**, *78*, 3530–3538. [CrossRef]
27. Pecellín, M. Structure and virulence of Legionella pneumophila populations from freshwater systems in Germany and Middle East. Ph.D. Thesis, Technische Universität Braunschweig, Braunschweig, Germany, October 2016.
28. Clarke, K.R.; Somerfield, P.J.; Gorley, R.N. Testing of null hypotheses in exploratory community analyses: Similarity profiles and biota-environment linkage. *J. Exp. Mar. Biol. Ecol.* **2008**, *366*, 56–69. [CrossRef]
29. Whiley, H.; Taylor, M. Legionella detection by culture and qPCR: Comparing apples and oranges. *Crit. Rev. Microbiol.* **2014**, *42*, 65–74. [CrossRef]
30. Bonetta, S.; Ferretti, E.; Balocco, F.; Carraro, E. Evaluation of Legionella pneumophilacontamination in Italian hotel water systems by quantitative real-time PCR and culture methods. *J. Appl. Microbiol.* **2010**, *108*, 1576–1583. [CrossRef]
31. Douterelo, I.; Boxall, J.B.; Deines, P.; Sekar, R.; Fish, K.E.; Biggs, C.A. Methodological approaches for studying the microbial ecology of drinking water distribution systems. *Water Res.* **2014**, *65*, 134–156. [CrossRef]
32. Leoni, E.; De Luca, G.; Legnani, P.; Sacchetti, R.; Stampi, S.; Zanetti, F. Legionella waterline colonization: Detection of Legionella species in domestic, hotel and hospital hot water systems. *J. Appl. Microbiol.* **2005**, *98*, 373–379. [CrossRef]
33. Borella, P.; Montagna, M.T.; Stampi, S.; Stancanelli, G.; Romano-Spica, V.; Triassi, M.; Marchesi, I.; Bargellini, A.; Tatò, D.; Napoli, C.; et al. Legionella Contamination in Hot Water of Italian Hotels. *Appl. Environ. Microbiol.* **2005**, *71*, 5805–5813. [CrossRef]
34. Shalash, I.; Ghanem, M. Hydrochemistry of the Natuf drainage basin in Ramallah area/West Bank. *Environ. Earth Sci.* **2007**, *55*, 359–367. [CrossRef]
35. Pierre, D.M.; Baron, J.L.; Ma, X.; Sidari, F.P.; Wagener, M.M.; Stout, J.E. Water Quality as a Predictor of Legionella Positivity of Building Water Systems. *Pathogens* **2019**, *8*, 295. [CrossRef]
36. Stout, J.E.; Yu, V.L.; Yee, Y.C.; Vaccarello, S.; Diven, W.; Lee, T.C. Legionella pneumophila in residential water supplies: Environmental surveillance with clinical assessment of Legionnaire's disease. *Epidemiol. Infect.* **1992**, *109*, 49–57.
37. Deng, J.; Brettar, I.; Luo, C.; Auchtung, J.M.; Konstantinidis, K.T.; Rodrigues, J.L.M.; Höfle, M.; Tiedje, J.M. Stability, genotypic and phenotypic diversity ofShewanella balticain the redox transition zone of the Baltic Sea. *Environ. Microbiol.* **2014**, *16*, 1854–1866. [CrossRef]
38. Deng, J.; Auchtung, J.M.; Konstantinidis, K.T.; Brettar, I.; Höfle, M.G.; Tiedje, J.M. Genomic Variations Underlying Speciation and Niche Specialization of Shewanella baltica. *MSystems* **2019**, *4*, e00560-19. [CrossRef] [PubMed]
39. Sharaby, Y.; Rodríguez-Martínez, S.; Oks, O.; Pecellin, M.; Mizrahi, H.; Peretz, A.; Brettar, I.; Höfle, M.G.; Halpern, M. Temperature-Dependent Growth Modeling of Environmental and Clinical Legionella pneumophila Multilocus Variable-Number Tandem-Repeat Analysis (MLVA) Genotypes. *Appl. Environ. Microbiol.* **2017**, *83*, e03295-16. [CrossRef]
40. Jaber, L.; Amro, M.; Abu Tair, H.; Bahader, S.A.; AlAlam, H.; Butmeh, S.; Abu Hilal, D.; Brettar, I.; Höfle, M.G.; Bitar, D.M. Comparison of in situ sequence type analysis of Legionella pneumophila in respiratory tract secretions and environmental samples of a hospital in East Jerusalem. *Epidemiol. Infect.* **2018**, *146*, 2116–2121. [CrossRef]

41. Hamilton, K.A.; Haas, C.N. Critical review of mathematical approaches for quantitative microbial risk assessment (QMRA) of Legionella in engineered water systems: Research gaps and a new framework. *Environ. Sci. Water Res. Technol.* **2016**, 2, 599–613. [CrossRef]
42. Edwards, J.; Keegan, A.; Fallowfield, H.; Ross, K. Uncertainties associated with assessing the public health risk from Legionella. *Front. Microbiol.* **2014**, 5, 501. [CrossRef]

**Publisher's Note:** MDPI stays neutral with regard to jurisdictional claims in published maps and institutional affiliations.

*pathogens*

MDPI

*Article*

# Water Quality as a Predictor of *Legionella* Positivity of Building Water Systems

**David Pierre [1], Julianne L. Baron [1], Xiao Ma [1], Frank P. Sidari III [1], Marilyn M. Wagener [2] and Janet E. Stout [1,3,*]**

1 Special Pathogens Laboratory, Pittsburgh, PA 15219, USA; dpierre@specialpathogenslab.com (D.P.); jbaron@specialpathogenslab.com (J.L.B.); maxiaosean1219@gmail.com (X.M.); fsidari@specialpathogenslab.com (F.P.S.III)

2 School of Medicine, University of Pittsburgh, Pittsburgh, PA 15261, USA; mmw5@pitt.edu

3 Department of Civil and Environmental Engineering, University of Pittsburgh, Pittsburgh, PA 15261, USA

* Correspondence: jstout@specialpathogenslab.com; Tel.: +1-(412)-281-5335

Received: 26 November 2019; Accepted: 11 December 2019; Published: 13 December 2019

**Abstract:** Testing drinking water systems for the presence of *Legionella* colonization is a proactive approach to assess and reduce the risk of Legionnaires' disease. Previous studies suggest that there may be a link between *Legionella* positivity in the hot water return line or certain water quality parameters (temperature, free chlorine residual, etc.) with distal site *Legionella* positivity. It has been suggested that these measurements could be used as a surrogate for testing for *Legionella* in building water systems. We evaluated the relationship between hot water return line *Legionella* positivity and other water quality parameters and *Legionella* colonization in premise plumbing systems by testing 269 samples from domestic cold and hot water samples in 28 buildings. The hot water return line *Legionella* positivity and distal site positivity only demonstrated a 77.8% concordance rate. Hot water return line *Legionella* positivity compared to distal site positivity had a sensitivity of 55% and a specificity of 96%. There was poor correlation and a low positive predictive value between the hot water return line and distal outlet positivity. There was no correlation between *Legionella* distal site positivity and total bacteria (heterotrophic plate count), pH, free chlorine, calcium, magnesium, zinc, manganese, copper, temperature, total organic carbon, or incoming cold-water chlorine concentration. These findings suggest that hot water return line *Legionella* positivity and other water quality parameters are not predictive of distal site positivity and should not be used alone to determine the building's *Legionella* colonization rate and effectiveness of water management programs.

**Keywords:** *Legionella*; distal site positivity; hot water return line; chlorine; HPC; temperature; water management

## 1. Introduction

*Legionella* is considered an opportunistic human pathogen and these bacteria have been found in up to 70% of building water systems [1]. In recent molecular studies, *Legionella pneumophila* was isolated from 38% of buildings sampled including 42% of residences and 35% of office buildings [2], and in 47% of all taps in a study of buildings and homes in the United States [3]. *Legionella* colonization of potable water systems can pose a public health risk, especially for immunocompromised individuals [1,4].

Several organizations including ASHRAE (formerly known as the American Society of Heating, Refrigerating and Air-Conditioning Engineers), the World Health Organization, the American Industrial Hygiene Association, and the Centers for Disease Control and Prevention recommend the creation of water management programs aimed at preventing the growth and spread of *Legionella* and other waterborne pathogens [5–8].

Testing water for the presence of *Legionella* is the most direct means of determining whether the building water system is colonized by *Legionella* [9,10] and its usefulness has been discussed in multiple technical guidelines [7,11,12]. The correlation with disease risk has been well established in healthcare facilities [13–16], but risk has also been demonstrated in hotels and other commercial properties. Rather than recommend testing for the bacteria, some guidelines and standards have suggested that building design or physical and chemical properties of the water can be used as predictors of risk or to demonstrate that water management programs have effectively controlled the growth and spread of *Legionella* [5,8].

For example, ASHRAE selected certain physical properties of buildings as requisite characteristics for requiring a water management program [5]. This included building height (greater than 10 stories including below grade), which had previously been found to have an increased presence of *Legionella* in the buildings' water heaters [17,18]. The Centers for Disease Control and Prevention (CDC) and the Centers for Medicare and Medicaid Services (CMS) recommend that temperature, pH, and free chlorine be tested at numerous outlets when doing a *Legionella* risk assessment [8,19]. Others have suggested that the temperature of the hot water in the recirculation line of a building [20] or other water quality parameters could predict the presence or absence of *Legionella* at the distal outlets (faucets and showers) [20–23].

There is a problem with these recommendations. There is little data to support them. If such monitoring is to be performed and relied upon as part of risk assessments and water management programs, the expectation is that this information will have some relationship to either the presence or absence of *Legionella*.

It is important that we better understand these assumptions. Therefore, we performed a large-scale field investigation to evaluate the presence of *Legionella* in premise plumbing systems in 28 buildings in New York City, San Francisco, and New Jersey. The objective of the study was to (1) evaluate the potential of using hot water return line *Legionella* positivity as an indicator of distal site *Legionella* colonization risk in these buildings and (2) evaluate the correlation between water quality parameters and the presence of *Legionella* in water systems.

## 2. Results

### 2.1. *Legionella* Positivity Correlation

A total of 269 samples were cultured for *Legionella* from domestic cold and hot water samples in 28 different buildings and from 45 recirculating hot water systems. *Legionella* was cultured from 65/269 (24.2%) samples from 15/28 (53.6%) buildings sampled. The hot water return line sample was positive in 12/45 (26.7%) systems (Figure 1). Positive distal sites (faucets) were observed in 20/45 (44.4%) of the hot water systems. *L. pneumophila* was the only species of *Legionella* isolated from these water samples. There was a trend towards larger buildings having more distal site positivity, however this was not statistically significant ($p = 0.06$).

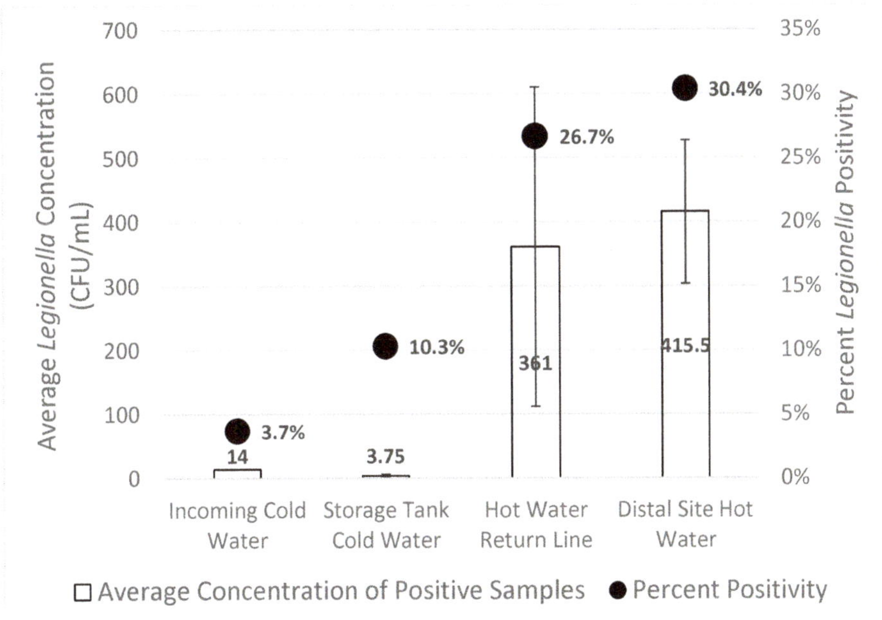


**Figure 1.** Percent *Legionella* positivity and concentration was highest in distal site hot water and hot water return lines. The bars represent the standard error of the mean for the average *Legionella* concentration.

*Legionella* was isolated from at least one distal site in 91.7% (11/12) of the hot water systems that also had *Legionella* isolated from the hot water return line, with an average distal site positivity of 83.3% ± 8.7% (Figure 2). *Legionella* was isolated from at least one distal site in 27.3% (9/33) of the hot water systems with a *Legionella* negative hot water return line, with an average distal site positivity of 13.1% ± 4.3% (Figure 2). In 35 of the 45 sampled hot water systems, there was agreement between *Legionella* distal site positivity and hot water return line *Legionella* positivity, resulting in a 77.8% concordance rate.

We then analyzed if hot water return line *Legionella* positivity was able to correctly predict whether the distal sites would be positive or negative for *Legionella*. Hot water return line positivity was related to distal site positivity ($p = 0.002$), with a sensitivity of only 55% (11/20) and a specificity of 96% (24/25). However, when hot water return line positivity was used as a screening tool for distal site positivity, the positive predictive value was 91.7% and the negative predictive value was only 72.7%. The average distal site concentration of *Legionella* in systems with a positive hot water return was 483.5 ± 147.4 CFU/mL, versus 20.7 ± 8.4 CFU/mL in negative hot water returns ($p < 0.003$) (Figure 2).

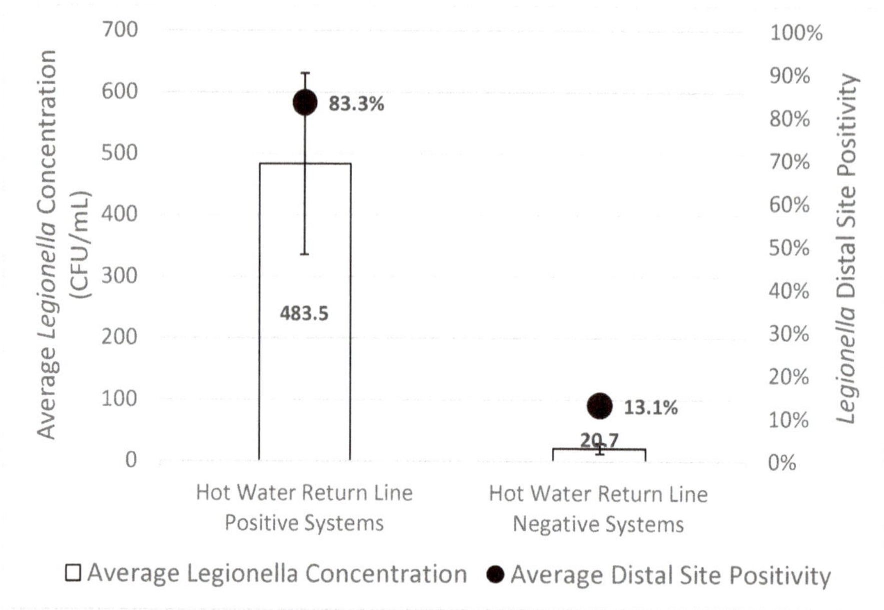


**Figure 2.** *Legionella* distal site positivity and concentration was highest for hot water systems with *Legionella* positive hot water return lines. The bars represent the standard error of the mean for average *Legionella* concentration.

## 2.2. Heterotrophic Plate Count (HPC) and Chemical Parameter Correlation

All samples collected were also cultured for heterotrophic plate count (HPC) bacteria (Table 1). HPC concentrations in hot water samples ranged from 3 CFU/mL to 2,100,000 CFU/mL. Statistical analysis showed no correlation between distal site HPC concentration and *Legionella* distal site positivity ($p = 0.788$) (Figure 3). The best-fit linear regression line demonstrates that HPC concentration explains only 0.68% of the variance in *Legionella* distal site positivity ($R^2 = 0.0068$).

**Table 1.** Average concentrations (±standard deviation) of total bacteria (heterotrophic plate count—HPC) and physicochemical parameters.

| | Incoming Cold Water | Storage Tank Cold Water | Hot Water Return | Distal Site Hot Water |
|---|---|---|---|---|
| HPC (CFU/mL) | 18,954 ± 71,478.3 | 7254 ± 20,138.3 | 11,360.4 ± 25,094.9 | 92,125.4 ± 221,547.2 |
| Temperature (°F, 30 s flush) | NT | NT | NT | 109.36 ± 21.85 |
| Temperature (°F, 1 min flush) | 63.12 ± 9.22 | 62.66 ± 9.76 | 122.80 ± 17.85 | 119.95 ± 73.22 |
| pH | 7.29 ± 0.78 | 7.17 ± 0.69 | 6.97 ± 0.67 | 6.93 ± 0.63 |
| Free Cl (mg/L) | 0.34 ± 0.24 | 0.23 ± 0.16 | 0.07 ± 0.13 | 0.05 ± 0.09 |
| Fe (mg/L) | NA | NA | NA | NA |
| Ca (mg/L) | 7.16 ± 4.41 | 6.56 ± 1.58 | 6.39 ± 1.19 | 6.61 ± 3.09 |
| Mg (mg/L) | 1.66 ± 1.09 | 1.46 ± 0.46 | 1.72 ± 1.80 | 1.79 ± 1.81 |
| Zn (mg/L) | 0.03 ± 0.04 | 0.02 ± 0.03 | 0.05 ± 0.09 | 0.04 ± 0.06 |
| Pb (mg/L) | NA | NA | NA | NA |
| Mn (mg/L) | 0.02 ± 0.03 | 0.02 ± 0.02 | 0.03 ± 0.12 | 0.03 ± 0.11 |
| Cu (mg/L) | 0.09 ± 0.20 | 0.05 ± 0.06 | 0.30 ± 0.51 | 0.19 ± 0.23 |
| TOC (mg/L) | NT | NT | 1.85 ± 0.26 | NT |

NA: 48.7% samples were with Fe concentration below detection limit of 0.03 mg/L, detectable Fe concentration ranged from 0.03 to 4.6 mg/L; 76.2% samples were with Pb concentration below detection limit of 0.001 mg/L; Pb concentrations in detectable samples ranged from 0.001 to 0.63 mg/L. NT: not tested. TOC = Total Organic Carbon.

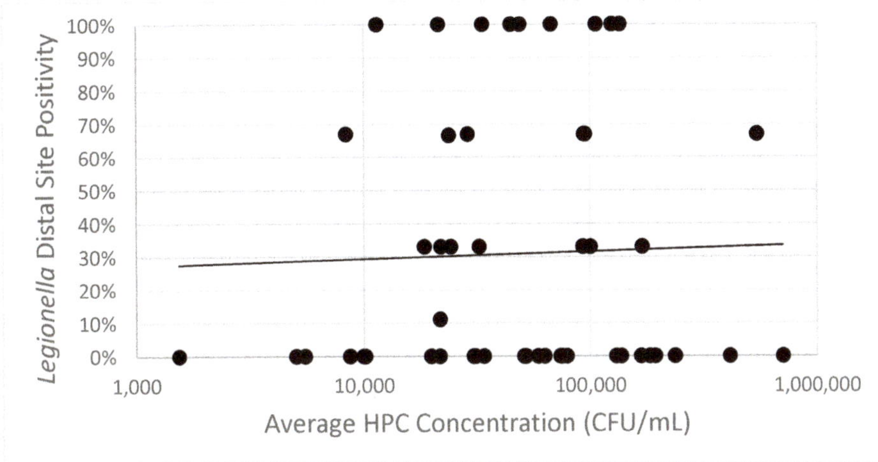


**Figure 3.** *Legionella* distal site positivity and average distal site HPC concentration are not significantly associated. Average HPC concentration is represented on a logarithmic scale x-axis. A line of best fit has been added to show the relationship between HPC concentration and distal site positivity.

The water from distal outlets was analyzed for pH, free chlorine, calcium (Ca), magnesium (Mg), zinc (Zn), manganese (Mn), copper (Cu), and TOC. These results were analyzed for correlation with distal site positivity (Table 1). None of the measured parameters were shown to have a correlation with *Legionella* distal site positivity ($p$ values > 0.05). Hot water return line pH, free chlorine, calcium, magnesium, zinc, manganese, and copper also were not correlated with *Legionella* distal site positivity ($p$ values > 0.05). No comparisons could be made between either distal site or hot water return line iron (Fe) or lead (Pb) concentrations and *Legionella* distal site positivity because the concentrations were below the lower detection limit of the test method in 49% and 77% of samples for iron and lead, respectively.

### 2.3. Temperature Correlation

Distal site temperatures averaged 119.95 °F after a 1-min flush. Distal site temperature was not statistically related to the distal site *Legionella* positivity ($p = 0.170$). Distal sites with no *Legionella* recovered trended towards higher hot water return temperatures, with an average 9.7 °F higher temperature. However, this was not statistically significant ($p = 0.0687$).

When using 124 °F, the minimum recommended return circulation temperature [24], as a threshold for *Legionella* distal site positivity, there was an association between the two values ($p = 0.013$) (Figure 4). This hot water return temperature threshold value had a sensitivity of 65% (13/20) and a specificity of 72% (18/25) for determining the *Legionella* distal site positivity. However, the positive predictive value was only 65% (13/20) and the negative predictive value was 72% (18/25). This recommended temperature threshold is a poor screening test for distal site positivity (receiver operating characteristic (ROC) curve area = 0.68). In the buildings with a hot water return temperature > 124 °F, 28% (7/25) were still positive for *Legionella* in distal sites.

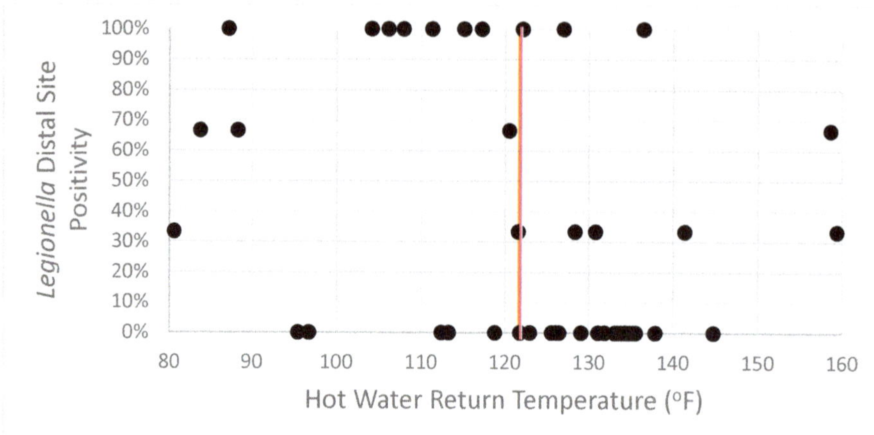


**Figure 4.** *Legionella* distal site positivity was seen in systems with hot water return line temperatures above and below the recommended threshold. The red vertical line represents the 124 °F threshold value.

### 2.4. Incoming Cold-Water Chlorine Concentration

Incoming cold-water chlorine concentration was analyzed to determine if the concentration could be used as a predictor of *Legionella* distal site positivity. Incoming cold-water chlorine measurements were available from 20 buildings. Using 0.5 mg/L as the threshold for the acceptable level of free residual chlorine found in drinking water [25], there was no correlation between incoming chlorine concentration and *Legionella* positivity ($p = 0.582$) (Figure 5). This 0.5 mg/L incoming chlorine threshold fails as a screening tool for *Legionella* distal site positivity (ROC curve area = 0.47).

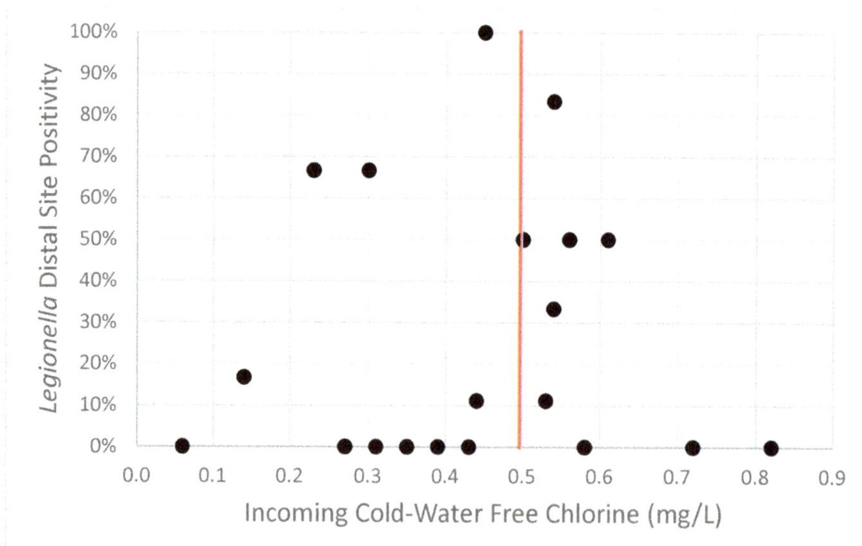


**Figure 5.** *Legionella* distal site positivity was seen in buildings with incoming cold-water free chlorine above and below the recommended threshold. The red vertical line represents the 0.5 mg/L threshold value.

## 3. Discussion

Public health agencies and some guidance documents recommend monitoring temperature and water quality parameters as part of *Legionella* risk assessments and water management programs. A reasonable assumption in following these recommendations is that there is some relationship to *Legionella* presence or absence.

However, few studies have been conducted to evaluate these relationships and to determine if any of these surrogate measurements can substitute for *Legionella* sampling in assessing risk or effectiveness of control measures [20–23] and there is a growing interest for a more efficient sampling approach for water system *Legionella* testing [20,26]. In this study we measured *Legionella* positivity, heterotrophic plate count bacteria, and other water physicochemical parameters, including temperature, free chlorine concentration, pH, calcium, magnesium, total organic carbon, iron, zinc, lead, manganese, and copper concentrations. We also evaluated the correlation between hot water return line *Legionella* positivity or these other water quality parameters to determine if any of these relationships were predictive of *Legionella* distal site positivity.

Our analysis showed that the concordance rate between hot water return line *Legionella* positivity and distal site *Legionella* positivity was 77.8%. This was similar to a previous report of 79% concordance between the hot water recirculation loop and distal sites [20]. In the current study, we further determined the sensitivity and specificity of using hot water return line *Legionella* positivity as a screening tool to predict distal site *Legionella* positivity. The low sensitivity (55%) indicates a low probability of finding distal site *Legionella* colonization based on the hot water return line *Legionella* positivity alone. In many cases, hot water return lines that yielded no *Legionella* had positive distal sites in that system.

Studies have linked the presence of *Legionella* in building water systems to water physicochemical parameters such as trace elements concentration, pH, and temperature. A significant association between *Legionella* presence and concentrations of Mn, Zn, and Fe was reported previously [22]. In another report, iron was significantly higher (average 1.43 mg/L) in *Legionella* positive public building water samples compared to *Legionella* negative samples (average 0.09 mg/L) [27]. This association was also seen with residential water systems [28]. In the present study, statistical analysis of the correlation between *Legionella* positivity and Fe as well as Pb concentration was not possible, because 49% and 77% of total samples had Fe and Pb concentrations lower than the lower detection limit (0.06 and 0.002 mg/L, respectively). Copper concentrations of > 0.05 mg/L have been associated with a lower risk of *Legionella* colonization [29].

We observed no statistically significant correlation between distal site *Legionella* positivity and HPC, temperature, pH, free chlorine (incoming cold water and distal site), Ca, Mg, Zn, Mn, Cu, and hot water return TOC. We found HPC concentration to be a poor predictor of *Legionella* positivity. HPC concentration was only able to explain 0.68% of the variation in *Legionella* distal site positivity. These results are consistent with a previous report that also showed the lack of correlation between total bacterial counts, measured by HPC, and *Legionella* colonization [23].

*Legionella* negative hot water systems trended towards higher temperature, Ca concentration, and lower hot water return TOC concentration, however, these were not statistically significant. In contrast to one other study, we found no correlation between *Legionella* colonization and manganese in building water systems [30]. *Legionella* positive systems trended towards higher Mn concentration on average, although this relationship was also not statistically significant.

From our experience, buildings > 10 stories high often have multiple centralized hot water systems installed to serve different building zones. The complexity of these centralized hot water systems lead to favorable environments for *Legionella* colonization, such as increased water age, favorable temperatures, and a lack of disinfectant residual [2,18,27,31,32]. Previous studies have shown a correlation between the building size and *Legionella* growth. These studies have indicated that larger buildings (>10 stories) and those with centralized hot water systems are more likely to support *Legionella* growth [17,18].

We did not see a statistical relationship in our study, between building size and *Legionella* positivity, however our data suggest the need for further investigation with larger data sets.

The risk of acquiring Legionnaires' disease has been previously associated with high levels of distal site *Legionella* positivity (>30%) [1,9,13–15]. Localized *Legionella* colonization at the point-of-use such as faucets and shower heads had been frequently observed and linked to the risk of susceptible individuals [33–35] and would serve as a patient–water system interaction point. Monitoring a facility's water system for *Legionella* colonization involves sampling distal hot water outlets regularly, which may necessitate the collection of numerous samples, especially for large facilities with multiple hot water systems [6,20].

Based on our study, sampling and culturing only the hot water return lines for *Legionella* presence demonstrates a low sensitivity of identifying *Legionella* colonization and therefore Legionnaires' disease risk. Similarly, the measured water quality parameters were not predictors of *Legionella* distal site positivity. Hot water temperature or incoming cold-water chlorine thresholds, 124 °F and 0.5 mg/L, respectively, also did not serve as good screening tools for *Legionella* colonization. In facilities with high risk residents, such as hospitals and long-term care facilities, a more conservative approach of direct sampling of at least 10 distal sites is recommended [9,36,37]. Based on our results, we recommend that in lower risk facilities, such as commercial or administrative buildings, sampling at least three distal sites and the hot water return should be done for routine surveillance in each hot water system. If positive samples are found, a more thorough examination of the extent and location of colonization may be warranted especially in office buildings where *Legionella* has been found to persist regardless of building age [2]. In ASHRAE Standard 188, an important part of any water management program is to ensure that there is validation of the program's efficacy. This is to ensure the water management program is controlling identified hazardous conditions, specifically the risk of *Legionella* growth and spread. Our results demonstrate that these surrogate measurements cannot be used to validate the control of *Legionella* risk at a facility because they are not predictive of the presence or absence of *Legionella* species.

## 4. Materials and Methods

### *4.1. Sample Collection and Onsite Water Quality Parameter Measurements*

Bulk water samples were collected from 28 buildings in New York City (25 buildings), San Francisco (two buildings), and New Jersey (one building) from March to September 2015. The sampled buildings included commercial buildings ranging from 5 to 57 floors. Samples were collected from the incoming cold-water per building, cold-water storage tank per cold water system, three hot water distal outlets per hot water system (near, mid and far), and the hot water return line from each hot water system.

Cold water and hot water return line samples were collected after a 1-min flush. A 1 L sample with sodium thiosulfate for microbiological analyses and a 250 mL sample preserved with nitric acid for metal analyses (iron (Fe), copper (Cu), lead (Pb), zinc (Zn), calcium (Ca), magnesium (Mg), and manganese (Mn)) were collected. Hot water distal outlet samples were collected and treated as above, but the 1 L microbiological sample was taken prior to flushing. Additionally, for hot water return line samples, two 50 mL vials with hydrochloric acid preservative were collected for total organic carbon (TOC) testing. Measurements for temperature, pH, and free chlorine were conducted onsite after sample collection.

Temperature, pH, and free chlorine residual concentration were measured on-site at the time of sample collection using a digital thermometer, portable Hach 900 colorimeter, and Oakton Acorn pH meter following the manufacturer's protocols. Samples for microbiological analyses were shipped on ice overnight to Special Pathogens Laboratory (Pittsburgh, PA, USA) and samples for metal analyses and total organic carbon testing were shipped to ALS Environmental (Middletown, PA, USA).

### 4.2. Microbiological Analyses

*Legionella* culture was conducted using buffered charcoal yeast extract agar (BCYE) (Remel, Lenexa, KS, USA) and selective media supplemented with glycine, vancomycin, and polymyxin B (DGVP) [38] using a modified ISO method [39,40]. Heterotrophic plate count (HPC) bacteria culture was performed using R2A agar (Remel, Lenexa, KS, USA) following standard method 9215B [41]. Culture media plates were prepared in-house with dehydrated media as noted above.

### 4.3. Statistical Analyses

For categorical data (building size, dichotomized threshold variables) *Legionella* positivity was compared using the Chi square test. Logistic models were used to evaluate *Legionella* positivity (presence/absence) and continuous variables (HPC, temperature, pH, free chlorine, calcium, magnesium, zinc, manganese, copper, and total organic carbon). A receiver operating characteristic (ROC) curve and the area under the curve (AUC) were generated to evaluate the utility of hot water return line temperature (dichotomized by recommended threshold) for predicting *Legionella* positivity. All statistics were performed using Stata version 13.0 (Stata Corp, College Station, TX, USA). A linear regression, and resulting $R^2$ value, was generated to evaluate the utility of HPC concentration for predicting *Legionella* positivity using Microsoft Excel.

## 5. Conclusions

Water quality measurements, including hot water return line *Legionella* positivity, total bacterial counts, temperature, and other physicochemical parameters, have previously been sought or suggested as alternative approaches to determine the *Legionella* risk for a building's water system instead of directly culturing the system. We found a concordance rate of only 77.8% between hot water return *Legionella* positivity and distal site *Legionella* positivity. Additionally, using hot water return line positivity as a predictor for *Legionella* distal site positivity had a sensitivity of only 55% and a specificity of 96%. There was no significant correlation between *Legionella* positivity and any water quality parameter (HPC, temperature, incoming cold-water chlorine, or physicochemical concentrations) tested. Neither hot water return line *Legionella* positivity nor other water quality parameters are suitable as a surrogate or stand-alone replacement for sampling and culturing distal sites for *Legionella* colonization in building water systems, especially in facilities with higher-risk populations.

**Author Contributions:** Conceptualization, F.P.S.III and J.E.S.; formal analysis, J.L.B., X.M., and M.M.W.; investigation, D.P. and X.M.; writing—original draft preparation, D.P., J.L.B. and X.M.; writing—review and editing, D.P., J.L.B., X.M., F.P.S.III, M.M.W., and J.E.S.; visualization, D.P. and J.L.B.; supervision, F.P.S.III and J.E.S.

**Funding:** This research received no external funding.

**Conflicts of Interest:** The authors declare no conflict of interest.

## References

1. Stout, J.E.; Yu, V.L.; Muraca, P.; Joly, J.; Troup, N.; Tompkins, L.S. Potable Water as a Cause of Sporadic Cases of Community-Acquired Legionnaires' Disease. *N. Engl. J. Med.* **1992**, *326*, 151–155. [CrossRef] [PubMed]
2. Donohue, M.J.; King, D.; Pfaller, S.; Mistry, J.H. The sporadic nature of Legionella pneumophila, Legionella pneumophila Sg1 and Mycobacterium avium occurrence within residences and office buildings across 36 states in the United States. *J. Appl. Microbiol.* **2019**, *126*, 1568–1579. [CrossRef] [PubMed]
3. Donohue, M.J.; O'Connell, K.; Vesper, S.J.; Mistry, J.H.; King, D.; Kostich, M.; Pfaller, S. Widespread Molecular Detection of Legionella pneumophila Serogroup 1 in Cold Water Taps across the United States. *Environ. Sci. Technol.* **2014**, *48*, 3145–3152. [CrossRef] [PubMed]
4. Anaissie, E.J.; Penzak, S.R.; Dignani, M.C. The hospital water supply as a source of nosocomial infections: A plea for action. *Arch. Intern. Med.* **2002**, *162*, 1483–1492. [CrossRef]
5. ASHRAE. ANSI/ASHRAE Standard 188-2018. In *Legionellosis: Risk Management for Building Water Systems*; American Society of Heating, Refrigerating and Air-Conditioning Engineers: Atlanta, GA, USA, 2018.

6. World Health Organization (WHO). *Legionella and the Prevention of Legionellosis*; WHO: Geneva, Switzerland, 2007.
7. AIHA. *Recognition, Evaluation, and Control of Legionella in Building Water Systems*; American Industrial Hygiene Association: Falls Church, VA USA, 2015.
8. Centers for Disease Control and Prevention (CDC). *Developing a Water Management Program to Reduce Legionella Growth & Spread in Buildings—A Practical Guide to Implementing Industry Standards*; US Department of Health and Human Services: Atlanta, GA, USA, 2017.
9. Stout, J.E.; Yu, V.L. Environmental culturing for Legionella: Can we build a better mouse trap? *Am. J. Infect. Control* **2010**, *38*, 341–343. [CrossRef]
10. Stout, J.E.; Muder, R.R.; Mietzner, S.; Wagener, M.M.; Perri, M.B.; DeRoos, K.; Goodrich, D.; Arnold, W.; Williamson, T.; Ruark, O.; et al. Role of environmental surveillance in determining the risk of hospital-acquired legionellosis: A national surveillance study with clinical correlations. *Infect. Control Hosp. Epidemiol. Off. J. Soc. Hosp. Epidemiol. Am.* **2007**, *28*, 818–824. [CrossRef]
11. Kozak, N.A.; Lucas, C.E.; Winchell, J.M. Identification of Legionella in the Environment. In *Legionella: Methods and Protocols*; Buchrieser, C., Hilbi, H., Eds.; Humana Press: Totowa, NJ, USA, 2013; pp. 3–25. [CrossRef]
12. ASTM. *D5952-08(2015), Standard Guide for the Inspection of Water Systems for Legionella and the Investigation of Possible Outbreaks of Legionellosis (Legionnaires' Disease or Pontiac Fever)*; ASTM International: West Conshohocken, PA, USA, 2015.
13. Best, M.; Yu, V.L.; Stout, J.; Goetz, A.; Muder, R.R.; Taylor, F. Legionellaceae in the hospital water supply–epidemiological link with disease and evaluation of a method of control of nosocomial Legionnaires' disease and Pittsburgh pneumonia. *Lancet* **1983**, *2*, 307–310. [CrossRef]
14. Kool, J.L.; Bergmire-Sweat, D.; Butler, J.C. Hospital characteristics associated with colonization of water systems by Legionella and risk of nosocomial legionnaires' disease: A cohort study of 15 hospitals. *Infect. Control Hosp. Epidemiol.* **1999**, *20*, 798–805. [CrossRef]
15. Sabria, M.; Modol, J.M.; Garcia-Nunez, M.; Reynaga, E.; Pedro-Botet, M.L.; Sopena, N. Environmental cultures and hospital-acquired Legionnaires' disease: A 5-year prospective study in 20 hospitals in Catalonia, Spain. *Infect. Control Hosp. Epidemiol. Off. J. Soc. Hosp. Epidemiol. Am.* **2004**, *25*, 1072–1076. [CrossRef]
16. Boccia, S.; Laurenti, P.; Borella, P.; Moscato, U.; Capalbo, G.; Cambieri, A.; Amore, R.; Quaranta, G.; Boninti, F.; Orsini, M.; et al. Prospective 3-year surveillance for nosocomial and environmental Legionella pneumophila: Implications for infection control. *Infect. Control Hosp. Epidemiol. Off. J. Soc. Hosp. Epidemiol. Am.* **2006**, *27*, 459–465. [CrossRef]
17. Flannery, B.; Gelling, L.B.; Vugia, D.J.; Weintraub, J.M.; Salerno, J.J.; Conroy, M.J.; Stevens, V.A.; Rose, C.E.; Moore, M.R.; Fields, B.S.; et al. Reducing Legionella colonization in water systems with monochloramine. *Emerg. Infect. Dis.* **2006**, *12*, 588–596. [CrossRef] [PubMed]
18. Borella, P.; Montagna, M.T.; Romano-Spica, V.; Stampi, S.; Stancanelli, G.; Triassi, M.; Neglia, R.; Marchesi, I.; Fantuzzi, G.; Tato, D.; et al. Legionella infection risk from domestic hot water. *Emerg. Infect. Dis.* **2004**, *10*, 457–464. [CrossRef] [PubMed]
19. Wright, D.R.; Centers for Medicare & Medicaid Services. Requirement to Reduce Legionella Risk in Healthcare Facility Water Systems to Prevent Cases and Outbreaks of Legionnaires' Disease (LD). In *S&C 17-30-Hospitals/CAHs/NHs*; Department of Health & Human Services: Baltimore, MD, USA, 2017.
20. Ditommaso, S.; Giacomuzzi, M.; Gentile, M.; Moiraghi, A.R.; Zotti, C.M. Effective environmental sampling strategies for monitoring *Legionella* spp contamination in hot water systems. *Am. J. Infect. Control* **2010**, *38*, 344–349. [CrossRef]
21. Bédard, E.; Fey, S.; Charron, D.; Lalancette, C.; Cantin, P.; Dolcé, P.; Laferrière, C.; Déziel, E.; Prévost, M. Temperature diagnostic to identify high risk areas and optimize Legionella pneumophila surveillance in hot water distribution systems. *Water Res.* **2015**, *71*, 244–256. [CrossRef] [PubMed]
22. Bargellini, A.; Marchesi, I.; Righi, E.; Ferrari, A.; Cencetti, S.; Borella, P.; Rovesti, S. Parameters predictive of Legionella contamination in hot water systems: Association with trace elements and heterotrophic plate counts. *Water Res.* **2011**, *45*, 2315–2321. [CrossRef]
23. Duda, S.; Baron, J.L.; Wagener, M.M.; Vidic, R.D.; Stout, J.E. Lack of correlation between Legionella colonization and microbial population quantification using heterotrophic plate count and adenosine triphosphate bioluminescence measurement. *Environ. Monit. Assess.* **2015**, *187*, 393. [CrossRef]

24. ASHRAE. ASHRAE Guideline 12-2000. In *Minimizing the Risk of Legionellosis Associated with Building Water Systems*; American Society of Heating, Refrigerating and Air-Conditioning Engineers: Atlanta, GA, USA, 2000.
25. World Health Organization (WHO). *Guidelines for Drinking-Water Quality*; WHO: Geneva, Switzerland, 2011.
26. Wang, H.; Bédard, E.; Prévost, M.; Camper, A.K.; Hill, V.R.; Pruden, A. Methodological approaches for monitoring opportunistic pathogens in premise plumbing: A review. *Water Res.* **2017**, *117*, 68–86. [CrossRef]
27. Edagawa, A.; Kimura, A.; Doi, H.; Tanaka, H.; Tomioka, K.; Sakabe, K.; Nakajima, C.; Suzuki, Y. Detection of culturable and nonculturable Legionella species from hot water systems of public buildings in Japan. *J. Appl. Microbiol.* **2008**, *105*, 2104–2114. [CrossRef]
28. Stout, J.E.; Yu, V.L.; Yee, Y.C.; Vaccarello, S.; Diven, W.; Lee, T.C. Legionella pneumophila in residential water supplies: Environmental surveillance with clinical assessment for Legionnaires' disease. *Epidemiol. Infect.* **1992**, *190*, 49–57.
29. Leoni, E.; De Luca, G.; Legnani, P.P.; Sacchetti, R.; Stampi, S.; Zanetti, F. Legionella waterline colonization: Detection of Legionella species in domestic, hotel and hospital hot water systems. *J. Appl. Microbiol.* **2005**, *98*, 373–379. [CrossRef]
30. Kehres, D.G.; Maguire, M.E. Emerging themes in manganese transport, biochemistry and pathogenesis in bacteria. *Fems Microbiol. Rev.* **2003**, *27*, 263–290. [CrossRef]
31. Rhoads, W.J.; Pruden, A.; Edwards, M.A. Convective Mixing in Distal Pipes Exacerbates Legionella pneumophila Growth in Hot Water Plumbing. *Pathogens* **2016**, *5*, 29. [CrossRef]
32. Serrano-Suarez, A.; Dellunde, J.; Salvado, H.; Cervero-Arago, S.; Mendez, J.; Canals, O.; Blanco, S.; Arcas, A.; Araujo, R. Microbial and physicochemical parameters associated with Legionella contamination in hot water recirculation systems. *Environ. Sci. Pollut. Res.* **2013**, *20*, 5534–5544. [CrossRef]
33. Halabi, M.; Wiesholzer-Pittl, M.; Schoberl, J.; Mittermayer, H. Non-touch fittings in hospitals: A possible source of *Pseudomonas aeruginosa* and *Legionella* spp. *J. Hosp. Infect.* **2001**, *49*, 117–121. [CrossRef]
34. Hosein, I.K.; Hill, D.W.; Tan, T.Y.; Butchart, E.G.; Wilson, K.; Finlay, G.; Burge, S.; Ribeiro, C.D. Point-of-care controls for nosocomial legionellosis combined with chlorine dioxide potable water decontamination: A two-year survey at a Welsh teaching hospital. *J. Hosp. Infect.* **2005**, *61*, 100–106. [CrossRef]
35. Kusnetsov, J.; Torvinen, E.; Perola, O.; Nousiainen, T.; Katila, M.L. Colonization of hospital water systems by legionellae, mycobacteria and other heterotrophic bacteria potentially hazardous to risk group patients. *APMIS Acta Pathol. Microbiol. Immunol. Scand.* **2003**, *111*, 546–556. [CrossRef]
36. Allegheny County Health Department. *Approaches to Prevention and Control of Legionella Infection in Allegheny County Health Care Facilities*, 2nd ed.; Allegheny County Health Department: Pittsburgh, PA, USA, 1997; pp. 1–15.
37. Clancy, C.M.; Department of Veterans Affairs; Veterans Health Administration. *VHA Directive 1061: Prevention of Healthcare-Associated Legionella Disease and Scald Injury from Potable Water Distribution Systems*; Veterans Health Administration: Washington, DC, USA, 2014.
38. Vickers, R.M.; Brown, A.; Garrity, G.M. Dye-containing buffered charcoal yeast extract medium for the differentiation of members of the family Legionellaceae. *J. Clin. Microbiol.* **1981**, *13*, 380–382.
39. ISO. Water Quality—Detection and Enumeration of *Legionella*. In *International Organization for Standardization*; ISO 11731: Geneva, Switzerland, 1998; pp. 1–16.
40. ISO. *11731-2:2004 Water Quality Detection and Enumeration of Legionella Part 2: Direct Membrane Filtration Method for Waters with Low Bacterial Counts*; ISO: Geneva, Switzerland, 2004.
41. Eaton, E.; Rice, E.; Baird, R. (Eds.) *Standard Methods for the Examination of Water and Waste Water*; American Public Health Association: Washington, DC, USA, 2005.

*pathogens*

MDPI

*Article*

# Evaluation of GVPC and BCYE Media for *Legionella* Detection and Enumeration in Water Samples by ISO 11731: Does Plating on BCYE Medium Really Improve Yield?

**Maria Scaturro [1], Elisa Poznanski [2], Mariarosaria Mupo [2], Paola Blasior [2], Margit Seeber [2], Anna-Maria Prast [2], Elisa Romanin [2], Antonietta Girolamo [1], Maria Cristina Rota [1], Antonino Bella [1], Maria Luisa Ricci [1,*,†] and Alberta Stenico [2,*,†]**

1 Department of Infectious Diseases, Istituto Superiore di Sanità, Viale Regina Elena 299, 00161 Rome, Italy; maria.scaturro@iss.it (M.S.); antonietta.girolamo@iss.it (A.G.); mariacristina.rota@iss.it (M.C.R.); antonino.bella@iss.it (A.B.)

2 Biological Laboratory, Provincial Agency for the Environment and Climate Protection, 39055 Laives (Bolzano), Italy; elisa.poznanski@provincia.bz.it (E.P.); mariarosaria.mupo@provincia.bz.it (M.M.); paola.blasior@provincia.bz.it (P.B.); margit.seeber@provincia.bz.it (M.S.); anna-maria.prast@provincia.bz.it (A.-M.P.); elisa.romanin@provincia.bz.it (E.R.)

* Correspondence: marialuisa.ricci@iss.it (M.L.R.); alberta.stenico@provincia.bz.it (A.S.); Tel.: +39-06-4990-2856 (M.L.R.); +39-04-7195-0431 (A.S.)

† These authors contributed equally to this work.

Received: 28 August 2020; Accepted: 15 September 2020; Published: 16 September 2020

**Abstract:** *Legionella spp* are the causative agents of Legionnaires' diseases, which is a pneumonia of important public health concern. Ubiquitous freshwater and soil inhabitants can reach man-made water systems and cause illness. *Legionella* enumeration and quantification in water systems is crucial for risk assessment and culture examination is the gold standard method. In this study, *Legionella* recovery from potable water samples, at presumably a low concentration of interfering microorganisms, was compared by plating on buffered charcoal yeast extract (BCYE) and glycine, vancomycin, polymyxin B, cycloheximide (GVPC) *Legionella* agar media, according to the International Standard Organization (ISO) 11731: 2017. Overall, 556 potable water samples were analyzed and 151 (27.1%) were positive for *Legionella*. *Legionella* grew on both BCYE and GVPC agar plates in 85/151 (56.3%) water samples, in 65/151 (43%) on only GVPC agar plates, and in 1/151 (0.7%) on only BCYE agar plates. In addition, GVPC medium identified *Legionella* species other than *pneumophila* in six more samples as compared with the culture on BCYE. Although the medians of colony forming units per liter (CFU/L) detected on the BCYE and GVPC agar plates were 2500 and 1350, respectively ($p$-value < 0.0001), the difference did not exceed one logarithm, and therefore is not relevant for *Legionella* risk assessment. These results make questionable the need to utilize BCYE agar plates to analyze potable water samples.

**Keywords:** *Legionella*; Legionnaires' disease; culture; BCYE and GVPC media

## 1. Introduction

*Legionella* is a water-born pathogen widely spread in man-made water systems, responsible for a severe pneumonia and a flu-like illness, named Legionnaires' disease (LD) and the Pontiac fever, respectively. Overall, at the present time, 62 *Legionella* spp. have been identified and less than a half were pathogenic, however, *Legionella pneumophila* was surely the most frequently found in LD cases.

After its first isolation following the large outbreak in Philadelphia in 1976, *Legionella* has become an opportunistic pathogen of major concern, because of a worldwide increasing number of both sporadic cases and outbreak events [1–4]. Outbreak investigations have widely demonstrated that the most frequent sources of infection are water systems of different buildings, such as hotels or hospitals and, specifically, showers, cooling towers, and spa pools [5,6].

The timely identification of the source of an infection is of great importance to prevent clusters or outbreaks and culture examination is the gold standard for the analyses of water samples. Although molecular methods have been demonstrated to be highly sensitive and specific, as well as able to detect all *Legionella* species and serogroups, they remain impracticable for *Legionella* enumeration because they detect DNA of both living and dead bacteria [7,8].

The *Legionella* laboratory isolation is of great relevance for further deeper molecular investigations, in order to characterize clinical and environmental strains and identify the source of infection [9]. Furthermore, according to the European guidelines for *Legionella* [10], a quantitative evaluation of the contamination of water systems due to *Legionella* can be determined only by a culture, even though a culture has been demonstrated to have some drawbacks [11]. The counting of colony forming units per liter (CFU/L) is a crucial step for risk assessment and, as a consequence, to decide the right control measures to be adopted. The fastidious growth requirements of *Legionella*, the overgrowth of other bacteria, as well as the medium required by the specifically adopted culture method can affect the results of the analysis and determine a variable range of *Legionella* concentrations.

Culture methods are generally performed according to standards, such as the international standard organization (ISO), recognized by each country's accreditation body [12]. In particular, the ISO 11731 is the most used and it has recently been updated, replacing the previous ones published in 1998 and 2004 (ISO 11731: 2017). Chemical formulation of culture media, as well as the acid and heat treatments, required by the ISO 11731 for the enumeration of *Legionella* in water samples, may have different effects on the recovery of *Legionella*, independent of the manufacturer of the commercially available media [13,14].

The updated ISO 11731 introduced the utilization of the following three media: the buffered charcoal yeast extract (BCYE) agar; the BCYE with selective supplements (BCYE + AB), containing polymixin B, sodium cefazolin, and pimaricin; and the highly selective Modified Wadowsky Yee (MWY) agar or, as an alternative, the glycine, vancomycin, polymyxin B, cycloheximide (GVPC) agar. The choice of the selective medium to be used is linked to the specific potential bacterial contamination of water samples. Potable waters and any other water samples with background microorganisms must be analyzed using selective media, with the capability of reducing background microorganisms. In the ISO 11731: 2017, GVPC and MWY are both considered to be equally efficient for *Legionella* recovery. It has been demonstrated that MWY was the best medium for isolating *Legionella pneumophila* from potable water samples and GVPC was the most effective for reducing additional microbial flora [15,16]. According to ISO 11731: 2017, the decision matrix in Annex J shows that plating on BCYE agar is specifically required for potable water samples when the enumeration of *Legionella* is determined by the following methods: (i) direct plating without any concentration and treatment, (ii) membrane filtration and direct placing of the membrane filter on culture media, and (iii) membrane filtration followed by washing procedure.

In this study, the *Legionella* recovery from potable water samples, concentrated by filtration with washing procedure, was determined and the recovery after plating on BCYE and GVPC agar plates was compared.

## 2. Material and Methods

### 2.1. Water Samples and Culture

Overall, 556 water samples were collected from accommodation sites, hospitals, and private homes and, according to the ISO 11731: 2017, they were classified as belonging to the identified Matrix

A, being water samples expected at low concentration of interfering microorganisms. Sampling was performed at different sampling points (shower, faucet, boiler, etc.) of accommodation sites and hospitals according to protocols reported in Italian guidelines [17]. Briefly, an instant water sample was collected to simulate exposure by a user, without flaming and disinfecting the outlet, and without running water. The temperature was measured immediately before filling the one-liter bottle.

The BCYE and GVPC (Oxoid, Thermo Fisher Diagnostics Limited, Cheshire, UK) agar plates and reagents were prepared according to ISO 11731: 2017. Legionella CYE agar base, BCYE-α growth supplement, and GVPC selective supplement were purchased from Oxoid, Thermo Fisher Diagnostics Limited, Cheshire, UK. For each lot of both media, a quality control was carried out according to ISO 11133: 2014 [18]. The reference material Easy-tab Reference Material (LGC, Bury, UK) was utilized for performance testing of the two media, and always resulted within the declared range. Selectivity of GVPC was qualitative determined according to ISO 11133: 2014. Ringer solution was used to wash polycarbonate membranes using a vortex mixer.

For each sample, a volume of one liter was collected and it was concentrated 200 times by filtration, using 0.22 µm polycarbonate membrane, followed by the washing procedure of the filter. After filtration the membrane was placed in a screw cap sterile container with 5 mL of diluted Ringer's solution. The membrane was washed by shaking vigorously for at least 2 min using a vortex mixer. The concentrated sample was divided into the following three aliquots, according to ISO 11731: 2017: one ml was heat treated, one ml was acid treated, and the remaining 3 mL were untreated. Then, one hundred µL of each aliquot were placed on both BCYE and GVPC agar plates and incubated at 37 °C, for ten days. The plates were checked after four or five days, and after 10 days. According to the concentration procedure, the detection limit of our method was 50 CFU/L.

The laboratory that analyzed the samples is accredited for the detection and enumeration of *Legionella* according to ISO 11731: 2017, by the Italian national accreditation body (Accredia).

### 2.2. Statistical Analysis

The McNemar's test was used to compare frequency on paired data. The concordance between media was evaluated using the Kappa test (K < 0.20 = "poor", 0.20–0.40 = "fair", 0.40–0.60 = "moderate", 0.60–0.80 = "good", and 0.80–1.00 = "very good"). Specificity and sensitivity, as well positive and negative predictive values (PPV and NPV, respectively), and 95% confidence intervals (CI) for both media were calculated, considering the BCYE as a reference medium.

All statistical analyses were performed by Stata software version 11.2 (Stata Corp, College Station, TX, USA).

## 3. Results

All of the 556 water samples were analyzed by culture and, overall, 151 (27.1%) were positive for Legionella, of which 65 (43%) grew on only GVPC, 85 (56.3%) grew on both GVPC and BCYE, and one (0.7%) grew on only BCYE. (Table 1). The difference of the results obtained analyzing the samples by the two media was significant (McNemar's test, $p < 0.0001$).

**Table 1.** Recovery of *Legionella* by using buffered charcoal yeast extract (BCYE) and glycine, vancomycin, polymyxin B, cycloheximide (GVPC) media.

| | | GVPC | | |
|---|---|---|---|---|
| | | **Negative** | **Positive** | **Total** |
| **BCYE** | Negative | 405 | 65 | 470 |
| | Positive | 1 | 85 | 86 |
| | Total | 406 | 150 | 556 |

McNemar's test *p*-value <0.0001.

The sensitivity, specificity, positive predictive value (PPV), and negative predictive value (NPV) of GVPC vs. BCYE media are shown in Table 2. GVPC demonstrated a greater sensitivity and a good specificity as compared with BCYE medium. The Cohen's Kappa coefficient calculated on these data provided a value of 0.65, indicating a good quality of agreement ($p < 0.0001$) (Table 2).

**Table 2.** GVPC vs. BCYE sensitivity, specificity, positive predictive value (PPV), negative predictive value (NPV), concordance, Kappa value, and *p*-value.

| Comparison | % (95% CI) |
|---|---|
| Sensitivity | 98.8 (97.9–99.7) |
| Specificity | 86.2 (83.3–89.0) |
| PPV | 56.7 (52.5–60.8) |
| NPV | 99.7 (99.3–100.2) |
| Concordance | 88.1 |
| Kappa value (*p*-value) | 0.65 (<0.0001) |

CI, confidence interval; PPV, positive predictive value; NPV, negative predictive value.

Considering the 85 samples that were positive on both GVPC and BCYE media, 11 samples showed fewer colony forming units per liter (CFU/L) on GVPC than on BCYE (Table 3). However, the differences of CFU/L found between the two media were never higher than 1 log (*p*-value = 0.0388).

**Table 3.** *Legionella* CFU/L range detected on samples positive on both BCYE and GVPC media.

| | | GVPC | | | Total |
|---|---|---|---|---|---|
| | CFU/L | 50–1000 | 1050–10,000 | >10,000 | |
| BCYE | 50–1000 | 30 | 3 | 0 | 33 |
| | 1050–10,000 | 5 | 33 | 0 | 38 |
| | >10,000 | 0 | 6 | 8 | 14 |
| | Total | 35 | 42 | 8 | 85 |

Figure 1 shows the distribution of *L. pneumophila* serogroups and *Legionella* species found on GVPC and BCYE agar plates. *Legionella* species were detected in 19 samples by GVPC and in 13 samples by BCYE. Considering the range of the colony forming unit (CFU)/L detected, the data proved that, among the 65 samples positive on only GVPC, there were 51 samples (78.4%) showing a range between 50 and 1000, highlighting the high efficiency of GVPC in isolating low bacterial counts. The calculated medians of CFU/L detected on BCYE and GVPC plates were equal to 2500 (interquartile range = 5500) on BCYE and 1350 (interquartile range = 3950) on GVPC, and the difference between the two media was significant (*p*-value < 0.0001).

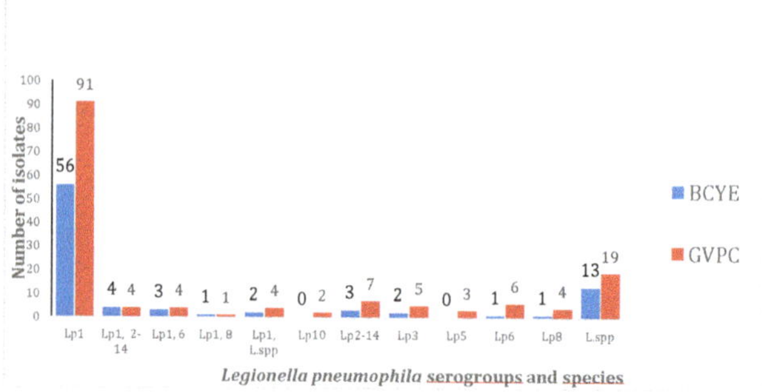


**Figure 1.** Distribution of *Legionella pneumophila* serogroups and *Legionella* species (BCYE n = 86 vs. GVPC n = 150).

## 4. Discussion

In this study, potable water samples were analyzed according to the ISO1173: 2017, in order to compare the *Legionella* recovery obtained by plating on BCYE and GVPC agar plates. We observed that GVPC was more efficient in detecting *Legionella* than BCYE medium. Indeed, 43% of the overall positive samples were detected on only GVPC agar plates and, in addition, the positivity of water samples at low bacterial counts, corresponding to 78.4% of the total positive samples, was determined only by using this medium. This finding has significant relevance especially when, in specific water systems as hospital wards or thermal waters, the absence or a strong containment of Legionella contamination must be guaranteed, due to possible exposure by people at increased risk of acquiring LD. In a previous study, it was demonstrated that a much greater yield of *Legionella* spp. was obtained by plating on BCYE than on MWY agar plates, and a significantly higher number of CFU of both *Legionella pneumophila* and non-pneumophila was counted on BCYE as compared with MWY [19]. On the contrary, Leoni et al. demonstrated the significantly higher yield on GVPC and MWY than on BCYE medium, in combination with the technique used of direct inoculum or pretreatment with acid or heat [18]. Furthermore, other studies have shown no statistically significant differences between BCYE and GVPC media in recovering *Legionella* in water samples [14,20]. In this study, *Legionella* recovery was determined by comparing the CFU/L counted in BCYE and GVPC, and although significantly higher CFU/L were found in BCYE than in GVPC agar plates, the difference did not exceed one log. Therefore, it was not relevant for *Legionella* risk assessment of drinking water systems.

Furthermore, the possibility of improving the isolation of *Legionella* non-*pneumophila* species by plating the water samples on BCYE was not confirmed. Indeed, *Legionella* species other than *pneumophila* were detected more on GVPC than on BCYE.

In conclusion, these results cast doubt on the advantages of analyzing water samples using only BCYE, as required by ISO 11731: 2017. Further investigations by analyzing a larger number of water samples should be conducted to confirm these data, which, if confirmed, would bring enormous benefits, saving time and money, especially for laboratories that carry out monitoring activities and analyze hundreds of samples daily.

**Author Contributions:** Conceptualization, A.S.; Data curation, E.P., M.M., P.B., A.B., and A.S.; Formal analysis, E.P., M.M., P.B., M.S. (Margit Seeber), E.R., and A.-M.P.; Investigation, E.P., A.G., M.S. (Margit Seeber), E.R., and A.-M.P.; Methodology, M.M., E.P. and P.B.; Software, A.B.; Supervision, A.S.; Validation, A.B. and M.L.R.; Visualization, A.G.; Writing—Original draft, M.S. (Maria Scaturro); Writing—Review and editing, M.C.R. and M.L.R. All authors have read and agreed to the published version of the manuscript.

**Funding:** This research received no external funding.

**Conflicts of Interest:** The author declares no conflict of interest.

## References

1. Fraser, D.W.; Tsai, T.R.; Orenstein, W.; Parkin, W.E.; Beecham, H.J.; Sharrar, R.G.; Harris, J.; Mallison, G.F.; Martin, M.S.; McDade, J.E.; et al. Legionnaires' disease: Description of an epidemic of pneumonia. *N. Engl. J. Med.* **1977**, *297*, 1189–1197. [CrossRef] [PubMed]
2. Hamilton, K.A.; Prussin, A.J., 2nd; Ahmed, W.; Haas, C.N. Outbreaks of Legionnaires' Disease and Pontiac Fever 2006–2017. *Curr. Environ. Health Rep.* **2018**, *5*, 263–271. [CrossRef] [PubMed]
3. Legionnaires' Disease–Annual Epidemiological Report for 2017. Surveillance Report 17 January 2019. Available online: https://www.ecdc.europa.eu/en/publications-data/legionnaires-disease-annual-epidemiological-report-2017 (accessed on 16 September 2020).
4. Rota, M.C.; Caporali, M.G.; Bella, A.; Scaturro, M.; Giannitelli, S.; Ricci, M.L. Rapporto annuale sulla legionellosi in italia nel 2018. *Not. Ist. Super. Sanità* **2019**, *32*, 7–13.
5. Orkis, L.T.; Harrison, L.H.; Mertz, K.J.; Brooks, M.M.; Bibby, K.J.; Stout, J.E. Environmental sources of community-acquired legionnaires' disease: A review. *Int. J. Hyg. Environ. Health* **2018**, *221*, 764–774. [CrossRef] [PubMed]

6. De Filippis, P.; Mozzetti, C.; Amicosante, M.; D'Alò, G.L.; Messina, A.; Varrenti, D.; Giammattei, R.; Di Giorgio, F.; Corradi, S.; D'Auria, A.; et al. Occurrence of *Legionella* in showers at recreational facilities. *J. Water Health* **2017**, *15*, 402–409. [CrossRef] [PubMed]
7. Collins, S.; Stevenson, D.; Walker, J.; Bennett, A. Evaluation of *Legionella* real-time PCR against traditional culture for routine and public health testing of water samples. *J. Appl. Microbiol.* **2017**, *122*, 1692–1703. [CrossRef] [PubMed]
8. Lee, J.V.; Lai, S.; Exner, M.; Lenz, J.; Gaia, V.; Casati, S.; Hartemann, P.; Lück, C.; Pangon, B.; Ricci, M.L.; et al. An international trial of quantitative PCR for monitoring *Legionella* in artificial water systems. *J. Appl. Microbiol.* **2011**, *110*, 1032–1044. [CrossRef] [PubMed]
9. Phin, N.; Parry-Ford, F.; Harrison, T.; Stagg, H.R.; Zhang, N.; Kumar, K.; Lortholary, O.; Zumla, A.; Abubakar, I. Epidemiology and clinical management of Legionnaires' disease. *Lancet. Infect. Dis.* **2014**, *14*, 1011–1021. [CrossRef]
10. European Technical Guidelines for the Prevention, Control and Investigation, of Infections Caused by Legionella Species" June 2017. Available online: https://www.ecdc.europa.eu/en/publications-data/european-technical-guidelines-prevention-control-and-investigation-infections (accessed on 16 September 2020).
11. Collins, S.; Jorgensen, F.; Willis, C.; Walker, J. Real-time PCR to supplement gold-standard culture-based detection of *Legionella* in environmental samples. *J. Appl. Microbiol.* **2015**, *119*, 1158–1169. [CrossRef] [PubMed]
12. *ISO-ISO 11731: 2017-Water quality—Enumeration of Legionella*; ISO: Geneva, Switzerland, 2017.
13. Lück, P.C.; Igel, L.; Helbig, J.H.; Kuhlisch, E.; Jatzwauk, L. Comparison of commercially available media for the recovery of *Legionella* species. *Int. J. Hyg. Environ. Health* **2004**, *207*, 589–593. [CrossRef] [PubMed]
14. Lee, T.C.; Vickers, R.M.; Yu, V.L.; Wagener, M.M. Growth of 28 *Legionella* species on selective culture media: A comparative study. *Clin. Microbiol.* **1993**, *31*, 2764–2768. [CrossRef] [PubMed]
15. Reinthaler, F.F.; Sattler, J.; Schaffler-Dullnig, K.; Weinmayr, B.; Marth, E. Comparative study of procedures for isolation and cultivation of *Legionella pneumophila* from tap water in hospitals. *J. Clin. Microbiol.* **1993**, *31*, 1213–1216. [CrossRef] [PubMed]
16. Edelstein, P.H. Comparative study of selective media for isolation of *Legionella pneumophila* from potable water. *J. Clin. Microbiol.* **1982**, *16*, 697–699. [CrossRef] [PubMed]
17. Linee Guida per La Prevenzione Ed Il Controllo Della Legionellosi. 2015. Available online: http://www.salute.gov.it/imgs/C_17_pubblicazioni_2362_allegato.pdf (accessed on 14 September 2020).
18. *ISO-ISO 11133: 2014-Microbiology of Food, Animal Feed and Water—Preparation, Production, Storage and Performance Testing of Culture Media*; Amendments 1:2018 and 2:2020 included; ISO: Geneva, Switzerland, 2014.
19. Ditommaso, S.; Gentile, M.; Giacomuzzi, M.; Zotti, C.M. Recovery of Legionella species from water samples using an internal method based on ISO 11731: Suggestions for revision and implementation. *Diagn. Microbiol. Infect. Dis.* **2011**, *70*, 200–206. [CrossRef]
20. Leoni, E.; Legnani, P.P. Comparison of selective procedures for isolation and enumeration of Legionella species from hot water systems. *J. Appl. Microbiol.* **2001**, *90*, 27–33. [CrossRef] [PubMed]

*pathogens*

MDPI

*Article*

# Sensitivity and Selectivity of Two Commercially Available Media for *Legionella* spp. Recovery from Environmental Water Samples

**Savina Ditommaso *, Monica Giacomuzzi, Gabriele Memoli, Jacopo Garlasco and Carla M. Zotti**

Department of Public Health and Pediatrics, University of Turin, 10126 Turin, Italy; monica.giacomuzzi@unito.it (M.G.); gabriele.memoli@unito.it (G.M.); jacopo.garlasco@unito.it (J.G.); carla.zotti@unito.it (C.M.Z.)

* Correspondence: savina.ditommaso@unito.it; Tel.: +39-0116705841; Fax: +39-0116705881

Received: 11 May 2020; Accepted: 22 June 2020; Published: 29 June 2020

**Abstract:** The quality control of culture media used for *Legionella* spp. isolation and enumeration is paramount to achieve a satisfactory degree of comparability among water testing results from different laboratories. Here, we report on a comparative assessment of the sensitivity and selectivity of MWY and BCYEα media supplied by two different manufacturers (i.e., Xebios Diagnostics GmbH and Oxoid Ltd) for the detection of *Legionella* spp. from environmental water samples. Even though our analysis showed an excellent agreement between the recovery rates of the four media tested (90.5%), the quantitative recovery of *Legionella* spp. colonies using Xebios media was significantly greater than that achieved by Oxoid media ($P$ = 0.0054). Furthermore, the sensitivity of detection was significantly higher when samples were plated on MWY $^{\text{Xebios}}$ agar ($P$ = 0.0442), while the selectivity of MWY appeared to be the same regardless of the manufacturer. Furthermore, MWY$^{\text{Xebios}}$ agar favored the growth of much larger colonies compared to those observed on MWY$^{\text{Oxoid}}$ agar. Finally, MWY$^{\text{Xebios}}$ medium enhanced the recovery of non-*pneumophila Legionella* species. Collectively, our findings demonstrate that quality control is crucial to ensure high selectivity and sensitivity of the culture media used for the detection and enumeration of *Legionella* spp. from environmental water resources. As water remediation measures strictly depend on *Legionella* spp. recovery, culture protocol standardization, as well as quality control of the culture media, is essential to achieve intra- and interlaboratory reproducibility and accuracy.

**Keywords:** *Legionella* spp.; culture media; environmental monitoring; quality control

## 1. Introduction

The most common protocol for environmental surveillance of *Legionella* spp. involves the use of buffered charcoal yeast extract agar enriched with 1 g/L alpha-ketoglutarate agar (BCYEα). Although this method has a proven record of effectively isolating and enumerating *Legionella* species from environmental or clinical specimens, its sensitivity and selectivity are often hampered by the presence of contaminating flora in the water samples, which may influence the final count of *Legionella* spp. due to overgrowth or inhibition. Therefore, culture on selective agar media—i.e., BCYEα supplemented with glycine, vancomycin, polymyxin B, cycloheximide (GVPC) or with glycine, polymyxin B, vancomycin, anisomycin, bromothymol blue, and bromocresol purple (MWY)—that is capable of inhibiting most non-*Legionellaceae* bacteria is the preferred solution to isolate *Legionella* spp. from environmental specimens [1–3].

Given that the recovery of *Legionella* spp. strictly depends on the type of agar being used, quality-assured culture media for water testing are key to consumer safety. However, the overall performance of commercially available nonselective (BCYEα) and selective (MWY and GVPC) media for *Legionella* spp. isolation has only been partially addressed. In this regard, the last studies on the quality assurance of these media date back to 2004 [4] and 2010 [5].

According to the literature, these types of media are very difficult to prepare, store, and test as minor differences in pH, cation content, and agar composition can heavily influence growth rates, plating efficiency, and colony formation [6–8]. Thus, the quality control of culture media used for *Legionella* spp. detection is now, more than ever, essential to achieve a satisfactory degree of comparability among water-testing results from different laboratories.

For over two decades, our group has been conducting *Legionella* spp. testing in numerous hospitals and health facilities in the Piedmont region of Italy. Furthermore, since 2007, our laboratory has been a permanent member of the External Quality Assessment (EQA) scheme for *Legionella* spp. isolation from water, which relies on the use of the BCYEα and MWY media manufactured by Oxoid Ltd. (Basingstoke, UK), with satisfactory z-scores of performance throughout. In particular, by conducting parallel seeding experiments, we have previously found that BCYEα allows a higher yield and recovery rate of *Legionella* spp. positive samples (93%) compared to that obtained with MWY (78%). Based on these findings, we were the first in 2011 [9] to recommend the use of BCYEα as a nonselective medium, in addition to MWY or GVPC, for optimal detection of *Legionella* spp. in environmental water samples. The combined use of one selective and one nonselective media for improved *Legionella* spp. detection was later on incorporated in the second edition of ISO 11731 [10].

Besides the recovery rate, we have also been actively involved in evaluating the sensitivity and selectivity of nonselective vs. selective media for the isolation and enumeration of *Legionella* spp., an investigation that has been more recently extended to BCYEα-AB media (Ditommaso et al., unpublished data).

In this study, we report on a comparative assessment between the sensitivity and selectivity of the aforementioned media from Oxoid Ltd. (Basingstoke, UK) and the corresponding media manufactured by Xebios Diagnostics GmbH (Düsseldorf, Germany) for the detection of *Legionella* spp. from environmental water samples.

## 2. Results

### 2.1. Overall Results

A total of 148 water samples were cultured on MWY and BCYEα agar produced by two different manufacturers (i.e., Xebios and Oxoid). Bacteria were isolated from 70/148 samples (47.3%) using either type of medium, showing comparable levels of *Legionella* spp. detection. Specifically, we detected 64 (43.2%) and 62 (41.9%) positive cultures using Xebios and Oxoid mediums, respectively, with excellent agreement between the two brands (90.5%; $\kappa = 0.807$).

According to the Wilcoxon test analysis performed on 148 results obtained with each medium, the detection sensitivity increased when the samples were plated on Xebios medium ($P = 0.0054$, median difference between log-counts: $\Delta = 0.192$, $CI_{95\%}$: 0.055, 0.394). Consistently, the 56 concordant positive samples also displayed the highest counts on Xebios agar plates ($P = 0.0006$, $\Delta = 0.159$, $CI_{95\%}$: 0.068, 0.295).

### 2.2. Results Relative to Medium Type

To further investigate the differences in the recovery of *Legionella* spp., we assessed the sensitivity and selectivity of Oxoid vs. Xebios alpha-ketoglutarate agar (BCYEα) and bromocresol purple (MWY) in detecting *Legionella* spp. from 148 environmental water samples (Table 1).

**Table 1.** Relative sensitivity and selectivity of BCYEα and MWY media from Xebios or Oxoid for *Legionella* spp. isolation.

| | BCYEα $^{Oxoid}$ | MWY $^{Oxoid}$ | BCYEα $^{Xebios}$ | MWY $^{Xebios}$ |
|---|---|---|---|---|
| **Number (%) plates growing *Legionella*** | 44 (29.7) | 52 (35.1) | 49 (33.1) | 62 (41.9) |
| **Sensitivity [a]** | (63) | (74) | (70) | (78) |
| **Number (%) plates growing microorganism other than *Legionella*** | 21 (14.2) | 70 (47.3) | 18 (12.2) | 70 (47.3) |
| **Selectivity [b]** | (14) | (47) | (12) | (47) |

[a] Sensitivity was calculated by comparing the number of positive plates for a given medium with the cumulative yield of *Legionella* spp. from all four media (n = 70). [b] Selectivity for each media was defined as the number of plates suppressing the growth of organisms that were not *Legionella* spp. over the total number of plates (n = 148).

The detection sensitivity was significantly higher in samples plated on MWY$^{Xebios}$ agar compared to that of samples grown on MWY$^{Oxoid}$ agar (McNemar test: $P = 0.0042$). No difference was found between BCYEα$^{Oxoid}$ and BCYEα$^{Xebios}$ media (McNemar test: $P = 0.03588$); no difference in suppressing non-*Legionella* bacteria (i.e., selectivity) was found between the two brands of MWY agar.

A Wilcoxon signed-rank test comparison (Table 2) revealed a significant difference in the number of *Legionella* spp. colonies, not only between BCYEα$^{Xebios}$ and MWY$^{Xebios}$ media but also between MWY$^{Oxoid}$ and MWY$^{Xebios}$ media. In either case, the recovery of *Legionella* spp. was significantly higher when the samples were plated on MWY$^{Xebios}$ agar.

**Table 2.** Wilcoxon signed-rank test analysis of *Legionella* spp. counts (CFU/L) obtained with four different types of mediums.

| | **Xebios Media vs. Oxoid Media** | | ***P*-Value** | **Δ ($CI_{95\%}$)** |
|---|---|---|---|---|
| **(a)** | BCYEα $^{Oxoid}$ | MWY $^{Oxoid}$ | 0.476 | |
| **(b)** | BCYEα $^{Xebios}$ | MWY $^{Xebios}$ | 0.0005 | Δ = 0.515 ($CI_{95\%}$: 0.182, 1.023) [MWY > BCYE] |
| **(c)** | BCYEα $^{Oxoid}$ | BCYEα $^{Xebios}$ | 0.0905 | |
| **(d)** | MWY $^{Oxoid}$ | MWY $^{Xebios}$ | 0.0014 | Δ = 0.618 ($CI_{95\%}$: 0.268, 0.981) [Xebios > Oxoid] |

Comparison between corresponding media from the two companies (a,b) and between different types of medium from each company (c,d). Δ is the median of the differences of log-counts, where the left column is the reference.

All data were disaggregated according to medium and manufacturer. Table 3 classifies the samples according to the presence or absence of *Legionella* spp. in MWY and BCYE media from Oxoid (Table 3a) or Xebios (Table 3b). There is a greater agreement between the two Xebios media (47/49) in comparison with that between the two Oxoid media (34/44).

**Table 3.** Assessment of *Legionella* spp. recovery according to culture medium and manufacturer.

| | | **(a) OXOID** | | | | **(b) XEBIOS** | | | | **(c) EXPECTED XEBIOS [from (a) with Deming-Stephan Method]** | | |
|---|---|---|---|---|---|---|---|---|---|---|---|---|
| | | **BCYEα** | | | | **BCYEα** | | | | **BCYEα** | | |
| | | **Pos (n)** | **Neg (n)** | **Total** | | **Pos (n)** | **Neg (n)** | **Total** | | **Pos (n)** | **Neg (n)** | **Total** |
| **MWY** | **Pos (n)** | 34 | 18 | 52 | **Pos (n)** | 47 | 15 | 62 | **Pos (n)** | 40.22 | 21.78 | 62 |
| | **Neg (n)** | 10 | 86 | 96 | **Neg (n)** | 2 | 84 | 86 | **Neg (n)** | 8.78 | 77.22 | 86 |
| | **Total** | 44 | 104 | 148 | **Total** | 49 | 99 | 148 | **Total** | 49 | 99 | 148 |

Pos = positive, Neg = negative; (b) Agreement = 81.1%; κ = 0.57; (c) Agreement = 88.5%; κ = 0.757.

Pearson's chi-squared test comparison between Oxoid and Xebios results, after a Deming–Stephan adjustment, revealed a statistically significant difference between the two brands ($\chi^2 = 9.0824$, $P = 0.0026$; Table 3c).

The number of *Legionella* spp. colonies (CFU/L) growing or not on BCYEα and MWY is shown in Tables 4 and 5.

**Table 4.** Number of *Legionella* spp. colonies (CFU/L) appearing on different plates: BCYEα positive results vs. MWY negative results.

| Oxoid | | | | | | Xebios | | | | | |
|---|---|---|---|---|---|---|---|---|---|---|---|
| Samples | BCYEα | MWY | *L.* spp. | B. Flora on BCYEα | B. Flora on MWY | Samples | BCYEα | MWY | *L.* spp. | B. Flora on BCYEα | B. Flora on MWY |
| 1 | 50 | 0 | Lp3 | 1+ | 0 | 1 | 50 | 0 | Lp3 | 0 | / |
| 2 | 50 | 0 | Lp1 | 1+ | 0 | 2 | 500 | 0 | Lp1 | 1+ | / |
| 3 | 150 | 0 | Lp6 | 1+ | 0 | | | | | | |
| 4 | 200 | 0 | L.spp | 0 | 0 | | | | | | |
| 5 | 400 | 0 | L.spp | 0 | 0 | | | | | | |
| 6 | 600 | 0 | L.spp | 1+ | 0 | | | | | | |
| 7 | 1050 | 0 | L.spp | 0 | 0 | | | | | | |
| 8 | 2200 | 0 | L.spp | 2+ | 0 | | | | | | |
| 9 | 2350 | 0 | L.spp | 0 | 0 | | | | | | |
| 10 | 3400 | 0 | L.spp | 0 | 0 | | | | | | |

Background (B.) flora was measured by semiquantitative counting: four categories were determined according to the visual density of colonies spread onto the plate, where zero is no background flora and 3+ is massive contamination (see supplementary materials). Lp 1 = *L. pneumophila* serogroups 1; Lp 3 = *L. pneumophila* serogroups 3; Lp 6 = *L. pneumophila* serogroups 6; L.spp. = *Legionella* spp. non-*pneumophila*.

**Table 5.** Number of *Legionella* spp. colonies (CFU/L) appearing on different plates: BCYEα negative results vs. MWY positive results.

| Oxoid | | | | | | Xebios | | | | | |
|---|---|---|---|---|---|---|---|---|---|---|---|
| Samples | BCYEα | MWY | *L.* spp. | B. Flora on BCYEα | B. Flora on MWY | Samples | BCYEα | MWY | *L.* spp. | B. Flora on BCYEα | B. Flora on MWY |
| 1 | 0 | 50 | Lp6 | 3+ | 0 | 1 | 0 | 50 | Lp6 | 3+ | 0 |
| 2 | 0 | 50 | L.spp | 2+ | 0 | 2 | 0 | 50 | Lp3, L.spp | 1+ | 1+ |
| 3 | 0 | 50 | Lp6 | 1+ | 0 | 3 | 0 | 50 | L.spp | 1+ | 0 |
| 4 | 0 | 50 | Lp6 | 3+ | 0 | 4 | 0 | 50 | Lp6 | 3+ | 0 |
| 5 | 0 | 50 | Lp6 | 3+ | 0 | 5 | 0 | 100 | Lp6 | 2+ | 0 |
| 6 | 0 | 100 | Lp1 | 1+ | 0 | 6 | 0 | 500 | Lp7-14 | 3+ | 0 |
| 7 | 0 | 150 | Lp6 | 2+ | 0 | 7 | 0 | 500 | Lp7-14 | 3+ | 0 |
| 8 | 0 | 200 | Lp7-14 | 3+ | 0 | 8 | 0 | 800 | Lp6, L.spp | 2+ | 0 |
| 9 | 0 | 300 | Lp7-14 | 3+ | 0 | 9 | 0 | 1200 | Lp7-14 | 3+ | 1+ |
| 10 | 0 | 300 | Lp714 | 3+ | 0 | 10 | 0 | 1300 | Lp714 | 3+ | 1+ |
| 11 | 0 | 300 | Lp7-14 | 3+ | 0 | 11 | 0 | 1500 | Lp7-14 | 3+ | 1+ |
| 12 | 0 | 300 | Lp6 | 3+ | 0 | 12 | 0 | 1700 | Lp6, L.spp | 3+ | 0 |
| 13 | 0 | 300 | Lp6 | 3+ | 0 | 13 | 0 | 2150 | Lp3, Lp6 | 3+ | 1+ |
| 14 | 0 | 500 | Lp7-14 | 3+ | 0 | 14 | 0 | 7700 | Lp6 | 3+ | 1+ |
| 15 | 0 | 600 | Lp7-14 | 3+ | 0 | 15 | 0 | 8300 | Lp6 | 3+ | 0 |
| 16 | 0 | 1500 | Lp3, Lp6 | 3+ | 0 | | | | | | |
| 17 | 0 | 4800 | Lp2-14 | 3+ | 0 | | | | | | |
| 18 | 0 | 15000 | Lp6 | 3+ | 3+ | | | | | | |

Background (B.) flora was measured by semiquantitative counting: four categories were determined according to the visual density of colonies spread onto the plate, where zero is no background flora and 3+ is massive contamination (see supplementary materials). Lp 1 = *L. pneumophila* serogroups 1; Lp 3 = *L. pneumophila* serogroups 3; Lp 6 = *L. pneumophila* serogroups 6; Lp 2–14 = *L. pneumophila* serogroups 2–14; Lp 7–14= *L. pneumophila* serogroups 7–14; L.spp = *Legionella* spp. non-*pneumophila*.

The observation that some water samples were *Legionella* spp. positive on BCYEα medium but negative on MWY agar (7/10 cultures, samples 4–10, Table 4) indicates that the use of selective medium can affect the recovery of non-*Legionella pneumophila* species. Cell concentrations from these water samples ranged from $2.0 \times 10^2$ to $3.4 \times 10^3$ CFU/L. These results only refer to samples cultured on Oxoid media (Table 4).

Table 5 shows that high levels of background flora can challenge the results: on BCYEα agar, the results were affected by the presence of concomitant background flora that prevented the growth or masked the observation of *Legionella* colonies in BCYEα agar from both manufacturers. No qualitative data are available for these aquatic bacteria. Contaminating non-*Legionella* bacteria were rarely recovered on MWY agar from either manufacturer.

The influence of the media on the detection time was also assessed. Typically, antibiotics added to the selective media suppress the accompanying flora but at the same time slow down the growth of the target organism. The average colony size of *Legionella* spp. cultured on Xebios media was greater than that of *Legionella* spp. plated on Oxoid medium, especially for bacteria plated onto MWY agar. With regard to colony count, the number of *Legionella* spp. was higher on $MWY^{Xebios}$ agar compared to $MWY^{Oxoid}$. As for the Oxoid media, even though the supplemented antibiotics suppressed the accompanying flora, it slowed down excessively the growth of the target organisms. No particular differences in colony size or count were observed between the two brands of BCYEα agar, as already shown in Table 2 (see also the supplementary materials).

### 3. Discussion

Several factors may hinder exact *Legionella* spp. quantification in environmental samples: (i) differences in the polycarbonate membrane characteristics (e.g., pore size, batches, fragility, crinkling and electrostatic interactions), (ii) different washing procedures to remove trapped bacteria from the membrane (e.g., shaker/vortex, ultrasound, finger and thumb scraping, or heat or acid treatment), which favor the detection of the microorganism but at the same time may reduce its concentration, and (iii) the choice of the culture medium [11–14].

As for the latter, the parameters affecting its quality are the following: (1) type and quantity of nutrients, (2) redox potential (Eh)—both after preparation and during incubation, (3) initial pH and buffering capacity, (4) water activity, and (5) type and activity of the antimicrobial agents—these can either be supplemented, already be present in the medium components, or accidentally form due to preparation errors, such as excessive heating [6,15]. Further evidence has highlighted several other deficiencies of those selective media that rely on a delicate balance between productive and selective mechanisms [4,8,16,17].

The quality of culture media has a dramatic effect on *Legionella* spp. recovery and counts. To evaluate the contribution of culture media, we checked relative recoveries of *Legionella* spp. from 148 environmental water samples because the response obtained from media plated with collection strains (i.e., quality control protocol) may vary when wild bacterial strains are present. It is known, for example, that virulent *L. pneumophila* are especially salt-sensitive, and that spontaneous mutations in stock strains may result in salt resistance [18]. Consequently, testing culture media using stock strains of bacteria may not always be a valid approach [5].

Although several different medium formulations are routinely used to detect *Legionella* spp. from environmental samples [10], there is paucity of studies assessing and comparing their abilities in growing *Legionella* spp. [4,5,9]. Since the quality of the culture medium strongly influences *Legionella* spp. detection and enumeration due to the presence of contaminating flora, here, we have assessed the recovery rates of *Legionella* spp. from hospital water samples using two different brands (i.e., Xebios and Oxoid) of either nonselective BCYEα or selective MWY medium.

Our analysis shows an excellent agreement between the recovery rates of the media from both companies (90.5%). Nonetheless, the quantitative recovery of *Legionella* colonies using Xebios media is significantly greater than that achieved by Oxoid media. Furthermore, the sensitivity of detection is significantly higher when samples are plated on $MWY^{Xebios}$ agar, while the selectivity of MWY appears to be the same regardless of the manufacturer. Moreover, there is a greater agreement between the two Xebios media compared to that between the two Oxoid media. Additionally, differences in colony size were apparent for the different agars (see supplementary materials). Specifically, the $MWY^{Xebios}$ agar favored the growth of much larger colonies compared to $MWY^{Oxoid}$, and enhanced the recovery of non-*Legionella pneumophila* species.

As we used four different batches over a one-year period of study, it is highly unlikely that batch-to-batch variability may have played a role in the performance differences that we observed. We hypothesize that other factors such as the presence in the medium of toxic compounds (e.g., metals), growth-promoting factors, or high gel strengths may have influenced the growth and colony size on different types of medium, as previously shown for *L. pneumophila* on BCYEα agar [19].

Collectively, these results highlight significant differences between the performances of media from different manufacturers. Despite generating the same number of positive cultures, the Xebios media generally yielded greater numbers of *Legionella* spp. and larger colony sizes, allowing easier detection. Thus, the use of Xebios culture media is indicated to achieve the highest sensitivity and selectivity when detecting environmentally sampled *L. pneumophila*.

## 4. Methods

### 4.1. Water Samples

Media were tested using environmental samples obtained from hospital building waterlines. One liter of sample was collected from each site in sterile one-liter plastic bottles. A sodium thiosulphate (100 mg/L) solution was added to the samples to neutralize free chlorine in treated water supplies.

### 4.2. Culture Media

To distinguish between *Legionella* spp. and non-*Legionella* bacteria, two different media were used: 0.1% BCYEα agar and BCYEα agar supplemented with 3 g/L glycine, 50,000 IU of polymyxin B, 0.001 g/L of vancomycin, 0.08 g/L of anisomycin, 0.01 g/L of bromothymol blue, and 0.01 g/L of bromocresol purple (MWY). The formulations of both media conform to ISO 11731 [10].

Commercially available agar plates were purchased by Oxoid Ltd. (Basingstoke, UK) or Xebios Diagnostics GmbH (Düsseldorf, Germany). For each batch supplied, each manufacturer provided us with detailed quality control information (e.g., type of bacteria, pH, colony morphology, selectivity, recovery and expected results). The microbiological performance test was carried out in accordance with ISO 11133:2014 requirements [20]: for BCYEα agar, colony count of positive strains was ≥70% for each inoculum (i.e., productivity); for MWY agar, colony count of positive strains was ≥50% for each inoculum (i.e., selectivity).

### 4.3. Laboratory Procedure

Briefly, the water samples were concentrated 100-fold by filtration through a 0.2-μm polycarbonate filter (Millipore, Billerica, MA, USA). The filter membrane was aseptically placed in one of the bottom corners inside the stomacher bag and rubbed with the finger and thumb of one hand for 1 min with 10 mL Page solution (pH 6.8) to detach the bacteria. A 0.2-mL volume of the concentrated sample was spread on duplicate plates of MWY or BCYEα agar. The plates were incubated at 36 °C in a humid 2.5% $CO_2$ chamber and examined after 3, 6, and 10 days of incubation. Suspected colonies were subcultured on blood and BCYEα agar.

### 4.4. Identification of Legionella spp.

The presence of background flora was measured through semiquantitative counting. According to the visual density of the colonies spread onto the plate, four categories were determined, where zero represented no background flora and 3+ massive contamination (see supplementary materials). Colonies grown on MWY or BCYEα agar were subsequently identified by agglutination test (*Legionella* latex test; Oxoid). This test allows the separate identification of *L. pneumophila* Serogroup 1 and Serogroups 2 to 14 and detection of seven other species of *Legionella* (polyvalent). Colonies identified as *L. pneumophila* Serogroup 2 to 14 were further tested with *Legionella* agglutination latex reagents (Pro-Lab Diagnostics, Richmond Hill, Canada), which are intended for the identification of a single *L. pneumophila* sero group. Colonies not identified by the agglutination test were tested by polymerase chain reaction (in-house PCR) for the detection of the genus *Legionella*. This PCR assay utilizes specific primers to amplify the 16S rRNA gene of *Legionella* spp. [21]. The plate showing the highest number of confirmed colonies was used to estimate the number of *Legionella* spp. in the original sample (report).

Concentrations of *Legionella* spp. in water samples are expressed as colony forming units per liter (CFU/l). According to this method, the lower limit of detection (LOD) is 50 CFU/L.

### 4.5. Statistical Analysis

Agreement between the final reports (i.e., Oxoid and Xebios) and the two media (i.e., MWY and BCYEα) was assessed by two-by-two contingency tables, through Cohen's κ coefficient. Comparison between *Legionella* spp. counts obtained by Xebios and Oxoid media was performed by a Wilcoxon signed-rank test.

To quantify the difference between counts on different media, our analyses took into account the decimal logarithm of counts observed. The median difference was chosen as the estimator of the actual difference for each comparison, as indicated by Wilcoxon's nonparametric analysis.

The two 2 × 2 tables were compared using the Deming–Stephan method [22,23]. Specifically, the first table (Oxoid counts, see Table 3a) was transformed through the algorithm developed by Mosteller et al. [24] in order to obtain another table (Table 3c). These could be considered as the expected values for the Xebios table according to the results of the Oxoid count. Therefore, values in Table 4c were compared to the actual observed Xebios counts (Table 3b), with Pearson's chi-squared test.

The sensitivity of *Legionella* spp. detection was calculated by comparing the number of positive plates for a given medium with the cumulative yield of *Legionella* spp. from all four media. Selectivity for each method was defined as the number of plates that suppressed the growth of organisms that were not *Legionella* spp. Sensitivity and selectivity were compared using the exact McNemar test.

All statistical analyses were performed using the statistical software R ("stats" package, version 3.6.3) [25]. For all analyses, the level of significance was set at $\alpha = 0.05$.

## 5. Conclusions

In conclusion, our data demonstrate that the quality of culture media is crucial in determining the level of *Legionella* spp. colonization in hospital water systems. As water remediation measures are based on quantitative *Legionella* spp. data obtained by culturing environmental samples, culture protocol standardization, as well as accurate quality control of the culture media, is essential to achieve intra- and interlaboratory reproducibility and accuracy.

Given the public health risk from *Legionella* spp., it is important that all water-testing laboratories carefully consider the following aspects: (1) there can be variability in *Legionella* spp. detection due to different types and brands of medium—of note, this variation can also be observed among different batches from the same supplier; (2) the medium should be purchased from a reputable company and fully validated in-house, with the inclusion of appropriate controls; (3) when switching to a different medium manufacturer, extensive validation should be performed in order to determine whether the new medium is "fit for purpose".

**Supplementary Materials:** The following are available online at http://www.mdpi.com/2076-0817/9/7/523/s1. File: images of plate with background flora; paired images (BCYEα and MWY agar by two manufacturers) obtained during incubation period of inoculated plates.

**Author Contributions:** Conceptualization, S.D. and M.G.; methodology, S.D. and M.G.; formal analysis, M.G. and G.M.; data curation, M.G. and J.G.; funding acquisition, C.M.Z.; writing original draft preparation, S.D. and M.G.; writing, review, and editing, S.D., M.G., J.G., and C.M.Z. All authors have read and agreed to the published version of the manuscript.

**Funding:** This work was supported by a grant from the Department fund ZOTC_CT_PREST_18_01, Department of Public Health and Pediatrics, University of Turin.

**Conflicts of Interest:** None of the authors declare to have a conflict of interest that could influence or bias the content of the paper.

## References

1. International Organization for Standardization. *ISO 11731: 1998 Water Quality—Detection and Enumeration of Legionella*; International Organization for Standardization: Geneva, Switzerland, April 1998.
2. Centers for Disease Control and Prevention (CDC). *Legionnaires Disease Procedures Manual for Recovery | Legionnaires*; CDC: Atlanta, GA, USA, 2005.
3. Health Protection Agency (HPA). *Detection and Enumeration of Legionella Species by Filtration and Centrifugation*; National Standards Method W 12; HPA: London, UK, 2006.
4. Lück, P.C.; Igel, L.; Helbig, J.H.; Kuhlisch, E.; Jatzwauk, L. Comparison of commercially available media for the recovery of Legionella species. *Int. J. Hygiene Environ. Health* **2004**, *207*, 589–593. [CrossRef] [PubMed]
5. Edelstein, P.H.; Edelstein, M.A.C. Comparison of the plating efficiencies and shelf lives of three different commercial buffered charcoal yeast extract media supplemented with alpha-ketoglutaric acid. *J. Clin. Microbiol.* **2010**, *48*, 1882–1883. [CrossRef] [PubMed]
6. Mossel, D.A.; Bonants-Van Laarhoven, T.M.; Ligtenberg-Merkus, A.M.; Werdler, M.E. Quality assurance of selective culture media for bacteria, moulds and yeasts: An attempt at standardization at the international level. *J. Appl. Bacteriol.* **1983**, *54*, 313–327. [CrossRef]
7. Feeley, J.C.; Gibson, R.J.; Gorman, G.W.; Langford, N.C.; Rasheed, J.K.; Mackel, D.C.; Baine, W.B. Charcoal-yeast extract agar: Primary isolation medium for Legionella pneumophila. *J. Clin. Microbiol.* **1979**, *10*, 437–441. [CrossRef] [PubMed]
8. Edelstein, P.H. Improved semiselective medium for isolation of Legionella pneumophila from contaminated clinical and environmental specimens. *J. Clin. Microbiol.* **1981**, *14*, 298–303. [CrossRef] [PubMed]
9. Ditommaso, S.; Gentile, M.; Giacomuzzi, M.; Zotti, C.M. Recovery of Legionella species from water samples using an internal method based on ISO 11731: Suggestions for revision and implementation. *Diagn. Microbiol. Infect. Dis.* **2011**, *70*, 200–206. [CrossRef] [PubMed]
10. International Organization for Standardization. *ISO 11731: 2017 Water Quality—Enumeration of Legionella*; International Organization for Standardization: Geneva, Switzerland, 2017.
11. Boulanger, C.A.; Edelstein, P.H. Precision and accuracy of recovery of Legionella pneumophila from seeded tap water by filtration and centrifugation. *Appl. Environ. Microbiol.* **1995**, *61*, 1805–1809. [CrossRef] [PubMed]
12. Ta, A.C.; Stout, J.E.; Yu, V.L.; Wagener, M.M. Comparison of culture methods for monitoring Legionella species in hospital potable water systems and recommendations for standardization of such methods. *J. Clin. Microbiol.* **1995**, *33*, 2118–2123. [CrossRef]
13. Brenner, K.P.; Rankin, C.C. New screening test to determine the acceptability of 0.45-micron membrane filters for analysis of water. *Appl. Environ. Microbiol.* **1990**, *56*, 54–64. [CrossRef]
14. Smith, L.; Carroll, K.; Mottice, S. Comparison of membrane filters for recovery of legionellae from water samples. *Appl. Environ. Microbiol.* **1993**, *59*, 344–346. [CrossRef] [PubMed]
15. Mossel, D.A. Food microbiology: An authentic academic discipline with substantial potential benefits for science and society. *J. Assoc. Anal. Chem.* **1991**, *74*, 1–13. [CrossRef]
16. Wadowsky, R.M.; Yee, R.B. Glycine-containing selective medium for isolation of Legionellaceae from environmental specimens. *Appl. Environ. Microbiol.* **1981**, *42*, 768–772. [CrossRef] [PubMed]
17. Edelstein, P.H. Comparative study of selective media for isolation of Legionella pneumophila from potable water. *J. Clin. Microbiol.* **1982**, *16*, 697–699. [CrossRef] [PubMed]
18. Hammer, B.K.; Swanson, M.S. Co-ordination of legionella pneumophila virulence with entry into stationary phase by ppGpp. *Mol. Microbiol.* **1999**, *33*, 721–731. [CrossRef] [PubMed]
19. Miyamoto, H.; Yamamoto, H.; Arima, K.; Fujii, J.; Maruta, K.; Izu, K.; Shiomori, T.; Yoshida, S. Development of a new seminested PCR method for detection of Legionella species and its application to surveillance of legionellae in hospital cooling tower water. *Appl. Environ. Microbiol.* **1997**, *63*, 2489–2494. [CrossRef] [PubMed]
20. International Organization for Standardization. *ISO 11133: 2014 Microbiology of Food, Animal Feed and Water—Preparation, Production, Storage and Performance Testing of Culture Media*; International Organization for Standardization: Geneva, Switzerland, 2014.
21. Edelstein, P.H.; Edelstein, M.A. Comparison of different agars used in the formulation of buffered charcoal yeast extract medium. *J. Clin. Microbiol.* **1991**, *29*, 190–191. [CrossRef] [PubMed]

22. Deming, W.E.; Stephan, F.F. On a Least Squares Adjustment of a Sampled Frequency Table When the Expected Marginal Totals are Known. *Ann. Math. Stat.* **1940**, *11*, 427–444. [CrossRef]
23. Altham, P.M.E.; Ferrie, J.P. Comparing Contingency Tables Tools for Analyzing Data from Two Groups Cross-Classified by Two Characteristics. *Hist. Methods J. Quant. Interdiscip. Hist.* **2007**, *40*, 3–16. [CrossRef]
24. Mosteller, F. Association and Estimation in Contingency Tables. *J. Am. Stat. Assoc.* **1968**, *63*, 1–28. [CrossRef]
25. R Development Core Team. *A Language and Environment for Statistical Computing*; R Foundation for Statistical Computing: Vienna, Austria, 2019.

pathogens

MDPI

Article

# Performance of Legiolert Test vs. ISO 11731 to Confirm *Legionella pneumophila* Contamination in Potable Water Samples

**Maria Scaturro [1], Matteo Buffoni [1], Antonietta Girolamo [1], Sandra Cristino [2], Luna Girolamini [2], Marta Mazzotta [2], Maria Antonietta Bucci Sabattini [3], Cristina Maria Zaccaro [3], Leonarda Chetti [3], Microbiology Arpa Novara Laboratory [4], Antonino Bella [1], Maria Cristina Rota [1] and Maria Luisa Ricci [1,*]**

1 Department Infectious Diseases, National Institute of Health, Viale Regina Elena 299, 00161 Rome, Italy; maria.scaturro@iss.it (M.S.); buffoni.matteo@gmail.com (M.B.); antonietta.girolamo@iss.it (A.G.); antonino.bella@iss.it (A.B.); mariacristina.rota@iss.it (M.C.R.)

2 Department of Biological, Geological, and Environmental Sciences, University of Bologna, via San Giacomo 12, 40126 Bologna, Italy; sandra.cristino@unibo.it (S.C.); luna.girolamini2@unibo.it (L.G.); marta.mazzotta2@unibo.it (M.M.)

3 Reference Laboratory for Environmental Control, Health Protection Agency, Bologna, Emilia Romagna Region, Via Trachini, 17, 40138 Bologna, Italy; mbucci@arpae.it (M.A.B.S.); cmzaccaro@arpae.it (C.M.Z.); lchetti@arpae.it (L.C.)

4 Reference Laboratory for Environmental Control, Health Protection Agency, Novara, Piemonte Region, Viale Roma 7/e, 28100 Rome, Italy; legionella@arpa.piemonte.it

* Correspondence: marialuisa.ricci@iss.it

Received: 31 July 2020; Accepted: 21 August 2020; Published: 23 August 2020

**Abstract:** Detection and enumeration of *Legionella* in water samples is of great importance for risk assessment analysis. The plate culture method is the gold standard, but has received several well-known criticisms, which have induced researchers to develop alternative methods. The purpose of this study was to compare *Legionella* counts obtained by the analysis of potable water samples through the plate culture method and through the IDEXX liquid culture Legiolert method. *Legionella* plate culture, according to ISO 11731:1998, was performed using 1 L of water. Legiolert was performed using both the 10 mL and 100 mL Legiolert protocols. Overall, 123 potable water samples were analyzed. Thirty-seven (30%) of them, positive for *L. pneumophila*, serogroups 1 or 2–14 by plate culture, were used for comparison with the Legiolert results. The Legiolert 10 mL test detected 34 positive samples (27.6%) and the Legiolert 100 mL test detected 37 positive samples, 27.6% and 30% respectively, out of the total samples analyzed. No significant difference was found between either the Legiolert 10 mL and Legiolert 100 mL vs. the plate culture ($p$ = 0.9 and $p$ = 0.3, respectively) or between the Legiolert 10 mL and Legiolert 100 mL tests ($p$ = 0.83). This study confirms the reliability of the IDEXX Legiolert test for *Legionella pneumophila* detection and enumeration, as already shown in similar studies. Like the plate culture method, the Legiolert assay is also suitable for obtaining isolates for typing purposes, relevant for epidemiological investigations.

**Keywords:** *Legionella*; Legiolert; ISO 11731; plate culture; potable water samples

## 1. Introduction

*Legionella* is a genus consisting of fastidious waterborne pathogens responsible for a severe form of pneumonia named Legionnaires' disease (LD) and for a flu-like infection known as Pontiac fever (PF) [1]. *Legionella* is widespread in natural freshwater environments, where it can be found free-living or intracellularly in hosts such as amoebae [2]. Among the 62 species known to date, *L. pneumophila* is the species most frequently found in cases of infection, amounting in 2018 to approximately 94.1% of the culture-confirmed LD cases in notified in EU/EEA (European Legionnaires' disease Surveillance Network annual meeting 2019, unpublished data). However, just under half of the known species cause illness and, in several countries such as New Zealand, soil-born *L. longbeachae* is the primary cause of LD [1]. Infection is acquired through inhalation of contaminated aerosols produced by various man-made water systems, such as showers, spa pools, fountains and cooling towers of air conditioning systems [3]. When *Legionella* colonizes the water systems, it often finds favorable conditions for growth, such as temperatures between 25 °C and 45 °C or the presence of biofilm, reaching high concentrations and becoming a serious risk for human health. After the first LD outbreak occurred in Philadelphia in 1976, numerous other outbreaks and sporadic cases have been reported worldwide [4–10]. In Italy in 2018, the incidence of Legionnaires' disease was 4.9 cases per million inhabitants, 2964 notified cases [11], and in the European network for Legionnaires' disease surveillance, Italy ranked first in terms of the number of reported cases [12]. In addition, in 2018 two important outbreaks occurred in Italy that required increased environmental monitoring ([13]; data unpublished).

Most European countries have adopted a preventive approach, implementing actions for prevention and control of *Legionella* contamination. Monitoring *Legionella* contamination of potable water systems is of paramount importance for risk assessment. To this end, the plate culture method, performed using specific media (buffered charcoal yeast extract, BCYE), usually supplemented with different combinations of antimicrobial selective substances, is considered the gold standard for detection and enumeration of *Legionella* in water samples [14]. Culture can also be performed in accordance with ISO 11731:2017, an updated norm which replaced both ISO 11731:1998 (used in this study) and ISO 11731-2:2004. [15–17]. Although plate culture methods are specific for *Legionella*, they have high variability in enumeration, are time-consuming and require significant experience in recognizing *Legionella* colonies [18]. In addition, the enumeration of a *Legionella* concentration may be under-estimated due to the inability to detect viable but not-culturable bacteria or *Legionella* within amoebae [1]. Molecular methods based on the detection of *Legionella* DNA have been demonstrated to be highly specific and sensitive, are able to discriminate between species and serogroup and can detect viable but nonculturable bacteria, but are not considered fully suitable to enumerate *Legionella* in water samples because they are unable to reliably discriminate whether DNA detected is from live or dead organisms [19,20]. A promising alternative method is the Legiolert test (IDEXX Laboratories, Westbrook, ME, USA), a liquid culture method based on bacterial enzyme detection technology, which determines the most probable number (MPN) of exclusively *L. pneumophila* species present in water samples. The presence of *L. pneumophila* is visualized through the utilization of a substrate present in the Legiolert reagent. In published studies, Legiolert has shown equal performance to traditional plate culture methods, providing results in seven days with simplified sample preparation and analysis [21].

In this study, the detection and enumeration of *L. pneumophila* were determined in both 10 mL and 100 mL of potable water samples using the Legiolert method. Data obtained were compared with those obtained by the traditional plate culture method, performed according to the ISO 11731:1998 using 1 L of potable water, in order to evaluate the possibility of using the Legiolert method as a valid alternative to traditional plate culture. The results of this study are encouraging for the adoption of the Legiolert method for *L. pneumophila* enumeration in water samples.

## 2. Results

Overall, 123 potable water samples were analyzed. Fifty-one of them were positive by plate culture. Among these, 37 (30%) were *L. pneumophila*, typed as serogroups 1 or 2–14 positive, 14 (11.4%) were *Legionella* non-*pneumophila* and 58.6% were negative. Since the Legiolert test is designed for specificity for *L. pneumophila* only, the 14 samples that were positive with non-*pneumophila Legionella* species samples were excluded from further statistical comparisons with Legiolert results.

Legiolert test results derived from 10 mL water samples showed that 34/123 (27.6%) samples were positive for *L. pneumophila*, whereas results from 100 mL water samples detected 37/123 (30%)

positive samples, consistent with plate culture results. Among the 37 plate culture positive samples, four showed very low *Legionella* concentrations (50, 150, 300 and 600 CFU/L) and these were found negative by the Legiolert test in both 10 and 100 mL water samples.

As per ISO 17994:2014 [22], eight samples that were positive with the Legiolert 100 mL test and exceeded the MPN count by Legiolert (too numerous to count, TNTC) were excluded from calculations. Thus, Legiolert data were analyzed for each of the 28 or 33 positive samples from the 100 mL and 10 mL tests, respectively, and were compared with plate culture data. Data analysis did not show significant differences in *Legionella pneumophila* detection between the two Legiolert protocols ($p = 0.83$), or between either Legiolert test (Legiolert 10 mL, $p = 0.82$; Legiolert 100 mL, $p = 0.2$) and the plate culture method performed according to ISO 11731:1998 (Table 1).

**Table 1.** Comparison of Legiolert 10 mL and Legiolert 100 mL tests and plate culture according to ISO 17994:2014.

| Comparison | Mean Relative Difference | 95% CI | | N. | Mean ± SD | *p*-Value |
|---|---|---|---|---|---|---|
| Legiolert 10 mL vs. Legiolert 100 mL | 0.13 | −0.59 | 0.33 | 22 | 6.97 ± 2.10<br>7.10 ± 1.79 | 0.8322 |
| Legiolert 10 mL vs. Plate Culture | −0.11 | −1.58 | 1.35 | 33 | 6.94 ± 3.57<br>7.05 ± 3.94 | 0.9042 |
| Legiolert 100 mL vs. Plate Culture | −0.73 | −1.86 | 0.39 | 28 | 5.86 ± 3.02<br>6.59 ± 2.83 | 0.3528 |

## 3. Discussion

Legiolert is characterized by very easy and rapid sample preparation, with the additional advantages of avoiding the need for large sampling volumes, membrane filtration, treatments, plating, colony isolation and additional confirmation or identification. Furthermore, the Legiolert test reduces the time required to obtain confirmed results (seven days, rather than 10 or more days required by the plate culture method). In this study, potable water samples were analyzed by both plate culture and Legiolert methods and no significant differences were found when comparing results. Furthermore, the results of the analyses carried out using only 10 mL of water samples showed that the Legiolert test was equally reliable using 10 mL of water as using 100 mL.

Although the plate culture method is the gold standard for the detection and enumeration of *Legionella* in water samples, different laboratories may choose to follow different procedures, depending on the expected *Legionella* concentration in the samples they process, or even for economic reasons, affecting the reliability of the data when the same sample is analyzed by different laboratories. The plate culture method should be performed by accredited laboratories, according to norms recognized by the country's accreditation body. Among the methods, there are the ISO 11731:1998 method or the ISO 11731-2:2004 method (both of which have since been replaced by ISO 11731:2017); the 2007 American public health association (APHA) method; the Association Française de Normalisation (AFNOR) method NF T90-431:2018; or the U.S. Centers for Disease Control and Prevention (CDC) method [23–25]. For ISO 11731:2017 [15], depending on the matrix to be analyzed, the user may select from four methods, four treatments and four selective culture media, for a total of 14 possible procedural scenarios. Regardless of the method used, plate culture involves many steps and significant time requirements. *Legionella* monitoring, as part of the risk assessment analysis, concerns many different buildings such as hospitals, hotels, public offices and, in the near future, according to the revision of the European directive concerning potable water requirements, every potable water system. In this study, although we analyzed potable water samples according to ISO 11731:1998, the differences between this method and ISO 11731:2017 did not affect the recovery of *Legionella pneumophila*, since the plating of the samples on BCYE medium, as suggested by ISO 11731:2017, would only improve the recovery of *Legionella* non-*pneumophila* species, which were excluded from this investigation. The MWY medium, although not included in ISO 11731:1998, is suggested by ISO11731:2017, and was adopted for this study because it has been known for many years to be the best for the recovery of *Legionella* from drinking water samples [26].

The newly drafted European Drinking Water Directive seems to have taken into consideration the emergence of many of these newer methods in Annex III part A, leaving to the national bodies the opportunity to choose the methods they find most appropriate for the purposes they specify [27]. In addition, the Legiolert method has recently been NF (Norme Francaise)validated by AFNOR certification and also included in the UK's Blue Book of validated test methods [28]. For many laboratories, the inclusion of testing in the Drinking Water Directive might lead to a large amount of work, time and financial expense. The Legiolert method may positively affect some of these difficulties, as well as those linked to the management of large volumes of water samples required for analysis by plate culture methods.

This study represents a confirmation of the reliability of the Legiolert method compared with the plate culture method, supporting conclusions from previous studies that documented the consistency of Legiolert for potable and non-potable water samples, analyzed according to ISO 11731:2004, which employs the filtration of 100 mL of water and the acid-treatment of filters which are directly placed on selective agar plates [16,29,30].

One limitation of the Legiolert method is that it is designed to detect only *Legionella pneumophila*, whereas other species remain undetectable. *Legionella pneumophila* is the most common species responsible for LD cases in Europe and, for this reason, in a few regions, such as France, Belgium, and the province of Quebec, Canada, it was decided to monitor only *Legionella pneumophila*, whereas in other countries, there is still a great debate on this matter.

Three fundamental factors can be identified in favor of monitoring exclusively for *Legionella pneumophila*. The first is risk—*L. pneumophila* is the species almost always cited in clinical cases and outbreaks and is the species most commonly found in the environment; the second is that laboratories may save time, human resources and money, and they can employ those saved resources to analyzing additional samples or locations, instead of identifying other *Legionella* species, which represent a much lower health risk; the third is that routinely monitoring only for the most pathogenic species of a bacteria is already an established practice. For example, *Pseudomonas aeruginosa* is routinely monitored, rather than all species of *Pseudomonas*.

At the same time, it is well known that other *Legionella* species are pathogenic to humans, although they represent a fraction of infections, with the exception of *Legionella longbeachae*, which is found in soil, rather than water, and is mostly detected in Australia and New Zealand. However *Legionella longbeachae* is beginning to be isolated also in EU/EEA, representing the 2.5% of isolated species in 2018 while other known and unknown species of *Legionella* were detected only in 3.3% of notified cases (European Legionnaires' disease Surveillance Network annual meeting 2019, unpublished data). It should be noted that the identification of species other than *L. pneumophila* suffers from extensive use of the urinary antigen, which exclusively detects *Legionella pneumophila* serogroup 1, and from the medium used for the isolation of *Legionella*, which has historically been optimized for *Legionella pneumophila*. Therefore, many cases caused by other species might not be detected even by culture for this reason. Until a suitable medium for growing other *Legionella* species is developed, a routine PCR test in diagnosing human specimens, capable of distinguishing between *Legionella pneumophila* and other species, should be adopted in order to identify the real burden of Legionnaires' disease, as already demonstrated in a few countries [31–37]. The results of these studies will be able to confirm the real incidence of infections caused by other *Legionella* species and consequently to address the choices on what should be the focus of monitoring in the environment.

The imminent introduction of the new drinking water legislation concerning the monitoring of an increasing number of water systems, however, will probably lead to streamlined choices aimed at reducing health risk by researching the most pathogenic and prevalent species present in the environment. Despite this, it must still be considered that for specific countries where other species of *Legionella*, such as *Legionella longbeachae*, are the prevalent in specific non-water matrices such as compost, and are an increasing cause of LD cases, Legiolert should not be utilized [1,31,35,37].

Concerning the enumeration of *Legionella pneumophila*, in this study, the most probable number did not provide any count in four of the 123 water samples, which instead tested positive by plate culture, though they were at lower concentrations of 50, 150, 300 and 650 CFU/L, respectively. For any Legiolert test, the limit of detection is 1 MPN, independent of the analyzed volume—a limit low enough to theoretically match that of plate culture. During this study, we used 100 CFU/L as the limit of detection for plate culture. We therefore suppose that the four samples which were negative in Legiolert but positive in plate culture could potentially be due to experimental errors. They could be, for example, faint colors of the wells not recognized as positive by the users. Unfortunately, these samples or isolates could not be tested again, as *Legionella* colonies were not kept for further investigations for detectability using the Legiolert test. The same bias can be considered for the eight TNTC samples tested by the 100 mL Legiolert protocol, which could have been included in the analysis if the original sample had been diluted or run with the 10 mL Legiolert protocol. No samples were found positive by Legiolert and negative by plate culture.

Although the data obtained showed that the two methods were comparable, a higher number of water samples at low *Legionella pneumophila* concentrations should perhaps be analyzed in order to assess any possible limitations with the Legiolert test.

In conclusion, Legiolert may be considered a valuable test for the detection and enumeration of *Legionella pneumophila* in potable water samples, and it can be used as a valid alternative to the traditional plate culture methods, especially considering the simplified protocol and the ability to employ smaller sample volumes to obtain the same quantification. Finally, it can be extremely useful when it is known that there is a prevalence of *Legionella pneumophila* in the water system under investigation.

## 4. Material and Methods

Over a 4-month period, 123 potable water samples were collected from hospitals, health care facilities for elderly people and industries located in 8 cities of the northern and central regions of Italy.

### *4.1. Enumeration of Legionella pneumophila and Legionella Spp. by ISO 11731:1998*

Water samples were collected in 2-L bottles (1.5 L collected) and, after proper mixing, 1L of each sample was analyzed by the culture method according to ISO 11731:1998. Water samples were collected according to the protocol contained in the Italian guidelines for *Legionella* [38] and were stored at 5 °C ± 3 °C until they were delivered (within 24 h) to the Italian reference laboratory for *Legionella*, where all samples were analyzed.

The sample was filtered through 0.2-µm polycarbonate membranes and the membranes were transferred to 10 mL of the same sampled water and were solubilized by vortexing. From the concentrate, three MWY (Modified Wadowsky-Yee, Oxoid, Thermo Fisher Diagnostics Limited, Cheshire, UK) agar plates were inoculated by spread plating: one plate with 0.2 mL of the concentrated sample, one with 0.2 mL of the concentrated sample pre-treated with acid and one with 0.2 mL of the concentrated sample pre-treated by heating to 50 °C ± 1 °C for X 30 ± 2 min in a water bath. All the plates were then incubated at 36 °C ± 1 °C, with 2.5% $CO_2$ for ten days. Presumptive *Legionella* colonies were confirmed by sub-culturing at least five colonies on BCYE agar plates with and without L-cysteine. The latex agglutination test (DR0800, Oxoid, Thermo Fisher Diagnostics Limited, Cheshire, UK) was utilized to obtain species information.

This study was carried out prior to the publication of the revision of the ISO 11731:2017 norm and the ISO 11731:1998 procedure was therefore applied. Regardless, the ISO 11731:1998 procedure adopted is still included in the new ISO 11731:2017 and is applicable for use with potable water samples, particularly when no information about the range of the *Legionella* concentration is known.

### 4.2. Enumeration of L. pneumophila by Legiolert/Quanti-Tray/Legiolert

The Legiolert test detects *Legionella pneumophila* through bacterial enzyme detection technology, which utilizes a substrate present in the Legiolert reagent in a liquid culture to reveal the presence of *L. pneumophila*. Generally, 100 mL of the culture is analyzed and results are received in 7 days. Any turbidity and/or brown color greater than the negative control indicates positivity. Enumeration is based on MPN. In this study Legiolert was performed using both 100 mL and 10 mL of the original 1.5 L water sample collected. Each aliquot was processed and analyzed following the procedure outlined in the Legiolert instructions, using the Quanti-tray/Legiolert device. Quanti-trays were incubated for 7 days at 39 °C +/−0.5 °C in a humidified environment.

### 4.3. Statistical Analyses

Data from plate culture and Legiolert testing were statistically analyzed by using Student's *t*-test and with relative difference according to ISO 17994:2014 [22].

**Author Contributions:** Conceptualization, M.L.R.; Data curation, M.S., M.B., S.C., M.A.B.S. and M.A.N.L.; Formal analysis, M.S., M.B., A.G. and A.B.; Investigation, M.B., A.G., S.C., L.G., M.M., M.A.B.S., C.M.Z., L.C. and M.A.N.L.; Data analysis, A.B.; Supervision, M.L.R. and M.S.; Writing–original draft, M.S.; Writing–review & editing, M.L.R. and M.C.R. All authors have read and agreed to the published version of the manuscript.

**Funding:** This work was supported by IDEXX Laboratories, Inc. (ISS-IDEXX grant N.W92)_.

**Conflicts of Interest:** The authors declare no conflict of interest. The sponsors had no role in the design, execution, interpretation or writing of the study.

## References

1. Phin, N.; Parry-Ford, F.; Harrison, T.; Stagg, H.R.; Zhang, N.; Kumar, K.; Lortholary, O.; Zumla, A.; Abubakar, I. Epidemiology and clinical management of Legionnaires' disease. *Lancet Infect. Dis.* **2014**, *14*, 1011–1021. [CrossRef]
2. Rowbotham, T.J. Preliminary report on the pathogenicity of *Legionella pneumophila* for freshwater and soil amoebae. *J. Clin. Pathol.* **1980**, *33*, 1179–1183. [CrossRef] [PubMed]
3. Orkis, L.T.; Harrison, L.H.; Mertz, K.J.; Brooks, M.M.; Bibby, K.J.; Stout, J.E. Environmental sources of community-acquired legionnaires' disease: A review. *Int. J. Hyg. Environ. Health* **2018**, *221*, 764–774. [CrossRef] [PubMed]
4. Fraser, D.W.; Tsai, T.R.; Orenstein, W.; Parkin, W.E.; Beecham, H.J.; Sharrar, R.G.; Harris, J.; Mallison, G.F.; Martin, M.S.; McDade, J.E.; et al. Legionnaires' disease: Description of an epidemic of pneumonia. *N. Engl. J. Med.* **1977**, *297*, 1197–1203. [CrossRef] [PubMed]
5. Den Boer, J.W.; Yzerman, E.P.; Schellekens, J.; Lettinga, K.D.; Boshuizen, H.C.; Van Steenbergen, J.E.; Bosman, A.; Van den Hof, S.; Van Vliet, H.A.; Peeters, M.F.; et al. A large outbreak of Legionnaires' disease at a flower show, the Netherlands, 1999. *Emerg. Infect. Dis.* **2002**, *8*, 37–43. [CrossRef] [PubMed]
6. Fernández, J.A.; López, P.; Orozco, D.; Merino, J. Clinical study of an outbreak of Legionnaire's disease in Alcoy, Southeastern Spain. *Eur. J. Clin. Microbiol. Infect. Dis.* **2002**, *21*, 729–735.
7. Lucas, K.D.; Wheeler, C.; McLendon, P.; Leistikow, B.N.; Mohle-Boetani, J.C. Outbreak of Legionnaires' disease associated with cooling towers at a California state prison, 2015. *Epidemiol. Infect.* **2018**, *146*, 297–302. [CrossRef]
8. Hamilton, K.A.; Prussin, A.J., 2nd; Ahmed, W.; Haas, C.N. Outbreaks of Legionnaires' disease and pontiac fever 20062-017. *Curr. Environ. Health Rep.* **2018**, *5*, 263–271. [CrossRef]
9. Wüthrich, D.; Gautsch, S.; Spieler-Denz, R.; Dubuis, O.; Gaia, V.; Moran-Gilad, J.; Hinic, V.; Seth-Smith, H.M.; Nickel, C.H.; Tschudin-Sutter, S.; et al. Air-conditioner cooling towers as complex reservoirs and continuous source of *Legionella pneumophila* infection evidenced by a genomic analysis study in 2017, Switzerland. *Euro Surveill.* **2019**, *24*. [CrossRef]
10. Scaturro, M.; Fontana, S.; Crippa, S.; Caporali, M.G.; Seyler, T.; Veschetti, E.; Villa, G.; Rota, M.C.; Ricci, M.L. An unusually long-lasting outbreak of community-acquired Legionnaires' disease, 20052-008, Italy. *Epidemiol. Infect.* **2015**, *143*, 2416–2425. [CrossRef]

11. Rota, M.C.; Caporali, M.G.; Bella, A.; Scaturro, M.; Giannitelli, S.E.; Ricci, M.L. Rapporto annuale sulla legionellosi in italia nel 2018. *Not. Ist. Super Sanità* **2019**, *32*, 7–13.
12. Legionnaires' Disease Annual Epidemiological Report for 2017. Surveillance Report. Available online: www.ecdc.europa.eu (accessed on 22 August 2020).
13. Faccini, M.; Russo, A.G.; Bonini, M.; Tunesi, S.; Murtas, R.; Sandrini, M.; Senatore, S.; Lamberti, A.; Ciconali, G.; Cammarata, S.; et al. Large community-acquired Legionnaires' disease outbreak caused by *Legionella pneumophila* serogroup 1, Italy, July to August 2018. *Euro Surveill.* **2018**, *23*, 1800032. [CrossRef] [PubMed]
14. Ta, A.C.; Stout, J.E.; Yu, V.L.; Wagener, M.M. Comparison of culture methods for monitoring Legionella species in hospital potable water systems and recommendations for standardization of such methods. *J. Clin. Microbiol.* **1995**, *33*, 2118–2123. [CrossRef] [PubMed]
15. ISO. *ISO 11731,2017, Water Quality—Enumeration of Legionella*; ISO: Geneva, Switzerland, 2017.
16. ISO. *ISO 11731, 2004, Water Quality—Detection and Enumeration of Legionella—Part 2: Direct Membrane Filtration Method for Waters with Low Bacterial Counts*; ISO: Geneva, Switzerland, 2004.
17. ISO. *ISO 11731, 1998, Water Quality—Detection and Enumeration of Legionella*; ISO: Geneva, Switzerland, 1998.
18. Lucas, C.E.; Taylor, T.H., Jr.; Fields, B.S. Accuracy and precision of Legionella isolation by US laboratories in the ELITE program pilot study. *Water Res.* **2011**, *45*, 4428–4436. [CrossRef] [PubMed]
19. Scaturro, M.; Fontana, S.; Dell'eva, I.; Helfer, F.; Marchio, M.; Stefanetti, M.V.; Cavallaro, M.; Miglietta, M.; Montagna, M.T.; De Giglio, O.; et al. A multicenter study of viable PCR using propidium monoazide to detect Legionella in water samples. *Diagn. Microbiol. Infect. Dis.* **2016**, *85*, 283–288. [CrossRef] [PubMed]
20. Lee, J.V.; Lai, S.; Exner, M.; Lenz, J.; Gaia, V.; Casati, S.; Hartemann, P.; Lück, C.; Pangon, B.; Ricci, M.L.; et al. An international trial of quantitative PCR for monitoring Legionella in artificial water systems. *J. Appl. Microbiol.* **2011**, *110*, 1032–1044. [CrossRef] [PubMed]
21. Rech, M.M.; Swalla, B.M.; Dobranic, J.K. Evaluation of legiolert for quantification of *Legionella pneumophila* from non-potable water. *Curr. Microbiol.* **2018**, *75*, 1282–1289. [CrossRef]
22. ISO. *ISO 17994, 201, Requirements for the Comparison of the Relative Recovery of Microorganisms by Two Quantitative Methods*; ISO: Geneva, Switzerland, 2014.
23. American Public Health Association (APHA). 9260 J. *Legionella*. In *Standard 9260—Detection of Pathogenic Bacteria*; American Public Health Association: Washington, DC, USA, 2007.
24. AFNOR. *NF T90-431,2017, Qualité De L'eau—Recherche et Dénombrement De Legionella Spp et De Legionella Pneumophila—Méthode Par Ensemencement Direct et Après Concentration Par Filtration Sur Membrane Ou Centrifugation*; AFNOR: Denis Cedex, France, 2017.
25. U.S. Centers for Disease Control and Prevention. *Developing a Water Management Program to Reduce Legionella Growth & Spread in Buildings: A Practical Guide to Implementing Industry Standards*; U.S. Centers for Disease Control and Prevention: Atlanta, GA, USA, 2016.
26. Edelstein, P.H. Comparative study ofselective media forisolationof *Legionella pneumophila* from potable water. *J. Clin. Microbiol.* **1982**, *16*, 697–699. [CrossRef]
27. Proposal for a Directive of the European Parliament and of the Council on the Quality of Water Intended for Human Consumption (Recast)—Political Agreement. Available online: https://eur-lex.europa.eu/legal-content/EN/TXT/PDF/?uri=CONSIL:ST_6060_2020_REV_1&from=EN (accessed on 22 August 2020).
28. The Determination of Legionella Bacteria in Waters and Other Environmental Samples (2020)—Part 2—Culture Methods for Their Detection and Enumeration. Available online: http://www.standingcommitteeofanalysts.co.uk/library/Full%20numbered%20set%20of%20books/269.pdf (accessed on 22 August 2020).
29. Spies, K.; Pleischl, S.; Lange, B.; Langer, B.; Hübner, I.; Jurzik, L.; Luden, K.; Exner, M. Comparison of the Legiolert™/Quanti-Tray®MPN test for the enumeration of *Legionella pneumophila* from potable water samples with the German regulatory requirements methods ISO 11731-2 and ISO 11731. *Int. J. Hyg. Environ. Health* **2018**, *221*, 1047–1053. [CrossRef]
30. Barrette, I. Comparison of legiolert and a conventional culture method for detection of *Legionella pneumophila* from Cooling Towers in Québec. *J. AOAC Int.* **2019**, *102*, 1235–1240. [CrossRef]
31. Von Baum, H.; Ewig, S.; Marre, R.; Suttorp, N.; Gonschior, S.; Welte, T.; Lück, C. Competence network for community acquired pneumonia study group community-acquired Legionella pneumonia: New insights

from the German competence network for community acquired pneumonia. *Clin. Infect. Dis.* **2008**, *46*, 1356–1364. [CrossRef]

32. Chen, D.J.; Procop, G.W.; Vogel, S.; Yen-Lieberman, B.; Richter, S.S. Utility of PCR, culture, and antigen detection methods for diagnosis of legionellosis. *J. Clin. Microbiol.* **2015**, *53*, 3474–3477. [CrossRef]
33. Cristovam, E.; Almeida, D.; Caldeira, D.; Ferreira, J.J.; Marques, T. Accuracy of diagnostic tests for Legionnaires' disease: A systematic review. *J. Med. Microbiol.* **2017**, *66*, 485–489. [CrossRef] [PubMed]
34. Avni, T.; Bieber, A.; Green, H.; Steinmetz, T.; Leibovici, L.; Paul, M. Diagnostic accuracy of PCR alone and compared to urinary antigen testing for detection of Legionella spp: A systematic review. *J. Clin. Microbiol.* **2016**, *54*, 401–411. [CrossRef] [PubMed]
35. Priest, P.C.; Slow, S.; Chambers, S.T.; Cameron, C.M.; Balm, M.N.; Beale, M.W.; Blackmore, T.K.; Burns, A.D.; Drinković, D.; Elvy, J.A.; et al. The burden of Legionnaires' disease in New Zealand (LegiNZ): A national surveillance study. *Lancet Infect. Dis.* **2019**, *19*, 770–777. [CrossRef]
36. Pletz, M.W.; Ewig, S.; Welte, T. PCR-based routine diagnostics uncover hidden burden of Legionnaires' disease. *Lancet Infect. Dis.* **2019**, *19*, 681–683. [CrossRef]
37. Cameron, R.L.; Pollock, K.G.J.; Lindsay, D.S.J.; Anderson, E. Comparison of *Legionella longbeachae* and *Legionella pneumophila* cases in Scotland; implications for diagnosis, treatment and public health response. *J. Med. Microbiol.* **2016**, *65*, 142–146. [CrossRef]
38. Linee Guida per La Prevenzione Ed Il Controllo Della Legionellosi. 2015. Available online: http://www.salute.gov.it/imgs/C_17_pubblicazioni_2362_allegato.pdf (accessed on 22 August 2020).

*pathogens*

MDPI

*Review*

# *Legionella pneumophila* and Protozoan Hosts: Implications for the Control of Hospital and Potable Water Systems

**Muhammad Atif Nisar, Kirstin E. Ross, Melissa H. Brown, Richard Bentham and Harriet Whiley ***

College of Science and Engineering, Flinders University, GPO Box 2100, Adelaide 5001, Australia; Muhammadatif.Nisar@flinders.edu.au (M.A.N.); Kirstin.Ross@flinders.edu.au (K.E.R.); Melissa.Brown@flinders.edu.au (M.H.B.); Richard.Bentham@flinders.edu.au (R.B.)

* Correspondence: Harriet.Whiley@flinders.edu.au; Tel.: +61-8-7221-8580

Received: 21 March 2020; Accepted: 14 April 2020; Published: 15 April 2020

**Abstract:** *Legionella pneumophila* is an opportunistic waterborne pathogen of public health concern. It is the causative agent of Legionnaires' disease (LD) and Pontiac fever and is ubiquitous in manufactured water systems, where protozoan hosts and complex microbial communities provide protection from disinfection procedures. This review collates the literature describing interactions between *L. pneumophila* and protozoan hosts in hospital and municipal potable water distribution systems. The effectiveness of currently available water disinfection protocols to control *L. pneumophila* and its protozoan hosts is explored. The studies identified in this systematic literature review demonstrated the failure of common disinfection procedures to achieve long term elimination of *L. pneumophila* and protozoan hosts from potable water. It has been demonstrated that protozoan hosts facilitate the intracellular replication and packaging of viable *L. pneumophila* in infectious vesicles; whereas, cyst-forming protozoans provide protection from prolonged environmental stress. Disinfection procedures and protozoan hosts also facilitate biogenesis of viable but non-culturable (VBNC) *L. pneumophila* which have been shown to be highly resistant to many water disinfection protocols. In conclusion, a better understanding of *L. pneumophila*-protozoan interactions and the structure of complex microbial biofilms is required for the improved management of *L. pneumophila* and the prevention of LD.

**Keywords:** *Legionella pneumophila*; protozoa; *Vermamoeba*; *Acanthamoeba*; potable water; hospital water; water disinfection; legionellosis

## 1. Introduction

*Legionella pneumophila* is an opportunistic pathogen associated with community-acquired and nosocomial infections. It is the causative agent of legionellosis, which includes Legionnaires' disease (LD), a severe atypical pneumonia infection, and Pontiac fever, an acute "flu-like" illness [1]. Globally, the incidence of LD has been increasing. In Europe, the number of notified cases increased from 4921 in 2011 to 11,343 in 2018 [2]. In the US, the number of notified LD cases has increased from 2301 in 2005 [3] to 7104 in 2018 [4], a 300% increase. Globally, the fatality rate of LD ranges from 2.2–10.3%, with the lowest in Singapore and the highest in European countries [5]. In nosocomial outbreaks the fatality rate can reach up to 48% [6–8].

The genus *Legionella* is comprised of 60 species and 80 distinct serogroups [9]. Globally, *L. pneumophila* is the primary aetiological agent of LD. In Europe and the US, *L. pneumophila* serogroup (SG1) is responsible for 70–92% reported cases [8]. According to WHO, 20–30% infections are caused by other *L. pneumophila* serogroups and only 5–10% are caused by other *Legionella* species (*L. micdadei*,

*L. bozemanii*, *L. dumoffii* and *L. longbeachae*) [10]. However, unlike rest of the world, in Australia and New Zealand, *L. longbeachae* is associated with ≈ 50% reported cases of legionellosis [11,12].

*L. pneumophila* is ubiquitous in manufactured water systems [10] and in the USA has been identified as the primary cause of all potable water related outbreaks [13]. Manufactured water systems, building plumbing systems, recreational water, cooling towers and humidifiers are major sources of *L. pneumophila* [10]. Inside these plumbing structures, *Legionella* and protozoan hosts are incorporated within biofilms. Factors like water stagnation, higher levels of organic carbon and moderate temperatures can increase the rate of biofilm formation [14,15]. Transmission occurs through inhalation or aspiration of contaminated aerosols or water [16]. *L. pneumophila* maintains long term contamination of manufactured water systems through its growth within protozoan hosts, association with biofilms and disinfectant resistance or tolerance [17,18]. Freshwater amoebae are the natural eukaryotic hosts of *Legionella*; whereas, humans are considered accidental hosts [19]. In the human body, *Legionella*–contaminated aerosols are inhaled into the lungs and phagocytosed by alveolar macrophages. The alveolar macrophages behave like amoebae hosts and facilitate the intracellular division and propagation of *Legionella*, resulting in LD [20,21].

Understanding the interactions between *L. pneumophila* and protozoan hosts is essential to inform water treatment and risk management strategies for the prevention of LD. Protozoan hosts play an important role in the ability of *L. pneumophila* to survive exposure to physiochemical and environmental stresses. Protozoans facilitate the intracellular replication and packaging of live bacterial cells in the stress resistant membrane bound infectious export vesicles [22,23]. The cysts of cyst-forming amoebae provide a protective shelter from prolonged environmental stress [24]. There are numerous reports describing existence of *L. pneumophila* harboring within protozoans from thermally-, chemically-, and UV radiation-treated potable water supplies and storage reservoirs [25]. Protozoan hosts and environmental stress may facilitate the genesis of highly resistant and potentially infectious viable but non-culturable (VBNC) *L. pneumophila* [26,27]. Importantly, water storage facilities and distribution networks of many countries have been shown to be highly contaminated with protozoans that may act as hosts for *L. pneumophila* (>0–4500 cell/L cell density) [28].

This systematic literature review collated studies which detected *L. pneumophila* in association/connection with protozoan hosts from hospital or municipal potable water distribution systems and discusses this relationship under diverse environmental conditions. The effectiveness of different physical and chemical water treatment methods to control the *L. pneumophila* and its protozoan hosts is described and implications for the control and management of these water distribution systems is explored.

## 2. Results

One thousand two hundred and seventy abstracts were obtained from the Web of Science and SCOPUS. After applying the described criteria (see Figure 1 and the Materials and Methods section), 29 research manuscripts discussing *L. pneumophila* and its protozoan hosts in hospital and potable water systems were included in the study (Table 1). Potential protozoan hosts playing crucial role(s) in the *L. pneumophila* life-cycle and living in both types of water systems are compiled in Table 2. These protozoan hosts have the potential to provide an appropriate habitat for replication and survival of *L. pneumophila*.

The articles from hospital settings showed that *L. pneumophila* Serogroup 1 (hereafter SG1) is the most common serogroup causing infection in USA and European countries. Globally, SG1 is also associated with community acquired legionellosis [29,30]. However, a limitation was that most municipal potable water supply studies did not characterize the *L. pneumophila* serogroups. To investigate the different *L. pneumophila*-protozoan interactions, some studies used co-isolation and co-culturing techniques or PCR. Other approaches included techniques like scanning electron microscopy or DVC-FISH to demonstrate the fate of internalized bacteria. The electron microscope studies conducted in hospital settings found that *L. pneumophila* SG1 is able to multiply inside *Echinamoeba exudans* [31] and *Vermamoeba vermiformis* (formerly *Hartmannella vermiformis*) [32]. Likewise, PCR-based examination of potable water also demonstrated the presence of *L. pneumophila* inside *V. vermiformis* [33]. Another study used DVC-FISH to detect intracellular *L. pneumophila* inside *Acanthamoeba* and *V. vermiformis* from a potable water supply [34]. Other studies (mentioned in the Table 1), demonstrated the co-existence of free-living *L. pneumophila* and protozoan hosts, but did not characterize the specific interaction or fate of internalized bacteria. The systematic literature review identified a more diverse number of potential protozoan hosts from hospitals compared with municipal potable water systems. This could be due to the more diverse dynamics of hospital water distributions systems (Table 2). The hosts identified in the hospital settings consisted of three phyla, five classes and twelve genera, whereas the hosts isolated from potable water consisted of only two phyla, three classes and five genera. Two genera of Amoebozoa namely, *Vermamoeba* and *Acanthamoeba*, are frequently reported from both types of facilities as potential hosts. Available literature demonstrated that non-cyst-forming and ciliated protozoans can also be potential hosts for *L. pneumophila*. Most of the studies were designed specifically to explore the interactions between *L. pneumophila* - *Vermamoeba*/*Acanthamoeba*, and the diversity and the role of other possible protozoans were not investigated.

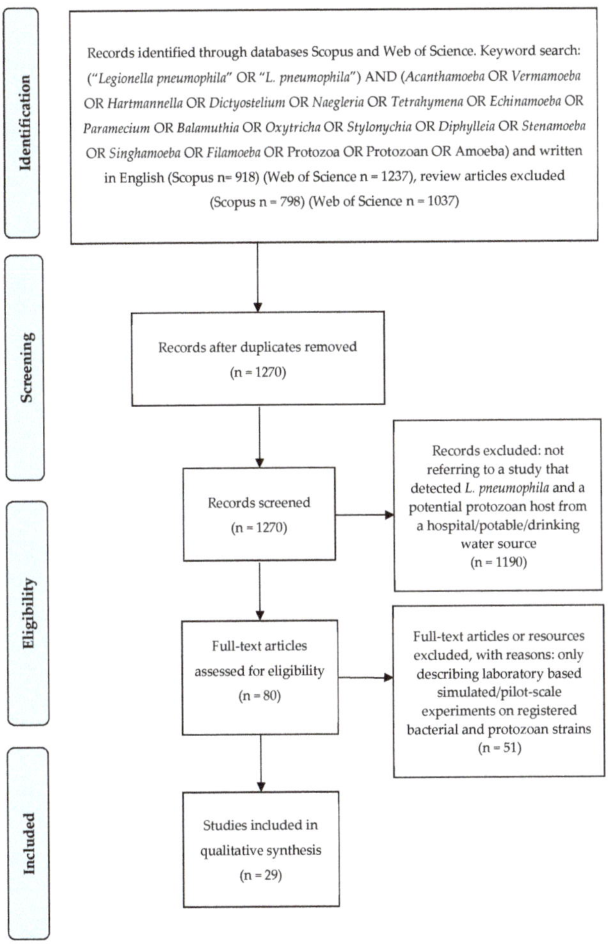


**Figure 1.** Overview of search methods and articles inclusion and exclusion criteria.

**Table 1.** Potential protozoan hosts of *Legionella pneumophila* isolated from and hospital and potable water systems.

| Isolation Source (Temperature at Time of Sampling) | Water Treatment Method | *L. pneumophila* | | Potential Protozoan Host | | Comments | Country of Origin (Sampling Site) | Reference |
|---|---|---|---|---|---|---|---|---|
| | | Identification Method | Serogroup Sequence-Type | Genus/Species | Identification Method | | | |
| **Hospital Settings** | | | | | | | | |
| Hot (45–52 °C) water tanks | - | Culturing, co-culture assay and serological identification | SG1 | *Hartmannella cantabrigiensis* *Vermamoeba vermiformis* *Echinamoeba exudans* | Culturing, light and transmission electron microscopy | Nosocomial legionellosis investigation | USA | [31] |
| Potable water sites (39–40 °C) | - | Culturing and monoclonal antibody based serotyping | SG1 | *Acanthamoeba hatchetti* *Hartmannella cantabrigiensis* *Vermamoeba vermiformis* *Vahlkampfia* *Filamoeba nolandi* *Comandonia operculata* *Paravahlkampfia ustiana* | Culturing and light microscopy | Nosocomial legionellosis investigation Thermal treatment (70 °C) and chlorination (1.5–2.0 mg/L) controlled the bacteria for 6 months but not amoebae. The treatment reduced incidence of legionellosis | South Dakota, USA | [35] |
| Cooling tower, humidifier, hot water tank and supply | - | Culturing and co-culture assay | - | *Vermamoeba vermiformis* *Naegleria* | Culturing, light and transmission electron microscopy | - | Paris, France | [32] |
| Hot (39–60 °C) and cold water supply | - | Culturing (ODR: $1 \times 10^3$–$9.7 \times 10^4$ CFU/L), direct fluorescent antibody and monoclonal antibody based serotyping | SG1 SG5 | *Hartmannella* (Hartmannellidae/ limax amoebae) | Culturing and light microscopy | Post nosocomial outbreak surveillance | Halifax, Nova Scotia, Canada | [36] |

**Table 1.** *Cont.*

| Isolation Source (Temperature at Time of Sampling) | Water Treatment Method | *L. pneumophila* | | Potential Protozoan Host | | Comments | Country of Origin (Sampling Site) | Reference |
|---|---|---|---|---|---|---|---|---|
| | | Identification Method | Serogroup Sequence-Type | Genus/Species | Identification Method | | | |
| Organ transplant unit hot (mean 56.2 °C) and cold water (mean 16.6 °C) supplies | - | Culturing and serological assay | SG1 | *Acanthamoeba*<br>*Hartmannella*<br>*Echinamoeba*<br>*Vahlkampfia*<br>*Tetrahymena*<br>*Vannella* | Culturing and light microscopy | Population density of amoebae was greater in hot water supplies than cold water supplies<br>Along amoebae other diverse eukaryotic microbes were detected as well | UK | [37] |
| Water supplies | Thermal treatment (60 and 70 °C) | Culturing (*Legionella* ODR: 2.89–6.74 × $10^5$ CFU/L), co-culture, latex agglutination, indirect and immunofluorescence assays, and PFGE | SG1<br>SG2 | *Acanthamoeba*<br>*Vahlkampfia*<br>*Mayorella* | Culturing and light microscopy | Thermal treatment (70 °C) only controlled bacterial contamination for 3 months<br>SG1 is more thermotolerant than SG2 at 60 °C | Germany | [38] |
| Water network system (mean 56 °C) | - | Amoebae co-culture assay, PCR and sequencing | - | *Vermamoeba vermiformis* | Culturing, PCR and sequencing | Detection of thermotolerant *Vermamoeba vermiformis* | Lausanne, Switzerland | [39] |
| Water distribution system (18.9–32.6 °C) | Chlorine dioxide treatment<br>Thermal treatment (<50 °C) | Culturing (ODR: *L. pneumophila* SG1: 1 × $10^2$–3.5 × $10^4$ CFU/L and *L. pneumophila* SG2-14: 1 × $10^2$–4 × $10^4$ CFU/L) and latex agglutination assay | SG1<br>SG2-14 | *Acanthamoeba*<br>*Hartmannella* | Culturing and light microscopy | - | Messina, Italy | [40] |
| Tap water | Chloramine (1.93 ± 1.04 mg/L) treatment | Culturing (protocol: ISO 11731-2:2004, LOD: 1 CFU/100 mL, ODR: 100–1.4 × $10^5$ ± 1.3 × $10^5$ CFU/L), qPCR (LOD: 5 GU, LOQ: 25 GU, *Legionella* ODR: 100–$10^9$ gu/L) and EMA-qPCR | ST269 | *Acanthamoeba polyphaga* | Culturing, light microscopy and PCR | | Italy | [27] |

**Table 1.** *Cont.*

| Isolation Source (Temperature at Time of Sampling) | Water Treatment Method | *L. pneumophila* | | Potential Protozoan Host | | Comments | Country of Origin (Sampling Site) | Reference |
|---|---|---|---|---|---|---|---|---|
| | | Identification Method | Serogroup Sequence-Type | Genus/Species | Identification Method | | | |
| Cold (14.9 °C) and warm (45.1 °C) potable water | Thermal treatment, chlorination (hypochlorates, chloramine), bacterial filters and chlorine dioxide treatment | Culturing (protocols: ISO 11731:1998 and ISO 11731-2:2004, LOD: 1 CFU/100 mL, ODR: 0–3 × $10^3$ CFU/100 mL) and MALDI-TOF MS | - | *Acanthamoeba* *Vermamoeba vermiformis* | Culturing and light microscopy | - | Bratislava, Slovakia | [41] |
| Cold water system (20–27.3 °C) | Chlorine contents 0.01–0.32 mg/L | qPCR (protocol: ISO/TS 12869:2012, LOD: 5 GU, LOQ: 25 GU, ODR: 2.7–3.8 × $10^2$ gu/L) | - | *Acanthamoeba* *Vermamoeba vermiformis* | Culturing and light microscopy | | Johannesburg, South Africa | [42] |
| Dental unit waterlines | $H_2O_2$ treatment (occasionally) | Heterotrophic plate counts, culturing (protocol: ISO 11731-2:2004, LOD: 1 CFU/100 mL, ODR: 0–2700 CFU/L) and agglutination test | - | *Vermamoeba vermiformis* | Culturing, light microscopy, PCR and sequencing | | Italy | [43] |
| **Potable Water System** | | | | | | | | |
| Unchlorinated water supplies (9.5–13.5 °C) | - | qPCR | - | *Acanthamoeba* *Acanthamoeba polyphaga* *Vermamoeba vermiformis* | qPCR (LOD: 1 cell/reaction), T-RFLP, cloning and sequencing | Along amoebae other diverse eukaryotic microbes were detected as well | Netherlands | [44] |
| Ground water supplies (5–39 °C) | Aeration, lime stone, granular activated carbon slow sand and rapid sand filtration, ozonisation and pellet softening | Culturing, biofilm batch test and qPCR | - | *Acanthamoeba* *Vermamoeba vermiformis* | 18S rDNA sequencing, PCR, T-RFLP and sequencing | Along amoebae other eukaryotic microbes were detected as well | Netherlands | [45] |

**Table 1.** *Cont.*

| Isolation Source (Temperature at Time of Sampling) | Water Treatment Method | *L. pneumophila* | | Potential Protozoan Host | | Comments | Country of Origin (Sampling Site) | Reference |
|---|---|---|---|---|---|---|---|---|
| | | Identification Method | Serogroup Sequence-Type | Genus/Species | Identification Method | | | |
| Water supplies (mean 30 °C) | Reverse osmosis, distillation (82%), chlorination (<0.005–0.2 mg/L), dolomite, limestone and granular activated carbon filtration, fluoride addition (0.3–0.7 mg/L), UV treatment (7.5–35.99 mJ/cm$^2$) | Culturing (LOD: 250 CFU/L, *Legionella* ODR: 2.5 × $10^2$–2.5 × $10^5$ CFU/L) and latex agglutination assay | - | *Acanthamoeba* *Vermamoeba vermiformis* *Echinamoeba exundans* *Echinamoeba thermarum* *Neoparamoeba* | qPCR (LOD: 2 C/L, ODR: *Acanthamoeba* < 2–56 C/L and *V. vermiformis* < 2–1670 C/L) | - | Caribbean islands, Leeward Antilles | [46] |
| Water distribution systems (mean 37.3 ± 8.4 °C) | Chloramine treatment (Chlorine contents 1.8 mg/L), flocculation, sedimentation, and dual-medium filtration | Culturing, qPCR (LOQ: 1–10 copies/reaction, maximum ODR: 13.7 ± 5.1 gc/mL) and T-RFLP | - | *Acanthamoeba* *Vermamoeba vermiformis* | qPCR (LOQ: 1–10 copies/reaction, maximum ODR: *Acanthamoeba* 6.8 ± 2.9 gc/mL and *V. vermiformis* 7.1 × $10^4$ ± 4.4 × $10^3$ gc/mL) | High concentration of chloramine is unable to disinfect water | Southwest Virginia, USA | [47] |
| Water treatment plant (7–21 °C) | - | Multiplex PCR | - | *Vermamoeba vermiformis* | Culturing, light microscopy, PCR and sequencing | Amoebae were frequently detected at 17 °C | Aragon, Spain | [33] |
| Water treatment facility (25 ± 3.4–28.2 ± 1.1 °C) | - | PCR (*Legionella* ODR: 1.2 × $10^4$–2.4 × $10^5$ gc/L) and sequencing | - | *Acanthamoeba* *Vermamoeba vermiformis* *Naegleria* | Culturing, PCR, qPCR (ODR: *Acanthamoeba* 2.1 × $10^2$–7.7 × $10^2$ gc/L and *Naegleria* 7.6 × $10^2$–9.4 × $10^2$ gc/L) and sequencing | - | Kaoping River, Taiwan | [48] |

**Table 1.** *Cont.*

| Isolation Source (Temperature at Time of Sampling) | Water Treatment Method | *L. pneumophila* | | Potential Protozoan Host | | Comments | Country of Origin (Sampling Site) | Reference |
|---|---|---|---|---|---|---|---|---|
| | | Identification Method | Serogroup Sequence-Type | Genus/Species | Identification Method | | | |
| Sediments of municipal water storage tank (2.2–28.9 °C) | Chlorination (<4 mg/L) | qPCR (LOD: 2 CE/reaction, *Legionella* ODR: 51 ± 114–7.98 × $10^4$ ± 2.49 × $10^4$ CE/g), cloning and sequencing | SG1 | *Acanthamoeba* *Vermamoeba vermiformis* | qPCR (LOD: 2 CE/reaction, ODR: *Acanthamoeba* 22 ± 50–391 ± 243 CE/g and *V. vermiformis* 17 ± 23 CE/g), cloning and sequencing | - | Northeast, East Coast, Midwest, South and West Coast, USA | [49] |
| Water distribution system | - | qPCR (LOD: 2 CE/reaction, *Legionella* ODR: 2 ± 4–391 ± 17 CE/L), cloning and sequencing | - | *Acanthamoeba* *Acanthamoeba castellanii* *Vermamoeba vermiformis* | qPCR (LOD: 2 CE/reaction, ODR: *Acanthamoeba* 1 ± 2–16 ± 2 * CE/L and *V. vermiformis* 1 ± 1–9 ± 11 * CE/L), cloning and sequencing | - | USA | [50] |
| Domestic water systems (mean 20.6 ± 3.8 °C) | - | Culturing, co-culture assay, PCR and sequencing | - | *Vermamoeba vermiformis* | Culturing, light microscopy, PCR and sequencing | - | Geneva, Lausanne and Sion, Switzerland | [51] |
| Sediments of water storage tank | - | qPCR (ODR: 25 ± 51–300 ± 38 gn/g) and NGS | - | *Acanthamoeba* *Vermamoeba vermiformis* | qPCR (ODR: *Acanthamoeba* 3–7 gn/g, *V. vermiformis* 99 ± 43–120 ± 60 gn/g) and NGS | - | Ohio, West Virginia and Texas, USA | [52] |
| Potable water | Polyaluminium chloride coagulation, sedimentation, sand and biologically activated carbon filtration and chlorination | qPCR (LOQ: 1–10 copy/reaction, minimum ODR: 3.5 log(gc)/mL) | - | *Acanthamoeba* *Vermamoeba vermiformis* | qPCR (LOD: 1–10 copy/reaction, minimum ODR: 2 log(gc)/mL for *V. vermiformis* and 4 log(gc)/mL for *Acanthamoeba*) and sequencing | Antibiotics (sulfadiazine and ciprofloxacin) promote growth of both bacterium and amoebae | Northern China | [53] |

**Table 1.** *Cont.*

| Isolation Source (Temperature at Time of Sampling) | Water Treatment Method | *L. pneumophila* | | Potential Protozoan Host | | Comments | Country of Origin (Sampling Site) | Reference |
|---|---|---|---|---|---|---|---|---|
| | | Identification Method | Serogroup Sequence-Type | Genus/Species | Identification Method | | | |
| Potable water | Polyaluminium chloride coagulation, sedimentation, sand and biologically activated carbon filtration, chlorination and ozonisation | qPCR (LOQ: 1–10 copies/reaction, minimum ODR ≈ 1 log(gc)/g) | - | *Acanthamoeba* *Naegleria* | qPCR (LOQ: 1–10 copies/reaction, minimum ODR: ≈ 0.5 log(gc)/g for *Naegleria* and ≈ 1 log(gc)/g for *Acanthamoeba*) | Combined chlorination and ozonisation are effective than chlorination only | Northern China | [54] |
| Potable water | Coagulation, ozonisation, pellet softening, granular activated carbon filtration, rapid and slow sand filtration | Heterotrophic plate counts, culturing (protocol: NEN 6275, LOD: 1 log(CFU)/cm$^2$) epifluorescence microscopy, bioluminescence assay, PCR and sequencing | - | *Vermamoeba vermiformis* | qPCR (ODR: 0.7–384 CE/cm$^2$) | - | Netherlands | [55] |
| Residential secondary water supply systems (13.9 ± 4.0–17.4 ± 2.9 °C) | Chloramine treatment (Chlorine contents 0.19–0.89 mg/L) | qPCR (LOQ: 10 copies/reaction, maximum ODR: ≈ $10^2$ gc/mL) and sequencing | - | *Acanthamoeba* *Vermamoeba vermiformis* | qPCR (LOQ: 10 copies/reaction, ODR: $10^1$–$10^3$ gc/mL for both *Acanthamoeba* and *V. vermiformis*) and sequencing | - | Shanghai, China | [56] |
| Water treatment facility | Coagulation, sedimentation, chlorination, ozonisation, granular activated carbon and sand filtration | qPCR (LOQ: 10 copies/reaction, minimum ODR: $10^2$ log(gc)/mL) and sequencing | - | *Vermamoeba vermiformis* | qPCR (LOQ: 10 copies/reaction) and sequencing | Sand filtration after granular activated carbon treatment improves water quality | Southeast China | [57] |

**Table 1.** *Cont.*

| Isolation Source (Temperature at Time of Sampling) | Water Treatment Method | *L. pneumophila* | | Potential Protozoan Host | | Comments | Country of Origin (Sampling Site) | Reference |
|---|---|---|---|---|---|---|---|---|
| | | Identification Method | Serogroup Sequence-Type | Genus/Species | Identification Method | | | |
| Water from private wells after flood | - | Culturing (protocol: ISO 11731, LOD: 1 CFU/100 mL) and qPCR (LOQ: 9.5 gc/mL, maximum ODR: 52.4 gc/mL) | - | *Naegleria fowleri* | qPCR (ODR: 11–610 gc/mL) | - | Louisiana, USA | [58] |
| Potable water | - | Culturing and DVC-FISH | - | *Acanthamoeba* *Vermamoeba vermiformis* | Culturing and PCR | - | Valencia, Spain | [34] |

*Vermamoeba vermiformis* was previously known as *Hartmannella vermiform*. *Paravahlkampfia ustiana* was previously known as *Vahlkampfia ustiana*. ODR: Observed detection range, the amount of bacteria/amoebae/DNA experimentally determined from the samples; CFU/L: colony forming unit/liter; PFGE: pulsed-field gel electrophoresis; PCR: polymerase chain reaction; ISO: International organization for standardization; MALDI-TOF MS: matrix assisted laser desorption ionization-time of flight mass spectrometry; qPCR: quantitative PCR; gu/L: genome unit/liter; LOQ: limit of quantification; LOD: limit of detection; EMA-qPCR: ethidium monoazide-qPCR; T-RFLP: terminal-restriction fragment length polymorphism; C/L: cells/liter; gc/mL: gene copy/milliliter; gc/L: gene copy/liter; CE/reaction: cell equivalent/reaction; CE/g: cell equivalent/gram; CE/L: cell equivalent/liter; * CE/L: cyst equivalent/liter; gn/g: genome copy number/gram; gc/g: gene copy/gram; NGS: next generation sequencing; NEN: Nederlands normalisatie instituut; CE/cm$^2$: cell equivalent/cm$^2$; DVC-FISH: direct viable count combined with fluorescence in situ hybridization.

**Table 2.** Taxonomic description of potential protozoan hosts.

| Hospital Settings | Potable Water System |
|---|---|
| **Phylum: Amoebozoa** | **Phylum: Amoebozoa** |
| **Class: Tubulinea**<br>Genera: *Vermamoeba, Echinamoeba, Hartmannella, Filamoeba* | **Class: Tubulinea**<br>Genera: *Vermamoeba, Echinamoeba* |
| **Class: Discosea**<br>Genera: *Acanthamoeba, Comandonia, Mayorella, Vannella* | **Class: Discosea**<br>Genera: *Acanthamoeba, Neoparamoeba* |
| **Class: Heterolobosea**<br>Genera: *Vahlkampfia, Paravahlkampfia* | |
| **Phylum: Percolozoa** | **Phylum: Percolozoa** |
| **Class: Heterolobosea**<br>Genus: *Naegleria* | **Class: Heterolobosea**<br>Genus: *Naegleria* |
| **Phylum: Ciliophora** | |
| **Class: Oligohymenophorea**<br>Genus: *Tetrahymena* | |

In the studies identified, diverse physical and chemicals methods were used to disinfect the hospital and municipal potable water systems. Chlorination (<0.05–<4 mg/L) using different chlorine compounds was frequently reported as being used in both settings. Protozoans and *L. pneumophila* could still be isolated from both hospital and municipal potable water systems despite chlorination (<0.05–<4 mg/L), and/or ozonisation and thermal (<50–70 °C) disinfection protocols being in place. Importantly, several studies from hospital settings reported regular outbreaks of legionellosis. This represents a failure of existing disinfection protocols. The systematic literature review revealed that *L. pneumophila–Acanthamoeba/Vermamoeba* were extensively co-isolated from chlorinated and thermally treated water. This demonstrates the potential tolerance of *L. pneumophila* and protozoan hosts to survive under a wide range of disinfection conditions.

## 3. Discussion

The studies identified in this review have demonstrated the failure of many common disinfection protocols to achieve long term elimination of *L. pneumophila* from hospital and potable water supplies when protozoan hosts are present [35,38] (as mentioned in Table 1). This long term survival could be attributed to association with biofilms, inherent tolerance of *L. pneumophila* to high temperature and chemical disinfectants, and constant reseeding from source water [59]. However, perhaps the most interesting and undervalued relationship is the interactions with protozoan hosts. The studies identified (Table 1) are from 14 different countries, which demonstrates the need for further research to understand the *L. pneumophila*–protozoan interaction under different environmental conditions found globally. Proper management of legionellosis requires a better understanding of *L. pneumophila*–protozoan interaction, the diversity of protozoan hosts in hospital and potable water systems and the role of the host in bacterial survival under different environmental conditions.

### *3.1. Implications for the Control of L. pneumophila*

Numerous studies have demonstrated the presence of *L. pneumophila* in disinfected water supplies [60,61]. An important factor enabling *L. pneumophila* survival in the built environment is its interaction with a protozoan host [62–64] (as mentioned in Table 3). Thermal treatment is one of the most common methods used to disinfect hospitals and building water supplies. In the USA [35], Germany [38] and Slovakia [41], thermal disinfection was adopted for management of nosocomial outbreaks of legionellosis. This strategy was unable to maintain water control for a long period of time [35,38] (as mentioned in Table 1). Rhoads et al. [64] reported that *L. pneumophila* associated with *V. vermiformis* can tolerate thermal (58 °C) treatment, and this disinfection protocol is unable to reduce microbial load in water. Published evidence suggests *Legionella* associated with *Acanthamoeba* are more thermos-tolerant and can survive at even higher temperatures ranging from 68–93 °C [63]. According to Steinert et al. [38] members of *L. pneumophila* SG1 are more thermo-tolerant than SG2. This is significant given the high number of legionellosis cases associated with *L. pneumophila* SG1.

As per WHO guidelines [65], 0.2 mg/L of free residual chlorine at point of delivery is recommended in potable water for disinfection. A pilot scale study conducted by Muraca et al. [66] demonstrated that 4 to 6 mg/L chlorine treatment for 6 h resulted in 5–6 log reduction of *L. pneumophila*. It was also observed that the efficacy of chlorine against *Legionella* was enhanced at 43 °C. However, at high temperatures a continuous flow of chlorine was required to overcome thermal decomposition. In vitro studies demonstrated higher level of tolerance to free chlorine (up to >50 mg/L) when bacteria are associated with host *Acanthamoeba* cysts [67]. According to Kool et al. [68], water disinfection with monochloramine resulted in a reduction of nosocomial LD outbreaks in USA. However, other studies have shown that some strains of *L. pneumophila* can tolerate high levels of monochloramine disinfection (17 mg-min/L for 3 log reduction) [69]. Donlan et al. [70] reported that *L. pneumophila* associated with amoebae in biofilm are less susceptible to chlorine and monochloramine treatment. It is also reported that monochloramine disinfection in hospital settings results in transformation of *L. pneumophila* vegetative cells to VBNC state [27].

According to Walker et al. [71] chlorine dioxide can effectively control *L. pneumophila* from hospital water system. In vitro studies demonstrated that 0.4 mg-min/L residual chlorine dioxide achieved a 3 log reduction of *L. pneumophila*. However, this procedure was not effective for amoebae associated *L. pneumophila* [69]. According to Schwartz et al. [72] *Legionella* biofilms on polyvinyl chloride, polyethylene and stainless steel materials can tolerate chlorine dioxide treatment. Muraca et al. [66] conducted a pilot scale study and reported that 1–2 mg/L residual concentration $O_3$ treatment for 5 h resulted in 5 log reduction of *L. pneumophila*. However, half-life of ozone in water is very short, so it is difficult to maintain residual concentration in water supplies. According to Wang et al. [54], if chlorination and ozonisation is used in combination, it can target both *L. pneumophila* and its host protozoans effectively. In combination both treatments effectively eliminated planktonic *L. pneumophila* and free living *Naegleria* from water, whereas this combination could only reduce the population of *Acanthamoeba* (≈0.9 $log_{10}$ gene copies/g). In comparison to chlorination alone, this combination method significantly reduced the population of *L. pneumophila* (≈3 $log_{10}$ gene copies/g) and host amoebae (≈3 $log_{10}$ *Naegleria* gene copies/g and ≈6.1 $log_{10}$ *Acanthamoeba* gene copies/g) co-existing in biofilms.

UV irradiation is another method of disinfection. These radiations harbor strong genotoxic attributes. Cervero-Arago et al. [73] demonstrated that 5–6 mJ/cm$^2$ UV dose was sufficient to achieve 4 log reduction *L. pneumophila* population. According to Muraca et al. [66] 30 mJ/cm$^2$ UV rays treatment for 20 min resulted in 5 log reduction of *L. pneumophila*. However, continued exposure to same fluence rate for 6 h unable to eliminate all culturable *L. pneumophila* (1–2 × $10^2$ CFU/mL). Schwartz et al. [72] reported that *Legionella* biofilms on stainless steel, polyvinyl chloride and polyethylene surfaces can tolerate UV treatment. It was also reported that amoebae associated *L. pneumophila* can tolerate much higher doses of UV rays [73].

### 3.2. Protozoan Host Control Strategies

Protozoans present in water supplies play an important role in *L. pneumophila* survival and resistance against disinfection protocols. Interesting, it has also been suggested that some protozoans infected by *L. pneumophila* have increased resistance to disinfection procedures compared to those uninfected [74]. As such, an understanding of protozoan disinfectant resistance and *L.pneuophila*–protozoan interactions is essential for the improved management of manufactured water systems. According to Loret et al. [75], common water chemical disinfection protocols, i.e., ozonisation (0.5 mg/L), chlorination (free chlorine 2 mg/L), electro-chlorination (free chlorine 2 mg/L), monochloramine (free chlorine 2 mg/L), chlorine dioxide (0.5 mg/L) and $Cu^{+}/Ag^{+}$ ions (0.5/0.001 mg/L) treatments, are unable to completely eliminate amoebae cysts hosting *Legionella* from water supplies (Table 3). These methods appear to be only effective against the free living amoebae population, as they are not feasible for targeting biofilm-associated amoebae [76]. The non-standardized approach to evaluating disinfection limits is one of the gaps in knowledge raised in the discussion section.

In vitro studies have shown 1 mg/L chlorine is sufficient to inhibit the growth of *Acanthamoeba*, *Vermamoeba* and *Vahlkampfia* trophozoites. Importantly, after two hours exposure, chlorine produced complete die-off of trophozoites [77]. According to Kuchta et al. [78] 2–4 mg/L chlorine treatment for 30 min can completely inactivate *Vermamoeba* trophozoites. Whereas, trophozoites of some strains of *Hartmannella* required 15 mg-min/L chlorine treatment for only 2 log reduction [79]. Mogoa et al. [80] reported that *Acanthamoeba* trophozoites exposed to 5 mg/L chlorine for 30 s resulted in a 3 log population reduction. It was also demonstrated that in *Acanthamoeba*, chlorination induces various cellular changes including reduction in cell size and alterations in cellular permeability. Dupuy et al. [79] noticed that *Acanthamoeba* trophozoites treated with 28 mg/L chlorine for 1 min only resulted in a 2 log reduction. In comparison with uninfected *Acanthamoeba* trophozoites, *L. pneumophila* infected *Acanthamoeba* trophozoites were more resistant against sodium hypochlorite (1024 mg/L) treatment [74].

Generally, inactivation of *Acanthamoeba* and *Vermamoeba* cysts required 5 mg/L chlorine, whereas for *Vahlkampfia* 2 mg/L chlorine treatment. It is important to note that cysts of *Acanthamoeba* were found highly resistant and only a 2 log reduction was noticed after eight hours exposure [77]. It was also reported that *Acanthamoeba* cysts can tolerate 100 mg/L of chlorine for 10 min [81]. According to Dupuy et al. [79] treatment of *Acanthamoeba* cysts with 856 mg-min/L results in only 2 log reduction. Loret et al. [82] reported that to achieve 4 log reduction for *Acanthamoeba polyphaga* cysts 3500 mg-min/L chlorine treatment is required. Likewise certain strains of *Hartmannella* cysts can tolerate high dose of chlorine (2 log reduction by 156 mg-min/L) [79]. Exposure of *Vermamoeba* cysts to 15 mg/L of chlorine for 10 min was lethal and resulted in complete inactivation [83].

Unlike *Acanthamoeba* and *Vermamoeba*, trophozoites and cysts of *Naegleria* were found sensitive to available disinfection protocols. *Naegleria* trophozoites were sensitive to 0.79 mg/L chlorine treatment for 30 min [84], whereas cysts were inactivated after exposure to 1.5 mg/L chlorine for 1 h [85]. Dupuy et al. [79] reported that chlorine treatment of *Naegleria* trophozoites with 5 mg-min/L resulted in only 2 log reduction and cysts can tolerate much higher levels of chlorine (29 mg-min/L for 2 log reduction). In potable water *Naegleria fowleri* associated with biofilms was able to tolerate 20 mg/L chlorine for 3 h [86].

In comparison to chlorine, chloramine is regarded as more stable disinfectant and capable to penetrate complex biofilms [68]. Dupuy et al. [79] suggested that instead of chlorine, monochloramine is effective chemical disinfectants against trophozoites and cysts of *Acanthamoeba*, *Vermamoeba* and *Naegleria*. It is possible that monochloramine harbors greater penetrating power than chlorine and easily enter in trophozoites and cysts. According to Mogoa et al. [87] monochloramine specifically targets the cell surface of *Acanthamoeba*. Dupuy et al. [79] identified that 352 mg-min/L monochloramine exposure resulted in 2 log reduction of *Acanthamoeba* cysts. Goudot et al. [88] noticed that 4–17 mg/L monochloramine exposure for 1 min only resulted in 2 log reduction of both planktonic and biofilm associated *Naegleria*. According to Dupuy et al. [79] to achieve 2 log reduction of *Hartmannella* trophozoites and cysts 12 mg-min/L and 34 mg-min/L monochloramine dose is required, respectively. Although in vitro studies demonstrate that higher concentration of chlorine-based disinfectants can inhibit the proliferation of protozoans; however, it can corrode the plumbing system pipes.

Chlorine dioxide has been shown to easily penetrate into amoeba trophozoites and cysts and specifically promotes cytoplasmic vacuolization in *Acanthamoeba* [87]. However the efficacy of chlorine dioxide varies from amoeba strains. The cyst form of some *Acanthamoeba* strains have been demonstrated to be highly tolerant to chlorine dioxide (35 mg-min/L for 2 log reduction) [79]. Loret et al. [82] demonstrated that an 80 mg-min/L dose of chlorine dioxide is required to achieve 4 log reduction of *Acanthamoeba polyphaga* cysts. Importantly, most studies were designed to investigate the effect of disinfection procedures on amoeba and there are limited studies on *L. pneumophila*-amoebae interactions during disinfection.

Ozonisation is an effective method of water disinfection. According to Cursons et al. [84], a dose of ozone 6.75 mg/L (0.08 mg/L residual level after 30 min) was sufficient to kill 99.9% (3 log reduction) trophozoites of *Acanthamoeba* and *Naegleria*. However, biofilm associated *Acanthamoeba*, *Hartmannella*, and *Vahlkampfia* were always found resistant to such treatments [76]. Loret et al. [82] demonstrated that 10 mg-min/L ozone dose resulted in 3 log reduction of *Acanthamoeba* trophozoites, however cysts retained viability.

Thermal treatment is a common physical disinfection protocol used for potable water supplies. According to Chang [89] trophozoites of *Naegleria* can survive at 55 °C for 15 min, whereas cysts can tolerate 65 °C for 3 min. *Vermamoeba* trophozoites and cysts have been shown to be completely inactivated by exposure to 60 °C for 30 min [78,83]. Thermal treatment of *Acanthamoeba* trophozoites and cysts at 65 °C for 10 min resulted in full inactivation [90]. Loret et al. [82] demonstrated that thermal treatment of *Acanthamoeba polyphaga* cysts at 65 °C for 120 min resulted in 5 log reduction. However, Storey et al. [81] reported that *Acanthamoeba castellanii* cysts are thermally stable and retain viability at 80 °C for 10 min. It has also been reported that thermal treatment can enhance the efficiency of chlorination. Although at high temperature (50 °C) the solubility of chlorine gas in water decreases significantly and very corrosive to pipe work, but its amoebicidal activity increases slightly [69].

UV treatment is another method of disinfection recommended by WHO. As per recommendation in 10 mJ/cm$^2$ dose is sufficient for 99.9% (3 log) inactivation of protozoans like *Giardia* and *Cryptosporidium* [65]. According to Cervero-Arago et al. [73] to achieve 3 log reduction of *V. vermiformis* trophozoites 26 mJ/cm$^2$ UV dose was required, whereas 76.2 mJ/cm$^2$ for cysts. It was also noticed that exposure to 72.2 mJ/cm$^2$ irradiance resulted in 3 log reduction of *Acanthamoeba* trophozoites [73]. Aksozek et al. [91] reported viability of *Acanthamoeba castellanii* cysts after exposure to high doses of UV rays (800 mJ/cm$^2$). According to Sarkar and Gerba [92] to achieve 4 log reduction of *Naegleria fowleri* trophozoites and cysts 24 mJ/cm$^2$ and 121 mJ/cm$^2$ UV irradiance is required, respectively. A pilot scale study conducted by Langmark et al. [93] demonstrated inability of UV irradiation to reduce biofilm associated amoebae. In contrast with other protozoans, members of the *Acanthamoeba* genera are more resistant to both chemical and physical disinfection protocols.

As per water quality guidelines of WHO [65], 41 mg-min/L chlorine at 25 °C OR 1000 mg-min/L monochloramine at 15 °C OR 7.3 mg-min/L chlorine dioxide 25 °C OR 0.63 mg-min/L $O_3$ at 15 °C OR 10 mJ/cm$^2$ UV rays, treatments are required for inactivation of pathogenic protozoan (reference protozoa *Giardia*), as mentioned earlier in this section protozoans facilitating growth of *L. pneumophila* can thrive in these conditions (Table 3).

So far, studies have investigated the efficacy of water disinfection protocols against *Acanthamoeba*, *Hartmannella*, *Naegleria* and *Vermamoeba*. However, there are numerous other waterborne cyst-forming, non-cyst forming and ciliated protozoans which support the proliferation of *L. pneumophila*. Therefore, there is a need for more research and a standardized approach to evaluating disinfection protocol(s) that target both *L. pneumophila* and potential protozoan hosts. According to our literature survey, the effectiveness of available disinfection protocols depends upon the species, strain and cellular state of protozoans, as well as the type of disinfection technique and exposure time.

### *3.3. Detection Methods*

The most commonly used methods to investigate potential *L. pneumophila* protozoan hosts are co-culture and co-isolation assays [19]. The co-culture assay is widely used in the laboratory to study *Legionella*-protozoan interactions. In this method, *Legionella* is allowed to grow in a protozoan host and fate of bacterium is determined microscopically [94]. In vitro laboratory studies showed that *Acanthamoeba* [95] and *Tetrahymena* [96] allow intracellular replication and packaging of live *L. pneumophila* into export vesicles. Other protozoan genera; *Balamuthia* [97], *Dictyostelium* [98],

*Echinamoeba* [31], *Naegleria* [99], *Paramecium* [100], and *Vermamoeba* [32], facilitate intracellular replication of *L. pneumophila*. The second method is used to detect naturally co-existing *Legionella*-protozoans from environment, but microscopically it is very difficult to find protozoans containing *Legionella* in the natural environment [101]. As an alternative approach, a sample is screened for the presence of both *Legionella* and protozoan hosts. Generally, samples are screened by PCR [102,103], fluorescence in situ hybridization [104], classical culturing techniques and microscopy [105,106]. These methods are good for screening environmental samples but are unable to delineate the underlying interactions between *Legionella* and host protozoans. Nowadays, PCR based protocols are widely used to detect *L. pneumophila* and protozoan hosts from engineered water systems. In comparison to classical culturing methods, these protocols are rapid and highly sensitive. However, most of the nucleic acid-based protocols are unable to differentiate viable and dead organisms. Propidium monoazide-PCR or ethidium monoazide-PCR are modified nucleic acid detection protocols to enumerate the live bacteria [107,108] and protozoan hosts [109,110]. To estimate burden of *L. pneumophila* and protozoan hosts in water distribution system, it is necessary to measure the quantity of alive and dead organisms regularly. This literature review demonstrates that *Vermamoeba* and *Acanthamoeba* are predominant hosts of *L. pneumophila* in the context of hospital and potable water systems. Many cyst-forming, non-cyst forming and ciliated protozoans have been found associated with *L. pneumophila* and are identified as potential hosts; however, in vitro co-culture assays and microscopic studies are required for confirmation and characterization of this interaction.

During stress (i.e., thermal, nutrient, chemical and radiation), *L. pneumophila* can enter into a VBNC state. After the end of such a stress period, in presence of a suitable host or favorable environmental conditions, the VBNC state can transform back into metabolically active cellular state [111]. Importantly, the underlying mechanisms of resuscitation from VBNC are not yet well understood. However, as the VBNC form is by definition a non-culturable state, classical microbiology culturing techniques cannot be used to monitor viability. Thus, in vitro co-culture assays can be used to resuscitate VBNC in the laboratory [74]. Alternative approaches to analyze VBNC are the analysis of membrane integrity and molecular screening [112]. There are also studies that have shown that intracellular replication of *L. pneumophila* induces VBNC state. According to Buse et al. [26] transformation of *V. vermiformis* trophozoites into cysts promotes biogenesis of VBNC *L. pneumophila*. Therefore, the interaction with protozoan hosts may also affect the ability to monitor the efficacy of disinfection protocols against *L. pneumophila*, because the bacteria may be in the VBNC form. Available literature only discusses disinfection protocols, which target culturable *L. pneumophila*. To our knowledge, there are limited studies investigating the effectiveness of disinfection protocols to eliminate VBNC *L. pneumophila*. It is our suggestion to adopt membrane integrity and in vitro co-culture assays to evaluate the disinfection procedure against VBNC *L. pneumophila*.

**Table 3.** Efficacy of available potable water disinfection protocols on *Legionella pneumophila* and host protozoans.

| Organisms | Disinfectant Dose | | | | | |
|---|---|---|---|---|---|---|
| | **Temperature (°C)** | **Chlorine (mg-min/L)** | **Monochloramine (mg-min/L)** | **Chlorine Dioxide (mg-min/L)** | **Ozone (mg-min/L)** | **UV Rays (mJ/cm²)** |
| ***Legionella pneumophila* studies** | | | | | | |
| *Legionella pneumophila* [1] | 70 °C [35,38] | 6 mg/L/6 h (5 log reduction) [66] | 17 (3 log reduction) [69] | 0.4 (3 log reduction) [69] | 1–2 mg/L/5 h (5 log reduction) [66] | 30 (5 log reduction) [2] [66] |
| ***Legionella pneumophila*–potential host protozoans coculture studies** | | | | | | |
| *Legionella pneumophila Acanthamoeba* coculture | 93 °C [3] [63] | >50 mg/L [67] | 23 (3 log reduction) [69] | 2.8 (3 log reduction) [69] | - | 10.8 (4 log reduction) [73] |
| *Legionella pneumophila Vermamoeba* coculture | 58 °C [64] | - | - | - | - | - |
| **Potential host protozoans studies** | | | | | | |
| *Acanthamoeba* (trophozoite) | 65 °C/10 min (inactivation) [90] | 28 (2 log reduction) [79] | 40 (2 log reduction) [69] | >5 (2 log reduction) [79] | 10 (3 log reduction) [82] | 72.2 (3 log reduction) [73] |
| *Acanthamoeba* (cyst) | 80 °C/10 min [81] | 3500 (4 log reduction) [82] | 352 (2 log reduction) [79] | 80 (4 log reduction) [82] | 15 (4 log reduction) [82] | 800 [91] |
| *Vermamoeba* (trophozoite) | 60 °C/5 min (4 log reduction) [113] | 2–4 mg/L/30 min inactivation) [78] | - | - | - | 26 (3 log reduction) [73] |
| *Vermamoeba* (cyst) | 60 °C/5 min (2 log reduction) [113] | 15 mg/L/10 min (inactivation) [83] | - | - | - | 76.2 (3 log reduction) [73] |
| *Hartmannella* (trophozoite) | 53 °C [114] | 15 (2 log reduction) [79] | 12 2 log reduction) [79] | 5 (2 log reduction) [79] | - | - |
| *Hartmannella* (cyst) | - | 156 (2 log reduction) [79] | 34 (2 log reduction) [79] | 1 (2 log reduction) [79] | - | - |

**Table 3.** *Cont.*

| Organisms | Disinfectant Dose | | | | | |
|---|---|---|---|---|---|---|
| | **Temperature (°C)** | **Chlorine (mg-min/L)** | **Monochloramine (mg-min/L)** | **Chlorine Dioxide (mg-min/L)** | **Ozone (mg-min/L)** | **UV Rays (mJ/cm$^2$)** |
| *Naegleria* (trophozoite) | 55 °C/15 min [89] | 5 (2 log reduction) [79] | 4–17 (2 log reduction) [88] | 1 (2 log reduction) [79] | 6.75 mg/L 30 min (3 log reduction) [84] | 24 (4 log reduction) [92] |
| *Naegleria* (cyst) | 65 °C/3 min [89] | 29 (2 log reduction) [79] | 13 (2 log reduction) [79] | 5.5 (2 log reduction) [115] | - | 121 (4 log reduction) [92] |
| *Vahlkampfia* (trophozoite) | - | 1 mg/L (inactivation) [77] | - | - | - | - |
| *Vahlkampfia* (cyst) | - | 2 mg/L/2 h (3 log reduction) [77] | - | - | - | - |

[1] Most of the studies focus on culturable bacteria, and non-culturable cells are not estimated. [2] No further bacterial inactivation possible, 1–2 × $10^2$ CFU/mL *L. pneumophila* remain stable. [3] Experiments conducted on *Legionella* sp.

## 4. Materials and Methods

The databases Scopus and Web of Science were searched for articles written in English containing the keywords ("*Legionella pneumophila*" OR "*L. pneumophila*") AND (*Acanthamoeba* OR *Vermamoeba* OR *Hartmannella* OR *Dictyostelium* OR *Naegleria* OR *Tetrahymena* OR *Echinamoeba* OR *Paramecium* OR *Balamuthia* OR *Oxytricha* OR *Stylonychia* OR *Diphylleia* OR *Stenamoeba* OR *Singhamoeba* OR *Filamoeba* OR Protozoa OR Protozoan OR Amoeba). The above search terms were modified from the review conducted by Boamah et al. [19]. Figure 1 presents the systematic approach to article inclusion or exclusion. Articles were screened by reading the titles and abstracts and initially excluded if they did not refer to a study that detected *L. pneumophila* and a potential protozoan host from a hospital or potable/drinking water source. Articles were then read in full and excluded if they only described laboratory based simulated or pilot-scale experiments on registered bacterial and protozoan strains.

## 5. Conclusions

Protozoans present in potable water play an important role in *L. pneumophila* survival. Further research is needed to better understand *L. pneumophila*-protozoan interactions and the implications for the prevention of Legionnaires' disease. To achieve long term disinfection of a water system the control protocols need to be effective against potential hosts harboring *L. pneumophila*. Additionally, an understanding of the mechanisms of VBNC state transformation, and the role of protozoans in this, is needed to effectively evaluate the efficacy of disinfection techniques.

**Author Contributions:** M.A.N. and H.W. drafted and edited the manuscript, H.W., K.E.R., M.H.B. and R.B. corrected and contributed to the manuscript. All authors have read and agreed to the published version of the manuscript.

**Funding:** This research received no external funding.

**Conflicts of Interest:** The authors declare no conflict of interest.

## References

1. Cunha, B.A.; Burillo, A.; Bouza, E. Legionnaires' disease. *Lancet* **2016**, *387*, 376–385. [CrossRef]
2. The European Legionnaires' Disease Surveillance Network; The European Centre for Disease Prevention and Control. Surveillance Atlas of Infectious Diseases. Available online: https://atlas.ecdc.europa.eu/public/index.aspx (accessed on 4 October 2019).
3. Smith, P.; Moore, M.; Alexander, N.; Hicks, L.; O'Loughlin, R. Surveillance for travel-associated Legionnaires disease-United States, 2005–2006. *Mmwr-Morb Mortal Wkly. Rep.* **2007**, *56*, 1261–1263.
4. National Notifiable Diseases Surveillance System; Centers for Disease Control and Prevention. *National Notifiable Infectious Diseases: Weekly Tables*; National Notifiable Diseases Surveillance System; Centers for Disease Control and Prevention: Atlanta, GA, USA, 2018.
5. Phin, N.; Parry-Ford, F.; Harrison, T.; Stagg, H.R.; Zhang, N.; Kumar, K.; Lortholary, O.; Zumla, A.; Abubakar, I. Epidemiology and clinical management of Legionnaires' disease. *Lancet Infect. Dis.* **2014**, *14*, 1011–1021. [CrossRef]
6. O'Mahony, M.C.; Stanwell-Smith, R.E.; Tillett, H.E.; Harper, D.; Hutchison, J.G.; Farrell, I.D.; Hutchinson, D.N.; Lee, J.V.; Dennis, P.J.; Duggal, H.V.; et al. The Stafford outbreak of Legionnaires' disease. *Epidemiol. Infect.* **1990**, *104*, 361–380. [CrossRef] [PubMed]
7. Soda, E.A. Vital Signs: Health care–associated Legionnaires' disease surveillance data from 20 states and a large metropolitan area-United States, 2015. *Mmwr-Morb Mortal Wkly. Rep.* **2017**, *66*, 584–589. [CrossRef]
8. Mercante, J.W.; Winchell, J.M. Current and emerging *Legionella* diagnostics for laboratory and outbreak investigations. *Clin. Microbiol. Rev.* **2015**, *28*, 95–133. [CrossRef]
9. Miyashita, N.; Higa, F.; Aoki, Y.; Kikuchi, T.; Seki, M.; Tateda, K.; Maki, N.; Uchino, K.; Ogasawara, K.; Kiyota, H.; et al. Distribution of Legionella species and serogroups in patients with culture-confirmed *Legionella pneumonia*. *J. Infect. Chemother* **2020**. [CrossRef]
10. Bartram, J.; Chartier, Y.; Lee, J.V.; Pond, K.; Surman-Lee, S. *Legionella and the Prevention of Legionellosis*; World Health Organization: Geneva, Switzerland, 2007.

11. NNDSS Annual Report Working Group. Australia's notifiable disease status, 2015: Annual report of the national notifiable diseases surveillance system. *Commun. Dis. Intel. (2018)* **2019**, *43*. [CrossRef]
12. Graham, F.F.; White, P.S.; Harte, D.J.; Kingham, S.P. Changing epidemiological trends of legionellosis in New Zealand, 1979–2009. *Epidemiol. Infect.* **2012**, *140*, 1481–1496. [CrossRef]
13. Benedict, K.M. Surveillance for waterborne disease outbreaks associated with drinking water—United States, 2013–2014. *Mmwr.-Morb. Mortal Wkly. Rep.* **2017**, *66*, 1216. [CrossRef]
14. Abdel-Nour, M.; Duncan, C.; Low, D.E.; Guyard, C. Biofilms: The stronghold of *Legionella pneumophila*. *Int. J. Mol. Sci* **2013**, *14*, 21660. [CrossRef]
15. Liu, Z.; Lin, Y.E.; Stout, J.E.; Hwang, C.C.; Vidic, R.D.; Yu, V.L. Effect of flow regimes on the presence of *Legionella* within the biofilm of a model plumbing system. *J. Appl. Microbiol.* **2006**, *101*, 437–442. [CrossRef]
16. Prussin, A.J., 2nd; Schwake, D.O.; Marr, L.C. Ten questions concerning the aerosolization and transmission of *Legionella* in the built environment. *Build. Environ.* **2017**, *123*, 684–695. [CrossRef]
17. Ashbolt, N.J. Environmental (Saprozoic) Pathogens of Engineered Water Systems: Understanding Their Ecology for Risk Assessment and Management. *Pathogens* **2015**, *4*, 390. [CrossRef]
18. Falkinham, J.O.; Pruden, A.; Edwards, M. Opportunistic premise plumbing pathogens: Increasingly important pathogens in drinking water. *Pathogens* **2015**, *4*, 373. [CrossRef]
19. Boamah, D.K.; Zhou, G.; Ensminger, A.W.; O'Connor, T.J. From many hosts, one accidental pathogen: The diverse protozoan hosts of *Legionella*. *Front. Cell Infect. Microbiol.* **2017**, *7*, 477. [CrossRef]
20. Newton, H.J.; Ang, D.K.; van Driel, I.R.; Hartland, E.L. Molecular pathogenesis of infections caused by *Legionella pneumophila*. *Clin. Microbiol. Rev.* **2010**, *23*, 274–298. [CrossRef]
21. Oliva, G.; Sahr, T.; Buchrieser, C. The Life Cycle of *L. pneumophila*: Cellular Differentiation Is Linked to Virulence and Metabolism. *Front. Cell Infect. Microbiol.* **2018**, *8*, 3. [CrossRef]
22. Bouyer, S.; Imbert, C.; Rodier, M.H.; Hechard, Y. Long-term survival of *Legionella pneumophila* associated with *Acanthamoeba castellanii* vesicles. *Environ. Microbiol.* **2007**, *9*, 1341–1344. [CrossRef]
23. Koubar, M.; Rodier, M.H.; Garduno, R.A.; Frere, J. Passage through *Tetrahymena tropicalis* enhances the resistance to stress and the infectivity of *Legionella pneumophila*. *Fems Microbiol. Lett.* **2011**, *325*, 10–15. [CrossRef]
24. Jjemba, P.K.; Johnson, W.; Bukhari, Z.; LeChevallier, M.W. Occurrence and control of *Legionella* in recycled water systems. *Pathogens* **2015**, *4*, 470. [CrossRef] [PubMed]
25. Kim, B.R.; Anderson, J.E.; Mueller, S.A.; Gaines, W.A.; Kendall, A.M. Literature review–efficacy of various disinfectants against *Legionella* in water systems. *Water Res.* **2002**, *36*, 4433–4444. [CrossRef]
26. Buse, H.Y.; Donohue, M.J.; Ashbolt, N.J. *Hartmannella vermiformis* inhibition of *Legionella pneumophila* cultivability. *Microb. Ecol.* **2013**, *66*, 715–726. [CrossRef]
27. Casini, B.; Baggiani, A.; Totaro, M.; Mansi, A.; Costa, A.L.; Aquino, F.; Miccoli, M.; Valentini, P.; Bruschi, F.; Lopalco, P.L.; et al. Detection of viable but non-culturable *Legionella* in hospital water network following monochloramine disinfection. *J. Hosp. Infect.* **2018**, *98*, 46–52. [CrossRef]
28. Thomas, J.M.; Ashbolt, N.J. Do free-living amoebae in treated drinking water systems present an emerging health risk? *Environ. Sci. Technol.* **2011**, *45*, 860–869. [CrossRef]
29. Yu, V.L.; Plouffe, J.F.; Pastoris, M.C.; Stout, J.E.; Schousboe, M.; Widmer, A.; Summersgill, J.; File, T.; Heath, C.M.; Paterson, D.L.; et al. Distribution of *Legionella* species and serogroups isolated by culture in patients with sporadic community-acquired legionellosis: An international collaborative survey. *J. Infect. Dis.* **2002**, *186*, 127–128. [CrossRef]
30. Beaute, J.; Zucs, P.; de Jong, B.; European Legionnaires' Disease Surveillance Network. Legionnaires disease in Europe, 2009–2010. *Euro Surveill.* **2013**, *18*, 20417. [CrossRef]
31. Fields, B.S.; Sanden, G.N.; Barbaree, J.M.; Morrill, W.E.; Wadowsky, R.M.; White, E.H.; Feeley, J.C. Intracellular multiplication of *Legionella pneumophila* in amoebae isolated from hospital hot water tanks. *Curr. Microbiol.* **1989**, *18*, 131–137. [CrossRef]
32. Nahapetian, K.; Challemel, O.; Beurtin, D.; Dubrou, S.; Gounon, P.; Squinazi, F. The intracellular multiplication of *Legionella pneumophila* in protozoa from hospital plumbing systems. *Res. Microbiol.* **1991**, *142*, 677–685. [CrossRef]
33. Garcia, A.; Goni, P.; Cieloszyk, J.; Fernandez, M.T.; Calvo-Begueria, L.; Rubio, E.; Fillat, M.F.; Peleato, M.L.; Clavel, A. Identification of free-living amoebae and amoeba-associated bacteria from reservoirs and water treatment plants by molecular techniques. *Environ. Sci. Technol.* **2013**, *47*, 3132–3140. [CrossRef]

34. Moreno, Y.; Moreno-Mesonero, L.; Garcia-Hernandez, J. DVC-FISH to identify potentially pathogenic *Legionella* inside free-living amoebae from water sources. *Environ. Res.* **2019**, *176*, 108521. [CrossRef]
35. Breiman, R.F.; Fields, B.S.; Sanden, G.N.; Volmer, L.; Meier, A.; Spika, J.S. Association of shower use with Legionnaires' disease-possible role of ameobas. *JAMA-J. Am. Med. Assoc.* **1990**, *263*, 2924–2926. [CrossRef]
36. Bezanson, G.; Burbridge, S.; Haldane, D.; Yoell, C.; Marrie, T. Diverse populations of *Legionella pneumophila* present in the water of geographically clustered institutions served by the same water reservoir. *J. Clin. Microbiol.* **1992**, *30*, 570–576. [CrossRef]
37. Patterson, W.J.; Hay, J.; Seal, D.V.; McLuckie, J.C. Colonization of transplant unit water supplies with *Legionella* and protozoa: Precautions required to reduce the risk of legionellosis. *J. Hosp. Infect.* **1997**, *37*, 7–17. [CrossRef]
38. Steinert, M.; Ockert, G.; Luck, C.; Hacker, J. Regrowth of *Legionella pneumophila* in a heat-disinfected plumbing system. *Zent. Bakteriol.* **1998**, *288*, 331–342. [CrossRef]
39. Thomas, V.; Herrera-Rimann, K.; Blanc, D.S.; Greub, G. Biodiversity of amoebae and amoeba-resisting bacteria in a hospital water network. *Appl. Environ. Microbiol.* **2006**, *72*, 2428–2438. [CrossRef]
40. Lagana, P.; Caruso, G.; Piccione, D.; Gioffre, M.E.; Pino, R.; Delia, S. *Legionella* spp., amoebae and not-fermenting Gram negative bacteria in an Italian university hospital water system. *Ann. Agric. Environ. Med.* **2014**, *21*, 489–493. [CrossRef]
41. Trnkova, K.; Kotrbancova, M.; Spalekova, M.; Fulova, M.; Boledovicova, J.; Vesteg, M. MALDI-TOF MS analysis as a useful tool for an identification of *Legionella pneumophila*, a facultatively pathogenic bacterium interacting with free-living amoebae: A case study from water supply system of hospitals in Bratislava (Slovakia). *Exp. Parasitol* **2018**, *184*, 97–102. [CrossRef] [PubMed]
42. Muchesa, P.; Leifels, M.; Jurzik, L.; Barnard, T.G.; Bartie, C. Detection of amoeba-associated *Legionella pneumophila* in hospital water networks of Johannesburg. *S. Afr J. Infect. D* **2018**, *33*, 72–75. [CrossRef]
43. Spagnolo, A.M.; Sartini, M.; Cave, D.D.; Casini, B.; Tuvo, B.; Cristina, M.L. Evaluation of microbiological and free-living protozoa contamination in dental unit waterlines. *Int. J. Environ. Res. Public Health* **2019**, *16*, 2468. [CrossRef]
44. Valster, R.M.; Wullings, B.A.; Bakker, G.; Smidt, H.; van der Kooij, D. Free-living protozoa in two unchlorinated drinking water supplies, identified by phylogenic analysis of 18S rRNA gene sequences. *Appl. Environ. Microbiol.* **2009**, *75*, 4736–4746. [CrossRef]
45. Valster, R.M.; Wullings, B.A.; van der Kooij, D. Detection of protozoan hosts for *Legionella pneumophila* in engineered water systems by using a biofilm batch test. *Appl. Environ. Microbiol.* **2010**, *76*, 7144–7153. [CrossRef]
46. Valster, R.M.; Wullings, B.A.; van den Berg, R.; van der Kooij, D. Relationships between free-living protozoa, cultivable *Legionella* spp., and water quality characteristics in three drinking water supplies in the Caribbean. *Appl. Environ. Microbiol.* **2011**, *77*, 7321–7328. [CrossRef]
47. Wang, H.; Edwards, M.; Falkinham, J.O., 3rd; Pruden, A. Molecular survey of the occurrence of *Legionella* spp., *Mycobacterium* spp., *Pseudomonas aeruginosa*, and amoeba hosts in two chloraminated drinking water distribution systems. *Appl. Environ. Microbiol.* **2012**, *78*, 6285–6294. [CrossRef]
48. Ji, W.T.; Hsu, B.M.; Chang, T.Y.; Hsu, T.K.; Kao, P.M.; Huang, K.H.; Tsai, S.F.; Huang, Y.L.; Fan, C.W. Surveillance and evaluation of the infection risk of free-living amoebae and *Legionella* in different aquatic environments. *Sci. Total Environ.* **2014**, *499*, 212–219. [CrossRef]
49. Lu, J.; Struewing, I.; Yelton, S.; Ashbolt, N. Molecular survey of occurrence and quantity of *Legionella* spp., *Mycobacterium* spp., *Pseudomonas aeruginosa* and amoeba hosts in municipal drinking water storage tank sediments. *J. Appl. Microbiol.* **2015**, *119*, 278–288. [CrossRef]
50. Lu, J.; Struewing, I.; Vereen, E.; Kirby, A.E.; Levy, K.; Moe, C.; Ashbolt, N. Molecular Detection of *Legionella* spp. and their associations with *Mycobacterium* spp., *Pseudomonas aeruginosa* and amoeba hosts in a drinking water distribution system. *J. Appl. Microbiol.* **2016**, *120*, 509–521. [CrossRef]
51. Lienard, J.; Croxatto, A.; Gervaix, A.; Levi, Y.; Loret, J.F.; Posfay-Barbe, K.M.; Greub, G. Prevalence and diversity of *Chlamydiales* and other amoeba-resisting bacteria in domestic drinking water systems. *New Microbes New Infect.* **2017**, *15*, 107–116. [CrossRef]
52. Qin, K.; Struewing, I.; Domingo, J.S.; Lytle, D.; Lu, J. Opportunistic pathogens and microbial communities and their associations with sediment physical parameters in drinking water storage tank sediments. *Pathogens* **2017**, *6*, 54. [CrossRef]

53. Wang, H.; Shen, Y.; Hu, C.; Xing, X.; Zhao, D. Sulfadiazine/ciprofloxacin promote opportunistic pathogens occurrence in bulk water of drinking water distribution systems. *Environ. Pollut.* **2018**, *234*, 71–78. [CrossRef]
54. Wang, H.; Hu, C.; Zhang, S.; Liu, L.; Xing, X. Effects of $O_3/Cl_2$ disinfection on corrosion and opportunistic pathogens growth in drinking water distribution systems. *J. Environ. Sci. (China)* **2018**, *73*, 38–46. [CrossRef]
55. Van der Kooij, D.; Veenendaal, H.R.; Italiaander, R.; van der Mark, E.J.; Dignum, M. Primary colonizing Betaproteobacteriales play a key role in the growth of *Legionella pneumophila* in biofilms on surfaces exposed to drinking water treated by slow sand filtration. *Appl. Environ. Microbiol.* **2018**, *84*, e01732–e01818. [CrossRef]
56. Li, H.; Li, S.; Tang, W.; Yang, Y.; Zhao, J.; Xia, S.; Zhang, W.; Wang, H. Influence of secondary water supply systems on microbial community structure and opportunistic pathogen gene markers. *Water Res.* **2018**, *136*, 160–168. [CrossRef]
57. Wang, H.; Xu, J.; Tang, W.; Li, H.; Xia, S.; Zhao, J.; Zhang, W.; Yang, Y. Removal efficacy of opportunistic pathogens and bacterial community dynamics in two drinking water treatment trains. *Small* **2019**, *15*, 1804436. [CrossRef]
58. Dai, D.J.; Rhoads, W.J.; Katner, A.; Strom, L.; Edwards, M.A.; Pruden, A.; Pieper, K.J. Molecular survey of *Legionella* and *Naegleria fowleri* in private well water and premise plumbing following the 2016 Louisiana flood. *Environ. Sci.-Wat. Res.* **2019**, *5*, 1464–1477. [CrossRef]
59. Lau, H.Y.; Ashbolt, N.J. The role of biofilms and protozoa in *Legionella* pathogenesis: Implications for drinking water. *J. Appl. Microbiol.* **2009**, *107*, 368–378. [CrossRef]
60. Lin, Y.S.; Stout, J.E.; Yu, V.L.; Vidic, R.D. Disinfection of water distribution systems for *Legionella*. *Semin. Respir. Infect.* **1998**, *13*, 147–159.
61. Whiley, H.; Bentham, R.; Brown, M.H. *Legionella* persistence in manufactured water systems: Pasteurization potentially selecting for thermal tolerance. *Front. Microbiol.* **2017**, *8*, 1330. [CrossRef]
62. Fields, B.S.; Benson, R.F.; Besser, R.E. *Legionella* and Legionnaires' disease: 25 years of investigation. *Clin. Microbiol. Rev.* **2002**, *15*, 506–526. [CrossRef]
63. Dobrowsky, P.H.; Khan, S.; Cloete, T.E.; Khan, W. Molecular detection of *Acanthamoeba* spp., *Naegleria fowleri* and *Vermamoeba (Hartmannella) vermiformis* as vectors for *Legionella* spp. in untreated and solar pasteurized harvested rainwater. *Parasit. Vectors* **2016**, *9*, 539. [CrossRef]
64. Rhoads, W.J.; Ji, P.; Pruden, A.; Edwards, M.A. Water heater temperature set point and water use patterns influence *Legionella pneumophila* and associated microorganisms at the tap. *Microbiome* **2015**, *3*, 67. [CrossRef]
65. World Health Organization. *Guidelines for Drinking-Water Quality*; WHO: Geneva, Switzerland, 2004.
66. Muraca, P.; Stout, J.E.; Yu, V.L. Comparative assessment of chlorine, heat, ozone, and UV light for killing *Legionella pneumophila* within a model plumbing system. *Appl. Environ. Microbiol.* **1987**, *53*, 447–453. [CrossRef]
67. Kilvington, S.; Price, J. Survival of *Legionella pneumophila* within cysts of *Acanthamoeba polyphaga* following chlorine exposure. *J. Appl. Bacteriol.* **1990**, *68*, 519–525. [CrossRef]
68. Kool, J.L.; Carpenter, J.C.; Fields, B.S. Effect of monochloramine disinfection of municipal drinking water on risk of nosocomial Legionnaires' disease. *Lancet* **1999**, *353*, 272–277. [CrossRef]
69. Dupuy, M.; Mazoua, S.; Berne, F.; Bodet, C.; Garrec, N.; Herbelin, P.; Menard-Szczebara, F.; Oberti, S.; Rodier, M.H.; Soreau, S.; et al. Efficiency of water disinfectants against *Legionella pneumophila* and *Acanthamoeba*. *Water Res.* **2011**, *45*, 1087–1094. [CrossRef]
70. Donlan, R.M.; Forster, T.; Murga, R.; Brown, E.; Lucas, C.; Carpenter, J.; Fields, B. *Legionella pneumophila* associated with the protozoan *Hartmannella vermiformis* in a model multi-species biofilm has reduced susceptibility to disinfectants. *Biofouling* **2005**, *21*, 1–7. [CrossRef]
71. Walker, J.T.; Mackerness, C.W.; Mallon, D.; Makin, T.; Williets, T.; Keevil, C.W. Control of *Legionella pneumophila* in a hospital water system by chlorine dioxide. *J. Ind. Microbiol.* **1995**, *15*, 384–390. [CrossRef]
72. Schwartz, T.; Hoffmann, S.; Obst, U. Formation of natural biofilms during chlorine dioxide and UV disinfection in a public drinking water distribution system. *J. Appl. Microbiol.* **2003**, *95*, 591–601. [CrossRef]
73. Cervero-Arago, S.; Sommer, R.; Araujo, R.M. Effect of UV irradiation (253.7 nm) on free *Legionella* and *Legionella* associated with its amoebae hosts. *Water Res.* **2014**, *67*, 299–309. [CrossRef]
74. Garcia, M.T.; Jones, S.; Pelaz, C.; Millar, R.D.; Abu Kwaik, Y. *Acanthamoeba polyphaga* resuscitates viable non-culturable *Legionella pneumophila* after disinfection. *Environ. Microbiol.* **2007**, *9*, 1267–1277. [CrossRef]
75. Loret, J.F.; Robert, S.; Thomas, V.; Cooper, A.J.; McCoy, W.F.; Levi, Y. Comparison of disinfectants for biofilm, protozoa and *Legionella* control. *J. Water Health* **2005**, *3*, 423–433. [CrossRef]

76. Thomas, V.; Bouchez, T.; Nicolas, V.; Robert, S.; Loret, J.F.; Levi, Y. Amoebae in domestic water systems: Resistance to disinfection treatments and implication in *Legionella* persistence. *J. Appl. Microbiol.* **2004**, *97*, 950–963. [CrossRef]
77. Critchley, M.; Bentham, R. The efficacy of biocides and other chemical additives in cooling water systems in the control of amoebae. *J. Appl. Microbiol.* **2009**, *106*, 784–789. [CrossRef]
78. Kuchta, J.M.; Navratil, J.S.; Shepherd, M.E.; Wadowsky, R.M.; Dowling, J.N.; States, S.J.; Yee, R.B. Impact of chlorine and heat on the survival of *Hartmannella vermiformis* and subsequent growth of *Legionella pneumophila*. *Appl. Environ. Microbiol.* **1993**, *59*, 4096–4100. [CrossRef]
79. Dupuy, M.; Berne, F.; Herbelin, P.; Binet, M.; Berthelot, N.; Rodier, M.H.; Soreau, S.; Hechard, Y. Sensitivity of free-living amoeba trophozoites and cysts to water disinfectants. *Int. J. Hyg. Environ. Health* **2014**, *217*, 335–339. [CrossRef]
80. Mogoa, E.; Bodet, C.; Legube, B.; Hechard, Y. *Acanthamoeba castellanii*: Cellular changes induced by chlorination. *Exp. Parasitol.* **2010**, *126*, 97–102. [CrossRef]
81. Storey, M.V.; Winiecka-Krusnell, J.; Ashbolt, N.J.; Stenstrom, T.A. The efficacy of heat and chlorine treatment against thermotolerant *Acanthamoebae* and *Legionellae*. *Scand. J. Infect. Dis.* **2004**, *36*, 656–662. [CrossRef]
82. Loret, J.; Jousset, M.; Robert, S.; Anselme, C.; Saucedo, G.; Ribas, F.; Martinez, L.; Catalan, V. Elimination of free-living amoebae by drinking water treatment processes. *Eur. J. Water Qual.* **2008**, *39*, 37–50. [CrossRef]
83. Fouque, E.; Hechard, Y.; Hartemann, P.; Humeau, P.; Trouilhe, M.C. Sensitivity of *Vermamoeba* (*Hartmannella*) *vermiformis* cysts to conventional disinfectants and protease. *J. Water Health* **2015**, *13*, 302–310. [CrossRef]
84. Cursons, R.T.; Brown, T.J.; Keys, E.A. Effect of disinfectants on pathogenic free-living amoebae: In axenic conditions. *Appl. Environ. Microbiol.* **1980**, *40*, 62–66. [CrossRef]
85. De Jonckheere, J.; van de Voorde, H. Differences in destruction of cysts of pathogenic and nonpathogenic *Naegleria* and *Acanthamoeba* by chlorine. *Appl. Environ. Microbiol.* **1976**, *31*, 294–297. [CrossRef] [PubMed]
86. Miller, H.C.; Wylie, J.; Dejean, G.; Kaksonen, A.H.; Sutton, D.; Braun, K.; Puzon, G.J. Reduced efficiency of chlorine disinfection of *Naegleria fowleri* in a drinking water distribution biofilm. *Environ. Sci. Technol.* **2015**, *49*, 11125–11131. [CrossRef] [PubMed]
87. Mogoa, E.; Bodet, C.; Morel, F.; Rodier, M.H.; Legube, B.; Hechard, Y. Cellular response of the amoeba *Acanthamoeba castellanii* to chlorine, chlorine dioxide, and monochloramine treatments. *Appl. Environ. Microbiol.* **2011**, *77*, 4974–4980. [CrossRef] [PubMed]
88. Goudot, S.; Herbelin, P.; Mathieu, L.; Soreau, S.; Banas, S.; Jorand, F.P. Biocidal efficacy of monochloramine against planktonic and biofilm-associated *Naegleria fowleri* cells. *J. Appl. Microbiol.* **2014**, *116*, 1055–1065. [CrossRef]
89. Chang, S.L. Resistance of pathogenic *Naegleria* to some common physical and chemical agents. *Appl. Environ. Microbiol.* **1978**, *35*, 368–375. [CrossRef]
90. Coulon, C.; Collignon, A.; McDonnell, G.; Thomas, V. Resistance of *Acanthamoeba* cysts to disinfection treatments used in health care settings. *J. Clin. Microbiol.* **2010**, *48*, 2689–2697. [CrossRef]
91. Aksozek, A.; McClellan, K.; Howard, K.; Niederkorn, J.Y.; Alizadeh, H. Resistance of *Acanthamoeba castellanii* cysts to physical, chemical, and radiological conditions. *J. Parasitol.* **2002**, *88*, 621–623. [CrossRef]
92. Sarkar, P.; Gerba, C.P. Inactivation of *Naegleria fowleri* by chlorine and ultraviolet light. *J. Am. Water Work Assoc.* **2012**, *104*, E173–E180. [CrossRef]
93. Langmark, J.; Storey, M.V.; Ashbolt, N.J.; Stenstrom, T.A. The effects of UV disinfection on distribution pipe biofilm growth and pathogen incidence within the greater Stockholm area, Sweden. *Water Res.* **2007**, *41*, 3327–3336. [CrossRef]
94. Dietersdorfer, E.; Cervero-Arago, S.; Sommer, R.; Kirschner, A.K.; Walochnik, J. Optimized methods for *Legionella pneumophila* release from its *Acanthamoeba* hosts. *BMC Microbiol.* **2016**, *16*, 74. [CrossRef]
95. Berk, S.G.; Ting, R.S.; Turner, G.W.; Ashburn, R.J. Production of respirable vesicles containing live *Legionella pneumophila* cells by two *Acanthamoeba* spp. *Appl. Environ. Microbiol.* **1998**, *64*, 279–286. [CrossRef] [PubMed]
96. Berk, S.G.; Faulkner, G.; Garduno, E.; Joy, M.C.; Ortiz-Jimenez, M.A.; Garduno, R.A. Packaging of live *Legionella pneumophila* into pellets expelled by *Tetrahymena* spp. does not require bacterial replication and depends on a Dot/Icm-mediated survival mechanism. *Appl. Environ. Microbiol.* **2008**, *74*, 2187–2199. [CrossRef] [PubMed]

97. Shadrach, W.S.; Rydzewski, K.; Laube, U.; Holland, G.; Ozel, M.; Kiderlen, A.F.; Flieger, A. *Balamuthia mandrillaris*, free-living ameba and opportunistic agent of encephalitis, is a potential host for *Legionella pneumophila* bacteria. *Appl. Environ. Microbiol.* **2005**, *71*, 2244–2249. [CrossRef] [PubMed]
98. Solomon, J.M.; Rupper, A.; Cardelli, J.A.; Isberg, R.R. Intracellular growth of *Legionella pneumophila* in *Dictyostelium discoideum*, a system for genetic analysis of host-pathogen interactions. *Infect. Immun.* **2000**, *68*, 2939–2947. [CrossRef] [PubMed]
99. Newsome, A.L.; Baker, R.L.; Miller, R.D.; Arnold, R.R. Interactions between *Naegleria fowleri* and *Legionella pneumophila*. *Infect. Immun.* **1985**, *50*, 449–452. [CrossRef] [PubMed]
100. Watanabe, K.; Nakao, R.; Fujishima, M.; Tachibana, M.; Shimizu, T.; Watarai, M. Ciliate *Paramecium* is a natural reservoir of *Legionella pneumophila*. *Sci. Rep.* **2016**, *6*, 24322. [CrossRef] [PubMed]
101. Kao, P.M.; Tung, M.C.; Hsu, B.M.; Hsu, S.Y.; Huang, J.T.; Liu, J.H.; Huang, Y.L. Differential *Legionella* spp. survival between intracellular and extracellular forms in thermal spring environments. *Environ. Sci. Pollut. Res. Int.* **2013**, *20*, 3098–3106. [CrossRef]
102. International Organization for Standardization. *ISO/TS12869:2019 Water Quality-Detection and Quantification of LEGIONELLA spp. and/or Legionella pneumophila by Concentration and Genic amplification by Quantitative Polymerase Chain Reaction (qPCR)*; International Organization for Standardization: Geneva, Switzerland, 2019.
103. Qvarnstrom, Y.; Visvesvara, G.S.; Sriram, R.; da Silva, A.J. Multiplex real-time PCR assay for simultaneous detection of *Acanthamoeba* spp., *Balamuthia mandrillaris*, and *Naegleria fowleri*. *J. Clin. Microbiol.* **2006**, *44*, 3589–3595. [CrossRef]
104. Whiley, H.; Taylor, M.; Bentham, R. Detection of Legionella species in potting mixes using fluorescent in situ hybridisation (FISH). *J. Microbiol. Methods* **2011**, *86*, 304–309. [CrossRef]
105. Health Protection Agency. *Isolation and Identification of Acanthamoeba Species*; Health Protection Agency: London, UK, 2004.
106. Standards Australia. *AS5132:2017 Waters-Examination for Legionella spp. Includeing Legionella Pneumophila-Using Concentration*; Standards Australia: Sydney, Australia, 2017.
107. Nocker, A.; Cheung, C.Y.; Camper, A.K. Comparison of propidium monoazide with ethidium monoazide for differentiation of live vs. dead bacteria by selective removal of DNA from dead cells. *J. Microbiol. Methods* **2006**, *67*, 310–320. [CrossRef]
108. Yanez, M.A.; Nocker, A.; Soria-Soria, E.; Murtula, R.; Martinez, L.; Catalan, V. Quantification of viable *Legionella pneumophila* cells using propidium monoazide combined with quantitative PCR. *J. Microbiol. Methods* **2011**, *85*, 124–130. [CrossRef] [PubMed]
109. Brescia, C.C.; Griffin, S.M.; Ware, M.W.; Varughese, E.A.; Egorov, A.I.; Villegas, E.N. *Cryptosporidium* propidium monoazide-PCR, a molecular biology-based technique for genotyping of viable *Cryptosporidium* oocysts. *Appl. Environ. Microbiol.* **2009**, *75*, 6856–6863. [CrossRef] [PubMed]
110. Fittipaldi, M.; Pino Rodriguez, N.J.; Adrados, B.; Agusti, G.; Penuela, G.; Morato, J.; Codony, F. Discrimination of viable *Acanthamoeba castellani* trophozoites and cysts by propidium monoazide real-time polymerase chain reaction. *J. Eukaryot. Microbiol.* **2011**, *58*, 359–364. [CrossRef]
111. Dietersdorfer, E.; Kirschner, A.; Schrammel, B.; Ohradanova-Repic, A.; Stockinger, H.; Sommer, R.; Walochnik, J.; Cervero-Arago, S. Starved viable but non-culturable (VBNC) *Legionella* strains can infect and replicate in amoebae and human macrophages. *Water Res.* **2018**, *141*, 428–438. [CrossRef] [PubMed]
112. Kirschner, A.K.T. Determination of viable *Legionellae* in engineered water systems: Do we find what we are looking for? *Water Res.* **2016**, *93*, 276–288. [CrossRef] [PubMed]
113. Cervero-Arago, S.; Rodriguez-Martinez, S.; Canals, O.; Salvado, H.; Araujo, R.M. Effect of thermal treatment on free-living amoeba inactivation. *J. Appl. Microbiol.* **2014**, *116*, 728–736. [CrossRef] [PubMed]
114. Rohr, U.; Weber, S.; Michel, R.; Selenka, F.; Wilhelm, M. Comparison of free-living amoebae in hot water systems of hospitals with isolates from moist sanitary areas by identifying genera and determining temperature tolerance. *Appl. Environ. Microbiol.* **1998**, *64*, 1822–1824. [CrossRef]
115. Chen, Y.-S.R.; Sproul, O.J.; Rubin, A.J. Inactivation of *Naegleria gruberi* cysts by chlorine dioxide. *Water Res.* **1985**, *19*, 783–789. [CrossRef]

*pathogens*

MDPI

*Article*

# Intracellular Behaviour of Three *Legionella pneumophila* Strains within Three Amoeba Strains, Including *Willaertia magna* C2c Maky

**Issam Hasni [1,2], Antoine Jarry [2], Benjamin Quelard [2], Antoine Carlino [2], Jean-Baptiste Eberst [2], Olivier Abbe [2] and Sandrine Demanèche [2,*]**

1 Microbes, Evolution, Phylogeny and Infection Department, Institut de Recherche pour le Développement IRD 198, Institut Hospitalo-Universitaire (IHU), Aix-Marseille Université, 13007 Marseille, France; issam.hasni@amoeba-biocide.com

2 R&D Department, Amoéba, 38 Avenue des Frères Montgolfier, 69680 Chassieu, France; antoinejarry93@gmail.com (A.J.); benjamin.quelard@amoeba-biocide.com (B.Q.); Antoine.CARLINO@amoeba-biocide.com (A.C.); Jean-Baptiste.EBERST@amoeba-biocide.com (J.-B.E.); Olivier.Abbe@amoeba-biocide.com (O.A.)

* Correspondence: sandrine.demaneche@amoeba-biocide.com

Received: 10 January 2020; Accepted: 5 February 2020; Published: 6 February 2020

**Abstract:** *Legionella pneumophila* is a facultative intracellular pathogen found in aquatic environments as planktonic cells within biofilms and as intracellular parasites of free-living amoebae such as *Acanthamoeba castellanii*. This pathogen bypasses the elimination mechanism to replicate within amoebae; however, not all amoeba species support the growth of *L. pneumophila*. *Willaertia magna* C2c Maky, a non-pathogenic amoeba, was previously demonstrated to possess the ability to eliminate the *L. pneumophila* strain Paris. Here, we study the intracellular behaviour of three *L. pneumophila* strains (Paris, Philadelphia, and Lens) within *W. magna* C2c Maky and compare this strain to *A. castellanii* and *W. magna* Z503, which are used as controls. We observe the intracellular growth of strain Lens within *W. magna* Z503 and *A. castellanii* at 22 °C and 37 °C. Strain Paris grows within *A. castellanii* at any temperature, while it only grows at 22 °C within *W. magna* Z503. Strain Philadelphia proliferates only within *A. castellanii* at 37 °C. Within *W. magna* C2c Maky, none of the three legionella strains exhibit intracellular growth. Additionally, the ability of *W. magna* C2c Maky to decrease the number of internalized *L. pneumophila* is confirmed. These results support the idea that *W. magna* C2c Maky possesses unique behaviour in regard to *L. pneumophila* strains.

**Keywords:** free-living amoebae; *Legionella*; biological biocide; cooling towers

## 1. Introduction

*Legionella pneumophila* is an aerobic, Gram-negative bacterium that causes Legionellosis, a severe form of pneumonia, following inoculation with contaminated aerosol [1]. This bacterial infection manifests as two clinical forms that include Legionnaires' disease, which is a life-threatening respiratory disease, and Pontiac fever, a milder self-limiting illness [2,3]. Among the sixteen currently identified serogroups of *L. pneumophila*, serogroup 1 is involved in the majority of infections [4,5]. This microorganism is ubiquitous throughout natural and artificial aquatic environments [6]. Legionellosis outbreaks are frequently related to contaminated water systems that produce aerosols, which occurs primarily within cooling towers [7]. Indeed, cooling towers provide ideal conditions for pathogen growth, as they frequently possess temperatures above 20 °C, at which *L. pneumophila* can proliferate [8–10].

Free-living amoebae (FLA) are ubiquitous protozoa that inhabit common aquatic environments and are frequently co-isolated with *L. pneumophila* in water cooling towers [11,12]. FLA are predatory and consume bacteria to facilitate their growth [13,14]; however, some bacteria such as *L. pneumophila* have evolved to avoid the phagolysosome fusion and can multiply within FLA, ultimately killing these amoebae before disseminating into the environment [9,15–17]. Furthermore, amoeba cysts can provide *L. pneumophila* with protection against unfavourable conditions and chemical treatments. Therefore, the association between FLA and this pathogen makes the control and monitoring of water-cooling towers difficult and makes eradication of *L. pneumophila* almost impossible [18,19].

Previous studies, however, have demonstrated that all FLAs do not exhibit the same behaviours when they come into contact with *L. pneumophila* strains. While *Acanthamoeba* sp. and *Vermamoeba* (formerly *Hartmannella*) *vermiformis* support the intracellular growth of *L. pneumophila*, the *Willaertia magna* strain C2c Maky has been demonstrated to eliminate the *L. pneumophila* serogroup 1 strain Paris ATCC 33152 [20], which is a virulent pathogen strain responsible for severe legionellosis epidemics in France [21]. *W. magna* C2c Maky is a free-living amoeba that is a member of the Vahlkampfiidae family [22]. This amoeba is a thermophilic FLA that is isolated from the water of thermal swimming pools (http://www.amoeba-biocide.com/en/page/learn-more-about-willaertia-magna-c2c-maky), and it has the capacity to grow at high temperatures (up to 44 °C) in xenic or axenic media. The living forms of this amoeba include a large trophozoite (50–100 μm) and a cyst (18–21 μm) form, and it can produce temporary flagella [22,23]. The lack of pathogenicity of this amoeba was demonstrated by cytotoxicity testing on human cells and was confirmed by genomic analysis [24]. According to these findings, the Amoéba company developed a natural biocide using *W. magna* C2c Maky to eliminate *L. pneumophila* as an alternative to chemical biocides (http://www.amoeba-biocide.com/en/page/revolutionary-biocide). The present study is performed to verify the elimination and the absence of the reservoir effect. Specifically, the decrease in the number of internal *L. pneumophila* and the absence of internal *L. pneumophila* multiplication within *W. magna* C2c Maky, when both microorganisms are co-cultured, is confirmed. The assay is performed by examining adhesion (the usual way of life for free-living amoeba) with three strains of *L. pneumophila* to assess the consistency of amoeba behaviour toward legionella strains. The assay lasts for one week and includes a daily count of intracellular *L. pneumophila* and amoebas by culture and Trypan blue staining, respectively. The behaviour of *W. magna* C2c Maky is compared to that of *W. magna* Z503 to determine if two amoeba strains of the same species have the same behavior. Moreover, it is compared to *A. castellanii*, an amoeba known to multiply amoeba-resistant bacteria such as the three *L. pneumophila* strains studied.

## 2. Results

### 2.1. *L. pneumophila Survival in Coculture Medium*

The survival of the three *L. pneumophila* strains in the calf serum-casein-yeast extract medium (SCYEM) was evaluated at 22 °C and 37 °C (Figure 1a,b). The survival of *L. pneumophila* Lens decreased to $2 \times 10^4$ CFU/mL and to 40 CFU/mL in SCYEM medium within 96 h at 22 °C and 37 °C, respectively. The survival of *L. pneumophila* Paris decreased to $7 \times 10^3$ CFU/mL and to 1 CFU/mL in SCYEM medium within 96 h at 22 °C and 37 °C, respectively. Finally, the survival of *L. pneumophila* Philadelphia decreased to $3 \times 10^3$ CFU/mL and to 2 CFU/mL in SCYEM medium within 96 h at 22 °C and 37 °C, respectively.

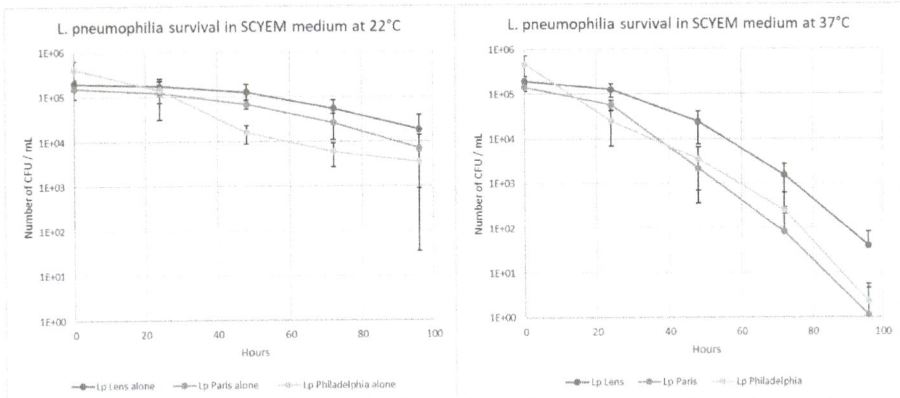


**(a)** **(b)**

**Figure 1.** *L. pneumophila* survival in coculture medium at 22 °C (**a**) and 37 °C (**b**). Results are expressed as the mean +/− 95% CI (Confidence Interval based on the standard error of the mean).

### 2.2. Amoeba Survival in coculture Medium

Survival of the three amoebas in the presence or in absence of bacteria was evaluated over 96 h at 22 °C and 37 °C in coculture medium (Figure 2a,b). The three amoeba strains could be maintained in SCYEM medium for 96 h in the presence or absence of bacteria at 22 °C and 37 °C with the exception of *A. castellanii* when co-cultivated with *L. pneumophila* strains. Found at the end of the experiment, the control condition of *A. castellanii* in the absence of bacteria was maintained at $2 \times 10^5$ cells/mL, while in the presence of *L. pneumophila* Lens, Paris, and Philadelphia, the amoeba number decreased to 556, 444 and 2333 cells/mL, respectively (Figure 2b). *A. castellanii* could not survive in the presence of the three *L. pneumophila* strains at 37 °C.

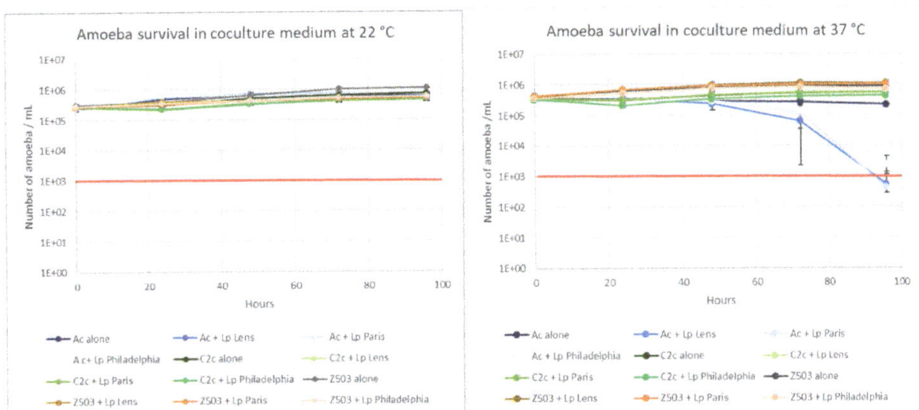


**(a)** **(b)**

**Figure 2.** Amoeba survival at 22 °C (**a**) and 37 °C (**b**) in coculture medium in the presence or absence of the three *L. pneumophila* strains (Lens, Paris, and Philadelphia). The red bar is the detection limit of the Malassez cell counting. Results are expressed as the mean +/− 95% CI (Confidence Interval based on the standard error of the mean).

### 2.3. Co-Culture Experiments

#### 2.3.1. *L. pneumophila* Lens co-cultivated with Amoeba Strains

The mean initial amount of amoeba-internalized bacteria at 22 °C was 16 ± 0.5% (16% in *A. castellanii*, 15% in *W. magna* C2c Maky, and 16% in *W. magna* Z503). Seen at 37 °C, a mean bacterial uptake of 20 ± 5.5% was observed (15% in *A. castellanii*, 26% in *W. magna* C2c Maky, and 18% in *W. magna* Z503).

A significant decrease ($p < 0.05$) in the number of intracellular *L. pneumophila* Lens per *W. magna* C2c Maky cell was observed after 24 h (5-fold and 10-fold reduction at 22 °C and 37 °C, respectively), while the level remained nearly constant for *A. castellanii* at 22 °C and 37 °C and for *W. magna* Z503 at 22 °C with no significant difference between $T_0$ and $T_0$ + 24 h ($p > 0.05$) (Figure 3). Occurring at $T_0$ + 96 h (Figure 3), the percentage of intracellular *L. pneumophila* Lens per *W. magna* C2c Maky cell was reduced by 48 ± 0.3% at 22 °C and 77 ± 1.2% at 37 °C, and an increase was observed for *W. magna* Z503 (9-fold at 22 °C and 5-fold at 37 °C) and *A. castellanii* (19-fold at 22 °C and 50,000-fold at 37 °C). Observed at 37 °C, a small number of *A. castellanii* cells were still alive ($5.6 \times 10^2 \pm 5.9 \times 10^2$ amoebas/mL), demonstrating that amoeba cell lysis occurred following the intracellular multiplication of *L. pneumophila* Lens.

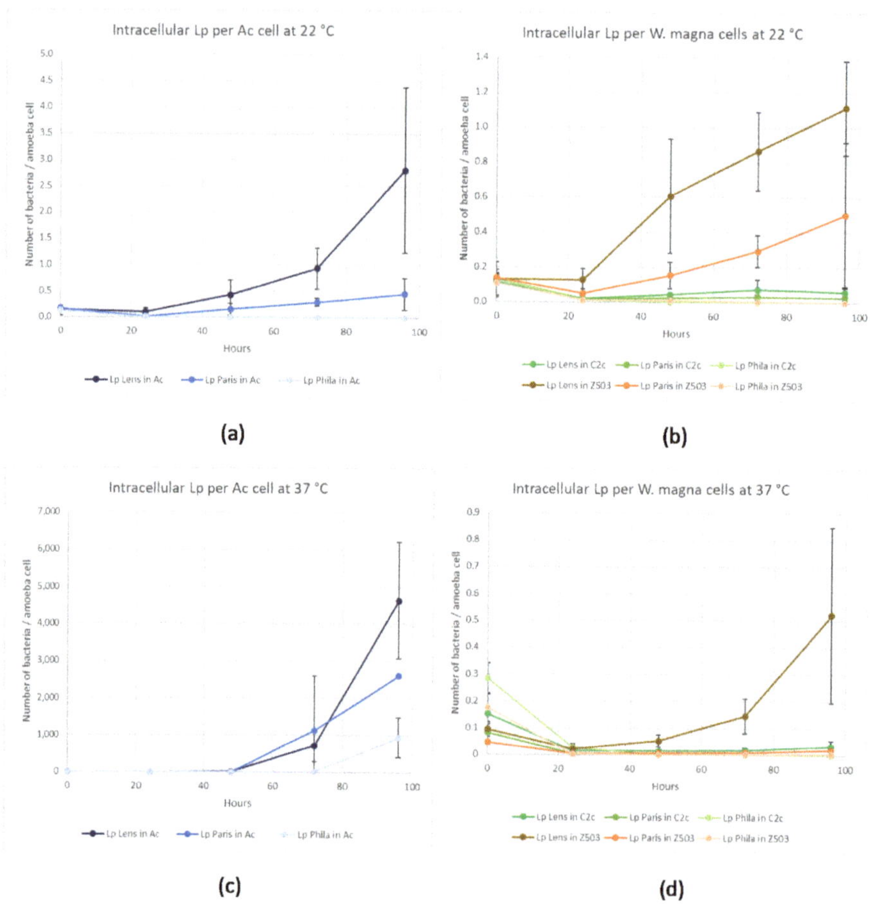


**Figure 3.** Comparison of the evolution of the number of intracellular *L. pneumophila* cells (Lens, Paris,

and Philadelphia) per amoeba cell (*A. castellanii*, *W. magna* C2c Maky, and *W. magna* Z503). Results are expressed as the mean +/− 95% CI (Confidence Interval based on the standard error of the mean). (**a**) *L. pneumophila* number per *A. castellanii* cell at 22 °C (n = 9 for Lp Lens and Paris, n = 15 for Lp Philadelphia); (**b**) *L. pneumophila* number per *A. castellanii* cell at 37 °C (n = 9); (**c**) *L. pneumophila* number per *W. magna* cell (C2c and Z503) at 22 °C (n = 9 for Lp Lens and Paris, n = 15 for Lp Philadelphia); (**d**) *L. pneumophila* number per *W. magna* cell (C2c and Z503) at 37 °C (n = 9).

Considering the number of *L. pneumophila* Lens at 22 °C and 37 °C, a significant increase ($p < 0.05$) was obtained when the bacterium was co-cultivated with *W. magna* Z503 and *A. castellanii*, and this was not observed when *L. pneumophila* Lens was cultivated alone or in the presence of *W. magna* C2c Maky (Figure 4a,b), demonstrating an intracellular multiplication of *L. pneumophila* Lens in *W. magna* Z503 and *A. castellanii* as the bacterium was unable to multiply by itself in the coculture medium (Figure 1a,b).

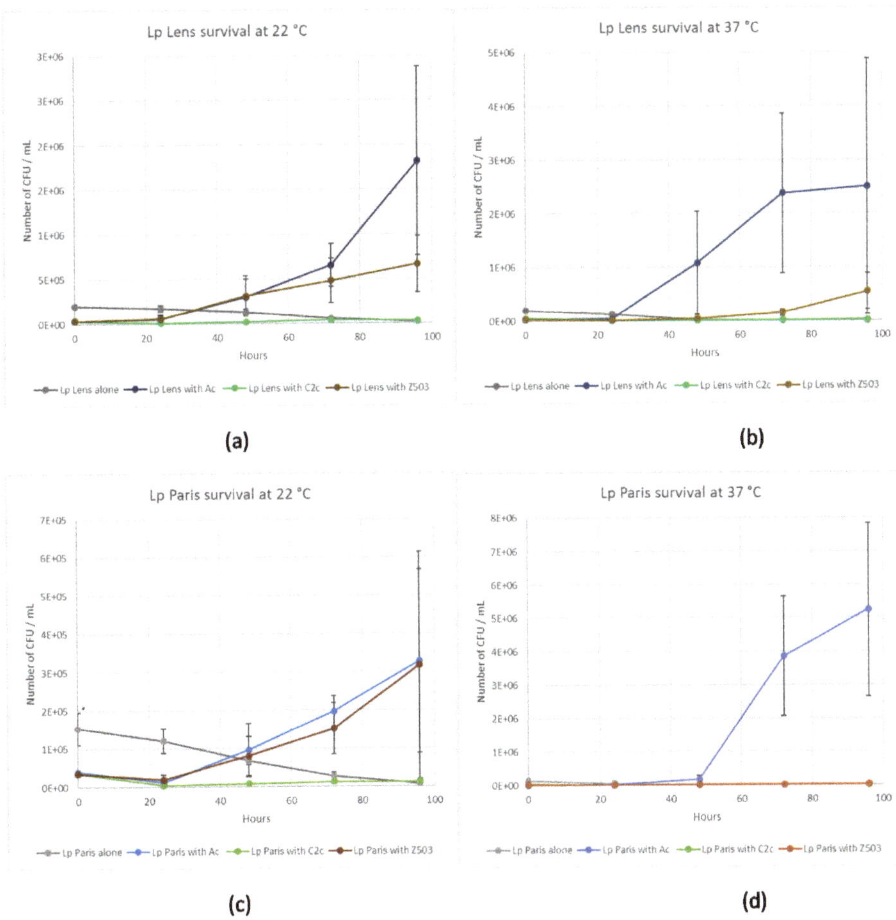


**Figure 4.** *Cont.*

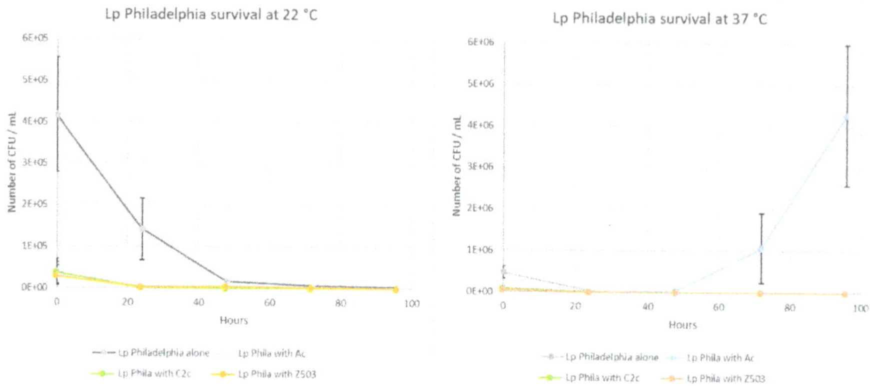


(e) (f)

**Figure 4.** Comparison of the evolution of the number of *L. pneumophila* cells in the presence or absence of amoeba cells (alone, or in presence of *A. castellanii*, *W. magna* C2c Maky, or *W. magna* Z503). Results are expressed as the mean +/− 95% CI (Confidence Interval based on the standard error of the mean). (**a**) *L. pneumophila* Lens at 22 °C (n = 9); (**b**) *L. pneumophila* Lens at 37 °C (n = 9); (**c**) *L. pneumophila* Paris at 22 °C (n = 9); (**d**) *L. pneumophila* Paris at 37 °C (n = 9); (**e**) *L. pneumophila* Philadelphia at 22 °C (n = 15); (**f**) *L. pneumophila* Philadelphia at 37 °C (n = 9).

### 2.3.2. *L. pneumophila* Paris Co-Cultivated with Amoeba Strains

Occurring at 22 °C, we reported a mean *L. pneumophila* Paris uptake by amoebas of 24 ± 1.5% (25% in *A. castellanii*, 23% in *W. magna* C2c Maky, and 23% in *W. magna* Z503). The initial mean amount of cells internalized by amoebas decreased to 14 ± 5.0% at 37 °C (9% in *A. castellanii*, 19% in *W. magna* C2c Maky and 13% in *W. magna* Z503).

A significant decrease of the number of intracellular *L. pneumophila* Paris per amoeba cell ($p < 0.05$) first was observed in the three amoebas after 24 h, with the exception of *A. castellanii* at 37 °C (8-fold for *W. magna* C2c Maky, 3-fold for *W. magna* Z503, and 9-fold for *A. castellanii* at 22 °C and 19-fold for *W. magna* C2c Maky, 11-fold for *W. magna* Z503, and 2-fold for *A. castellanii* at 37 °C) (Figure 3). This decrease was maintained until the end of the experiment ($T_0$ + 96 h) only by *W. magna* C2c Maky, and the percentage of intracellular *L. pneumophila* Paris per amoeba cell was reduced by 79 ± 2% at 22 °C and 98 ± 0.1% at 37 °C ($p < 0.05$). The opposite was observed for *W. magna* Z503 and *A. castellanii* at 22 °C and 37 °C, as the decrease measured after 24 h was not maintained. Seen at 48 h, the level of intracellular *L. pneumophila* Paris per amoeba cell began to increase until it reached 4-fold and 3-fold more bacteria per amoeba cell than that observed at $T_0$ for *W. magna* Z503 and *A. castellanii*, respectively at 22 °C. Observed at 37 °C for *W. magna* Z503, the number of intracellular *L. pneumophila* Paris per amoeba cell at $T_0$ + 96 h was 5-fold the ratio observed at 24 h, but it did not reach the initial ratio. Regarding *A. castellanii*, a strong increase was observed at both temperatures, and the initial ratio was slightly increased by 3-fold at 22 °C ($p > 0.05$) and strongly increased by 60,000-fold at 37 °C ($p < 0.05$). Furthermore, the correlation between the increase in *L. pneumophila* Paris and the low concentration of viable *A. castellanii* ($5.6 \times 10^2 \pm 5.9 \times 10^2$ cells/mL) after 96 h indicated that a high intracellular multiplication of *L. pneumophila* Paris occurred that was followed by a release of bacteria in the medium after *A. castellanii* death.

Considering the number of *L. pneumophila* Paris at 22 °C, a significant increase ($p < 0.05$) was obtained when the bacterium was co-cultured with *W. magna* Z503 and *A. castellanii*, and this was not observed when *L. pneumophila* Paris was cultured alone or in the presence of *W. magna* Z503 at 37 °C and

*W. magna* C2c Maky at both 22 °C and 37 °C (Figure 4c,d), demonstrating an intracellular multiplication of *L. pneumophila* Paris in *W. magna* Z503 and *A. castellanii* at 22 °C and only in *A. castellanii* at 37 °C as the bacterium was unable to multiply by itself in the coculture medium (Figure 1a,b).

#### 2.3.3. *L. pneumophila* Philadelphia Co-Cultivated with Amoeba Strains

The mean bacterial internalization by amoebas was 9 ± 1.1% (9% in *A. castellanii*, 10% in *W. magna* C2c Maky, and 7% in *W. magna* Z503) at 22 °C, and the initial amount of internalized cells by amoebas increased to 17 ± 3.8% (19% in *A. castellanii*, 20% in *W. magna* C2c Maky, and 13% in *W. magna* Z503).

Occurring at 22 °C, a rapid and significant ($p < 0.05$) decrease in the number of intracellular *L. pneumophila* per amoeba cell was observed within 24 h (20-fold for *A. castellanii*, 11-fold for *W. magna* C2c Maky, and 10-fold for *W. magna* Z503) in the three amoebas (Figure 3). Then, a slow but significant ($p < 0.05$) decrease continued until the death of more than 99% of intracellular *L. pneumophila* Philadelphia in all cases. Even if this decrease could be attributed to the bacterial death in the coculture medium, the experiment demonstrated the absence of intra-amoeba multiplication of *L. pneumophila* Philadelphia necessary for survival at 22 °C.

Occurring at 37 °C, a similar rapid decrease in the number of intracellular *L. pneumophila* per amoeba was observed within 24 h for all three amoebas (20-fold for *A. castellanii*, 10-fold for *W. magna* C2c Maky, and 92-fold for *W. magna* Z503). Then, differential behaviours were observed depending on the amoeba strains. Regarding *W. magna* C2c Maky, the significant decrease ($p < 0.05$) continued until the death of more than 99.99% of the intracellular *L. pneumophila* Philadelphia per amoeba cell (Figure 3d). Concerning *W. magna* Z503, a decrease also was observed up to 97% elimination of intracellular *L. pneumophila* Philadelphia per amoeba cell after 96 h ($p < 0.05$) (Figure 3d). To contrast, for *A. castellanii*, a significant increase ($p < 0.05$) in intracellular *L. pneumophila* Philadelphia per amoeba cell appeared after 48 h, demonstrating an intra-amoeba multiplication up to 2600-fold at the end point (Figure 3c).

Considering the number of *L. pneumophila* Philadelphia at 22 °C, a significant decrease ($p < 0.05$) was obtained in all cases (Figure 4e), while at 37 °C, a significant increase ($p < 0.05$) was observed when *L. pneumophila* Philadelphia was cultured in the presence of *A. castellanii* (Figure 4f). This demonstrated an intracellular multiplication of *L. pneumophila* Philadelphia *A. castellanii* at 37 °C, as the bacterium was unable to multiply by itself in SCYEM medium (Figure 1a,b).

### 2.4. *Microscopic Observations of Intracellular L. pneumophila Philadelphia at 37 °C*

Microscopic observations were performed at $T_0$, $T_0$ + 48 h, and $T_0$ + 96 h. Occurring at $T_0$, excess intracellular *L. pneumophila* Philadelphia bacteria were observed in the presence of the three amoebas (Figure 5A,D,G). Regarding *A. castellanii* at 48 h, a strong bacterial multiplication was observed (Figure 5B) which was not observed for both *W. magna* strains (Figure 5E,H). Occurring at 96 h, lysis of *A. castellanii* after intracellular bacterial multiplication was clearly evident (Figure 5C), and only a small amount of amoeba lysis could be observed for both *W. magna* strains (Figure 5F,I).

### 2.5. *Statistical Comparison of Amoeba Behavior*

Analysis of variance tests (ANOVA) were performed to determine if *W. magna* C2c Maky interacted with *L. pneumophila* in a significantly different manner compared to interactions with the two other amoebas.

Concerning the three bacterial strains, $T_0$ data obtained in the presence of the three amoebas were not statistically different at 22 °C ($p > 0.05$); however, at 37 °C, a significant difference in behaviour ($p < 0.05$) was detected at $T_0$.

Pairwise comparisons (Dunn test) established that at 72 h and 96 h at both temperatures and with the three legionella strains, *W. magna* C2c Maky behaviour was statistically different from that of the two other amoeba strains (Table 1). This significant difference was observed even after 24 h with strain Paris at both temperatures, and at 22 °C for strain Lens. Statistical tests provided evidence that *W. magna* C2c Maky behaved differently compared to *W. magna* Z503 and *A. castellanii* cells in the presence of *Legionella* strains.

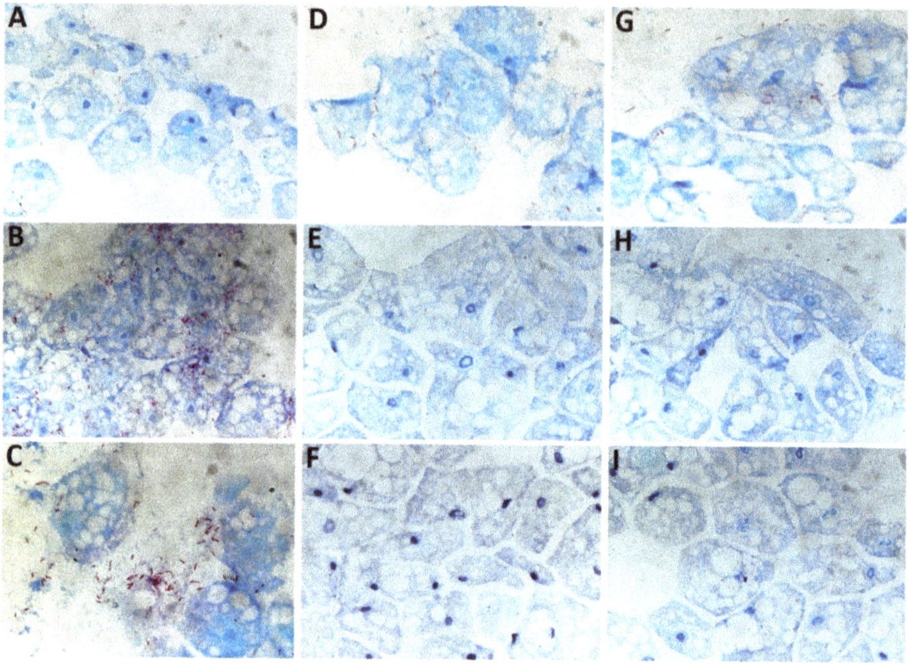


**Figure 5.** Optical microscopy observation using Gimenez staining of *A. castellanii* (**A–C**), *W. magna* C2c Maky (**D–F**), and *W. magna* Z503 (**G–I**) infected with *L. pneumophila* Philadelphia at 37 °C. Photos of the co-cultures were acquired at $T_0$ (**A,D,G**), $T_0$ + 48 h (**B,E,H**), and $T_0$ + 96 h (**C,F,I**).

**Table 1.** Statistical analysis of the behaviour of the three amoeba strains in the presence of the three *Legionella* strains at 22 °C and 37 °C. Significant differences for *W. magna* C2c Maky are highlighted in yellow.

| | 22 °C | | | | | 37 °C | | | | |
|---|---|---|---|---|---|---|---|---|---|---|
| ***L. pneumophila*** **Lens** | **T0** | **24 h** | **48 h** | **72 h** | **96 h** | **T0** | **24 h** | **48 h** | **72 h** | **96 h** |
| **With *A. castellanii*** | A | A | A | A | A | A | A | A | A | A |
| **With *W. magna* Z503** | A | A | A | A | A | AB | AB | A | A | A |
| **With *W. magna* C2c Maky** | A | B | B | B | B | A | B | B | C | B |
| ***L. pneumophila*** **Paris** | T0 | 24 h | 48 h | 72 h | 96 h | T0 | 24 h | 48 h | 72 h | 96 h |
| **With *A. castellanii*** | A | A | A | A | A | C | A | A | A | A |
| **With *W. magna* Z503** | A | A | A | A | A | B | A | B | B | B |
| **With *W. magna* C2c Maky** | A | B | B | B | B | A | B | C | C | C |
| ***L. pneumophila*** **Philadelphia** | T0 | 24 h | 48 h | 72 h | 96 h | T0 | 24 h | 48 h | 72 h | 96 h |
| **With *A. castellanii*** | A | A | A | AB | AB | AB | A | A | AB | AB |
| **With *W. magna* Z503** | A | A | A | A | A | B | B | B | A | A |
| **With *W. magna* C2c Maky** | A | A | A | B | B | A | A | B | C | C |

## 3. Discussion

This work explores the permissiveness of three amoeba strains regarding the intracellular multiplication of three pathogenic *L. pneumophila* strains under two temperature conditions (22 °C and

37 °C) that correspond to temperatures found in cooling towers in which *L. pneumophila* are known to replicate within certain strains of amoebae [10,25]. It is important to demonstrate that *W. magna* C2c Maky does not multiply *L. pneumophila* as we aim to propose it as a natural biocide to treat cooling towers.

The three *L. pneumophila* strains are a representative set of *L. pneumophila* serogroup 1 that is responsible for 95% of the legionellosis disease world-wide [5]. Strain Philadelphia is a clinical isolate that is historically responsible for the very first outbreak. It possesses gene traits that allow for multiplication in a number of hosts such as peripheral blood mononuclear cells, peritoneal macrophages, and *A. castellanii*, *A. polyphaga*, or *A. lenticulate* [26–29]. The Philadelphia strain is, according to the EN 13623 European standard, the only strain for which testing is required to validate a disinfectant against *Legionella* in Europe. *L. pneumophila* Lens was chosen because it was responsible for an outbreak in the north of France between November 2003 and January 2004 where 86 confirmed cases resulted in 17 deaths [30]. *L. pneumophila* Paris was chosen because, among the endemic strains of *L. pneumophila* serogroup 1, sequence type 1 (ST1) strains are among the most prevalent, particularly the ST1/Paris pulsotype. This endemic type was responsible for 8.2% of French culture-proven cases of Legionnaire's disease from 1995 through 2006. ST1/Paris pulsotype isolates also have been detected in clinical and environmental samples taken from several other countries around the world, including Switzerland, Italy, Spain, Sweden, the United States, Japan, Senegal, and Canada [21,30].

Our experiments demonstrate differential behaviours among amoeba species infected by the pathogenic bacteria. Compared to *A. castellanii* and *W. magna* Z503, the intracellular *L. pneumophila* are efficiently eliminated by *W. magna* C2c Maky at 22 °C and 37 °C. Indeed, the experiments report not only a non-replication, but also an elimination of the intracellular strains Lens, Paris and Philadelphia within *W. magna* C2c Maky. Furthermore, the coculture medium used in the survey is not adapted to the survival of the legionella bacteria, and they, therefore, must parasitize the amoebae to facilitate their own growth. Indeed, the experiments demonstrate that the three legionella strains were unable to remain at the inoculation level and began to die after 24 h (Figure 1). Although the medium is not adapted to *L. pneumophila* strains, it was chosen for the co-culture study because an increase of the bacterial number during the co-culture experiment necessarily indicates that the multiplication occurred within amoeba. The bacterial multiplication is observed both in *A. castellanii* and *W. magna* Z503, and it is not observed in *W. magna* C2c Maky. The assays reveal a multiplication of all legionella strains within *A. castellanii* at 37 °C and the intracellular multiplication of strain Lens and Paris at 22 °C. Indeed, the strain Philadelphia grows at 37 °C (Figure 3c) and does not multiply at 22 °C (Figure 3a) within *A. castellanii*. Based on this, these results suggest a behaviour that is influenced by the temperature conditions. Several previous studies revealed the effect of temperature on the relationship between *L. pneumophila* and free-living amoeba (FLA) [9,31,32]. *L. pneumophila* serogroup 1, for example, replicated in *A. castellanii* at 25 °C but were digested at temperatures below 20 °C [25]. Dupuy et al. assessed the ability of 12 amoeba strains of *Naegleria* sp., *Acanthamoeba* sp., and *Vermamoeba* sp. to support the multiplication of *L. pneumophila* Lens at various temperatures (25 °C, 30 °C and 40 °C), and they revealed a more efficient intracellular proliferation with increasing temperatures [33]. Additionally, we did not observe the same behaviour according to the different bacteria and amoeba strains used during our experiments. Indeed, the strain Lens replicates at 37 °C within *W. magna* strain Z503, but not in *W. magna* C2c Maky (Figure 3d). The co-culture at 22 °C of *W. magna* Z503 with *L. pneumophila* strain Paris and strain Lens reveals a multiplication of the bacteria; however, no replication is observed during co-culture with strain Philadelphia (Figure 3b). The difference in amoeba permissiveness has been highlighted previously, especially in regard to *Naegleria*, *Acanthamoeba*, *Vermamoeba* and *Micriamoeba tesseris* [9,34]. The non-replication of legionella within *W. magna* C2c Maky was previously observed with strain Paris [20]. Our study confirms this result, as the resistance of *W. magna* C2c Maky towards *L. pneumophila* Paris is illustrated by the observed significant decrease in the bacterial concentration after 4 days of co-culture at 22 °C and 37 °C (Figure 4c,d). Dey et al. [20], however, reported a moderate increase in strains Philadelphia and Lens within *W. magna* C2c at 37

°C while in our study the intracellular bacterial concentration significantly decreased in culture with *W. magna* C2c Maky at 22 °C and 37 °C. These differences can be explained by the protocol parameters used in the former study, particularly regarding the culture medium and elimination of extracellular bacteria. The authors used serum casein glucose yeast extract medium (SCGYEM) that was favourable to *L. pneumophila* survival, so bacteria were not forced to multiply into amoeba to survive. Additionally, Dey and co-workers did not eliminate extracellular bacteria by centrifugation, and the observed increase could be due to extracellular bacterial replication, such as that resulting from necrotrophic growth as previously demonstrated [35].

*W. magna C2c* Maky is demonstrated to possess a high efficiency for digesting the intracellular *L. pneumophila* cells in all strains used in this survey. The growth of *L. pneumophila* within amoebas is known to enhance the pathogenicity and invasion of *L. pneumophila* [15,36]; however, no intracellular bacterial replication is observed when we infect *W. magna* C2c Maky with *L. pneumophila* strains derived from a first co-culture that was thought to be more virulent (unpublished data).

The action on different *L. pneumophila* strains and the absence of internal proliferation support the fact that *W. magna* C2c Maky could be used as a biocide to combat *L. pneumophila* proliferation in cooling tower water. This observation is consistent with the control of legionella by *W. magna* C2c Maky observed in real conditions during field trials in functioning cooling towers (http://www.amoeba-biocide.com/sites/default/files/180711_cp_amoeba_us_positive_efficacy_field_test_en_vedf_0.pdf). The traditional method to control bacterial growth in cooling tower water is primarily based on the use of chemical biocides [37,38]. Indeed, the oxidizing agent chlorine is the most used product for cooling tower treatment [39]. The chemical biocide is efficient to prevent *L. pneumophila* proliferation, although some previous studies reported incomplete eradication of legionella from installations and progressive re-colonization within these systems within weeks or months [40,41]. Moreover, these chemical biocides are dangerous to the environment, they degrade the installation systems, and they require the application of other products such as anti-corrosive agents [42,43]. Described by Iervolino, treatment with another oxidizing agent ($H_2O_2$/Ag) was inadequate for legionella control, and, instead, it caused a rapid increase of one logarithmic unit [44]. Chemical biocide action also is not completely efficient against biofilms and amoeba cysts that can provide protection against disinfection treatment [16,17,45]. Finally, chemical biocides used in cooling towers can select *L. pneumophila* populations, and chemical biocides can promote resistance to biocides and to human health antibiotics [46,47].

To conclude, *W. magna* C2c Maky is not associated with any human or animal infection, and this is in agreement with the lack of pathogenicity demonstrated in vivo and suggested by genomic analysis [24,48]. This organism is likely a safe and efficient candidate for legionella control in cooling towers and could provide an alternative solution to chemical biocides.

## 4. Materials and Methods

### 4.1. Free-Living Amoebae Culture

*Willaertia magna* C2c Maky (ATCC® PTA-7824), *Willaertia magna* Z503 (ATCC® 50035), and *Acanthamoeba castellanii* (ATCC® 30010) were purchased from ATCC and cultivated according to their recommendation into 10 mL of modified PYNFH medium (ATCC medium 1034) in a T-25 tissue culture flask. Amoebae were then grown in cell factories in serum casein yeast extract medium (SCYEM) at 30 °C. SCYEM medium is derived from serum casein glucose yeast extract medium (SCGYEM) medium [49] and contained 10 g·L$^{-1}$ casein, 5 g·L$^{-1}$ yeast extract, 10% foetal calf Serum, 1.325 g·L$^{-1}$ $Na_2HPO_4$, and 0.8 g·L$^{-1}$ $KH_2PO_4$. After 72 h (during exponential phase), the cell factories were gently shaken, and the amoeba suspensions were transferred to 50 mL Falcon® tubes. Amoeba populations were then quantified using a Malassez haemocytometer cell counting chamber method (Thermo Fisher Scientific, France) with Trypan blue by mixing 100 µL of Trypan blue with 100 µL of amoeba sample. According to the results, the amoebae concentration in Falcon® tubes was then adjusted to $3 \times 10^5$ cells/mL by the addition of SCYEM. The amoebas were then washed twice in

SCYEM using centrifugation at 3000× *g* for 10 min, and the supernatants were then discarded. Amoeba populations were then re-quantified, and the amoeba suspensions were finally adjusted to 3 × $10^5$ cells/mL in 100 mL of SCYEM. A final quantification was performed to verify the concentration.

Each final solution of *W. magna* C2c Maky, *W. magna* Z503, and *A. castellanii* corresponded to working suspensions that were named $AWS_{C2C}$, $AWS_{Z503}$, and $AWS_{AC}$, respectively (Table 2).

**Table 2.** Preparation of the co-cultures.

| Co-Culture | AWS [1] Volume | BWS [2] Volume |
|---|---|---|
| *L.p.* Philadelphia + *W. magna* C2c Maky | 10 mL $AWS_{C2C}$ | 0.1 mL $BWS_{Phila}$ |
| *L.p.* Philadelphia + *W. magna Z503* | 10 mL $AWS_{Z503}$ | 0.1 mL $BWS_{Phila}$ |
| *L.p.* Philadelphia + *A. castellanii.* | 10 mL $AWS_{AC}$ | 0.1 mL $BWS_{Phila}$ |
| *L.p.* Paris + *W. magna* C2c Maky | 10 mL $AWS_{C2C}$ | 0.1 mL $BWS_{Paris}$ |
| *L.p.* Paris + *W. magna Z503* | 10 mL $AWS_{Z503}$ | 0.1 mL $BWS_{Paris}$ |
| *L.p.* Paris + *A. castellanii.* | 10 mL $AWS_{AC}$ | 0.1 mL $BWS_{Paris}$ |
| *L.p.* Lens + *W. magna* C2c Maky | 10 mL $AWS_{C2C}$ | 0.1 mL $BWS_{Lens}$ |
| *L.p.* Lens + *W. magna Z503* | 10 mL $AWS_{Z503}$ | 0.1 mL $BWS_{Lens}$ |
| *L.p.* Lens + *A. castellanii.* | 10 mL $AWS_{AC}$ | 0.1 mL $BWS_{Lens}$ |
| *Control L.p.* Philadelphia | 10 mL SCYEM | 0.1 mL $BWS_{Phila}$ |
| *Control L.p.* Paris | 10 mL SCYEM | 0.1 mL $BWS_{Paris}$ |
| *Control L.p.* Lens | 10 mL SCYEM | 0.1 mL $BWS_{Lens}$ |
| *Control W. magna* C2c Maky | 10 mL $AWS_{C2C}$ | 0 mL |
| *Control W. magna Z503* | 10 mL $AWS_{Z503}$ | 0 mL |
| *Control A. castellanii* | 10 mL $AWS_{AC}$ | 0 mL |

[1] AWS: Amoeba Working Solution at 3 × $10^5$ cells / mL; [2] BWS: Bacteria Working Solution at 3 × $10^7$ CFU / mL.

### 4.2. *Legionella Pneumophila Cultures*

*L. pneumophila* strain Philadelphia (ATCC 33152), *L. pneumophila* strain Lens (CIP 108280), and *L. pneumophila* strain Paris (CIP 107629) were grown on buffered charcoal yeast extract (BCYE) agar plates (Thermo Fisher Scientific, Dardilly, France) at 36 °C for 72 hours and then harvested by scraping, suspended in phosphate-buffered saline (PBS), centrifuged at 9500 xg for 10 min, and washed once in PBS. The supernatants were then discarded. The *L. pneumophila* suspensions were then diluted in PBS to obtain 3 × $10^7$ bacteria/mL.

The legionella final suspensions represented the bacterial stock working suspensions, and they were identified as $BWS_{Phila}$, $BWS_{Paris}$, and $BWS_{Lens}$ (Table 2).

### 4.3. *Bacterial Survival in the Coculture Medium (Control)*

The three control bacterial conditions were prepared as described in Table 2 by adding 10 mL of SCYEM to the 0.1 mL bacteria working solutions ($BWS_{Phila}$, $BWS_{Paris}$, or $BWS_{Lens}$) in 25 $cm^3$ flasks (Dutscher, Brumath, France) and incubated at 22 °C or 37 °C. This operation corresponded to the $T_0$ time point of the bacterial controls. Occurring at $T_0$, $T_0$ + 24 h, $T_0$ + 48 h, $T_0$ + 72 h, and $T_0$ + 96 h, 1 mL was sampled in each flask and then serially 10-fold diluted in SCYEM and plated on buffered charcoal yeast extract plates (BCYE) in triplicate. BCYE plates were incubated at 36 °C, and colony forming units (CFU) were counted after 5 days. Each condition was performed for three independent replicates and repeated three times (n = 9).

### 4.4. *Amoeba Survival in the coculture Medium (Control)*

The three amoeba working solutions ($AWS_{C2C}$, $AWS_{Z503}$, or $AWS_{AC}$) were prepared as described in Table 2 (10 mL of working solutions) and incubated at 22 °C or 37 °C in 25 $cm^3$ flasks. Occurring at $T_0$, $T_0$ + 24 h, $T_0$ + 48 h, $T_0$ + 72 h, and $T_0$ + 96 h, the flasks were gently shaken, and the numbers of amoeba cells were quantified using a haemocytometer cell counting chamber method with Trypan blue. Each condition was performed for three independent replicates and repeated three times (n = 9).

### *4.5. Co-Culture Assays*

Amoeba and bacterial working solutions were mixed in 25 $cm^3$ flasks by adding the required volume according to Table 1. To provide an example, 10 mL of *W. magna* C2c Maky at $3 \times 10^5$ cells/mL was mixed with 0.1 mL of *L. pneumophila* Lens at $3 \times 10^7$ CFU / mL. All flasks were left to stand for 2 h at 22 °C ± 2 °C or at 37 °C ± 2 °C to allow for amoebae/bacteria contact and the internalization of *L. pneumophila* into amoebae. After the 2-h contact process, each flask was gently shaken 10 times, and the suspension was transferred into a 15 mL Flacon® tube and centrifuged at 3000× *g* for 5 min. This step allowed for the removal of non-internalized (i.e., extracellular) *L. pneumophila* from the co-culture suspensions. The pellet was resuspended in 10 mL of sterile SCYEM, and the suspension was poured into a new 25 $cm^3$ flask and incubated at 22 °C ± 2 °C or at 37 °C ± 2 °C. This time point corresponded to the $T_0$ time point of the assay. Each condition was performed for three independent replicates and repeated three times (n = 9), with the exception of the co-culture with strain Philadelphia that was repeated four times at 22 °C (n = 15).

### *4.6. L. pneumophila and Amoeba Quantifications in Co-Culture Assays from $T_0$ to $T_0$ + 96 h*

Occurring at $T_0$, $T_0$ + 24 h, $T_0$ + 48 h, $T_0$ + 72 h, and $T_0$ + 96 h, a washing step was performed. The culture supernatant was removed from each flask and replaced by 10 mL of sterile SCYEM. This step was intended to remove extracellular *L. pneumophila* to allow for the detection of only intracellular bacteria. Each flask was gently shaken 10 times and an aliquot of 1 mL was sampled. Quantification of amoeba populations was performed using 0.1 mL of each aliquot utilizing a haemocytometer cell counting chamber method with Trypan blue. The remaining 0.9 mL were treated with Triton™ X-100 [31] at 0.02% v/v (final concentration) for 2 min to lyse amoebas and to recover the internal *L. pneumophila*. The sample was then serially 10-fold diluted in SCYEM and plated on BCYE plates in triplicate, with the exception of the undiluted conditions that were spread onto five plates when the number of *L. pneumophila* was intended to decrease below the detection limit. BCYE plates were incubated at 36 °C, and CFU were counted after 5 days.

### *4.7. Microscopic Observations in Co-Culture with L. pneumophila Philadelphia at 37 °C*

Co-cultures of *L. pneumophila* Philadelphia using the three amoeba strains at 37 °C were sampled from running experiments and stained by the Gimenez technique [50,51] at $T_0$, $T_0$ + 48 h, and $T_0$ + 96 h. Co-cultures (0.1 mL) were deposited onto glass slides by using a Shandon Cytospin 4 cytocentrifuge (Thermo Scientific, Illkirch-France) at 800× *g* for 10 min and then stained using the Gimenez technique. Briefly, each of the glass slides were stained with fuchsin solution for 3 min and washed with water. Then, the glass slides were stained with malachite green for 5–10 s and washed, and this step was repeated twice. Finally, the glass slides were allowed to dry at room temperature.

The observations were performed using a LEICA DM 2500 LED microscope (Leica Microsystemes SAS, Nanterre-France) under an ×100 oil immersion objective.

### *4.8. Statistical Analyses*

Statistical significance of co-culture studies was determined for 22 °C and 37 °C conditions through the use of analysis of variance (ANOVA) (Kruskal–Wallis test and multiple pair-wise comparison Dunn test).

**Author Contributions:** Conceptualization, S.D. and J.-B.E.; methodology, I.H., A.J., A.C., B.Q.; formal analysis, I.H. and S.D.; writing—original draft preparation; I.H., S.D. and J.-B.E.; writing—review and editing, S.D.; supervision, S.D. and O.A. All authors have read and agreed to the published version of the manuscript.

**Funding:** This research was funded by the French Government under the "Investissements d'avenir" (Investments for the Future) program managed by the Agence Nationale de la Recherche (ANR, French National Agency for Research), (reference: Méditerranée Infection 10-IAHU-03), by Région Provence-Alpes-Côte d'Azur and European funding FEDER PRIMI.

**Acknowledgments:** We are very grateful to B. La Scola for his advice throughout the completion of this study and for reviewing this article prior to publication.

**Conflicts of Interest:** The authors declare no conflict of interest.

## References


1. Stout, J.E.; Yu, V.L. Legionellosis. *N. Engl. J. Med.* **1997**, *337*, 682–687. [CrossRef] [PubMed]
2. Fraser, D.W.; Tsai, T.R.; Orenstein, W.; Parkin, W.E.; Beecham, H.J.; Sharrar, R.G.; Harris, J.; Mallison, G.F.; Martin, S.M.; McDade, J.E.; et al. Legionnaires' Disease. *N. Engl. J. Med.* **1977**, *297*, 1189–1197. [CrossRef] [PubMed]
3. Kaufmann, A.F.; McDade, J.E.; Patton, C.M.; Bennett, J.V.; Skaliy, P.; Feeley, J.C.; Anderson, D.C.; Potter, M.E.; Newhouse, V.F.; Gregg, M.B.; et al. Pontiac Fever: Isolation of the etiologic agent (legionella pneumophila) and demonstration of its mode of transmission. *Am. J. Epidemiol.* **1981**, *114*, 337–347. [CrossRef] [PubMed]
4. Benson, R.F.; Fields, B.S. Classification of the genus Legionella. *Semin. Respir. Infect.* **1998**, *13*, 90–99.
5. Phin, N.; Parry-Ford, F.; Harrison, T.; Stagg, H.R.; Zhang, N.; Kumar, K.; Lortholary, O.; ZumLa, A.; Abubakar, I. Epidemiology and clinical management of Legionnaires' disease. *Lancet Infect. Dis.* **2014**, *14*, 1011–1021. [CrossRef]
6. Borella, P.; Guerrieri, E.; Marchesi, I.; Bondi, M.; Messi, P. Water ecology of Legionella and protozoan: Environmental and public health perspectives. *Biotechnol. Annu. Rev.* **2005**, *11*, 355–380.
7. Hamilton, K.A.; Prussin, A.J.; Ahmed, W.; Haas, C.N. Outbreaks of Legionnaires' Disease and Pontiac Fever 2006–2017. *Curr. Environ. Health Rep.* **2018**, *5*, 263–271. [CrossRef]
8. Pine, L.; George, J.R.; Reeves, M.W.; Harrell, W.K. Development of a chemically defined liquid medium for growth of Legionella pneumophila. *J. Clin. Microbiol.* **1979**, *9*, 615–626.
9. Rowbotham, T.J. Preliminary report on the pathogenicity of Legionella pneumophila for freshwater and soil amoebae. *J. Clin. Pathol.* **1980**, *33*, 1179–1183. [CrossRef]
10. Yamamoto, H.; Sugiura, M.; Kusunoki, S.; Ezaki, T.; Ikedo, M.; Yabuuchi, E. Factors stimulating propagation of legionellae in cooling tower water. *Appl. Environ. Microbiol.* **1992**, *58*, 1394–1397. [CrossRef]
11. Barbaree, J.M.; Fields, B.S.; Feeley, J.C.; Gorman, G.W.; Martin, W.T. Isolation of protozoa from water associated with a legionellosis outbreak and demonstration of intracellular multiplication of Legionella pneumophila. *Appl. Environ. Microbiol.* **1986**, *51*, 422–424. [CrossRef] [PubMed]
12. Scheikl, U.; Sommer, R.; Kirschner, A.; Rameder, A.; Schrammel, B.; Zweimüller, I.; Wesner, W.; Hinker, M.; Walochnik, J. Free-living amoebae (FLA) co-occurring with legionellae in industrial waters. *Eur. J. Protistol.* **2014**, *50*, 422–429. [CrossRef] [PubMed]
13. Clarholm, M. Protozoan grazing of bacteria in soil—impact and importance. *Microb. Ecol.* **1981**, *7*, 343–350. [CrossRef] [PubMed]
14. Rodríguez-Zaragoza, S. Ecology of Free-Living Amoebae. *Crit. Rev. Microbiol.* **1994**, *20*, 225–241. [CrossRef]
15. Cirillo, J.D.; Falkow, S.; Tompkins, L.S. Growth of Legionella pneumophila in Acanthamoeba castellanii enhances invasion. *Infect. Immun.* **1994**, *62*, 3254–3261. [CrossRef]
16. Greub, G.; Raoult, D. Microorganisms Resistant to Free-Living Amoebae. *Clin. Microbiol. Rev.* **2004**, *17*, 413–433. [CrossRef]
17. Rowbotham, T.J. Isolation of Legionella pneumophila from clinical specimens via amoebae, and the interaction of those and other isolates with amoebae. *J. Clin. Pathol.* **1983**, *36*, 978–986. [CrossRef]
18. Kilvington, S.; Price, J. Survival of Legionella pneumophila within cysts of Acanthamoeba polyphaga following chlorine exposure. *J. Appl. Bacteriol.* **1990**, *68*, 519–525. [CrossRef]
19. Linder, J.W.-K. Ewert Free-living Amoebae Protecting Legionella in Water: The Tip of an Iceberg? *Scand. J. Infect. Dis.* **1999**, *31*, 383–385. [CrossRef]
20. Dey, R.; Bodennec, J.; Mameri, M.O.; Pernin, P. Free-living freshwater amoebae differ in their susceptibility to the pathogenic bacterium Legionella pneumophila. *FEMS Microbiol. Lett.* **2009**, *290*, 10–17. [CrossRef]
21. Aurell, H.; Etienne, J.; Forey, F.; Reyrolle, M.; Girardo, P.; Farge, P.; Decludt, B.; Campese, C.; Vandenesch, F.; Jarraud, S. Legionella pneumophila Serogroup 1 Strain Paris: Endemic Distribution throughout France. *J. Clin. Microbiol.* **2003**, *41*, 3320–3322. [CrossRef] [PubMed]

22. De Jonckheere, J.F.; Dive, D.G.; Pussard, M.; Vickerman, K. Willaertia Magna gen. nov. sp. nov. (Vahlkampfiidae), a Thermophilic Amoeba Found in Different Habitats. Available online: https://eurekamag.com/research/001/281/001281223.php (accessed on 6 February 2019).
23. Robinson, B.S.; Christy, P.E.; De Jonckheere, J.F. A temporary flagellate (mastigote) stage in the vahlkampfiid amoeba Willaertia magna and its possible evolutionary significance. *Biosystems* **1989**, *23*, 75–86. [CrossRef]
24. Hasni, I.; Chelkha, N.; Baptiste, E.; Mameri, M.R.; Lachuer, J.; Plasson, F.; Colson, P.; La Scola, B. Investigation of potential pathogenicity of Willaertia magna by investigating the transfer of bacteria pathogenicity genes into its genome. *Sci. Rep.* **2019**, *9*, 18318. [CrossRef] [PubMed]
25. Ohno, A.; Kato, N.; Sakamoto, R.; Kimura, S.; Yamaguchi, K. Temperature-Dependent Parasitic Relationship between Legionella pneumophila and a Free-Living Amoeba (Acanthamoeba castellanii). *Appl. Environ. Microbiol.* **2008**, *74*, 4585–4588. [CrossRef]
26. Casini, B.; Baggiani, A.; Totaro, M.; Mansi, A.; Costa, A.L.; Aquino, F.; Miccoli, M.; Valentini, P.; Bruschi, F.; Lopalco, P.L.; et al. Detection of viable but non-culturable legionella in hospital water network following monochloramine disinfection. *J. Hosp. Infect.* **2018**, *98*, 46–52. [CrossRef]
27. Arslan-Aydoğdu, E.Ö.; Kimiran, A. An investigation of virulence factors of Legionella pneumophila environmental isolates. *Braz. J. Microbiol. Publ. Braz. Soc. Microbiol.* **2018**, *49*, 189–199. [CrossRef]
28. García, M.T.; Jones, S.; Pelaz, C.; Millar, R.D.; Abu Kwaik, Y. Acanthamoeba polyphaga resuscitates viable non-culturable Legionella pneumophila after disinfection. *Environ. Microbiol.* **2007**, *9*, 1267–1277. [CrossRef]
29. Molmeret, M.; Jarraud, S.; Mori, J.P.; Pernin, P.; Forey, F.; Reyrolle, M.; Vandenesch, F.; Etienne, J.; Farge, P. Different growth rates in amoeba of genotypically related environmental and clinical Legionella pneumophila strains isolated from a thermal spa. *Epidemiol. Infect.* **2001**, *126*, 231–239. [CrossRef]
30. Cazalet, C.; Rusniok, C.; Brüggemann, H.; Zidane, N.; Magnier, A.; Ma, L.; Tichit, M.; Jarraud, S.; Bouchier, C.; Vandenesch, F.; et al. Evidence in the Legionella pneumophila genome for exploitation of host cell functions and high genome plasticity. *Nat. Genet.* **2004**, *36*, 1165–1173. [CrossRef]
31. Buse, H.Y.; Ashbolt, N.J. Differential growth of Legionella pneumophila strains within a range of amoebae at various temperatures associated with in-premise plumbing. *Lett. Appl. Microbiol.* **2011**, *53*, 217–224. [CrossRef]
32. Rowbotham, T.J. Pontiac fever, amoebae, and legionellae. *Lancet Lond. Engl.* **1981**, *1*, 40–41. [CrossRef]
33. Dupuy, M.; Binet, M.; Bouteleux, C.; Herbelin, P.; Soreau, S.; Héchard, Y. Permissiveness of freshly isolated environmental strains of amoebae for growth of Legionella pneumophila. *FEMS Microbiol. Lett.* **2016**, *363*, fnw022. [CrossRef] [PubMed]
34. Atlan, D.; Coupat-Goutaland, B.; Risler, A.; Reyrolle, M.; Souchon, M.; Briolay, J.; Jarraud, S.; Doublet, P.; Pélandakis, M. Micriamoeba tesseris nov. gen. nov. sp.: A new taxon of free-living small-sized Amoebae non-permissive to virulent Legionellae. *Protist* **2012**, *163*, 888–902. [CrossRef]
35. Temmerman, R.; Vervaeren, H.; Noseda, B.; Boon, N.; Verstraete, W. Necrotrophic growth of Legionella pneumophila. *Appl. Environ. Microbiol.* **2006**, *72*, 4323–4328. [CrossRef]
36. Richards, A.M.; Von Dwingelo, J.E.; Price, C.T.; Abu Kwaik, Y. Cellular microbiology and molecular ecology of *Legionella*-amoeba interaction. *Virulence* **2013**, *4*, 307–314. [CrossRef]
37. McDade, J.E. *Legionella and the Prevention of Legionellosis*; Bartram, J., Chartier, Y., Lee, J.V., Bond, K., Surman-Lee, S., Eds.; World Health Organization: Geneva, Switzerland, 2007; ISBN 978-92-4-156297-3.
38. Carducci, A.; Verani, M.; Battistini, R. Legionella in industrial cooling towers: Monitoring and control strategies. *Lett. Appl. Microbiol.* **2010**, *50*, 24–29. [CrossRef]
39. Canals, O.; Serrano-Suárez, A.; Salvadó, H.; Méndez, J.; Cervero-Aragó, S.; Ruiz de Porras, V.; Dellundé, J.; Araujo, R. Effect of chlorine and temperature on free-living protozoa in operational man-made water systems (cooling towers and hot sanitary water systems) in Catalonia. *Environ. Sci. Pollut. Res. Int.* **2015**, *22*, 6610–6618. [CrossRef]
40. Heimberger, T.; Birkhead, G.; Bornstein, D.; Same, K.; Morse, D. Control of nosocomial Legionnaires' disease through hot water flushing and supplemental chlorination of potable water. *J. Infect. Dis.* **1991**, *163*, 413. [CrossRef]
41. Thomas, V.; Bouchez, T.; Nicolas, V.; Robert, S.; Loret, J.F.; Lévi, Y. Amoebae in domestic water systems: Resistance to disinfection treatments and implication in Legionella persistence. *J. Appl. Microbiol.* **2004**, *97*, 950–963. [CrossRef]

42. Abdel-Wahab, A.; Batchelor, B. Chloride Removal from Recycled Cooling Water Using Ultra-High Lime with Aluminum Process. *Water Environ. Res.* **2002**, *74*, 256–263. [CrossRef]
43. Pagnier, I.; Merchat, M.; La Scola, B. Potentially pathogenic amoeba-associated microorganisms in cooling towers and their control. *Future Microbiol.* **2009**, *4*, 615–629. [CrossRef]
44. Iervolino, M.; Mancini, B.; Cristino, S. Industrial Cooling Tower Disinfection Treatment to Prevent Legionella spp. *Int. J. Environ. Res. Public. Health* **2017**, *14*, 1125. [CrossRef]
45. Farhat, M.; Moletta-Denat, M.; Frère, J.; Onillon, S.; Trouilhé, M.-C.; Robine, E. Effects of disinfection on Legionella spp., eukarya, and biofilms in a hot water system. *Appl. Environ. Microbiol.* **2012**, *78*, 6850–6858. [CrossRef]
46. Maillard, J.-Y. Resistance of Bacteria to Biocides. *Microbiol. Spectr.* **2018**, *6*. [CrossRef]
47. Wéry, N.; Bru-Adan, V.; Minervini, C.; Delgénes, J.-P.; Garrelly, L.; Godon, J.-J. Dynamics of Legionella spp. and bacterial populations during the proliferation of L. pneumophila in a cooling tower facility. *Appl. Environ. Microbiol.* **2008**, *74*, 3030–3037. [CrossRef]
48. Rivera, F.; Lares, F.; Gallegos, E.; Ramirez, E.; Bonilla, P.; Calderon, A.; Martinez, J.J.; Rodriguez, S.; Alcocer, J. Pathogenic amoebae in natural thermal waters of three resorts of Hidalgo, Mexico. *Environ. Res.* **1989**, *50*, 289–295. [CrossRef]
49. De Jonckheere, J. Use of an axenic medium for differentiation between pathogenic and nonpathogenic Naegleria fowleri isolates. *Appl. Environ. Microbiol.* **1977**, *33*, 751–757. [CrossRef]
50. Gimenez, D.F. Staining rickettsiae in yolk-sac cultures. *Stain Technol.* **1964**, *39*, 135–140. [CrossRef]
51. Greer, P.W.; Chandler, F.W.; Hicklin, M.D. Rapid demonstration of Legionella pneumophila in unembedded tissue. An adaptation of the Giménez stain. *Am. J. Clin. Pathol.* **1980**, *73*, 788–790. [CrossRef]

pathogens

MDPI

*Article*

# Mitigation of Expression of Virulence Genes in *Legionella pneumophila* Internalized in the Free-Living Amoeba *Willaertia magna* C2c Maky

**Rayane Mouh Mameri [1], Jacques Bodennec [2], Laurent Bezin [2] and Sandrine Demanèche [1,*]**

1 R&D Department, Amoéba, 69680 Chassieu, France; mameri77@yahoo.fr
2 Lyon Neuroscience Research Center CRNL UMR5292 U1028, University of Lyon, Univ Lyon 1, CNRS, Inserm, 69500 Bron, France; jacques.bodennec@univ-lyon1.fr (J.B.); laurent.bezin@univ-lyon1.fr (L.B.)
* Correspondence: sandrine.demaneche@amoeba-biocide.com; Tel.: +33-(04)-2669-1600

Received: 14 May 2020; Accepted: 3 June 2020; Published: 5 June 2020

**Abstract:** *Legionella pneumophila* is a human pathogen responsible for a severe form of pneumonia named Legionnaire disease. Its natural habitat is aquatic environments, being in a free state or intracellular parasites of free-living amoebae, such as *Acanthamoeba castellanii*. This pathogen is able to replicate within some amoebae. *Willaertia magna* C2c Maky, a non-pathogenic amoeba, was previously demonstrated to resist to *L. pneumophila* and even to be able to eliminate the *L. pneumophila* strains Philadelphia, Lens, and Paris. Here, we studied the induction of seven virulence genes of three *L. pneumophila* strains (Paris, Philadelphia, and Lens) within *W. magna* C2c Maky in comparison within *A. castellanii* and with the gene expression level of *L. pneumophila* strains alone used as controls. We defined a gene expression-based virulence index to compare easily and without bias the transcript levels in different conditions and demonstrated that *W. magna* C2c Maky did not increase the virulence of *L. pneumophila* strains in contrast to *A. castellanii*. These results confirmed the non-permissiveness of *W. magna* C2c Maky toward *L. pneumophila* strains.

**Keywords:** free-living amoebae (FLA); *Legionella pneumophila*; virulence genes; *Willaertia magna* C2c Maky

## 1. Introduction

The genus *Legionella* includes more than 60 species, but human infections that progress to a severe pneumonia, known as Legionnaires' disease, are most often caused by *Legionella pneumophila* [1,2]. This Gram-negative bacterium especially affects immunocompromised individuals after inhalation of Legionella-contaminated aerosols. *L. pneumophila* invades and replicates within alveolar macrophages and epithelial cells of the lungs, inducing a severe respiratory infection [1,3].

*L. pneumophila* is ubiquitous in natural, artificial, and industrial aquatic environments and is mostly nested in intracellular hosts, such as free-living amoebae (FLA). In 1980, Rowbotham first demonstrated the intracellular multiplication of *L. pneumophila* within *Acanthamoeba* spp. and *Naegleria* spp. [4]. Following this study, several reports described the replication of *Legionella* isolates from clinical samples within protozoa isolated from the presumed source of infection. Intracellular growth within protozoa was shown to increase the ability of *L. pneumophila* to infect human monocytes and to resist to chemical disinfectants, biocides, and antibiotics [5–7]. Inhalation of legionellae packaged in amoebae is associated with the induction of more severe clinical cases of legionellosis. The speculated link between the homing of legionella in amoebae and the increased virulence of legionella is supported by the publication of a mouse model of co-inhalation of *L. pneumophila* and *Hartmannella vermiformis*. It was shown to significantly enhance the intrapulmonary growth of *L. pneumophila*, with a greater mortality than that observed from inhalation of legionellae alone [8]. The intra-amoebae growth was demonstrated to enhance the ability of *L. pneumophila* to infect epithelial cells (100- to 1000-fold), murine macrophages (10- to 100-fold), human monocytes (100- to 1000-fold), and *Acanthamoeba castellanii* (10- to 100-fold) [9]. Moreover, *L. pneumophila* grown in *A. castellanii* displays enhanced infection in monocytes compared to buffered charcoal yeast extract (BCYE)-grown bacteria [10]. Within FLA and human macrophages, *L. pneumophila* cells are able to reroute the phagosome thanks to a functional Dot/Icm type 4 secretion system (T4SS) and the approximately 300 proteins it secretes and to induce the biogenesis of a legionella-containing vacuole (LCV) [11,12]. Bacterial replication occurs in LCV, evades fusion with lysosomes, and associates intimately with the host endoplasmic reticulum (ER), inducing the lysis of the host cells. The induction of apoptosis in the host cell is induced by a type IV-like secretion machinery [13]. The internalization of *L. pneumophila* into host cells, such as FLA, promotes not only its ability to survive and multiply but also to acquire and increase its virulence [14]. The *L. pneumophila* Dot/Icm-secreted effector PlcC/CegC1 was demonstrated to promote virulence [15], and the translocated Dot/Icm type IVB secretion system effector SdhA was demonstrated to be of crucial importance in infection processes [16].

Genes of the T4SS system have also been identified, with several other genes, as being responsible for the increased virulence in *L. pneumophila* once internalized by FLA, such as *A. castellanii* and *Vermamoeba vermiformis* [17]. In two studies from NJ Ashbolt's laboratory, about 30 transcripts of genes involved in bacterial metabolism, replication, and virulence have been investigated using reverse transcription quantitative polymerase chain reaction (RT-qPCR) in *L. pneumophila* Philadelphia after exposure to CuO nanoparticles (CuO-NPs) or synthetic gray water (Gw) for a period ranging from 3 to 48 h [18,19].

Here, we investigated the expression level by RT-qPCR of 7 genes (*htpX*, *icmE*, *lirR*, *ccmF*, *gacA*, *tatB*, and *lvrE*) that we have shown to be expressed by all three reference strains, *L. pneumophila* Paris, Philadelphia, and Lens, after co-incubation with two FLA: *A. castellanii* known to be permissive to the legionella multiplication [10], and *Willaertia magna* C2c Maky considered as non-permissive to the legionella multiplication [20]. We also evaluated transcript levels of these genes in *L. pneumophila* cultivated alone and harvested at the end of the growth exponential phase corresponding to the end of the replicative phase and the beginning of the virulence phase. The aim of this work was to evaluate the evolution of virulence of *L. pneumophila* strains after internalization into a non-permissive amoeba, such as *W. magna* C2c Maky. To facilitate the interpretation and comparison of gene expression, a GENE EXPRESSION-based index was developed.

## 2. Results

### 2.1. Virulence Gene Selection

Virulence genes of interest were selected based on previous studies [18,19] that investigated the potential increase in virulence gene expression in L. pneumophila Philadelphia exposed to CuO-NPs or synthetic Gw (Table S1). As our objective was to assess virulence gene expression in *L. pneumophila* internalized for three days in amoebae, we focused on genes whose expression was increased after 24–48 h of environmental exposure to CuO-NPs (*dotA*, *enhC*, *htpX*, *icmE*, and *pvcA*) or synthetic Gw (*lirR*, *ccmF*, *gacA*, *tatB*, *lvrB*, and *lvrE*). We excluded the genes *cegC1* and *sidF* displaying conflicting variations between the two environmental conditions (Table S1).

Among the 11 genes selected, we were not able to amplify *pvcA* cDNA in *L. pneumophila* Philadelphia using published primers [19]. We thus ran qPCR for the other 10 genes, using the primers listed in Table 1.

**Table 1.** The sequences of primer pairs used for qPCR amplification, the slope "a" of the calibration curve: Cq = a × Log[DNA]i + b and the efficiency "E" of the qPCR are given for the targeted cDNAs.

| Gene ID | Size (bp) | Forward Primer (5′->3′) | Reverse Primer (5′->3′) | a | E |
|---|---|---|---|---|---|
| *ccmF* | 149 | TGA ATA CAC AGG GCC GTG ATC TGA | ACT GGT TTC TAC TTT CCC TGC CCA | −3.369 | 1.98 |
| *dotA* | 81 | CTG AGA TGG ATA GGT GGT AGT C | TCT TAC TCT ACC TTT GGC TTC CTC | −3.473 | 1.94 |
| *enhC* | 438 | AAT GCT TTG TAT GCC CTC GG | CAT ATC AGC GCT TTG GCC ATC | −3.401 | 1.97 |
| *gacA* | 120 | TTT AAA CGA CGC GTC ACT TCC CAC | TGC AGA TGC TGA AAG TGG TGA GCA | −3.386 | 1.97 |
| *htpX* | 196 | ATT GAC TCT CAT GGT TGC CGT GCT | AGC CAT GTA TTC TCT GGT TCG GCT | −3.330 | 2.00 |
| *icmE* | 200 | GCT CAA ATC AAA GCT GCT CAG GCA | CCT GCG TTT GCT AAA TCC GCA TCA | −3.331 | 2.00 |
| *lirR* | 124 | CCA TGC TTA ATG CTC TCT ACC A | GGG TTG CTC CGC AAT TAA AC | −3.541 | 1.92 |
| *lvrB* | 99 | CAT TGG TGT ACT CTC GGT CTT C | AGC ACC ATG CAG AGC ATA C | −3.385 | 1.97 |
| *lvrE* | 128 | CCG TAA CAA GTG GGT GAT TCT | CAT TGC CCA ACA AAC CAT AGA C | −3.330 | 2.00 |
| *rpsL* | 132 | GAA AGC CTC GTG TGG ACG TA | CAA CCT TAC GCA TAG CTG AGT TA | −3.340 | 1.99 |
| *tatB* | 115 | ATT GTG TTT GGG CCA TCA AAG | CAT TGA GTT GTT GCT GCC AAA | −3.484 | 1.94 |

As expected, these primers allowed high quality amplification of the targeted cDNAs in *L. pneumophila* Philadelphia cultured for three days in BCYE plates, with the exception of *dotA* and *enhC* cDNAs. Indeed, the amplification of these two amplicons did not reach satisfactory criteria for either the melt curve analysis or the agarose gel electrophoresis of the end products, for the three *L. pneumophila* strains. We thus excluded these two genes from the final analysis. The primers used for the eight remaining genes produced good amplifications, with end products of the expected size (Figure 1). However, the expression of *lvrB* in *L. pneumophila* Paris was not detected in all conditions tested. It is possible that the *lvrB* gene differs in *L. pneumophila* Paris from the two other strains, in regions recognized by the primers used. To avoid any bias when comparing the three strains, *lvrB* was not included in the final list of the seven selected transcripts.

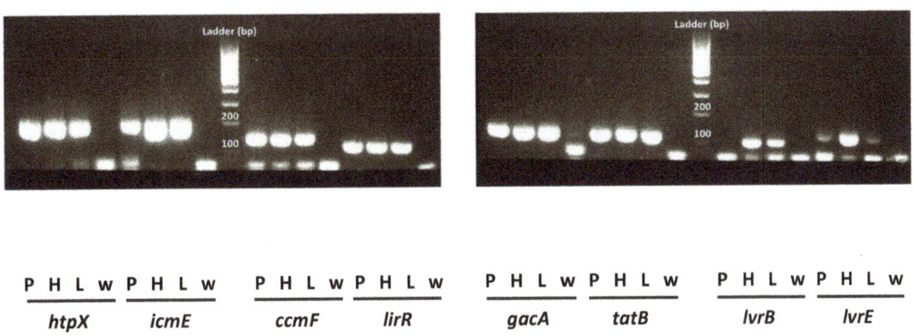


**Figure 1.** Amplicon size verification on 2% agarose gel for the three L. pneumophila strains. bp: base pair; P: strain Paris; H: strain Philadelphia; L: strain Lens; w: water. The ladder is the MassRuler Low Range DNA Ladder (Thermo Fisher Scientific, Lyon, France SM0383); the 100 and 200 bp bands are indicated.

### *2.2. Validation of rpsL as a Non-Acceptable Housekeeping Gene, in Which Transcript Level Should Be Stable between the Different Tested Conditions*

While real-time PCR is a quantitative method with high sensitivity and great reproducibility, to measure targeted DNAs over a wide range of concentrations, the RT of biological sample RNAs is a reaction that is difficult to calibrate. To normalize differences in RT efficiency across samples, the transcript level of a gene, mostly a housekeeping gene, is used as an internal control, whose expression is supposed to be invariant between various treatments or conditions. The ribosomal gene *rpsL* is a housekeeping gene commonly used for phylogenic analysis of *Legionella* [21] and has been used to normalize the reverse transcription of RNA extracted from *L. pneumophila* exposed to Cuo-NPs nanoparticles or synthetic Gw [18,19]. An alternative way to normalize the RT reaction between the different samples is to use a known quantity of synthetic RNA, added directly to the reaction mix, thereby allowing the same number of copies of this synthetic RNA to be added across all samples. As an external control, we used the so-called synthetic non-homologous standard mRNA (SmRNA) [22], of which we ensured that the primers used for the amplification of its cDNA do not amplify DNA sequences resulting from the RT of the endogenous RNAs of *L. pneumophila*.

Thus, using SmRNA as an external control, the expression of *rpsL* could be examined by RT-qPCR like any other gene. Using a graphical representation with an ordinate axis on a linear scale, we show that there is a high variability in the *rpsL* mRNA level, in particular in *L. pneumophila* Paris and *L. pneumophila* Lens (Figure S1). In a logarithmic representation of the ordinate axis, one can note that the *rpsL* mRNA level was almost identical between the three *Legionella* strains under the T′0 control condition. In addition, the *rpsL* mRNA level in the *L. pneumophila* Philadelphia strain remained unchanged under the various conditions tested. On the other hand, the maintenance for three days of the incubation of the *L. pneumophila* Paris and *L. pneumophila* Lens strains in the liquid medium (3D-FREE) caused an increase in the *rpsL* mRNA level, which was significant compared to the T′0 control in *L. pneumophila* Lens. The internalization of *L. pneumophila* Paris and *L. pneumophila* Lens in *A. castellanii* was also followed by a strong and significant increase in the *rpsL* mRNA level, not only compared to the T′0 condition but also compared to the T3D-FREE condition (*A. castellanii* only). Finally, the internalization of the three strains in *W. magna* C2c Maky did not lead to a significant increase in the *rpsL* mRNA level (Figure 2).

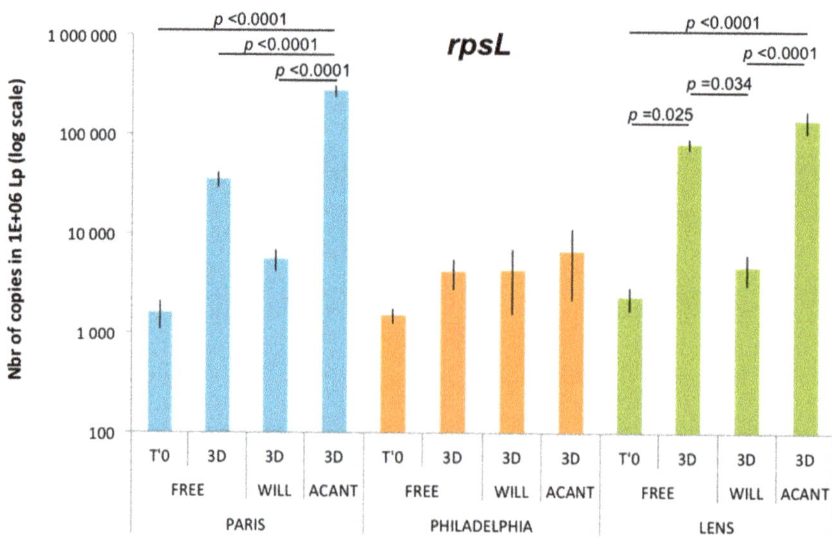


**Figure 2.** Level of *rpsL* transcript in the different conditions, expressed as the number of copies in $1 \times 10^6$ *L. pneumophila* (Lp) ± SD and displayed using a log scale. Abbreviations: FREE: *L. pneumophila* strains alone; WILL: *L. pneumophila* strains cocultured with *W. magna* C2c Maky; ACANTH: *L. pneumophila* strains cocultured with *A. castellanii*; T′0: reference transcript level; 3D: transcript level after 3 days. ANOVA 2: Factor 1: "*L. pneumophila* strain", $p < 0.0001$; Factor 2: "culture conditioned", $p < 0.0001$; Interaction Factor 1 × Factor 2, $p < 0.0001$.

Given that *rpsL* is not a gene whose expression is invariant under the conditions tested, we did not use it for the normalization of the RT reaction and instead used the SmRNA for all of the targeted genes.

### 2.3. Definition of a Gene Expression-Based Virulence Index

Studies aimed at investigating virulence gene expression usually analyze each gene separately, making it difficult to draw clear conclusions, especially when expression increases for some genes, decreases for others, and finally remains stable for the latest.

Since qPCR makes it possible to quantify cDNA copies in a sample, we eluded the above-mentioned issue by defining for each sample a virulence index, which is the sum of all virulence cDNAs quantified by qPCR.

However, in the calculation of the virulence index, we paid attention to avoid masking important variations for genes expressed at low levels in basal conditions by genes initially expressed at high levels. To this end, for each transcript, the cDNA copy number contained in a given sample has been expressed as a percentage of the averaged copy number measured in all samples.

## 2.4. Comparison of the Virulence Index of three L. pneumophila Strains

Firstly, we evaluated the "Gene Expression-based" virulence index of three strains of *L. pneumophila* after coincubation for three days within two amoebic species: *W. magna* C2c Maky and *A. castellanii*, (Figure 3A), based on the measurements of the transcript levels of the seven virulence genes selected: *ccmF* (Figure S2), *gacA* (Figure S3), *htpX* (Figure S4), *icmE* (Figure S5), *lirR* (Figure S6), *lvrE* (Figure S7), and *tatB* (Figure S8). These measures were performed after three days of coculture when the amount of intracellular *Legionella* was not significantly different between *W. magna* C2c Maky and *A. castellanii*.

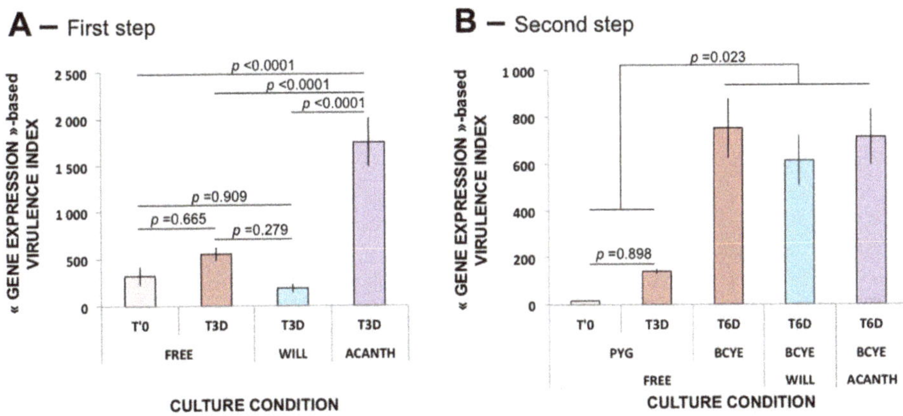


**Figure 3.** Measurement of the virulence index. (**A**) After 3 days in liquid medium; (**B**) after 3 days in liquid medium plus 3 days on buffered charcoal yeast extract (BCYE) plates. FREE: *L. pneumophila* strains alone; WILL: *L. pneumophila* strains cocultured with *W. magna* C2c Maky; ACANTH: *L. pneumophila* strains cocultured with *A. castellanii*; T'0: reference virulence index; T3D: virulence index after 3 days; T6D: virulence index after 6 days, 3 days in liquid medium, and 3 days on BCYE plates.

Secondly, after Day 3, every condition (*L. pneumophila* alone, *L. pneumophila* strains co-incubated with *W. magna* C2c Maky, and *L. pneumophila* strains co-incubated with *A. castellanii*) was seeded on BCYE plates for three additional days (until Day 6) to remove all amoeba traces, because amoebae are not able to survive on a BCYE plate, and to evaluate the fate of the virulence of *L. pneumophila* after their release from amoebae. Once *L. pneumophila* had grown on the BCYE plates, bacteria were harvested, and the virulence index was evaluated (Figure 3B) as above, based on the measurements of the transcript levels of the seven virulence genes selected: *ccmF* (Figure S9A), *gacA* (Figure S9B), *htpX* (Figure S10A), *icmE* (Figure S10B), *lirR* (Figure S11A), *lvrE* (Figure S11B), and *tatB* (Figure S12). As expected, the transcript levels varied differently between the different genes selected, making it difficult to draw a clear conclusion (Figures S2–S12).

A virulence index was thus calculated for each condition, and an ANOVA 2 revealed for Steps 1 and 2 (Figure S13A,B) that the three *L. pneumophila* strains behaved in the same way under all tested conditions. Their results were then pooled and averaged (Figure 3).

ANOVA 2 also revealed a statistical difference between culture conditions ($p < 0.0001$), and post-hoc analysis (Tukey's HSD test) showed a significant increase of the virulence genes after the internalization of *L. pneumophila* in *A. castellanii* compared to the controls containing *L. pneumophila* alone at Day 0 and Day 3 (Figure 3A). Conversely, a tendency (not significant) to decrease the level of virulence of the three *L. pneumophila* strains within *W. magna* C2c Maky was observed when compared to *L. pneumophila* alone (control) at Day 0 and Day 3 (Figure 3A). *L. pneumophila* strains internalized within *W. magna* C2c Maky exhibit a virulence index that is not statistically different from the controls at T'0 and T3D, demonstrating that *W. magna* C2c Maky did not increase the virulence of *L. pneumophila* strains in contrast to *A. castellanii*.

## 3. Discussion

These experiments brought evidence of a different behavior between *L. pneumophila* strains internalized by *W. magna* C2c Maky and strains internalized by *A. castellanii*. Indeed, the expression of virulence genes is reduced in *L. pneumophila* internalized into *W. magna* C2c Maky cells, while it is significantly increased in *L. pneumophila* internalized into *A. castellanii*.

These results confirm the non-permissiveness observed in *W. magna* C2c Maky towards *L. pneumophila*. Unlike *A. castellanii*, *W. magna* C2c Maky is able to internalize and digest *L. pneumophila* by phagocytosis [20,23]. These data suggest that *L. pneumophila* strains were unable to use their T4SS system to deflect the cellular machinery of *W. magna* C2c Maky. As a consequence, the phago-lysosomal fusion could happen and LCV could not be created, leading to the intracellular destruction of bacteria.

Considering FLA known to be permissive to *L. pneumophila* multiplication, such as *A. castellanii*, Buse et al. reported that the Dot/Icm T4SS system was responsible for the translocation of a large number of Legionella effectors in the host cells to promote infectivity [18]. These translocated effector genes are located in a hypervariable region of the *L. pneumophila* genome, and the plasminogen activator homologue of *L. pneumophila*, was strongly involved in *A. castellanii* intracellular growth [24,25]. The virulence associated genes *lvrB* and *lvrE* were also shown to be involved in the T4SS [21,26]. Another effector of the Dot/Icm substrate, SidF, was shown to be involved in the inhibition of programmed cell death in the host and anchoring of binding effectors to bacterial phagosomes [27,28]. However, Legionella *vir* homologues were shown to be not required for intracellular replication in amoeba or macrophage [21].

Surprisingly, *L. pneumophila* strains cultivated alone in liquid medium for three days and then incubated for three additional days on BCYE plates exhibit a great increase of the virulence index (Figure 3B), demonstrating an induction of the expression of virulence genes. ANOVA 2 revealed a statistical difference between culture conditions ($p < 0.0001$). Post-hoc analysis (Tukey's HSD test) showed a tendency of the virulence index to increase from Day 0 to Day 3 under the control conditions (not significant, $p = 0.898$) and a clear statistical difference from other conditions (Figure 3B), i.e., the control at Day 6, and the *L. pneumophila* strains internalized within *W. magna* C2c Maky and *A. castellanii* at Day 6 ($p = 0.023$). No statistical difference was observed at Day 6 between the control condition and the *L. pneumophila* strains co-cultivated with both amoebae. It seems that cultivation on BCYE plates erased the virulence index acquired during internalization in amoebae to confer a virulence index specific to BCYE culturing. Bacteria cultivated on BCYE plates were shown to be virulent, even if some isolates transiently lost their flagella. Indeed, some isolates characterized as aflagellate when harvested from BCYE agar were shown to be able to multiply in amoebae, and flagella were subsequently detectable by immunologic methods [29]. Moreover, Nowicki et al. have concluded that infection with aerosols of *L. pneumophila* coming from cultures on BCYE plates causes mortality in guinea pigs, showing that BCYE-cultivated *Legionella* are indeed virulent [30]. Consequently, the virulence index can only be assessed in conditions of liquid culture without sub-culturing on BCYE plates.

To conclude, we developed a tool that can be named the "GENE EXPRESSION-based Virulence Index" based on the quantitative measurement of cDNAs of 7 virulence genes (*ccmF*, *gacA*, *htpX*, *icmE*, *lirR*, *lvrE*, and *tatB*) using a calibrated RT and real-time PCR. Virulence genes were expressed at low levels in the three strains of *L. pneumophila* in the absence of amoebae, making it possible to determine a "GENE EXPRESSION-based Virulence Index" both at T0 and after 3 days in a liquid medium.

*L. pneumophila* strains followed a similar trend, since the "GENE EXPRESSION-based Virulence Index" between the three strains was not statistically different. After internalization in *W. magna* C2c Maky, there was no increase in the "GENE EXPRESSION-based Virulence Index" compared to the control conditions, since no statistical difference between the index determined after internalization in *W. magna* C2c Maky and the index determined in control *L. pneumophila* was observed for the three *L. pneumophila* strains. A huge and significant increase in "GENE EXPRESSION-based Virulence Index" was observed in the three tested *L. pneumophila* strains internalized in *A. castellanii* compared to *W. magna* C2c Maky, with a fold difference between the two amoebae reaching 9.4 (Virulence Index was 185 ± 44 in WILL condition and 1745 ± 257 in ACANTH condition). These data confirmed that *W. magna* C2c Maky cells resist *Legionella* strains, are able to phagocyte *Legionella* strains, and do not increase their virulence after internalization.

This index synthesized the expression of all the genes in one data for each condition, rendering the comparison easier. Moreover, a particular attention was brought to the genes that were expressed at low level by expressing the cDNA copy number contained in a given sample in percent of the averaged copy number measured in all samples. This GENE EXPRESSION-based Virulence Index could be used in many studies that focus on the variation of gene expression in a given function (here virulence, elsewhere inflammation, development, cell death, etc.) where many transcripts are analyzed. When different genes associated with the same function are compiled or integrated in the form of an index, it makes it easier to conclude about the overall evolution of these genes, a conclusion often made difficult when the individual variations do not always point in the same direction.

## 4. Materials and Methods

### 4.1. *L. pneumophila* Cultures

The three strains of *L. pneumophila* serogroup 1, Lens CIP 108 286, Paris CIP 107 629T, and Philadelphia ATCC 33152, were cultured at 36 °C for 3 days on buffered charcoal yeast extract (BCYE) agar plates (Thermo Fisher Scientific, Dardilly, France) before coculture experiments. Quantification of *L. pneumophila* was performed via Real-time qPCR using Biorad kits (Biorad, Hercules, CA, USA). Briefly, aliquots of cell suspension were sampled, and genomic DNA was extracted using the AquadienTM DNA extraction kit. *L. pneumophila* was quantified using the iQ-CheckTM L. pneumophila Real-time qPCR kit and a CFX96 Biorad thermocycler. The data were analyzed using the Bio-Rad CFX Manager Industrial Diagnostic Edition software (version 2.2). The whole procedure was designed for quantification of *L. pneumophila*, including intra-amoebic bacteria, and has been validated by AFNOR (Agence Française de Normalisation, the French standardization authority).

### 4.2. FLA Culture

The amoebae used in this study were *W. magna* C2c Maky ATCC PTA-7824 and *A. castellanii* ATCC 30010. Amoebae were grown at 30 °C for 3 days using adhesion culture on CF4 with serum casein glucose yeast extract medium (SCGYEM) [31]. Amoeba cells were maintained in the exponential growth phase by subculturing every 3 days. Quantification of amoeba populations was performed using 0.1 mL of each aliquot utilizing a hemocytometer cell counting chamber method with Trypan blue.

### 4.3. Coculture of *L. pneumophila* with Amoebae

Tubes containing 5 mL of Peptone Yeast Extract Glucose Broth (PYG) [32] were seeded with $5 \times 10^5$ *W. magna* C2c Maky or *A. castellanii* cells. At Day 0, the different strains of *L. pneumophila*, grown on BCYE plates, were suspended in sterile distilled water at $2.5 \times 10^8$ cells/mL, and inoculated into the amoebic cultures at a multiplicity of infection (MOI) of 50. Low-speed centrifugation (30 min at

1000× *g*) was used to initiate physical interaction between bacteria and amoebae, and then incubated at 36 °C for 1 h. To eliminate extracellular legionella, cocultures were treated for 2 h at 30 °C with 0.5 mg/mL of Penicillin and Streptomycin. Cells were then washed twice by, first, centrifuging assay tubes for 10 min at 1000× *g*, second, removing the supernatant, and, finally, adding 5 mL of fresh PYG medium preheated at room temperature. After the second wash, cocultures were incubated at 30 °C in PYG medium for 3 days (Figure S13A).

In a second step, coculture solutions were submitted to a mechanical cell lysis using a FastPrep®-24 instrument (MP Biomedicals, Illkirch-Graffenstaden, France) for 2 × 30 s at a speed of 5.0 in order to release legionella from the amoebae. Afterward, 100 µL of each treated coculture were deposited onto BCYE plates and incubated for 3 days at 36 °C (Figure S13B).

Control conditions consisted in legionella cultured in PYG medium in the absence of amoebae (FREE condition) for 3 days; for the second step, 100 µL of the latter culture were deposited onto BCYE plates for 3 additional days (Day 6), as described above.

### 4.4. Preservation of RNA Samples

Free legionella and cocultures (amoebae with internalized legionella) in liquid medium, or legionella colonies on BCYE plates were collected, rinsed in sterile osmosed water, and centrifuged for 5 min at 4 °C at 6000× *g*. The TRIzol® Max™ Bacterial RNA Isolation Kit (Thermo Fisher Scientific, Lyon, France) was used to improve the isolation of intact total RNA. The kit utilizes both the Max™ Bacterial Enhancement Reagent and TRIzol® Reagent to inactivate endogenous RNases and promote protein denaturing, improving RNA quality and integrity. After removal of the supernatant, 200 µL of pre-heated (95 °C) Max Bacterial Enhancement Reagent buffer was added on the pelleted cells and incubated for 5 min at 95 °C. Afterward, 1 mL of TRIzol Reagent was added, and the tubes were frozen in dry ice before being stored at −80 °C.

### 4.5. Total RNA Extraction

Biological material in TRIzol® was used for total RNA extraction according to the manufacturer's instruction and MIQE guidelines [33]. Total nucleic acid extracted was treated with Turbo DNA-free DNAse (Thermo Fisher Scientific, Lyon, France) as recommended by the manufacturer, to remove genomic DNA. Total RNA was then transferred into a nuclease-free tube and stored at −80 °C after the concentration was measured using a Biodrop spectrophotometer (Biodrop, Cambridge, UK).

### 4.6. Calibrated Reverse Transcription (cRT)

Briefly, 180 ng of purified total RNAs were used for cRT using random hexamers and Multiscribe™ reverse trancriptase in the presence of RNAse inhibitor (Thermo Fisher Scientific). A synthetic external and non-homologous Standard RNA (SmRNA) was used to control the quality of the RT and to normalize the reverse transcription of mRNAs of biological samples [34]. At the end of the RT, cDNAs were stored at −20 °C until further use. It should be noted that the Multiscribe® reverse transcriptase was not added in the RT master mix as recommended by the manufacturer. Indeed, thanks to the data obtained for the SmRNA, we discerned that the enzyme was rapidly altered in the master mix, thus significantly affecting the efficiency of reverse transcription across samples. This problem was resolved by preparing a master mix without the enzyme and then by adding the enzyme directly in each individual reaction tube.

### 4.7. Quantitative PCR (qPCR)

qPCR was performed using a Rotorgene Q (Qiagen, Courtaboeuf, France) and the Rotor-Gene SYBR Green PCR kit (Qiagen). All qPCR assays were run under the following conditions: 95 °C for 5 min, followed by 40 cycles of 5 s at 95 °C and 20–45 s (depending on the size of the amplicons) at 60 °C. The melting curve analysis was performed to evaluate the specificity of the DNA amplified during the PCR. The cycle of quantification Cq was determined at the intersection between the threshold line and the amplification curves when data were displayed as semi-logarithmic representation of the accumulated fluorescence versus cycle number. Cq was then transformed into a number of cDNA copies, according to standard curves composed of standards with known copies of cDNAs (1 to $10^8$ copies). These standard curves reliably indicated the method's detection sensitivity to targeted cDNAs (1 copy detected in 25% of cases) and the linear range of the quantification (from 10 to $10^8$ copies). Primer sequences, the slope of the standard curve established between the Cq and the initial cDNA concentration, as well as the efficiency of the qPCR of each targeted cDNAs, are presented in Table 1.

### 4.8. Calculation of the Virulence Index

As mentioned above (Section 2.3), the cDNA copy number contained in a given sample has been expressed in percent of the averaged copy number measured in all samples: 36 samples in each series of Task-1 (Lp Paris + Lp Philadelphia + Lp Lens) and 30 samples in Task-2 (Lp Paris + Lp Lens). Thus, at the end of each experiment series, and for each gene, the averaged "balanced" cDNA copy number calculated from all samples equaled 100. Finally, for a given sample, Gene expression-based Virulence Index was calculated as follows:

$$\text{Virulence} - \text{Index for sample A} = \sum_{n=1}^{7} \frac{\left(\text{cDNA copy nbr for gene}_{(n)}\right) \text{ in sample A} \times 100}{\text{average cDNA copy nbr for gene}_{(n)} \text{ in all samples}}$$

### 4.9. Data and Statistical Analysis

For each group of the first step, experiments were replicated once with $n = 3$ in each replicate, and the data obtained in each replicate were pooled; therefore, $n = 6$ in each group. For the second step, experiments were performed once with $n = 3$ in each group. For each group, data are expressed as mean ± SD of either the number of copies of the different cDNAs analyzed or the "gene expression"-based virulence index. Statistical significance for within-group comparisons was calculated by a two-way analysis of variance (ANOVA 2) with Tukey's post hoc test, using XLSTAT Software (version 19.4). The $p$ value of 0.05 defined the significance cut-off.

When the $p$ value for Factor 1 was below 0.05, meaning that differences between *L. pneumophila* strains were not statistically different, data for *L. pneumophila* strains were pooled together with "$n$ = n L. pneumophila Paris + n L. pneumophila Philadelphia + n L. pneumophila Lens."

For the first step, Factor 1 equals "L. pneumophila strain (Paris, Philadelphia, and Lens)" and Factor 2 equals "culture condition (T'0, T3D-FREE, T3D-WILL, T3D-ACANTH)." For the virulence index, because ANOVA 2 showed no statistical difference between the three strains ($p = 0.832$), values for the three strains were pooled together ($n = 18$).

For the second step, Factor 1 equals "L. pneumophila strain (Paris and Lens)" and Factor 2 equals "culture condition (T'0, T3D-FREE, T6D-FREE, T6D-WILL, T6D-ACANTH)." Strain Philadelphia was not analyzed in this second step because after internalization into both amoebae, it did not grow on BCYE plates in 2 out of 3 samples; because $n = 1$ for this condition, data from strain Philadelphia were not considered for the statistical analysis. Because ANOVA 2 showed no statistical difference between the two other strains ($p = 0.566$), values for both strains were pooled together ($n = 6$).

**Supplementary Materials:** The following are available online at http://www.mdpi.com/2076-0817/9/6/447/s1, Figure S1: Level of *rspL* transcript in the different conditions, expressed as the number of copies in $10^6$ *L. pneumophila* (Lp) ± SD, Figure S2: *ccmF* cDNA copy number measured in $10^6$ *L. pneumophila* for each strain and each culture condition, Figure S3: *gacA* cDNA copy number measured in $10^6$ *L. pneumophila* for each strain and each culture condition, Figure S4: *htpX* cDNA copy number measured in $10^6$ *L. pneumophila* for each strain and each culture condition, Figure S5: *icmE* cDNA copy number measured in $10^6$ *L. pneumophila* for each strain and each culture condition, Figure S6: *lirR* cDNA copy number measured in $10^6$ *L. pneumophila* for each strain and each culture condition, Figure S7: *lvrE* cDNA copy number measured in $10^6$ *L. pneumophila* for each strain and each culture condition, Figure S8: *tatB* cDNA copy number measured in $10^6$ *L. pneumophila* for each strain and each culture condition, Figure S9: *ccmF* and *gacA* cDNA copy number measured in $10^6$ *L. pneumophila* for each strain and each culture condition, Figure S10: *htpX* and *icmE* cDNA copy number measured in $10^6$ *L. pneumophila* for each strain and each culture condition, Figure S11: *lirR* and *lvrE* cDNA copy number measured in $10^6$ *L. pneumophila* for each strain and each culture condition, Figure S12: *tatB* cDNA copy number measured in $10^6$ *L. pneumophila* for each strain and each culture condition, Figure S13: Synoptic diagram of the experiments, Table S1: Fold changes of gene expression after 3-48h exposure of *L. pneumophila* Philadelphia to CuO nanoparticles (CuO-NPs) or synthetic Gray water (Gw).

**Author Contributions:** Conceptualization, R.M.M., J.B. and L.B.; methodology, R.M.M.; formal analysis, L.B.; writing—original draft preparation; R.M.M., S.D., and L.B.; writing—review and editing, S.D. and L.B.; supervision, L.B.; L.B. and S.D. contributed equally to the manuscript. All authors have read and agreed to the published version of the manuscript.

**Funding:** This research received no external funding.

**Acknowledgments:** The authors are grateful to Florian Lanet, Béatrice Georges and Florian Bénitière for their technical assistance in RT-qPCR and to Benjamin Quelard for his technical assistance for the coculture experiments

**Conflicts of Interest:** J.B. is the co-founder of Amoeba SA.

## References

1. Cunha, C.B.; Cunha, B.A. Legionnaire's Disease Since Philadelphia. *Infect. Dis. Clin. N. Am.* **2017**, *31*, 1–5. [CrossRef] [PubMed]
2. Khodr, A.; Kay, E.; Gomez-Valero, L.; Ginevra, C.; Doublet, P.; Buchrieser, C.; Jarraud, S. Molecular epidemiology, phylogeny and evolution of Legionella. *Infect. Genet. Evol.* **2016**, *43*, 108–122. [CrossRef] [PubMed]
3. Rossier, O.; Starkenburg, S.R.; Cianciotto, N.P. Legionella pneumophila Type II Protein Secretion Promotes Virulence in the A/J Mouse Model of Legionnaires' Disease Pneumonia. *Infect. Immun.* **2004**, *72*, 310–321. [CrossRef] [PubMed]
4. Rowbotham, T.J. Preliminary report on the pathogenicity of Legionella pneumophila for freshwater and soil amoebae. *J. Clin. Pathol.* **1980**, *33*, 1179–1183. [CrossRef]
5. Balczun, C.; Scheid, P. Free-Living Amoebae as Hosts for and Vectors of Intracellular Microorganisms with Public Health Significance. *Viruses* **2017**, *9*, 65. [CrossRef] [PubMed]
6. Berjeaud, J.-M.; Chevalier, S.; Schlusselhuber, M.; Portier, E.; Loiseau, C.; Aucher, W.; Lesouhaitier, O.; Verdon, J. Legionella pneumophila: The Paradox of a Highly Sensitive Opportunistic Waterborne Pathogen Able to Persist in the Environment. *Front. Microbiol.* **2016**, *7*, 21660. [CrossRef] [PubMed]
7. Borella, P.; Guerrieri, E.; Marchesi, I.; Bondi, M.; Messi, P. Water ecology of Legionella and protozoan: Environmental and public health perspectives. *Biotechnol. Annu. Rev.* **2005**, *11*, 355–380. [CrossRef]
8. Neumeister, B.; Reiff, G.; Faigle, M.; Dietz, K.; Northoff, H.; Lang, F. Influence of Acanthamoeba castellanii on Intracellular Growth of Different Legionella Species in Human Monocytes. *Appl. Environ. Microbiol.* **2000**, *66*, 914–919. [CrossRef]
9. Cirillo, J.D.; Falkow, S.; Tompkins, L.S. Growth of Legionella pneumophila in Acanthamoeba castellanii enhances invasion. *Infect. Immun.* **1994**, *62*, 3254–3261. [CrossRef]
10. Cirillo, J.D.; Cirillo, S.L.G.; Yan, L.; Bermudez, L.E.; Falkow, S.; Tompkins, L.S. Intracellular Growth in Acanthamoeba castellanii Affects Monocyte Entry Mechanisms and Enhances Virulence of Legionella pneumophila. *Infect. Immun.* **1999**, *67*, 4427–4434. [CrossRef]
11. Michard, C.; Doublet, P. Post-translational modifications are key players of the Legionella pneumophila infection strategy. *Front. Microbiol.* **2015**, *6*, 87. [CrossRef] [PubMed]
12. Prashar, A.; Terebiznik, M.R. Legionella pneumophila: Homeward bound away from the phagosome. *Curr. Opin. Microbiol.* **2015**, *23*, 86–93. [CrossRef] [PubMed]
13. Zink, S.; Pedersen, L.; Cianciotto, N.P.; Abu Kwaik, Y. The Dot/Icm Type IV Secretion System of Legionella pneumophila Is Essential for the Induction of Apoptosis in Human Macrophages. *Infect. Immun.* **2002**, *70*, 1657–1663. [CrossRef] [PubMed]
14. Gomes, T.S.; Gjiknuri, J.; Magnet, A.; Vaccaro, L.; Ollero, D.; Izquierdo, F.; Fenoy, S.; Hurtado, C.; Del Águila, C. The Influence of Acanthamoeba–Legionella Interaction in the Virulence of Two Different Legionella Species. *Front. Microbiol.* **2018**, *9*, 2962. [CrossRef] [PubMed]
15. Aurass, P.; Schlegel, M.; Metwally, O.; Harding, C.R.; Schroeder, G.; Frankel, G.; Flieger, A. The Legionella pneumophila Dot/Icm-secreted Effector PlcC/CegC1 Together with PlcA and PlcB Promotes Virulence and

Belongs to a Novel Zinc Metallophospholipase C Family Present in Bacteria and Fungi. *J. Boil. Chem.* **2013**, *288*, 11080–11092. [CrossRef] [PubMed]

16. Harding, C.R.; Stoneham, C.; Schuelein, R.; Newton, H.J.; Oates, C.V.; Hartland, E.L.; Schroeder, G.; Frankel, G. The Dot/Icm Effector SdhA Is Necessary for Virulence of Legionella pneumophila in Galleria mellonella and A/J Mice. *Infect. Immun.* **2013**, *81*, 2598–2605. [CrossRef] [PubMed]
17. Fontana, M.F.; Banga, S.; Barry, K.; Shen, X.; Tan, Y.; Luo, Z.-Q.; E Vance, R. Secreted Bacterial Effectors That Inhibit Host Protein Synthesis Are Critical for Induction of the Innate Immune Response to Virulent Legionella pneumophila. *PLoS Pathog.* **2011**, *7*, e1001289. [CrossRef]
18. Buse, H.Y.; Lu, J.; Ashbolt, N.J. Exposure to Synthetic Gray Water Inhibits Amoeba Encystation and Alters Expression of Legionella pneumophila Virulence Genes. *Appl. Environ. Microbiol.* **2015**, *81*, 630–639. [CrossRef]
19. Lu, J.; Struewing, I.; Buse, H.Y.; Kou, J.; Shuman, H.A.; Faucher, S.P.; Ashbolt, N.J. Legionella pneumophila Transcriptional Response following Exposure to CuO Nanoparticles. *Appl. Environ. Microbiol.* **2013**, *79*, 2713–2720. [CrossRef]
20. Hasni, I.; Jarry, A.; Quelard, B.; Carlino, A.; Eberst, J.-B.; Abbe, O.; Demanèche, S. Intracellular Behaviour of Three Legionella pneumophila Strains within Three Amoeba Strains, Including Willaertia magna C2c Maky. *Pathogens* **2020**, *9*, 105. [CrossRef]
21. Schroeder, G.; Petty, N.K.; Mousnier, A.; Harding, C.R.; Vogrin, A.J.; Wee, B.; Fry, N.K.; Harrison, T.G.; Newton, H.J.; Thomson, N.R.; et al. Legionella pneumophila Strain 130b Possesses a Unique Combination of Type IV Secretion Systems and Novel Dot/Icm Secretion System Effector Proteins. *J. Bacteriol.* **2010**, *192*, 6001–6016. [CrossRef] [PubMed]
22. Xuereb, B.; Bezin, L.; Chaumot, A.; Budzinski, H.; Augagneur, S.; Tutundjian, R.; Garric, J.; Geffard, O. Vitellogenin-like gene expression in freshwater amphipod Gammarus fossarum (Koch, 1835): Functional characterization in females and potential for use as an endocrine disruption biomarker in males. *Ecotoxicology* **2011**, *20*, 1286–1299. [CrossRef] [PubMed]
23. Dey, R.; Bodennec, J.; Mameri, M.O.; Pernin, P. Free-living freshwater amoebae differ in their susceptibility to the pathogenic bacteriumLegionella pneumophila. *FEMS Microbiol. Lett.* **2009**, *290*, 10–17. [CrossRef] [PubMed]
24. Vranckx, L.; De Buck, E.; Anné, J.; Lammertyn, E. Legionella pneumophila exhibits plasminogen activator activity. *Microbiology* **2007**, *153*, 3757–3765. [CrossRef]
25. Zusman, T.; Degtyar, E.; Segal, G. Identification of a Hypervariable Region Containing New Legionella pneumophila Icm/Dot Translocated Substrates by Using the Conserved icmQ Regulatory Signature. *Infect. Immun.* **2008**, *76*, 4581–4591. [CrossRef]
26. Cianciotto, N.P. Many substrates and functions of type II secretion: Lessons learned fromLegionella pneumophila. *Futur. Microbiol.* **2009**, *4*, 797–805. [CrossRef]
27. Banga, S.; Gao, P.; Shen, X.; Fiscus, V.; Zong, W.-X.; Chen, L.; Luo, Z.-Q. Legionella pneumophila inhibits macrophage apoptosis by targeting pro-death members of the Bcl2 protein family. *Proc. Natl. Acad. Sci. USA* **2007**, *104*, 5121–5126. [CrossRef]
28. Hsu, F.; Zhu, W.; Brennan, L.; Tao, L.; Luo, Z.-Q.; Mao, Y. Structural basis for substrate recognition by a unique Legionella phosphoinositide phosphatase. *Proc. Natl. Acad. Sci. USA* **2012**, *109*, 13567–13572. [CrossRef]
29. Bosshardt, S.C.; Benson, R.F.; Fields, B.S. Flagella are a positive predictor for virulence inLegionella. *Microb. Pathog.* **1997**, *23*, 107–112. [CrossRef]
30. Nowicki, M.; Bornstein, N.; Paucod, J.; Binder, P.; Fleurette, J. Effect of Culture Medium on Morphology and Virulence of Legionella pneumophila Serogroup 1. *Zent. Bakteriol. Mikrobiol. Hyg.* **1987**, *264*, 167–177. [CrossRef]
31. De Jonckheere, J. Use of an axenic medium for differentiation between pathogenic and nonpathogenic Naegleria fowleri isolates. *Appl. Environ. Microbiol.* **1977**, *33*, 751–757. [CrossRef] [PubMed]
32. Laskowski-Arce, M.A.; Orth, K. Acanthamoeba castellanii Promotes the Survival of Vibrio parahaemolyticus. *Appl. Environ. Microbiol.* **2008**, *74*, 7183–7188. [CrossRef] [PubMed]
33. Bustin, S.; Benes, V.; Garson, J.A.; Hellemans, J.; Huggett, J.F.; Kubista, M.; Mueller, R.; Nolan, T.; Pfaffl, M.W.; Shipley, G.L.; et al. The MIQE Guidelines: Minimum Information for Publication of Quantitative Real-Time PCR Experiments. *Clin. Chem.* **2009**, *55*, 611–622. [CrossRef] [PubMed]

34. Bezin, L.; Morales, A. Method of Calibration of Reverse Transcription Using a Synthetic Messenger RNA (SmRNA) as an Internal Control. Patent WO20044092414, 28 October 2004.

www.ingramcontent.com/pod-product-compliance
Lightning Source LLC
Chambersburg PA
CBHW061814210926
51850CB00006B/697

MDPI
St. Alban-Anlage 66
4052 Basel
Switzerland
Tel. +41 61 683 77 34
Fax +41 61 302 89 18
www.mdpi.com

*Pathogens* Editorial Office
E-mail: pathogens@mdpi.com
www.mdpi.com/journal/pathogens